Hydrogen-Transfer Reactions

Edited by
James T. Hynes,
Judith P. Klinman,
Hans-Heinrich Limbach,
Richard L. Schowen

1807–2007 Knowledge for Generations

Each generation has its unique needs and aspirations. When Charles Wiley first opened his small printing shop in lower Manhattan in 1807, it was a generation of boundless potential searching for an identity. And we were there, helping to define a new American literary tradition. Over half a century later, in the midst of the Second Industrial Revolution, it was a generation focused on building the future. Once again, we were there, supplying the critical scientific, technical, and engineering knowledge that helped frame the world. Throughout the 20th Century, and into the new millennium, nations began to reach out beyond their own borders and a new international community was born. Wiley was there, expanding its operations around the world to enable a global exchange of ideas, opinions, and know-how.

For 200 years, Wiley has been an integral part of each generation's journey, enabling the flow of information and understanding necessary to meet their needs and fulfill their aspirations. Today, bold new technologies are changing the way we live and learn. Wiley will be there, providing you the must-have knowledge you need to imagine new worlds, new possibilities, and new opportunities.

Generations come and go, but you can always count on Wiley to provide you the knowledge you need, when and where you need it!

William J. Pesce
President and Chief Executive Officer

Peter Booth Wiley
Chairman of the Board

Hydrogen-Transfer Reactions

Volume 1

Edited by
James T. Hynes, Judith P. Klinman,
Hans-Heinrich Limbach, Richard L. Schowen

WILEY-VCH Verlag GmbH & Co. KGaA

The Editors

Prof. James T. Hynes
Department of Chemistry and Biochemistry
University of Colorado
Boulder, CO 80309-0215
USA

Prof. Judith P. Klinman
Departments of Chemistry and
Molecular and Cell Biology
University of California
Berkeley, CA 94720-1460
USA

Département de Chimie
Ecole Normale Supérieure
24 rue Lhomond
75231 Paris
France

Prof. Hans-Heinrich Limbach
Institut für Chemie und Biochemie
Freie Universität Berlin
Takustrasse 3
14195 Berlin
Germany

Prof. Richard L. Schowen
Departments of Chemistry, Molecular
Biosciences, and Pharmaceutical Chemistry
University of Kansas
Lawrence, KS 66047
USA

Cover
The cover picture is derived artistically from the potential-energy profile for the dynamic equilibrium of water molecules in the hydration layer of a protein (see A. Douhal's chapter in volume 1) and the three-dimensional vibrational wavefunctions for reactants, transition state, and products in a hydride-transfer reaction (see the chapter by S.J. Benkovic and S. Hammes-Schiffer in volume 4).

Library of Congress Card No.: applied for

British Library Cataloguing-in-Publication Data
A catalogue record for this book is available from the British Library.

Bibliographic information published by Die Deutsche Bibliothek
Die Deutsche Nationalbibliothek lists this publication in the Deutsche Nationalbibliografie; detailed bibliographic data are available in the Internet at <http://dnb.d-nb.de>.

Typesetting Kühn & Weyh, Freiburg; Asco Typesetters, Hongkong
Printing betz-druck GmbH, Darmstadt
Bookbinding Litges & Dopf GmbH, Heppenheim
Cover Design Adam-Design, Weinheim

Printed in the Federal Republic of Germany.
Printed on acid-free paper.

ISBN: 978-3-527-30777-7

Foreword
The Remarkable Phenomena of Hydrogen Transfer

Ahmed H. Zewail*
California Institute of Technology
Pasadena, CA 91125, USA

Life would not exist without the making and breaking of chemical bonds - chemical reactions. Among the most elementary and significant of all reactions is the transfer of a hydrogen atom or a hydrogen ion (proton). Besides being a fundamental process involving the smallest of all atoms, such reactions form the basis of general phenomena in physical, chemical, and biological changes. Thus, there is a wide-ranging scope of studies of hydrogen transfer reactions and their role in determining properties and behaviors across different areas of molecular sciences.

Remarkably, this transfer of a small particle appears deceptively simple, but is in fact complex in its nature. For the most part, the dynamics cannot be described by a classical picture and the process involves more than one nuclear motion. For example, the transfer may occur by tunneling through a reaction barrier and a quantum description is necessary; the hydrogen is not isolated as it is part of a chemical bond and in many cases the nature of the bond, "covalent" and/or "ionic" in Pauling's valence bond description, is difficult to characterize; and the description of atom movement, although involving the local hydrogen bond, must take into account the coupling to other coordinates. In the modern age of quantum chemistry, much has been done to characterize the rate of transfer in different systems and media, and the strength of the bond and degree of charge localization. The intermediate bonding strength, directionality, and specificity are unique features of this bond.

* The author is currently the Linus Pauling Chair Professor of chemistry and physics and the Director of the Physical Biology Center for Ultrafast Science & Technology and the National Science Foundation Laboratory for Molecular Sciences at Caltech in Pasadena, California, USA. He was awarded the 1999 Nobel Prize in Chemistry.
Email: zewail@caltech.edu
Fax: 626.792.8456

Hydrogen-Transfer Reactions. Edited by J. T. Hynes, J. P. Klinman, H. H. Limbach, and R. L. Schowen

ISBN: 978-3-527-30777-7

The supreme example for the unique role in specificity and rates comes from life's genetic information, where the hydrogen bond determines the complementarities of G with C and A with T and the rate of hydrogen transfer controls genetic mutations. Moreover, the not-too-weak, not-too-strong strength of the bond allows for special "mobility" and for the potent hydrophobic/hydrophilic interactions. Life's matrix, liquid water, is one such example. The making and breaking of the hydrogen bond occurs on the picosecond time scale and the process is essential to keeping functional the native structures of DNA and proteins, and their recognition of other molecules, such as drugs. At interfaces, water can form ordered structures and with its amphiphilic character, utilizing either hydrogen or oxygen for bonding, determines many properties at the nanometer scale.

Hydrogen transfer can also be part of biological catalysis. In enzyme reactions, a huge complex structure is involved in bringing this small particle of hydrogen into the right place at the right time so that the reaction can be catalytically enhanced, with rates orders of magnitude larger than those in solution. The molecular theatre for these reactions is that of a very complex energy landscape, but with guided bias for specificity and selectivity in function. Control of reactivity at the active site has now reached the frontier of research in "catalytic antibody", and one of the most significant achievements in chemical synthesis, using heterogeneous catalysis, has been the design of site-selective reaction control.

Both experiments and theory join in the studies of hydrogen transfer reactions. In general, the approach is of two categories. The first involves the study of prototypical but well-defined molecular systems, either under isolated (microscopic) conditions or in complexes or clusters (mesoscopic) with the solvent, in the gas phase or molecular beams. Such studies over the past three decades have provided unprecedented resolution of the elementary processes involved in isolated molecules and en route to the condensed phase. Examples include the discovery of a "magic solvent number" for acid-base reactions, the elucidation of motions involved in double proton transfer, and the dynamics of acid dissociation in finite-sized clusters. For these systems, theory is nearly quantitative, especially as more accurate electronic structure and molecular dynamics computations become available.

The other category of study focuses on the nature of the transfer in the condensed phase and in biological systems. Here, it is not perhaps beneficial to consider every atom of a many-body complex system. Instead, the objective is hopefully to project the key electronic and nuclear forces which are responsible for behavior. With this perspective, approximate, but predictive, theories have a much more valuable outreach in applications than those simulating or computing bonding and motion of all atoms. Computer simulations are important, but for such systems they should be a tool of guidance to formulate a predictive theory. Similarly for experiments, the most significant ones are those that dissect complexity and provide lucid pictures of the key and relevant processes.

Progress has been made in these areas of study, but challenges remain. For example, the problem of vibrational energy redistribution in large molecules, although critical to the description of rates, statistical or not, and to the separation of intra and intermolecular pathways, has not been solved analytically, even in an approximate but predictive formulation. Another problem of significance concerns the issue of the energy landscape of complex reactions, and the question is: what determines specificity and selectivity?

This series edited by prominent players in the field is a testimony to the advances and achievements made over the past several decades. The diversity of topics covered is impressive: from isolated molecular systems, to clusters and confined geometries, and to condensed media; from organics to inorganics; from zeolites to surfaces; and, for biological systems, from proteins (including enzymes) to assemblies exhibiting conduction and other phenomena. The fundamentals are addressed by the most advanced theories of transition state, tunneling, Kramers' friction, Marcus' electron transfer, Grote-Hynes reaction dynamics, and free energy landscapes. Equally covered are state-of-the-art techniques and tools introduced for studies in this field and including ultrafast methods of femtochemistry and femtobiology, Raman and infrared, isotope probes, magnetic resonance, and electronic structure and MD simulations.

These volumes are a valuable addition to a field that continues to impact diverse areas of molecular sciences. The field is rigorous and vigorous as it still challenges the minds of many with the fascination of how the physics of the smallest of all atoms plays in diverse applications, not only in chemistry, but also in life sciences. Our gratitude is to the Editors and Authors for this compilation of articles with new knowledge in a field still pregnant with challenges and opportunities.

Pasadena, California *Ahmed Zewail*
August, 2006

Contents

Hydrogen-Transfer Reactions. Edited by J. T. Hynes, J. P. Klinman, H. H. Limbach, and R. L. Schowen

ISBN: 978-3-527-30777-7

Preface

As one of the simplest of chemical reactions, pervasive on this highly aqueous planet populated by highly aqueous organisms, yet still imperfectly understood, the transfer of hydrogen as a subject of scientific attention seems hardly to require defense. This claim is supported by the readiness with which the editors of this series of four volumes on *Hydrogen-transfer Reactions* accepted the suggestion that they organize a group of their most active and talented colleagues to survey the subject from viewpoints beginning in physics and extending into biology. Furthermore, forty-nine authors and groups of authors acceded, with alacrity and grace, to the request to contribute and have then supplied the articles that make up these volumes.

Our scheme of organization involved an initial division into physical and chemical aspects on the one hand, and biological aspects on the other hand (and one might well have said biochemical and biological aspects). In current science, such a division may provide an element of convenience but no-one would seriously claim the segregation to be either easy or entirely meaningful. We have accordingly felt quite entitled to place a number of articles rather arbitrarily in one or the other category. It is nevertheless our hope that readers may find the division adequate to help in the use of the volumes. It will be apparent that the division of space between the two categories is unequal, the physical and chemical aspects occupying considerably more pages than the biological aspects, but our judgment is that this distribution of space is proper to the subjects treated. For example, many of the treatments of fundamental principles and broadly applicable techniques were classified under physical and chemical aspects. But they have powerful implications for the understanding and use of the matters treated under biological aspects.

Within each of these two broad disciplinary categories, we have organized the subject by beginning with the simple and proceeding toward the complex. Thus the physical and chemical aspects appear as two volumes, volume1 on simple systems and volume 2 on complex systems. Similarly, the biological aspects appear as volume 3 on simple systems and volume 4 on complex systems.

Volume 1 then begins with isolated molecules, complexes, and clusters, then treats condensed-phase molecules, complexes, and crystals, and finally reaches

Hydrogen-Transfer Reactions. Edited by J. T. Hynes, J. P. Klinman, H. H. Limbach, and R. L. Schowen

ISBN: 978-3-527-30777-7

treatments of molecules in polar environments and in electronic excited states. Volume 2 reaches higher levels of complexity in protic systems with bimolecular reactions in solution, coupling of proton transfer to low-frequency motions and proton-coupled electron transfer, then organic and organometallic reactions, and hydrogen-transfer reactions in solids and on surfaces. Thereafter articles on quantum tunneling and appropriate theories of hydrogen transfer complete the treatment of physical and chemical aspects.

Volume 3 begins with simple model (i.e., non-enzymic) reactions for proton-transfer, both to and from carbon and among electronegative atoms, hydrogen-atom transfer, and hydride transfer, as well as the extension to small, synthetic peptides. It is completed by treatments of how enzymes activate C-H bonds, multiple hydrogen transfer reactions in enzymes, and theoretical models. Volume 4 moves then into enzymic reactions and a thorough consideration of quantum tunneling and protein dynamics, one of the most vigorous areas of study in biological hydrogen transfer, then considers several specific enzyme systems of high interest, and is completed by the treatment of proton conduction in biological systems.

While we do not claim any sort of comprehensive coverage of this large subject, we believe the reader will find a representative treatment, written by accomplished and respected experts, of most of the matters currently considered important for an understanding of hydrogen-transfer reactions. I am enormously grateful to James T. (Casey) Hynes and Hans-Heinrich Limbach, who saw to the high quality of the volumes on the physical and chemical aspects, and to Judith Klinman, who gave me a nearly free pass as her co-editor of the volumes on biological aspects. We are all grateful indeed to the authors who contributed their wisdom and eloquence to these volumes. It has been a very great pleasure to be assisted, encouraged, and supported at every turn by the outstanding staff of VCH-Wiley in Weinheim, particularly (in alphabetical order) Ms. Nele Denzau, Dr. Renate Dötzer, Dr. Tim Kersebohm, Dr. Elke Maase, Ms. Claudia Zschernitz, and – of course – Dr. Peter Gölitz.

Lawrence, Kansas, USA, September 2006 *Richard L. Schowen*

Preface to Volumes 1 and 2

These volumes together address the subject of the physical and chemical aspects of hydrogen transfer, volume 1 focusing on comparatively simple systems and volume 2 treating relatively more complex ones.

Volume 1 comprises three parts, commencing with Part I, dealing with hydrogen transfers of polyatomic molecules and complexes in relatively isolated conditions. In the first three contributions, the transfer is a coherent tunneling process rather than a rate process, characterized by "tunnel splittings" or delocalized hydrogen nuclei, for which electronic and vibrational spectroscopies are common and potent tools. The molecular systems discussed are malonaldehyde and tropolone (Redington, Ch. 1), carboxylic acid dimers (Havenith, Ch. 2) and strongly hydrogen-bonded systems such as $(H_2O...H...OH_2)^+$ (Asmis, Neumark and Bauman, Ch. 3). Kühn and Gonzales (Ch. 4) consider theoretically the more active role of infrared radiation in controlling hydrogen dissociation dynamics in e.g. OHF^-.

The five contributions of Part II focus on condensed matter. If the barriers are large, the hydrogen transfer becomes a rate process which may involve incoherent tunneling. Ceulemans (Ch. 5) examines proton abstraction by alkanes from strongly acidic alkane radical cations in inert matrices. Limbach (Ch. 5) follows the kinetics of single and multiple hydrogen and deuteron transfers in liquids and solids via NMR. Optical methods are applied by Douhal (Ch. 6) to systems embedded in a nanocavity, and embedded in liquids and polymer matrices by Waluk (Ch. 7), with a contrast to coherent hydrogen transfer in supersonic jets. Finally, Vener (Ch. 9) compares theory and experiment for anharmonic vibrations of strong hydrogen bonds in crystals.

Part III, comprising four chapters, commences the examination of hydrogen transfer – here proton transfer – in polar environments. The strong electrostatic proton-environment interaction guarantees incoherent rate phenomena. Kiefer and Hynes (Ch. 10) lay out the theoretical description for such reactions. The next three chapters exploit the greatly enhanced acidity of aromatic acids in the excited electronic state. Lochbrunner, Schriever and Riedle (Ch. 11) focus on the role of the motion of the groups between which the proton transfers, Pines and Pines (Ch. 12) thoroughly examine the insight to be gained from Förster cycle and free

Hydrogen-Transfer Reactions. Edited by J. T. Hynes, J. P. Klinman, H. H. Limbach, and R. L. Schowen

ISBN: 978-3-527-30777-7

energy analyses, while Tolbert and Solnstev (Ch. 13) pursue related themes for "super" photoacids in the concluding chapter of volume 1.

Volume 2 opens with Part IV dealing with hydrogen transfer in protic systems. Generally, a larger number of solvent molecules is involved, and hence multiple protons may be transferred. The first two chapters elucidate molecular details of proton transfer in solution via ultrafast infrared spectroscopy. Nibbering and Pines (Ch. 14) examine the transfer between acid-base pairs for the acid in the excited electronic state, while Elsaesser (Ch. 15) discusses coherent low frequency motions coupled to related proton transfers as well as in hydrogen-bonded complexes. The final two chapters in Part IV deal with proton transfer coupled to electron transfer, with Hammes-Schiffer (Ch. 16) expounding and illustrating the theory for these, while Hodgkiss, Rosenthal and Nocera (Ch. 17) discuss these reactions with a special emphasis on the connection to hydrogen atom transfer.

Part V, consisting of four chapters, opens with a discussion of the kinetics and mechanisms of proton abstraction from carbon in organic systems by Koch (Ch. 18) and then turns to a presentation by Williams (Ch. 19) on free energy relationships for proton transfer, as informed by various theoretical approaches. The final two chapters are devoted to hydrogen and dihydrogen mobility in the coordination sphere of transition metal complexes, where the transition from coherent to incoherent H-tunneling can be observed, with a review of the field given by Kubas in Ch. 20 and a discussion of insights from NMR studies presented by Buntkowsky and Limbach in Ch. 21.

In the first three of the five chapters of Part VI, hydrogen transfer is examined in assorted complex solids of importance in various applications: zeolites by Sauer in Ch. 22, fuel cells by Kreuer in Ch. 23 and ice bilayers by Aoki in Ch. 24. Attention is then turned to hydrogen transfer at metal surfaces in Ch. 25 by Christmann and in metals in Ch. 26 by Hempelmann and Skripov.

Volume 2 concludes in Part VII with contributions on the variational transition state theory approach to hydrogen transfer in various contexts (Truhlar and Garrett, Ch. 27), on experimental evidence of hydrogen atom tunneling in simple systems (Ingold, Ch. 28), and finally on a theoretical perspective for multiple hydrogen transfers (Smedarchina, Siebrand and Fernández-Ramos, Ch. 29).

JTH acknowledges the support of grant CHE-0417570 from the US National Science Foundation. HHL thanks the Deutsche Forschungsgemeinschaft, Bonn, and the Fonds der Chemischen Industrie, Frankfurt, for financial support.

Boulder and Paris, September 2006 — *James T. Hynes*
Berlin, September 2006 — *Hans-Heinrich Limbach*

List of Contributors to Volumes 1 and 2

Katsutoshi Aoki
Synchroton Radiation Research Center
Kansai Research Establishment
Japan Atomic Energy Research Institute
Kouto 1-1-1
Mikazuki-cho
Sayo-gun
Hyogo
Japan

Knut R. Asmis
Department of Molecular Physics
Fritz-Haber-Institut der Max-Planck-Gesellschaft
Faradayweg 4–6
14195 Berlin
Germany

Joel M. Bowman
Department of Chemistry and
Cherry L. Emerson Center for
Scientific Computation
Emory University
Dickey Drive
Atlanta, GA 30322
USA

Gerd Buntkowsky
Department of Chemistry
FSU Jena
Helmholtzweg 4
07743 Jena
Germany

Jan Ceulemans
Department of Chemistry
K.U. Leuven
Celestijnenlaan 200-F
3001 Leuven
Belgium

Klaus Christmann
Institut für Chemie und Biochemie
Physikalische und Theoretische Chemie
Freie Universität Berlin
Takustrasse 3
14195 Berlin
Germany

Abderrazzak Douhal
Departamento de Química Física
Sección de Químicas
Facultad de Ciencias del Medio Ambiente
Universidad de Castillo-La Mancha
Avda. Carlos III
S.N. 45071 Toledo
Spain

Hydrogen-Transfer Reactions. Edited by J. T. Hynes, J. P. Klinman, H. H. Limbach, and R. L. Schowen

ISBN: 978-3-527-30777-7

Thomas Elsaesser
Max-Born-Institut für Nichtlineare Optik
und Kurzzeitspektroskopie
Max-Born-Strasse 2A
12489 Berlin
Germany

Antonio Fernández-Ramos
Department of Physical Chemistry
Faculty of Chemistry
University of Santiago de Compostela
15706 Santiago de Compostela
Spain

Bruce C. Garrett
Chemical Sciences Division
Pacific Northwest National Laboratory
Richland, WA 99352
USA

Leticia González
Institut für Chemie und Biochemie
Freie Universität Berlin
Takustrasse 3
14195 Berlin
Germany

Sharon Hammes-Schiffer
Department of Chemistry
104 Chemistry Building
Pennsylvania State University
University Park, PA 16802-4615
USA

Rolf Hempelmann
Institute of Physical Chemistry
Saarland University
66123 Saarbrücken
Germany

Justin Hodgkiss
Department of Chemistry
Massachusetts Institute of Technology
77 Massachusetts Avenue
Cambridge, MA 02139-4307
USA

James T. Hynes
Department of Chemistry
and Biochemistry
Pacific Northwest National Laboratory
University of Colorado
Boulder, CO 80309-0215
USA
and
Ecole Normale Supérieure
CNRS UMR 8640 PASTEUR
Département de Chimie
24, rue Lhomond
75231 Paris
France

Keith U. Ingold
National Research Council
Ottawa, ON K1A 0R6
Canada

Philip M. Kiefer
Department of Chemistry and
Biochemistry
University of Colorado
Boulder, CO 80309-0215
USA
and
Ecole Normale Supérieure
CNRS UMR 8640 PASTEUR
Département de Chimie
24, rue Lhomond
75231 Paris
France

Heinz F. Koch
Department of Chemistry
Ithaca College
Ithaca, NY 14850
USA

Klaus-Dieter Kreuer
Max-Planck-Institut für
Festkörperforschung
Heisenbergstrasse 1
70569 Stuttgart
Germany

Gregory J. Kubas
Los Alamos National Laboratory
Chemistry Division
MS J514
Los Alamos, NM 87545
USA

Oliver Kühn
Institut für Chemie und Biochemie
Freie Universität Berlin
Takustrasse 3
14195 Berlin
Germany

Hans-Heinrich Limbach
Institut für Chemie und Biochemie
Freie Universität Berlin
Takustrasse 3
14195 Berlin
Germany

Stefan Lochbrunner
Department of Physics
Ludwig Maximillians University
Oettingenstrasse 67
80538 München
Germany

Daniel M. Neumark
Department of Chemistry
University of California
Berkeley, CA 94702
USA

Erik T. J. Nibbering
Max-Born-Institut für nichtlineare
Optik und Kurzzeitspektroskopie
Max-Born-Strasse 2A
12489 Berlin
Germany

Daniel G. Nocera
Department of Chemistry
Massachusetts Institute of Technology
77 Massachusetts
Cambridge, MA 02139-4307
USA

Dina Pines
Department of Chemistry
Ben-Gurion University of the Negev
P.O.B. 653
Beer Sheva 84105
Israel

Ehud Pines
Department of Chemistry
Ben-Gurion University of the Negev
P.O.B. 653
Beer Sheva 84105
Israel

Richard L. Redington
Department of Chemistry and
Biochemistry
Texas Tech University
Mail Stop 1061
Lubbock, TX 79409
USA

Eberhard Riedle
Department of Physics
Ludwig Maximilians University
Oettingenstrasse 67
80538 München
Germany

Joel Rosenthal
Department of Chemistry
Massachusetts Institute of Technology
77 Massachusetts Avenue
Cambridge, MA 02139-4307
USA

Joachim Sauer
Institut für Chemie
Humboldt-Universität zu Berlin
Unter den Linden 6
10099 Berlin
Germany

C. Schriever
Department of Physics
Ludwig Maximilians University
Oettingenstrasse 67
80538 München
Germany

Willem Siebrand
Steacie Institute for Molecular Sciences
National Research Council of Canada
Vorontsova pole 10
Ottawa, K1A 0R6
Canada

Alexander Skripov
Institute of Metal Physics
Urals Branch of the Academy of Sciences
Ekaterinburg 620219
Russia

Zorka Smedarchina
Steacie Institute for Molecular Sciences
National Research Council of Canada
Ottawa, K1A 0R6
Canada

Kyril M. Solntsev
School of Chemistry and Biochemistry
Georgia Institute of Technology
Atlanta, GA 30332-0400
USA

Laren M. Tolbert
School of Chemistry and Biochemistry
Georgia Institute of Technology
Atlanta, GA 30332-0400
USA

Donald G. Truhlar
Department of Chemistry
University of Minnesota
Minneapolis, MN 55455-0431
USA

Mikhail V. Vener
Department of Quantum Chemistry
Mendeleev University of Chemical Technology
Miusskaya Sq. 9
Moscow 125047
Russia

Jacek Waluk
Institute of Physical Chemistry
Polish Academy of Sciences
Kasprzaka 44/52
01-224 Warsaw
Poland

Ian H. Williams
Department of Chemistry
University of Bath
Bath BA2 7AY
UK

I Physical and Chemical Aspects

Part I
Hydrogen Transfer in Isolated Hydrogen Bonded Molecules, Complexes and Clusters

This first part is devoted to hydrogen transfers in small isolated molecular systems in the gas phase and jet streams. Coherent proton tunneling in malonaldehyde and tropolone in the electronic ground- and excited states is reviewed in Ch. 1 by Redington. These splittings, first observed by E. B. Wilson for malonaldehyde, depend on the vibrational and the electronic state. Often, only the difference of tunnel splittings between the initial and final spectrocopic state can be observed. In the case of formic acid dimers, however, the ground state splittings could be observed (Ch. 2 by Havenith). The challenge for the theoretical chemist is to reproduce experimental tunnel splittings from first principles. For this purpose, not only the Schrödinger equation for electrons but also for the nuclei has to be solved. Various theories are reviewed, e.g. the theory of Benderskii and of Wójcik which take multidimensional tunneling into account.

Malonaldehyde and tropolone exhibit hydrogen bonds of medium strength, releatively large barriers and OH-stretching frequencies and small tunnel splittings of the order of a few wave numbers or smaller. The results of novel gas phase vibrational spectroscopic experiments of strong hydrogen bonded systems such as bihalide anions, $(H_2O\cdot\cdot H\cdot\cdot OH_2)^+$ and $(HO\cdot\cdot H\cdot\cdot OH)^-$ etc. are reviewed by Asmis, Neumark, Bauman (Ch.3). These ions exhibit either low-barrier hydrogen bonds where tunnel splittings become larger than the barrier heights or no-barrier hydrogen bonds. The goal is to describe the proton motion by calculation of the strongly anharmonic vibrations from first principles. This task is a prerequisite before the well-known IR continua observed by Zundel in condensed matter will be really understood.

In Ch. 4, Kühn and González deals with the theory of ultrafast hydrogen transfer dynamics driven by tailored IR laser pulses. For small systems such as OHF^- pulse schemes can be devised, which either lead to a breaking of the OH or of the HF bond. For multidimensional hydrogen bond dynamics in medium sized mole-

Hydrogen-Transfer Reactions. Edited by J. T. Hynes, J. P. Klinman, H. H. Limbach, and R. L. Schowen

ISBN: 978-3-527-30777-7

cules anharmonic coupling and intramolecular vibrational energy redistribution constitute a serious challenge to any control attempt. The situation may be improved in alternative control schemes involving a properly timed UV excitation.

1
Coherent Proton Tunneling in Hydrogen Bonds of Isolated Molecules: Malonaldehyde and Tropolone

Richard L. Redington

1.1
Introduction

A series of articles reporting temperature dependent H and D isotope effects on the rates of certain enzymatic H transfer reactions show there is quantum tunneling by the H in these systems [1–7]. Theoretical considerations [4, 6, 7] infer vibrations within the enzyme–substrate complex, especially within the enzyme protein, develop transient geometrical configurations fomenting the tunneling process. Details of the promoting vibrations are still speculative, but it seems clear they work through transient reshaping of the potential energy surface (PES) barrier region, its width, energy maximum, overall contour, to provide "gated" impetus to H tunneling and to product escape.

Molecules hosting coherent H tunneling activity in the presence of complex vibrational structure are of immediate interest in the above context. They provide sharp data points probing specific vibrational couplings along the large amplitude tunneling coordinate, and they probe portions of the PES topography in detail. Studies on isotopomers and close chemical congeners provide clusters of data points reflecting systematic changes of the dynamical behavior. In this chapter research on the coherent tunneling properties of malonaldehyde, tropolone, and tropolone derivatives is surveyed. These are among the few currently known 10–15 atom molecules showing clear-cut spectral doublet structures signifying coherent tunneling properties. Interestingly, the equal double-minimum global PESs for these molecules, notably for S_0 tropolone, are also poised for demonstrations of tunneling activity by the heavy atoms. The presence of H tunneling in some enzymatic systems is newly recognized; tunneling processes by heavy atoms such as C, N, and O may also prove consequential. Zuev et al. [8] recently published low temperature rate data and theoretical-computational results showing the tunneling of C atom from a single quantum state during the ring expansion isomerization of 1-methycyclobutylfluorocarbene.

Hydrogen-Transfer Reactions. Edited by J. T. Hynes, J. P. Klinman, H. H. Limbach, and R. L. Schowen

ISBN: 978-3-527-30777-7

As illustrated in Fig. 1.1, the tautomerizations of tropolone (TRN, $C_7H_6O_2$) and malonaldehyde (MA, $C_3H_4O_2$) involve an O···HO → OH···O transfer accompanied by shifts of the skeletal atom positions as the bond characters interchange. Spectroscopic experiments performed on low pressure gaseous samples of TRN and MA experience minimal disruptive interactions with the environment, and spectral doublet structures observed for both molecules unequivocally reveal coherent quantum tunneling. MA has 21 and TRN has 39 internal coordinates available for coupling into their tunneling processes. MA, with a developing experimental base [9–21], holds a computational edge over TRN which currently stands alone in its size group for its catalog of spectroscopic tunneling structures [22–39]. With 15 atoms TRN is reasonably expected to model elements of the intramolecular dynamical behavior occurring in much larger molecules, and it is amenable to high level experimental and theoretical studies. The topography of its PES has saddle-point (SP) energy barriers just low enough, and tautomer-to-tautomer heavy atom excursions that are just short enough, to produce numerous resolvable vibrational state-specific coherent tunneling splittings. They are also observed in the ^{2}H [40, 41] and ^{18}O [39, 42, 43] isotopomers of gaseous TRN, and in simple chemical derivatives [44–57]. The numerous van der Waals complexes of tropolone are not discussed in this chapter.

The rotation–contortion–vibration electronic states of the nonrigid TRN and MA molecules are classified according to the G_4 molecular symmetry group using the permutation-inversion operations E, P, i, and P^* [58]. The G_4 group is iso-

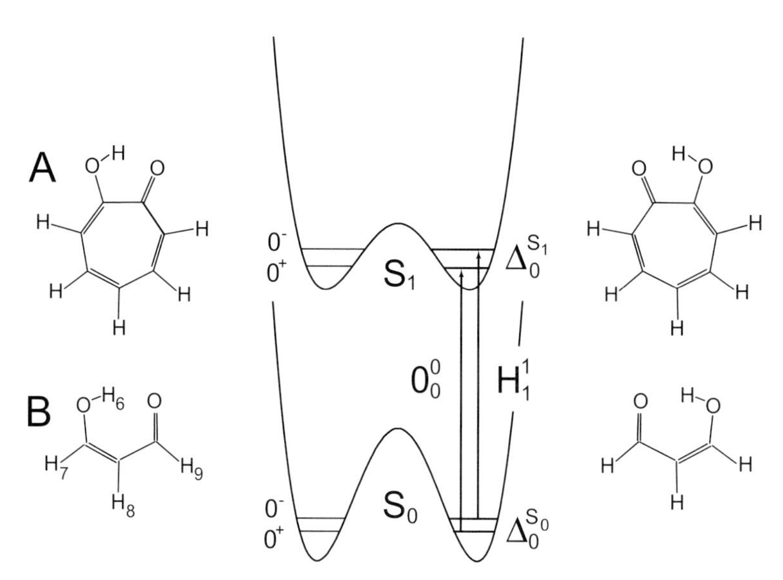

Figure 1.1 Symbolic representation of equal double-well potential energy functions for (A) tropolone and (B) malonaldehyde in the S_0 ground and lowest S_1 singlet electronic states. The spectral doublet of the S_1–S_0 zero-point transitions is shown. The doublet separation is $(H_1{}^1 - 0_0{}^0) = |\Delta_0{}^{S_1} - \Delta_0{}^{S_0}|$.

morphic to the C_{2v} point group and the energy states are conventionally labeled using the a_1, a_2, b_1, and b_2 irreducible representations of C_{2v}. The spectra observed for tropolone are complicated by many overlaps of the tunneling doubled fundamental, overtone, combination, and hot band vibrational transitions – the latter abundant at 25 °C. Many of the vibrational states are coupled by anharmonic resonance interactions affecting transition intensities more strongly than frequencies, and perturbing tunneling doublets [39].

The Schrödinger equation for a particle moving in a symmetrical one-dimensional (1D) double-well potential energy function (PEF) is readily solved for the lower energy levels and wavefunctions. The time-dependent wavefunctions produce probability density functions which coherently oscillate at a constant frequency between the two wells. In one's dreams multidimensional coordinate spaces of real molecules separate into simple 1D equations, and discovering the most effective real life coordinate separations and the most pertinent regions of the PES topography for complex molecules requires access to as much diverse and discriminating experimental information as possible. The status of research on coherent tunneling in the S_0 and S_1 electronic states of malonaldehyde and tropolone is discussed in Sections 1.2 and 1.3, respectively, with observations for several TRN derivatives sketched in Section 1.4, and a summary given in Section 1.5.

1.2 Coherent Tunneling Splitting Phenomena in Malonaldehyde

In 1976 Wilson's group published the first of several [9–12] studies on the microwave spectroscopy of the C, H, and O isotopomers of MA. These established its internally H bonded geometry, equal double-minimum nature of the PES, and coherent quantum tunneling activity. Their conclusions concerning the value of the zero-point (ZP) tunneling splitting, Δ_0, from precise observations of tunneling–rotation transitions were supported by their early far-infrared spectral measurements in the 20 cm^{-1} region. Subsequent precision studies in this wavelength region by Firth et al. [17] and by Baba et al. [14] showed $\Delta_0 = 21.583\ cm^{-1}$ for MA in its S_0 ground electronic state. Baughcum et al. [10] reported $\Delta_0[MA(OD)] = 2.915(4)\ cm^{-1}$ for the symmetrical $MA(D_6D_7D_9)$ isotopomer (see atom numbering in Fig. 1.1), and slightly smaller values for nine other MA(OD) isotopomers, six of them incorporating ^{13}C or ^{18}O atoms.

ZP tunneling splitting values for MA in excited electronic states have proved elusive. Seliskar and Hoffman [20] reported vibronic spectral structure for the weak $\tilde{A}\ ^1B_1 \leftarrow \tilde{X}\ ^1A_1$ (π^*–n) [$S_1 \leftarrow S_0$] absorption near 354 nm to have a spectral doublet separation (DS) of 7 cm^{-1}. Arias et al. [21] applied the absorbance-based degenerate four-wave mixing (DFWM) spectroscopic method to the $S_1 \leftarrow S_0$ transition. With higher intrinsic spectral resolution, and spectral clarification arising through the reduction of signal due to hot band transitions that is inherent in this nonlinear spectroscopic technique, they found DS = $|\Delta_0^{S_1} - \Delta_0^{S_0}| \sim 19\ cm^{-1}$, to place the $\tilde{A}\ ^1B_1$ zero-point tunneling splitting value at $\Delta_0^{S_1} \sim 2.5\ cm^{-1}$. The 19 cm^{-1}

DS value appears for the members of a 185 cm^{-1} vibronic progression, but details of the doublings for other vibrational states have not yet been reported. The strong quenching of $\Delta_0^{S_1}$ relative to $\Delta_0^{S_0}$ suggests the $\pi^*\leftarrow$n promotion increases the energy maximum and/or width of the effective PEF barrier [21].

Infrared spectra of gaseous MA and its deuterium isotopomers were reported at medium resolution by Seliskar and Hoffman [16] and by Smith et al. [15] and an overview at 1 cm^{-1} resolution is provided by Duan and Luckhaus [19]. At these modest resolutions the observed transition profiles include isolated PQR rotational envelopes as well as Q branch regions with multiple peaks suggesting doublets and probable hot band structures. Figure 1.2 is taken from Smith et al. [15] and shows peaks resolved in the 200–300 cm^{-1} region. Peaks #20 and #26 were assigned as representing the ν_{15} (nominal O···O stretch) and ν_{21} (out-of-plane skeletal deformation) fundamentals, respectively [15, 59, 60], with no specific tunneling doublet or hot band assignments made. No evidence for the presence of resolved spectral tunneling doublets was reported by Firth et al. [17] for Ar-isolated MA or by Chiavassa et al. [18] in their IR studies of MA isotopomers isolated

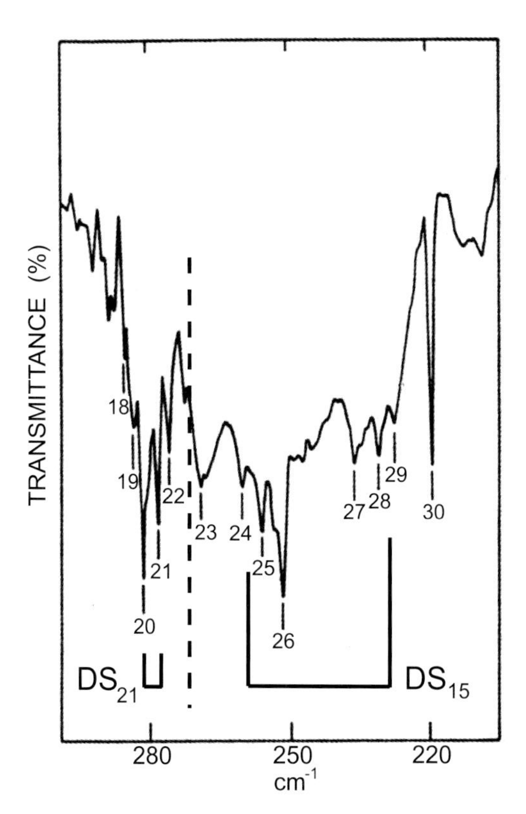

Figure 1.2 Infrared spectrum of MA(OH) in the 200–300 cm^{-1} region due to Smith et al. [15]. Spectral doublet separations DS_{21} and DS_{15} replace the single-peak assignments ν_{15} = peak 20 and ν_{21} = peak 26.

in Ar, Kr, and Xe matrices. Smith et al. [15] proposed a vibrational assignment for MA on the basis of their IR spectra for H and D isotopomers, comparative spectra for other substances, and a valence force field. Tayyari and Milani-Nejad [59] reassigned many fundamentals with the guidance of MP2/6-31G*//MP2/6-31G*(sp and p on H) computations performed on MA, MA(D_6D_8), and MA(D_7D_9). Alparone and Millefiori [60] affirmed the assignment through vibrational self-consistent field (VSCF) and correlation-corrected-VSCF computations introducing anharmonicities to estimate the fundamental isotopomer spectra.

Duan and Luckhaus [19] made a major experimental advance by using a Pb-salt diode laser to measure rotational transitions for ν_6 in jet-cooled samples prepared by seeding MA into He carrier gas with expansion through a pulsed slit nozzle. The clarified spectrum allowed the origin for the $\nu_6^+ \leftarrow \nu_0^+$ (type A) tunneling doublet component to be evaluated as 1594.08983(26) cm^{-1}. Further, the determination of 12 rotation and centrifugal distortion constants provides comparisons for corresponding ZP state constants determined by Baba et al. [14]. The analysis of weak IR transitions produced a selection of parameters for the $\nu_6^- \leftarrow \nu_0^-$ tunneling doublet component, to yield the remarkably small DS = $\Delta_6 - \Delta_0 = -0.03$ cm^{-1} value for the ν_6 fundamental (νC=O + in-phase (iph) νC=C + δCOH [60]). Duan and Luckhaus suggested the tunneling enhancing effect of symmetrical C=C—C=O motion may be cancelled by the tunneling damping effect of concomitant COH angle widening, and they noted the challenge to theoretical treatments of the tunneling dynamics in MA provided by this accurate new experimental result.

While Smith et al. [15] eschewed the assignment of tunneling doublets, Seliskar and Hoffman [16] noted the Q branch doublet separations implying DS_{20} = 6.3 cm^{-1} for ν_{20} at 384 cm^{-1} (out-of-plane ring deformation) and DS_{14} = 5.8 cm^{-1} for ν_{14} at 512 cm^{-1} (in-plane ring deformation). These $DS_V = |\Delta_V - \Delta_0|$ values suggest the upper state tunneling splittings Δ_{20} = 15.3 or 27.9 cm^{-1} and Δ_{14} = 15.8 or 27.4 cm^{-1}. At 1 cm^{-1} resolution [19] there is a single sharp Q peak for ν_6, and a single sharp Q peak for ν_{19} = 766 cm^{-1}. Isolated PQR envelopes with sharp Q spikes [19] reasonably imply $\Delta_V \sim \Delta_0$ for ν_{10} at 1260 cm^{-1}, ν_8 at 1346 cm^{-1}, and ν_7 at 1452 cm^{-1}. To extract Δ_V values from Fig. 1.2 the cluster of peaks to the blue of ~270 cm^{-1} (dotted line) is observed to be qualitatively similar to peaks shown for ν_{20} in the 370–410 cm^{-1} region (Fig. 7 of Ref. [15]). Assigning peak #20 = 282.2 cm^{-1} and #21 = 279.3 cm^{-1} as the sharp Q spikes of type C absorption profiles for the doublet components of the *out-of-plane* – i.e. ν_{21} instead of ν_{15} – vibration yields $DS_{21} = |\Delta_{21} - \Delta_0|$ = 2.9 cm^{-1}, Δ_{21} = 18.7 or 24.5 cm^{-1}, and ν_{21} = $1/2$(282.2+279.3) = 280.8 cm^{-1}. The value ν_{21} = 277.5 cm^{-1} is computed from the CC-VSCF treatment of Alparone and Millefiori [60] and the calculated minus observed value difference of –1.2% is similar to the –1.6% value found for ν_{20}.

The cluster of peaks in Fig. 1.2 to the red of ~270 cm^{-1} divide into similar groups with peaks #23 to #26, separated by 17.1 cm^{-1}, and peaks #27 to #30, separated by 16.6 cm^{-1}, spanning the high and low frequency groups, respectively. One of the two central peaks in each group is probably due to a type A Q spike of the Y polarized transition dipole, and the other probably to a hot band Q spike. The outer peaks are attributed to apparent band heads in the P and R branches.

Pending further data, the doublet component origins are represented using peaks #24 and #25 as $\frac{1}{2}(260.7 + 256.8) = 258.8$ cm^{-1} and peaks #28 and #29 as $\frac{1}{2}(231.9 + 228.5) = 230.2$ cm^{-1}. Then $DS_{15} = |\Delta_{15} - \Delta_0| = 28.6$ cm^{-1} yielding $\Delta_{15} =$ 7.0 or 50.2 cm^{-1}, and $\nu_{15} = 244.5$ cm^{-1}. The CC-VSCF computed estimate [60] for ν_{15} is 267.2 cm^{-1}, a value 9.3% higher than $\nu_{15} = 244.5$ cm^{-1}. For the original assignment the differences are 277.5 – 252 = 25.5 (10.1%) and 267.2 – 282 = –14.8 cm^{-1} (–5.2%).

The experimentally based Δ_V values for nine fundamentals are listed in column 3 of Table 1.1 and approximate descriptions of the vibrations by Alparone and Millefiori [60], whose computations fit the observed IR spectrum to 2 or 3%, are listed in the right-most column. In column 4 the Δ_V/Δ_0 ratios are listed next to theoretical ratios [61–64] discussed below. Of the nine experimental values only the four lower frequency skeletal displacements – two in-plane and two out-of-plane – appear to significantly impact the tunneling process. Experimental splittings remain unknown for the vibrations dominated by motion of the tunneling H (that is the νOH, δCOH, and γOH internal coordinates). The ν_{OH} stretching vibration is anharmonically coupled to many nearby overtone and combination vibrations and its doublet separation is concealed in the broad IR absorption profile. The vibrational compositions in Table 1.1 show that the fundamentals containing the γOH torsion and δCOH bending coordinates are strongly mixed with other internal coordinates.

Carrington and Miller [65] using 2D and Shida et al. [66] using 3D quantum chemical reaction surface models produced the first of many extant theoretical estimates for Δ_0 and vibrational state specific splittings Δ_V for MA. The differences of 2D and 3D computations are very noticeable and they are the exclusive topic of a recent article by Yagi et al. [67]. The theoretical approaches to MA vary from extensions of the cited early work, to that on full multidimensional instanton computations by Smedarchina et al. [68], to work on new techniques such as the short time propagation method being tested by Giese and Kühn [69]. Došlic and Kühn [70] studied modes in the OHO fragment in 4D computation. Tautermann et al. [71] developed an improved semiclassical method for determining the optimized multidimensional tunneling path between the straight line and minimum energy path routes. The ZP corrected PES was constructed using low-level energy points calculated using B3LYP/6-31G(d) density functional theory (DFT) methodology and this PES was then scaled to match a few high level G3(MP2) reference points. The procedure produced computed Δ_0 values for $(HF)_2$, MA, and TRN that were each in good agreement with experiment.

PES features are now routinely computed using quantum chemistry methods that must necessarily include estimates for the electron correlation energy. The computed SP energy values depend strongly on the level of theory used, as exemplified by the range 10.34 kcal mol^{-1} (RHF/6-31G**) to 2.86 kcal/mol [B3LYP/6-311+G(2d,p)] reported by Benderskii et al. [61]. Barone and Adamo [72] proposed 4.3 kcal mol^{-1} as the best value for the SP barrier for MA on the basis of high level comparative studies. Sadhukhan et al. [73] reported 3.9 kcal mol^{-1} for CCSD(T)/cc-pVDZ computation.

Yagi et al. [74] determined an *ab initio* PES for MA, based on 698 reference points in the full 21D vibrational space, in work made possible through the use of a modified Shepard interpolation method. For this PES they estimated $\Delta_0 = 13.9\ cm^{-1}$ by using the semiclassical trajectory method proposed by Makri and Miller [75]. The Δ_0 value can also be calculated using exact quantum dynamics methods, and Coutinho-Neto et al. [76] computed Δ_0 using the diffusion Monte Carlo based projection operator imaginary time spectral evolution method to obtain $\Delta_0 = 25.7 \pm 0.3\ cm^{-1}$. The multiconfigurational time-dependent Hartree approach yielded $\Delta_0 = 25\ cm^{-1}$ to about 10% accuracy. These results establish an independent benchmark comparison for the various semiclassical results. Mil'nikov and Nakamura [77] reformulated instanton theory to allow "practical" computations of multidimensional tunneling splittings which, with some refinements, was applied [78] to the 21D PES to obtain $\Delta_0 = 30.7\ cm^{-1}$.

The reformulated instanton theory introduces a variational procedure to minimize the classical action functional determining the instanton trajectory. Few iterations of the procedure are required to converge an initial approximate trajectory into the instanton. Because the required quantum chemical PES computations are focused on the instanton region of the PES, relatively few points (energy, gradient, Hessian) are required. The method allows the instanton obtained from a low level PES to be the starting point for determining the instanton generated from a higher level PES [77, 78]. Mil'nikov et al. [78] could address MA – with full 21D consideration – using *ab initio* PESs obtained at high levels of quantum chemistry. The new theoretical development produces smooth instanton paths (unlike results found for the PES obtained through the Shepard interpolation procedure). The reduced quantum chemistry work load allowed the variational instanton method to determine $\Delta_0 = 16.4\ cm^{-1}$ for MA at the CCSD(T)/(aug-)pVDZ level of PES computation. With recognition that full computation using the (aug-)cc-pVTZ basis would require more than 5 years of CPU time because of the Hessian calculation, a partial computation with the instanton pre-exponential B and S_1 factors transferred from MP2/cc-pVDZ computations yielded $\Delta_0 = 21.2\ cm^{-1}$ – very near the $21.6\ cm^{-1}$ experimental result. Similarly, $\Delta_0[MA(OD)] = 3.0\ cm^{-1}$ was calculated with $2.9\ cm^{-1}$ observed. Mil'nikhov et al. [78] estimate the error in the computed Δ_0 value for MA to be at most 10%. This work re-emphasizes the sensitivity of the Δ_0 computation to the PES determination, while taking advantage of the lower sensitivity of the B and S_1 pre-exponential factor computations to the quantum chemical method. The developmental work [77] included the demonstration that the utilized multidimensional path integral method and the multidimensional Wentzel–Kramers–Brillouin (WKB) method yield equivalent results. The variational instanton method has not yet been applied to the tunneling splittings of vibrational excited states of MA.

Benderskii et al. [61] developed a theory called the perturbative instanton approach (PIA) which they applied to MA in full 21D treatment. An important feature of their work is the expression of the carefully partitioned and symmetrized Hamiltonian in expansions using dimensionless variables including a semiclassical parameter called γ. The γ value allows scaling of the tunneling

Tab. 1.1 Comparisons of observed and computed fundamental tunneling splitting ratios Δ_V/Δ_0 for malonaldehyde.

Mode[a] (V)	Observed		Δ_V/Δ_0 ratios						Approx. description[a]
	ν_V [a]	Δ_V [b]	Obs.[c] 21.6	Bend.[d] 21.6	Sew.[e] 21.8	Meyer[f] 22.0	Tew[g] 38	Tew[g] 15	
ZP	21.583	21.583	1.00	1.00	1.00	1.00	1.00	1.00	
1	2856			2.96	1.56	7.89			νOH + νCH7
5	1655			1.03	1.10	0.89	1.11	1.60	νC=O + oph νC=C
6	1594	21.55	1.000	0.80	1.06	1.14	0.79	0.73	νC=O + iph νC=C + δOH
7	1452	~21.6	~1.00	1.10		1.03	0.76	1.87	δCH7 + δCH8 + νC–C
8	1346			23.3	1.26[e]	1.31	6.63	1.40	δOH + δCH9 + νC=O + iph νC=C
9	1358	~21.6	~1.00	1.26		1.49	1.05	1.53	δCH9 + δOH + νC–O
10	1260	~21.6	~1.00	1.76		2.14	0.61	1.53	δCH7 + δOH + νC–O
14	512	15.8[b]	0.73	0.91			0.76		δRING
		27.4	1.27		1.11	1.47		1.93	δRING
15	245[a]	7.0[b]	0.32			0.79			νO···O
		50.2	2.32	1.56	2.45		3.26	5.20	νO···O
16	1028			0.96		0.30	0.21	1.20	γCH7 + γCH9

Mode[a] (V)	Observed		Δ_V/Δ_0 ratios						Approx. description[a]
	ν_V[a]	Δ_V[b]	Obs.[c] 21.6	Bend.[d] 21.6	Sew.[e] 21.8	Meyer[f] 22.0	Tew[g] 38	Tew[g] 15	
17	981			0.86		0.89	0.58	0.27	γOH + iph γCH7 + γCH9
18	873			0.73		0.72	0.39	1.13, 0.67	γOH + oph γCH7
19	766	~21.6	~1.00	1.04		1.06	0.92	1.60	γCH8
20	384	15.3[b]	0.71				0.87		γRING
		27.9	1.29	1.05	1.08	1.08		1.47	γRING
21	281[a]	18.7[b]	0.87	0.83	0.91	0.67	0.53		γRING
		24.5	1.13					1.47	γRING

a Mode numbers, observed frequencies, and descriptions (oph = out of phase; iph = in phase) are from Table 1 of Ref. [60] – except reassigned frequencies for ν_{15} and ν_{21}. Computed frequencies [d,e,f,g] do not exactly match the observed values.
b IR data from Refs. [15,19]. Indeterminate doublet splittings $DS_V = |\Delta_V - \Delta_0|$ give two choices for the tunneling splittings Δ_V.
c ZP tunneling splittings Δ_0 (cm^{-1}) are included in the Δ_V/Δ_0 headings.
d Benderskii et al. [61]. $\Delta_8/\Delta_0 = 23.3$ because of resonance interaction. Δ_0 is fit to the exact experimental value.
e Sewell et al. [62]. Vibrations from the utilized PES correlate poorly with the observed midrange vibrations. The 1.26 ratio correlates with δOH type mode with 1452 cm^{-1} frequency.
f Meyer and Ha [63].
g Tew et al. [64]. The $\Delta_0 = 38$ and 15 cm^{-1} columns are for harmonic and anharmonic valleys, respectively. The 1358 and 1346 cm^{-1} vibrations are labeled ν_8 and ν_9, respectively, by Tew et al.

splitting Δ_0 and SP barrier height, which are quite sensitive to γ, as well as spectral densities and other quantities that are relatively insensitive to γ. The selected PES was scaled with γ chosen to duplicate the experimental $\Delta_0 = 21.6\ cm^{-1}$ splitting. The resulting SP energy is 4.30 kcal mol^{-1} – the same value chosen by Barone and Adamo [72] as the best value for MA on the basis of their comparative quantum chemical computations. Conversely, choosing γ to reproduce this *ab initio* SP energy value yields the observed Δ_0 splitting. The full 21D vibrational spectrum and 21 Δ_V values were computed in the PIA. In column 5 of Table 1.1 a selection of the resulting Δ_V/Δ_0 ratios are listed for comparison with columns containing the available experimental values and results from other computational procedures. At this stage, the appreciable dispersion of the computed PES topologies, vibrational spectra, and tunneling splitting values favors comparison of the results through ratios. Benderskii et al. [61] present a detailed description of the dynamics calculated for MA and several isotopomers. They refer to ν_8, observed at 1346 cm^{-1}, as the tunneling coordinate with ν_1 (OH stretching) and ν_{15} (O···O stretching) designated as promoting modes. The very large (500 cm^{-1}) computed splitting obtained for ν_8 is attributed to a resonance interaction with ν_1 – also stated to be resonantly coupled with two nearby CH stretching modes with computed tunneling splittings above 100 cm^{-1}. The values $\Delta_8/\Delta_0 = 23.3$, $\Delta_3/\Delta_0 = 7.55$, and $\Delta_2/\Delta_0 = 5.89$ are by far the largest ratios arising in this study.

Column 6 of Table 1.1 presents several Δ_V/Δ_0 ratios obtained from the full 21D classical trajectory/WKB computations of Sewell et al. [62]. The utilized PES is based on the 1983 valence force field, and early vibrational assignments of Smith et al. [15] – some of which differ from later work [59, 60]. Many of the computed splitting ratios are therefore not entered in Table 1.1. The PES was given Morse functions for the bond stretching coordinates. At 10.0 kcal mol^{-1} the SP energy is very high but the barrier is noticeably narrow in its upper reaches. The computed ZP tunneling splitting is $\Delta_0 = 21.8\ cm^{-1}$, fortuitously close to the 21.6 cm^{-1} observed value. The Δ_V/Δ_0 ratios listed in column 6 are those with the closest agreement to current understanding of the vibrational structure and, for whatever reason, they follow the lead of Δ_0 by showing quite good agreement with the experimental results.

Full 21D computations were produced using direct quantum chemical methods by Meyer and Ha [63] and by Tew et al. [64, 79]. Meyer and Ha developed a reference reaction path model applied using a PES with the SP barrier of 3.27 kcal mol^{-1}. They obtained $\Delta_0 = 22.0\ cm^{-1}$ for MA(OH), 3.8 cm^{-1} for MA(OD), and values for all 21 Δ_V splittings. Some of the resulting Δ_V/Δ_0 ratios are listed in column 7 of Table 1.1, where it is seen that a large splitting is predicted for ν_1 (OH stretching) and that damping of the ZP splitting is predicted for the low frequency H transfer coordinate ν_{15} (O···O stretching). The largest ratios are $\Delta_1/\Delta_0 = 7.89$, $\Delta_{10}/\Delta_0 = 2.14$, and $\Delta_{13}/\Delta_0 = 1.66$. Δ_V values for several modes are significantly larger than suggested by the experimental data. Tew et al. developed an internal coordinate reaction path Hamiltonian allowing the usual division of coordinates: a single large amplitude motion with 20 orthogonal normal modes. In their work the energy levels are solved by matrix diagonalization (that is, by variational configuration interaction computations that were performed on three increasingly inclusive approximations for the Hamiltonian.) They initially [79] applied the method to MA using a PES determined by DFT. This work was followed [64] by an article concentrating on the use of PESs determined using the MP2 approximation for electron correlation energies because of convergence inadequacies arising through the use of DFT. In columns 8 and 9 of Table 1.1 the listing of computed Δ_V/Δ_0 ratios is obtained from results for the highest level Hamiltonians using a computed harmonic PES valley (Δ_0 = 38 cm^{-1}, column 8) and a computed anharmonic PES valley (Δ_0 = 15 cm^{-1}, column 9). It is noted that the computations for the harmonic valley predict quenching of the ZP tunneling on exciting ν_{20} or ν_{21}, while for the anharmonic valley the computations predict enhancement instead. For the harmonic valley the three largest ratios are Δ_8/Δ_0 = 6.63, Δ_{15}/Δ_0 = 3.26, and Δ_{13}/Δ_0 = 2.82. For the anharmonic valley they are Δ_{15}/Δ_0 = 5.20, Δ_{13}/Δ_0 = 2.00, and Δ_{14}/Δ_0 = 1.93. Good convergence was obtained for the lower vibrational frequencies but converged Δ_V splittings could not be obtained for the OH/CH stretching frequencies.

1.3 Coherent Tunneling Phenomena in Tropolone

The first experimental evidence for coherent tunneling in the moderately large molecules of interest in this chapter was reported for TRN in 1972 by Alves and Hollas [22]. Vibronic absorption bands of gaseous TRN(OH) and TRN(OD) in the 370 nm region were recorded at relatively high spectral resolution to clearly show rovibronic fine structures with peak spacings of ~ 0.1 cm^{-1}. Observed band head peaks were systematically paired as spectral doublets [22, 23] with separations both larger and smaller than the value of 18.93 cm^{-1} determined for the ZP origin doublet. This important work on gaseous tropolone showed that readily measured coherent tunneling processes occur over at least the lower 700 cm^{-1} range of vibrational excited states in the $\tilde{A}$ $^1B_2$ (S_1) electronic state and it stimulated many experimental and theoretical studies on this system.

For TRN(OH) Alves and Hollas assigned the ZP doublet transitions labeled 0_0^0 (0^+–0^+) and H_1^1 (0^-–0^-) in Fig. 1.3 and determined their origins to be 27017.54 and 27036.47 cm^{-1}, respectively. The spectral doublet separation is DS = $|\Delta_0^{S_1} - \Delta_0^{S_0}|$ = 18.93 cm^{-1}; for TRN(OD) they found DS = 2.2 cm^{-1}. Independent data are required to determine the individual $\Delta_0^{S_1}$ and $\Delta_0^{S_0}$ values, and Alves and Hollas were limited to recognizing that $\Delta_0^{S_1}$ is 18.93 cm^{-1} larger than $\Delta_0^{S_0}$. In 1999 Tanaka et al. [32] used microwave spectroscopy to determine that (rounded to three digits) $\Delta_0^{S_0}$ = 0.974 cm^{-1} for gaseous tropolone. The untruncated ZP splitting, $\Delta_0^{S_0}$ = 29193.788(26) MHz, was determined using a single interaction parameter with six rotational and centrifugal distortion constants each for the symmetric (0^+) and antisymmetric (0^-) tunneling states. Stark effect measurements on the $3_{2,1} \leftarrow 2_{2,0}$ and $3_{2,2} \leftarrow 2_{2,1}$ rotational transitions show the average dipole moments of gaseous tropolone along the Z (longitudinal) axis in these rotational

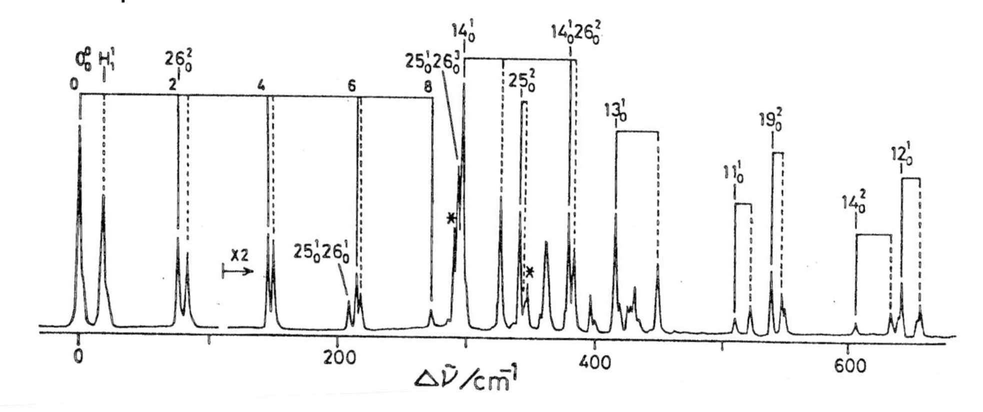

Figure 1.3 Vibrational state-specific doublets observed in the S_1–S_0 fluorescence excitation spectrum of jet-cooled TRN(OH) by Sekiya et al. [31].

states are 3.428(50) D for 0^+ and 3.438(50) D for 0^-. The analogous dipole moment component for MA is 2.59 D [10]. The microwave study verified beyond any doubt that tropolone possesses a symmetrical double-minimum PES in its S_0 electronic state, that the molecule undergoes coherent state-specific rotation-tunneling transitions, and that there are small but measurable differences between rotational constants of the 0^+ and 0^- states. The node in the 0^- tunneling wavefunction may cause the inequality $A^+ = 2743.0931(54) > A^- = 2742.7181(114)$ MHz [32].

A few years after Alves and Hollas studied the vibronic absorption spectrum of gaseous tropolone in a 1 m cell at room temperature, the application of pulsed UV laser spectroscopy to gaseous samples cooled in pulsed molecular beams became possible. Passing the carrier gas over tropolone crystals loaded into a pulsed nozzle at 45 °C produces excellent samples with rotational temperatures ~ 3 K and vibrational temperatures ~ 15 K. The initial works were by Tomioka et al. [26], Redington et al. [28], and Sekiya et al. [30]. With the elimination of hot bands and with rotational envelopes reduced to widths of about 2 cm^{-1} [28, 29], the S_1–S_0 fluorescence excitation spectrum is greatly simplified compared with the 25 °C absorption spectrum. In the best resolved fluorescence excitation data the ZP origin doublet components show peaks with central dips which, taken as the band origins, yield DS = 18.90 cm^{-1}, in good agreement with the value reported by Alves and Hollas.

The most recent publications on the S_1–S_0 transition of TRN(OH) are by Bracamonte and Vacarro [37, 38], who applied their polarization resolved DFWM spectroscopy method [37] to TRN vapor at its room temperature sublimation pressure. This remarkable nonlinear spectroscopic method simplifies the spectrum by allowing Q branch ($\Delta J = 0$) transitions to be observed with greatly diminished R and P branch ($\Delta J = \pm 1$) signal and, with rotation of a polarization element, *vice versa*. This selective nonlinear optical technique also produces a large reduction in the signal generated by hot bands, as noted in the above discussion for MA. However, with optical saturation the hot band Q branch signals are sufficiently strong that some of their band origins could be approximated [38]. With the S_0 ground state rotational parameters known from the work of Tanaka et al. [32], the S_1 state parameters, including the coherent tunneling splitting value $\Delta_0^{S_1} = 19.846(25)$ cm^{-1}, were accurately determined by Bracamante and Vaccaro [38]. The extra large inertial defects $\Delta I_0 +^{S_1} = -0.802(86)$ and $\Delta I_0 -^{S_1} = -0.882(89)$ amu Å^2 led them to conclude that the geometry of tropolone in the S_1 electronic state is slightly nonplanar.

Tropolone is easily deuterated and the S_1–S_0 fluorescence excitation spectrum of jet-cooled TRN(OD) was reported by Sekiya et al. [40]. The observed doublet separation, DS = 2 cm^{-1}, is near the 2.2 cm^{-1} value reported by Alves and Hollis [22], and about 10% of the value observed for TRN(OH). Presently the only available experimental estimate for the ZP tunneling splitting of S_0 TRN(OD) is $\Delta_0^{S_0} \leq 0.17$ cm^{-1} [80] obtained using a chloroform solvent and NMR spectroscopy to measure the deuteron spin–lattice relaxation time.

The S_1–S_0 fluorescence excitation and dispersed fluorescence spectra of TRN(OH) and TRN(OD) isolated in Ne matrices at ~ 3.5 K were studied by Rossetti and Brus [25] using UV pulsed laser excitation. Low temperature Ne matrix-isolation sampling is advantageous because, in principle, the sample monomers are dispersed in identical trapping sites resulting in a single sharp peak very near the gas-phase origin for each spectral transition. Rossetti and Brus found that the S_1–S_0 band origin transitions are doublets with the separations DS = 21±2 cm^{-1} for TRN(OH) and DS = 7±1 cm^{-1} for TRN(OD). These values are *larger* than the gas phase values by 2±2 cm^{-1} for TRN(OH) and 5±1 cm^{-1} for TRN(OD). According to theory [81], coupling a tunneling system to a bath of harmonic oscillators produces a damping of the tunneling splittings. The increased spectral doublet separations observed for Ne-isolated tropolone are explainable [82] through the interaction of nonplanar S_1 tropolone with an asymmetric Ne matrix trapping site.

The large observed inertial defects [38], and molecular orbital (MO) computations reported by Wójcik et al. [83], suggest the geometry of tropolone is probably slightly nonplanar in the S_1 state. For discussion it is assumed to be deformed towards a boat shape. (In the crystal state S_0 tropolone is boat-shaped with folding angles of 1 or 2° [84]). Tautomerization inverts the concavity of the TRN geometry to, for example, uncradle an initially cradled Ne atom in the asymmetric trapping site. The small reorientation of the TRN position causes minimal disturbance of the trapping site configuration, but it produces an offset between energy minima of the two tautomer configurations to distinguish the behavior of planar S_0 and slightly nonplanar S_1 tropolone in an asymmetric Ne matrix trapping site.

Hameka and de la Vega [85] presented a theoretical description of the ZP energy levels and probability density functions for a particle in a slightly asymmetric double-well PEF. They analyzed the model behavior in terms of two basic parameters: δ (the energy offset between the two PEF minima), and Δ (the tunneling splitting that occurs for the limiting equal double-well PEF – namely at $\delta = 0$). Equations presented by Hameka and de la Vega are easily combined [39, 82] into the equation $(^{as}\Delta)^2 = \delta^2 + \Delta^2$, where $^{as}\Delta$ is the splitting of the ZP energy levels in the PEF.

Except for the small nonplanarity of S_1 TRN, the S_0 and S_1 geometries, Ne matrix trapping sites, and (probably) fundamental tautomerization processes of TRN(OH) and TRN(OD) are similar. For gaseous TRN in the S_0 and S_1 states the PEF is symmetrical so that $\delta = 0$ and $\Delta = \Delta_0 = {}^{as}\Delta$. The observed IR and UV spectral data for Ne-isolated TRN suggest $\delta \sim 0$ for TRN in the S_0 state, and $\delta = 7{\pm}1$ cm^{-1} for TRN in the S_1 state. The latter value accounts for the comparative spectral doublet behavior of the S_1–S_0 origin bands of Ne-isolated and gaseous TRN(OH) and TRN(OD). The spectral data also suggest that the Ne trapping site environment damps the Δ tunneling splitting values by an amount d taken to be the same for the S_0 and S_1 states of both isotopomers (if $d > \Delta_V$, the Δ_V value is fully quenched). Dampings $1 \le d \le 2.5$ cm^{-1} therefore completely quench the TRN(OH) splitting $\Delta_0^{S_0} = 0.974$ cm^{-1}, but only mildly damp $\Delta_0^{S_1} = 19.864$ cm^{-1}. The value $d = 2.5$ cm^{-1} is the maximum value consistent with error bars of the observed spectral doublets.

The modeling computations of Makri and Miller [81] show that if a specific tunneling system is coupled to a high bath frequency the system tunneling splitting is only slightly damped, whereas if the coupling is to a low bath frequency the damping is strong. An analogous situation is to compare dampings generated by the couplings of a specific bath frequency to different tunneling systems, say high barrier system A with splitting Δ_0 and low barrier system B with splitting $10\Delta_0$. Compared to the bath oscillator the parameters of system A would be high [that is, relative to them the bath frequency would be low (say below the barrier) with strong damping of the splitting]. Similarly, the parameters of system B would be low [that is, relative to them the bath frequency would be high (say above the barrier) with weak damping of the splitting]. A specific Ne trapping site, with its fixed distribution of vibrational frequencies, can thus be envisioned as strongly damping (percentage wise) a tunneling system with a small Δ_0 value ($\Delta_0^{S_0}$ of TRN), and weakly damping (percentage wise) a tunneling system with a large Δ_0 value ($\Delta_0^{S_1}$ of TRN).

Fluorescence excitation [43] and IR absorption [39] spectra of ^{16,16}TRN(OH), ^{16,18}TRN(OH), and ^{18,18}TRN(OH) show that an asymmetric double-well PES occurs for gaseous ^{16,18}TRN in the S_0 and S_1 electronic states. This is due to unequal anharmonic vibrational energy contributions to the effective PES at the limiting ^{16}OH···^{18}O and ^{16}O···H^{18}O tautomer configurations. The lowered symmetry of ^{16,18}TRN(OH) facilitates the observation of four (rather than two) fluorescence excitation transitions [43]. Combination differences of the observed transitions show the ZP levels of ^{16,18}TRN(OH) are separated by 1.7 cm^{-1} in the S_0 state [compared to $\Delta_0^{S_0} = 0.974$ cm^{-1} for ^{16,16}TRN(OH)].

The fluorescence excitation spectra of jet-cooled TRN(OH) and TRN(OD) show vibrational transitions in the S_1 state reaching to about 700 cm^{-1}. The lowest frequency fundamental of TRN(OH) occurs at 39 cm^{-1} in the S_1 state, and at 110 cm^{-1} in the S_0 state. The MO-computed normal coordinate for the S_0 state suggests primary OCCO/skeletal twisting, rather than wagging, character. If its true nature is torsional, it should be identified as ν_{19} (a_2) rather than as ν_{26} (b_1). The low frequency values reflect the competition between drives favoring planar geometry (the internal H bond and π resonance interactions), and drives favoring nonplanar geometry (the relaxation of heptacyclic ring strain, and of *cis* OCCO alignment).

The fluorescence excitation spectrum of jet-cooled TRN(OH) as presented by Sekiya et al. [31] is shown in Fig. 1.3. A series of vibrational state-specific tunneling doublets (labeled 26_0^V) are seen to proceed to the blue of the S_1–S_0 origin doublet. These involve even numbered vibrational excitations of the 39 cm^{-1} mode in the S_1 state. The spectral doublet separations, 7.23, 4.72, 3.51, ··· cm^{-1}, are strongly damped from the 18.90 cm^{-1} value for the origin band. The figure also shows vibrationally enhanced doublet separations for ν_{13} (414.66, 446.7 cm^{-1}) and ν_{14} (295.82, 326.41 cm^{-1}) [28–31], which are a_1 skeletal deformation fundamentals with strong components of O···O stretching. The above vibrational state-specific doublet separations for the S_1 state are collected in column 3 Table 1.2, where they are compared with theoretical splittings computed by Wójcik et al. [83] as noted below.

Many of the spectral doublet separations observed for S_1 state vibrations are large and easily measured and, in principle, dispersion of the fluorescence excited from the S_1 states allows many S_0 vibrational states to be probed. The individual doublet components of the S_1–S_0 origin or other sufficiently strongly fluorescent doublet component can be excited with laser pulses to generate fluorescence transitions of the same parity terminating on S_0 vibrational states. Values for the frequencies of ν_{13}, ν_{14}, and other fundamental and excited state transitions are assigned [27, 31], but most S_0 state tunneling splittings Δ_V cannot be determined from the extant fluorescence data. Alves et al. [27] published a large body of dispersed single vibronic level fluorescence spectra excited from doublet components of gaseous tropolone at room temperature by a UV laser with a 1 cm^{-1} band width. The dispersed fluorescence spectra access many high lying S_0 state vibrations that may be basically unobservable to direct IR absorption spectroscopy. This work added greatly to understanding of the vibrational spectra in the S_1 and S_0 electronic states – especially concerning the lowest frequency modes.

Turning to spectral doublings observed in the IR spectra of TRN(OH) and TRN(OD), it was found that samples isolated in Ne matrices produce many more transitions than there are fundamental vibrations [24]. Of these, ten pairs of lines were originally assigned as apparent spectral doublets, but in later experimental and computational research [34, 86] virtually all of the extraneous lines were assignable as binary, and some ternary, overtone and combination transitions. Relative absorption intensities suggest the anharmonic resonance interactions couple many of the vibrational states, and several weak computed fundamentals of TRN(OH) and TRN(OD) have not yet been observed in IR spectroscopic experiments. Spectral tunneling doublet assignments in the Ne matrix data were made only for the OH(OD) stretching (DS = 19[4.8] cm^{-1}) and skeletal contortion (DS = 10.6[9.6] cm^{-1}) fundamentals. The absence of other assignable tunneling doublets in the IR spectra of Ne-isolated TRN is attributed to the above noted total damping of the gas phase coherent tunneling splittings $\Delta_V < 2.5$ cm^{-1} by couplings to the Ne matrix [82]. In Ar matrices the peaks show shifting, broadening, and sometimes structure relative to the Ne matrix data [24].

Table 1.2 Calculated energy splittings ($\Delta_0^{S_1}$) in cm^{-1} for the 3D model potentials for the Ã state of tropolone. (Table VII of Ref. [83].)

Band	CIS/6-31++G(d,p)	Exp.
$13^0 14^0$	19.6	20
$13^0 14^1$	24.2	31
$13^0 14^2$	29.0	29
$13^0 14^3$	34.0	
$13^0 14^4$	39.3	
$13^1 14^0$	20.4	33
$13^2 14^0$	21.2	
$13^3 14^0$	21.9	
$13^4 14^0$	22.5	
$25^0 26^0$	24.4	20
$25^0 26^1$	24.2	
$25^0 26^2$	24.0	8
$25^0 26^3$	23.8	
$25^0 26^4$	23.6	6
$25^0 26^5$	23.4	
$25^0 26^6$	23.3	5
$25^0 26^7$	23.1	
$25^0 26^8$	22.8	2
$25^1 26^0$	23.9	
$25^2 26^0$	23.4	5
$25^3 26^0$	22.9	
$25^4 26^0$	22.5	

Vibrations in the S_0 state of gaseous TRN(OH) were directly addressed in the 700–3500 cm^{-1} region using high resolution Fourier transform infrared (FTIR) absorption spectroscopy [35] on samples at 25 °C, the sublimation pressure ~0.01 Torr, and optical path length 32 m. Individual rotational transitions of tropolone are not resolved at a spectral resolution of 0.0025 cm^{-1}. However, sharp Q branch peaks arising in rotational contours of the FTIR absorptions mark approximate band origins for the out-of-plane vibrations (type C profiles), and for vibrations with intense transition dipoles paralleling the fictive C_2 axis (type A profiles). The observed sharp Q branch spikes led to 32 reported [35] spectral tunneling doublets in the region between 700 and 950 cm^{-1}. Additional doublets were observed at 0.1 cm^{-1} resolution at frequencies above 950 cm^{-1}, but the details have not yet been reported – exclusive of the CH/OH stretching region which is shown below to have a smooth contour with no resolved spectral doublets [35].

The CCOH torsion fundamental (ν_{22}) and two members of an associated progression of hot bands are shown in Fig. 1.4. The doublet separation for ν_{22} is $DS_{22} = |\Delta_{22} - \Delta_0| = 0.90$ cm^{-1}. Together with $\Delta_0 = 0.974$ cm^{-1} for the ZP state, this yields the strongly damped tunneling splitting $\Delta_{22} = 0.074$ cm^{-1} for this fundamental. The transitions in Fig. 1.4 have also been studied for the oxygen isotopomers ^{16,18}TRN(OH) and ^{18,18}TRN(OH) [39]. An analysis including the assumption that $^{16,16}\Delta_{22}/^{16,16}\Delta_0 = {}^{18,18}\Delta_{22}/^{18,18}\Delta_0$ for the oxygen isotopomer splittings in the ν_{22} COH torsion and ZP states contributes to first estimates for the isotopomer ZP tunneling splittings $^{18,18}\Delta_0 = 0.865$ cm^{-1} and $^{16,18}\Delta_0 = 0.920$ cm^{-1}. The ^{18}O vibrational isotope shifts of the ν_{22} transitions depend on normal isotopomer reduced mass values near 1.07 amu. In contrast, the ^{18}O isotope dependence of the isotopomer tunneling *splittings* for ν_{22} correlates with the large reduced mass values of the skeletal contortion fundamental [39]. This unique isotope effect supports the role for skeletal tunneling in the tautomerization process of tropolone in the S_0 electronic state.

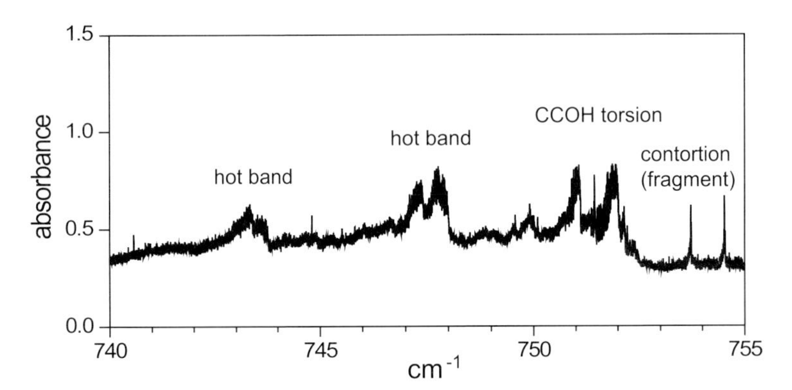

Figure 1.4 A 15 cm^{-1} overview of the high resolution FTIR absorption spectrum of gaseous TRN(OH) at 25 °C [35]. The spectral doublets are for the CCOH torsion fundamental and two members of its hot band progression. The ultra-narrow spikes are attributed, with other structure, to the high frequency doublet component of the skeletal contortion fundamental.

The OH/CH stretching region around 3100 cm^{-1} is of obvious importance to a discussion of coherent H tunneling in TRN. Sharp Q peaks are neither predicted nor observed for the OH stretching fundamental, although Fig. 1.5 shows peaks appear in the region. Vibrational assignments [34, 86] were proposed on the basis of (i) gas phase IR spectra at lower frequencies, (ii) moderately high level MO computations, (iii) two-laser fluorescence dip IR spectra (FDIRS) of jet-cooled TRN(OH) and TRN(OD) by Frost et al. [33], (iv) the IR spectra of Ne matrix-isolated samples [24], and (v) the results of independent theoretical computations of the spectral doublet for the OH stretching fundamental. The weak OH stretching fundamental assigned in Fig. 1.5 [3121, 3102 cm^{-1} in Ne-isolated TRN(OH)] occurs near the five CH stretching fundamentals (overlapped at 3063, 3030 cm^{-1}), with anharmonic resonance couplings to these and many (perhaps all) of the binary overtone and combination states present in the region. The broad gas phase absorption for TRN(OH) closely spans the limits set by these transitions (as

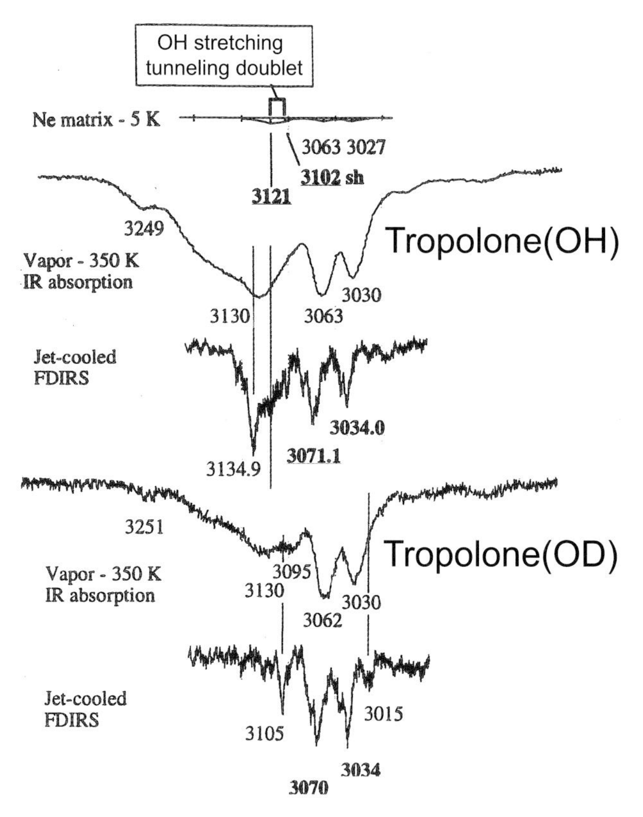

Figure 1.5 Infrared spectra of TRN(OH) and TRN(OD) in the 3100 cm^{-1} region [34]. In Ne-isolated TRN(OH) the fundamental OH stretching spectral doublet is at 3121 cm^{-1} and a shoulder at 3102 cm^{-1}. The doublet is not discerned in the gas phase IR absorption profile. The jet-cooled fluorescence dip IR spectra (FDIRS) are due to Frost et al. [33].

shown in Fig. 3 of Ref. [34].) Hot band absorptions are also present. The assignment of the OD stretching fundamental is 2344.8, 2340 cm^{-1} for Ne-isolated TRN(OD), with DS = 4.8 cm^{-1}. Because of damping by the Ne matrix, the spectral doublet separations for gaseous TRN(OH) and TRN(OD) would be ~ 2 cm^{-1} larger than the 19 and 5 cm^{-1} values observed in Ne-matrix isolation [82].

As noted in Section 1.2 for MA, the vibrational state-specific tunneling dynamics of TRN are addressed using PES features obtained through quantum chemical computations. The results of computing the C_{2v} saddle-point energy barrier maxima [86] for TRN at various levels of theory are shown in Table 1.3. The SP barriers computed using the larger basis sets with second order Møller–Plesset perturbation theory (MP2) for correlation energy are only about one quarter of the values obtained using the restricted Hartree–Fock (RHF) methodology with the 6-31G(d,p) basis set. Vener et al. [87] addressed the problem of coherent H tunneling in tropolone by approximating solutions to the 2D and 3D Schrödinger equations for the OH···O group using adiabatic separations of the high and low frequency degrees of freedom. The molecular PES was constructed using the 6-31G basis with added polarization functions for some atoms. Self-consistent field (SCF) and configuration interaction singles (CIS) methodologies were used for the S_0 and S_1 states, respectively. The computed vibrational state-specific tunneling splittings reflect the dynamics of H motion in a double-minimum PES with a high barrier maximum (~ 15.7 kcal mol^{-1}). The computational model produced enhanced tunneling splitting on exciting the O···O stretching coordinate and modest effects on exciting COH bending. Tunneling behavior on the excitation of vibrations beyond the OH···O group was not considered. Smedarchina et al. [88] used a modified semiclassical instanton approach to consider the ZP and excited state tunneling splittings of the S_0 and S_1 states. PES features were computed at the RHF/6-31G(d,p) level. Scaled by 0.9, the high computed adiabatic barrier reproduced the experimental ZP splitting for S_1 tropolone, and the splittings computed for 13 excited vibrational states yielded excellent to poor agreement with experiment. The instanton computations produced a large deuterium isotope effect, as observed.

Takada and Nakamura [89] studied S_0 tropolone using model 3D analytical PES functions based on coordinates for the OH stretch, a tunneling-promoting nominal O···O stretch, and a low frequency out-of-plane deformation mode. These were parametrized using optimized geometries for the tautomer and SP configurations computed at the MP2/6-31G(d,p) level, with refinement of the critical point energies at the MP4/6-31G(d,p) level. ZP energies for all coordinates, except the three defining the PES, were included using vibrational spectra computed at the MP2/6-31G(d,p) level. Tunneling eigenstates were evaluated by a numerical method. The high value of the C_{2v} SP barrier entering the PES produced good computed agreement with the observed $\Delta_0^{S_0}$ = 0.974 cm^{-1} tunneling splitting. Enhanced tunneling splittings were computed for O···O stretching excitations, with mildly damped splittings for excitations of the out-of-plane mode. In a very similar paper, Wójcik et al. [83] used model 2D and 3D analytical PESs for TRN in the S_1 electronic state that were parametrized by data computed at the CIS/6-

Tab. 1.3 MO-computed C_{2v} saddle-point potential energies for tropolone [86].

Level	Atoms	Basis	Energy (kcal mol^{-1})
MP2[a]	COHOC	6-311G(df,pd)	3.64
	5CH	6-311G(d,p)	
MP3	COHOC	6-311G(df,pd)	7.12
	5CH	6-311G(d,p)	
MP4(DQ)	COHOC	6-311G(df,pd)	7.88
	5CH	6-311G(d,p)	
MP4(SDQ)[b]	COHOC	6-311G(df,pd)	6.96
	5CH	6-311G(d,p)	
MP2	C_7OHO	6-311G(df,pd)	3.64
	5H	6-311G(d,p)	
MP2	COHOC	6-311++G(df,pd)	3.62
	5CH	6-311G(d,p)	
MP2	OHO	6-311G(2df,2pd)	4.57
	C_7H_5	6-311G(d,p)	
MP2	all	6-311G(d,p)	5.02
MP2	all	6-31++G(d,p)	5.57
RHF	all	6-31G(d,p)	15.62
MP2	all	6-31G(d,p)	5.29
MP3	all	6-31G(d,p)	9.22
MP4(DQ)	all	6-31G(d,p)	9.93
MP4(SDQ)	all	6-31G(d,p)	8.86

a MP2/GEN methodology is used for geometry optimizations and harmonic frequency computations.
b MP4(SDQ)/GEN methodology is used for points in the PES topography. The G3(MP2) computed SP energy is 7.2 kcal mol^{-1} [93].

31++G(d,p) level of theory (including the vibrational spectrum for the ZP correction). The tunneling splittings, accurately computed using a variational method, are shown with the experimental splittings for S_1 tropolone in Table 1.2 taken from the article.

Paz et al. [90] used MO-computed energies at modest theoretical levels to parametrize a 2D model PES and obtain tunneling splittings resembling the observed data. As they did for MA, Guo et al. [91] used a semiclassical approach on the tunneling splittings, with trajectory calculations, to compute Δ_V values for all 39 fundamentals of the S_0 state of TRN(OH). MO-computations were used to establish the PEF functions. With adjustment of the barrier to give the observed Δ_0 value, the two lowest frequency out-of-plane modes realistically damp the tunneling, several likely modes enhance the tunneling, and the Δ_{OH} for the OH stretch is 31 cm^{-1}. The scope of the various developing multidimensional theories is general and, while the computations on TRN lag those on MA, inroads on higher dimension computations are being made. Giese and Kühn [92] applied a multidimensional reaction surface approach to TRN using 4D and 12D versions of the reaction surface. The 4D surface included reduced normal modes favoring OH stretch, COH bend, the skeletal mode near 750 cm^{-1} presented [34, 86] as the tautomerization coordinate, and O···O stretching. The 12D model was used to compute IR spectra using the multiconfiguration time-dependent Hartree method. The spectrum provides theoretical justification for the small Δ_{OH} tunneling splitting value, and for the attribution of the broad OH stretching absorption of gaseous TRN to resonantly coupled binary combination modes as discussed above for Fig. 1.5. This work and investigations described above for TRN –and in the previous section for MA – give prominence to the lowest frequency in-plane mode (nominal O···O stretching) as the H transfer coordinate.

As already noted in Section 1.2 Tautermann et al. [71] obtained good agreement with experiment for the ZP splittings of MA. They also successfully applied their semiclassical method for finding the tunneling path to $(HF)_2$ and TRN [93], where the 7.2 kcal mol^{-1} quantum chemistry barrier for TRN is 50 to 100% lower than the barriers used for the computations on TRN discussed above. It is similar to the MP4/GEN results (footnote b of Table 1.3) used in the following discussion of the tautomerization of S_0 TRN that is based on an examination of the computed and spectroscopic data set for TRN(OH) and TRN(OD) [34–36, 86]. MP2/GEN (footnote a in Table 1.3) was used for geometry optimizations and computation of harmonic vibrational spectra, with MP4(SDQ)/GEN refinement of energies at critical points and other PES points of interest. The unscaled MP2/GEN computed vibrational spectra provided insight into sorting of the fundamental vibrations of TRN(OH) and TRN(OD) from the numerous binary (and occasional ternary) overtone and combination transitions resolved in the Ne matrix-isolation IR spectra [24, 34]. Rostkowska et al. [94] recently reported a normal mode analysis for the TRN monomer, and Wójcik et al. [95] studied normal modes of the dimer along with IR and Raman spectra of polycrystalline tropolone. To interpret the observed broadening of the OH absorption, they applied a model coupling the low frequency inter- and intramolecular O···O stretching modes to the OH and OD stretching vibrations.

The MP2/GEN optimized minimum energy path (MEP) [86] was found to be very long and to reach C_{2v} geometry at a high energy PES bifurcation point – the source of the energy ridge that divides the tautomers and hosts C_{2v} SP configura-

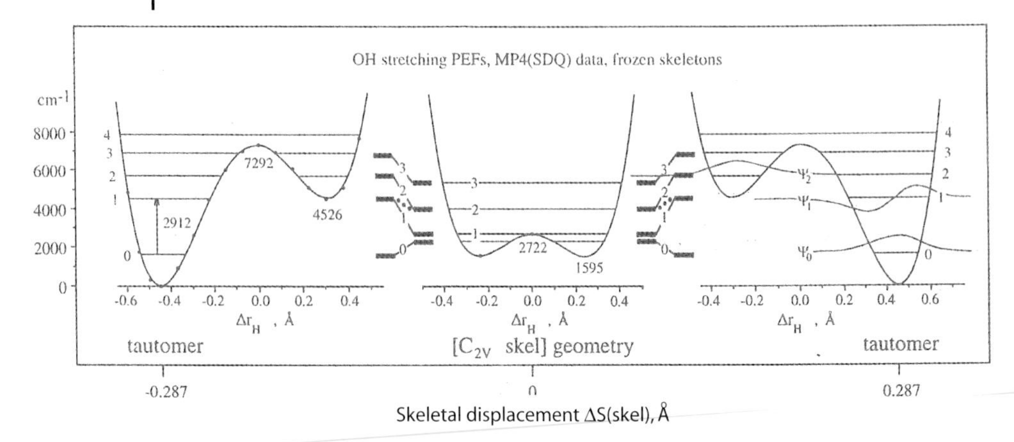

Figure 1.6 Dependence of the 1D potential energy function for OH stretching on the skeletal geometry [86]. PEFs are shown for the localized OH at the tautomer configurations, and for the delocalized H at the intermediate C_{2v} saddle-point configuration of the $C_7H_5O_2$ skeleton. Plotted as a function of the skeletal displacement ΔS(skel) (see also Fig. 1.7), the OH stretching eigenvalues E_1 and E_2 show an avoided crossing symbolized by dots between the levels correlated in the present figure.

tions. At geometries near the tautomer minima the H atom is found to be localized to one O atom as suggested in Fig. 1.6. The figure shows that with contortion of the 14-atom $C_7H_5O_2$ skeleton to C_{2v} configurations on the energy ridge the OH-stretching PEF is transformed into an equal double-minimum function allowing full delocalization of the H atom. The contortional displacement coordinate thus "vibrationally assists" H tunneling in S_0 tropolone which, in turn, allows completion of the skeletal tautomerization. The first examinations of this dynamical model are being made at the lowest possible descriptive level: informal adiabatic separation of the fast OH stretching and slow skeletal contortion motions, with quasiharmonic separation of the other 37 vibrations. Despite the disregarded kinetic and potential energy couplings, the informally separated 1D equations allow key features of the tunneling behavior to be outlined and compared with experimental data.

The quasiharmonic vibration spectrum is computed at the MP2/GEN level. The 1D OH stretching PEF (Fig. 1.6) is computed using MP4/GEN input points and its vibrational energies are solved numerically. The 1D PEF for the contortion is given quadratic-quartic-Gaussian functionality anchored by MP4/GEN level critical point energies supplemented with the 37 MP2/GEN computed quasiharmonic ZP or excited vibrational energies – and the numerically calculated, geometry-dependent, OH stretching vibrational energy. The Gaussian functionality of the PEF interpolates the vibrational energy contribution between the SP and tautomer geometries. The effective PEF for the tautomerization coordinate depends on the other-coordinate vibrational state. The state-specific ZP contortion levels in the upper and lower vibrational states of the spectroscopic transitions give first estimates for the spectral doublet separations as illustrated for an out-of-plane mode ν_X in Fig. 1.7.

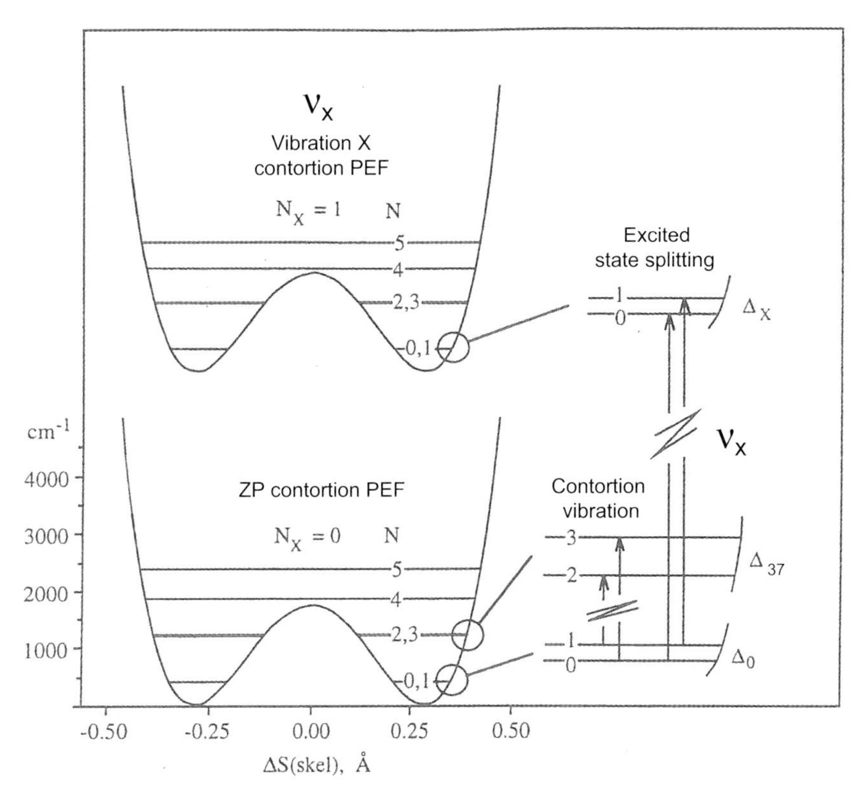

Figure 1.7 Vibrational state-specific effective potential energy functions for the skeletal contortion vibration [86]. Transitions at 754, 743.3 cm^{-1} for TRN(OH) isolated in a Ne matrix are assigned as the spectral doublet, and Δ_{37} is the upper state splitting, of the ν_{37} contortion vibration. The spectral doublet for a vibration ν_X, for example the CCOH torsion in Fig. 1.4, has the upper state vibrational state-specific tunneling splitting Δ_X. The spectral doublet components have the separation $DS_X = |\Delta_X - \Delta_0|$.

The model [34–36, 39, 86], labeled the {tunneling skeleton}{tunneling H atom} [TSTH] model for convenience, closely approximates the frequencies and doublet separations of the contortion fundamental as observed in Ne matrix isolation at 754, 743.4 cm^{-1} for TRN(OH) and 751, 741.4 cm^{-1} for TRN(OD). Peaks assigned to the higher – but not the lower – frequency component are discerned in the complex IR absorption profiles of gaseous TRN(OH). The observed ^{18}O isotope shifts, 3.3 cm^{-1} for ^{16,18}TRN(OH) and 6.9 cm^{-1} for ^{18,18}TRN(OH), are reproduced by the TSTH model. The MP2/GEN isotope shifts for the 754 cm^{-1} transition, if treated as a quasiharmonic mode, are about twice these values. The contortional reduced mass values are determined as $^{16,16}\mu = 7.25$, $^{16,18}\mu = 7.32$, and $^{18,18}\mu = 7.37$ amu, and these values also correlate the observed ^{18}O effects on the tunneling splittings for the COH torsion fundamental shown for ^{16,16}TRN(OH) in Fig. 1.4. The TSTH model approximates the frequencies and small tunneling splittings of the OH(OD) stretching fundamentals observed in Ne matrix isolation at 3121,

3102 cm^{-1} for TRN(OH) and 2344.8, 2340 cm^{-1} for TRN(OD). The successful computation of the OH and OD stretching frequencies depends on a curve-hopping at the major avoided crossing point lying in plots of the OH and OD stretching energies versus contortional displacement. Additional support for the TSTH model is discussed in the articles [34–36, 39, 86]. Spectroscopic experiments pending for tropolone include comprehensive high resolution FTIR spectra for TRN(OD) and its ^{18}O isotopomers – plus extension of the spectral range to the 300 cm^{-1} region for all TRN isotopomers.

1.4 Tropolone Derivatives

Sekiya and his collaborators [44–48] found spectral doublets in the S_1–S_0 excitation and dispersed fluorescence spectra of jet-cooled halotropolones that are modifications of the basic coherent tunneling splittings observed for TRN(OH) and TRN(OD). The symmetrical chlorotropolones showed enhancements of the spec-

Tab. 1.4 Tunneling splittings (cm^{-1}) in the chlorotropolones and bromotropolones [48].

Molecule	$\|\Delta_{v'} - \Delta_{0''}\|$			
	0_0^0	26_0^2	26_0^4	14_0^1
TRN(OH)	19	7	4	30
5-chloro-TRN(OH)	23	4	≤1	21
3,7-dichloro-TRN(OH)	45	7	≤1	53
3,5,7-trichloro-TRN(OH)	31	25	12	38
5-bromo-TRN(OH)	16	6	3	13[a]
3,7-dibromo-TRN(OH)	≤1			
TRN(OD)	2	≤1	≤1	11
5-chloro-TRN(OD)	2	≤1		
3,7-dichloro-TRN(OD)	8	1	≤1	6
3,5,7-trichloro-TRN(OD)	4	2	2	6
5-bromo-TRN(OD)	2	≤1		
3,7-dibromo-TRN(OD)	≤1			

a This 14_0^1 transition is only tentatively assigned.

tral doublets for the S_1–S_0 origin bands, while the bromotropolones showed damping. Spectral doublets are absent for the jet-cooled asymmetrical molecules because there is a large asymmetry offset of the energy minima of the double-well PESs [47]. Observations for the ZP, out-of-plane deformation, and nominal O···O stretching spectral doublets are shown in Table 1.4 [48]. The less pronounced consequences on the effective PESs, vibrational spectral doublets, and symmetries arising through the deuteration of the 3-, 4-, and 5- positions have been presented [41, 96].

The couplings between the OH···O and -NH_2 groups of the 5-amino-tropolone molecule have been of experimental and theoretical interest [52]. In part this is because the –NH_2 in the S_0 state may be pyramidal, and in the S_1 state it may be planar. The 5-hydroxy-tropolone molecule is planar with an asymmetric double-minimum PES, and part of its interest resides in the possibility of driving the tropolone tautomerization by laser pumping of the C(5)OH torsion vibration [49, 50, 97]. In the lowest energy configuration the two OH bonds point in opposing directions. Resolved spectral tunneling structures similar to those of the parent TRN(OH) and TRN(OD) molecules are observed for the aforementioned molecules, as well as others: 5-methyl-tropolone [53, 54], isopropyltropolones [55], and 5-phenyl-tropolone [56–57]. The effects of coupling the OH···O coordinates to the methyl internal rotation, or of coupling them to the low frequency phenyl torsional motions, are of interest.

1.5 Concluding Remarks

The zero-point tunneling splittings of S_0 malonaldehyde, $\Delta_0 = 21.583$ cm^{-1}, and S_0 tropolone, $\Delta_0 = 0.974$ cm^{-1}, plus values for S_0 MA isotopomers, are accurately known. The ZP splitting for TRN(OH) in the $\tilde{A}\,^1B_2$ (π^*–π), S_1, state is 19.846 cm^{-1} and that for MA(OH) in the $\tilde{A}\,^1B_1$ (π^*–n), S_1, state is ~ 2.5 cm^{-1}. A sizeable array of less precise vibrational state specific tunneling splittings is known for S_0 and S_1 TRN(OH). For S_0 MA(OH) this type of experimental evidence includes the value $\Delta_6 = 21.55$ cm^{-1} for ν_6 at 1594 cm^{-1}, several small doublet separations $DS_V \sim 0$ cm^{-1} implying $\Delta_V \sim \Delta_0$ in the midrange region, and doublet separations $DS_V = |\Delta_V - \Delta_0|$ of several cm^{-1} for each of the four fundamentals below 600 cm^{-1}. Spectral evidence clearly defining the splitting behavior and couplings for H motions in the OHO group of MA(OH) have not yet been recognized. The IR spectrum of S_0 TRN shows clear-cut tunneling doublets in the region below 950 cm^{-1}, evidence for small splittings in the midfrequency range, and DS = 19(5) cm^{-1} for OH(OD) stretching. All in all, the evidence supporting state-specific coherent tunneling behavior in these molecules is very strong and future experimental work can be expected to provide many accurately measured data points fully amenable to discriminating interpretive models and the discovery of novel intramolecular phenomena. Experimental effort, including overtone and combination spectra, clarifying the dynamical behavior of the OHO modes is a priority need. The ZP Δ_0 values for the S_0 MA and TRN isotopomers are reasonably reproduced by several different semiclassical and *ab initio* quantum chemical methods – in some cases remarkably well. The extant work shows the ZP tunneling splittings are very sensitive to the quality of the utilized *ab initio* input PES, and the scatter of results among the method-dependent computed splitting behavior verifies this is true for all energy states. The developments for coordinate systems and coordinate transformations for the kinetic and potential energy operators are well advanced and several full (21D) dimensional computations have been published for MA. The theoretical models use low-dimensioned reaction surface analysis for the large amplitude motions. PES expansions for the remaining orthogonalized, presumably tunneling inert, "bath" modes are truncated at the quadratic, and for some models the quartic, terms. It is important to have accurately computed energy values at distances well out from the expansion points, and the "Morsification" of the stretching coordinate expansions is an approach under examination [98]. Strong promoting modes for H tunneling are found to be sparse. The OH stretching PEF for S_0 TRN shown in Fig. 1.6 is an example with pronounced behavior of a type proposed for consideration of an enzyme–substrate complex [99]. Effects of the strong resonance couplings arising in some PES versions [61] are duly noted.

Acknowledgments

The author is very grateful for the help of T. E. Redington with this chapter.

References

1 Y. Cha, C. J. Murray, J. P. Klinman, *Science* **1989**, *243*, 1325.

2 T. Jonsson, M. H. Glickman, S. J. Sun, J. P. Klinman, *J. Am. Chem. Soc.* **1996**, *118*, 10319.

3 A. Kohen, R. Cannio, S. Bartolucci, J. P. Klinman, *Nature* **1999**, *399*, 496.

4 M. J. Knapp, J. P. Klinman, *Eur. J. Biochem.* **2002**, *269*, 3113.

5 J. Basran, M. J. Sutcliffe, N. S. Scrutton, *Biochemistry* **1999**, *38*, 3218.

6 M. J. Sutcliffe, N. S. Scrutton, *Eur. J. Biochem.* **2002**, *269*, 3096.

7 D. Antoniou, S. Caratzoulas, C. Kalyanaraman, J. S. Mincer, S. D. Schwartz, *Eur. J. Biochem.* **2002**, *269*, 3103.

8 P. S. Zuev, R. S. Sheridan, T. V. Albu, D. G. Truhlar, D. A. Hrovat, W. T. Borden, *Science* **2003**, *299*, 867.

9 W. F. Rowe Jr., R. W. Duerst, E. B. Wilson, *J. Am. Chem. Soc.* **1976**, *98*, 4021.

10 S. L. Baughcum, R. W. Duerst, W. F. Rowe, Z. Smith, E. B. Wilson, *J. Am Chem. Soc.* **1981**, *103*, 6296.

11 P. Turner, S. L. Baughcum, S. L. Coy, Z. Smith, *J. Am. Chem. Soc.* **1984**, *106*, 2265.

12 S. L. Baughcum, Z. Smith, E. B. Wilson, R. W. Duerst, *J. Am. Chem. Soc.* **1984**, *106*, 2260.

13 D. W. Firth, K. Beyer, M. A. Dvorak, S. W. Reeve, A. Grushow, K. R. Leopold, *J. Chem. Phys.* **1991**, *94*, 1812.

14 T. Baba, T. Tanaka, I. Morino, K. M. T. Yamada, K. Tanaka, *J. Chem. Phys.* **1999**, *110*, 4131.

15 Z. Smith, E. B. Wilson, R. W. Duerst, *Spectrochim. Acta, Part A* **1983**, *39*, 1117.

16 C. J. Seliskar, R. E. Hoffman, *J. Mol. Spectrosc.* **1982**, *96*, 146.

17 D. W. Firth, P. F. Barbara, H. P. Trommsdorff, *Chem. Phys.* **1989**, *136*, 349.

18 T. Chiavassa, P. Roubin, L. Pizzala, P. Verlaque, A. Allouche, F. Marinelli, *J. Phys. Chem.* **1992**, *96*, 10659.

19 C. Duan, D. Luckhaus, *Chem. Phys. Lett.* **2004**, *391*, 129.

20 C. J. Seliskar, R. E. Hoffman, *J. Mol. Spectrosc.* **1981**, *88*, 30.

21 A. A. Arias, T. A. W. Wasserman, P. H. Vaccaro, *J. Chem. Phys.* **1997**, *107*, 5617.

22 A. C. P. Alves, J. M. Hollas, *Mol. Phys.* **1972**, *23*, 927.

23 A. C. P. Alves, J. M. Hollas, *Mol. Phys.* **1973**, *25*, 1305.

24 R. L. Redington, T. E. Redington, *J. Mol. Spectrosc.* **1979**, *78*, 229.

25 R. Rossetti, L. E. Brus, *J. Chem. Phys.* **1980**, *73*, 1546.

26 Y. Tomioka, M. Ito, N. Mikami, *J. Phys. Chem.* **1983**, *87*, 4401.

27 A. C. P. Alves, J. M. Hollas, H. Musa, T. Ridley, *J. Mol. Spectrosc.* **1985**, *109*, 99.

28 R. L. Redington, Y. Chen, G. J. Scherer, R. W. Field, *J. Chem. Phys.* **1988**, *88*, 627.

29 R. L. Redington, R. W. Field, *Spectrochim. Acta, Part A* **1989**, *45*, 41.

30 H. Sekiya, Y. Nagashima, Y. Nishimura, *Bull. Chem. Soc. Jpn.* **1989**, *62*, 3229.

31 H. Sekiya, Y. Nagashima, Y. Nishimura, *J. Chem. Phys.* **1990**, *92*, 5761.

32 K. Tanaka, H. Honjo, T. Tanaka, H. Kohguchi, Y. Ohshima, Y. Endo, *J. Chem. Phys.* **1999**, *110*, 1969.

33 R. K. Frost, F. C. Hagemeister, C. A. Arrington, T. S. Zwier, K. D. Jordan, *J. Chem. Phys.* **1996**, *105*, 2595.

34 R. L. Redington, T. E. Redington, J. M. Montgomery, *J. Chem. Phys.* **2000**, *113*, 2304.

35 R. L. Redington, R. L. Sams, *J. Phys. Chem. A* **2002**, *106*, 7494.

36 R. L. Redington, R. L. Sams, *Chem. Phys.* **2002**, *283*, 135.

37 A. E. Bracamonte, P. H. Vaccaro, *J. Chem. Phys.* **2003**, *119*, 997.

38 A. E. Bracamonte, P. H. Vaccaro, *J. Chem. Phys.* **2004**, 120, 4638.

39 R. L. Redington, T. E. Redington, T. A. Blake, R. L. Sams, T. J. Johnson, *J. Chem. Phys.* **2005**, *122*, 224311.

40 H. Sekiya, Y. Nagashima, Y. Nishimura, *Chem. Phys. Lett.* **1989**, *160*, 581.

41 H. Sekiya, Y. Nagashima, T. Tsuji, Y. Nishimura, A. Mori, H. Takeshita, *J. Phys. Chem.* **1991**, *95*, 10311.

42 H. Sekiya, K. Sasaki, Y. Nishimura, Z.-H. Li, A. Mori, H. Takeshita, *Chem. Phys. Lett.* **1990**, *173*, 285.

43 R. L. Redington, T. E. Redington, M. A. Hunter, R. W. Field, *J. Chem. Phys.* **1990**, *92*, 6456.

44 H. Sekiya, K. Sasaki, Y. Nishimura, A. Mori, H. Takeshita, *Chem. Phys. Lett.* **1990**, *174*, 133.

45 T. Tsuji, H. Sekiya, Y. Nishimura, A. Mori, H. Takeshita, *J. Chem. Phys.* **1991**, *95*, 4801.

46 T. Tsuji, H. Sekiya, Y. Nishimura, R. Mori, A. Mori, H. Takeshita, *J. Chem. Phys.* **1992**, *97*, 6032.

47 T. Tsuji, H. Sekiya, S. Ito, H. Ujita, M. Habu, A. Mori, H. Takeshita, Y. Nishimura, *J. Chem. Phys.* **1993**, *98*, 6571.

48 H. Sekiya, T. Tsuji, S. Ito, A. Mori, H. Takeshita, Y. Nishimura, *J. Chem. Phys.* **1994**, *101*, 3464.

49 F. A. Ensminger, J. Plassard, T. S. Zwier, S. Hardinger, *J. Chem. Phys.* **1995**, *102*, 5260.

50 R. K. Frost, F. Hagemeister, D. Schleppenbach, G. Laurence, T. S. Zwier, *J. Phys. Chem.* **1996**, *100*, 16835.

51 T. Tsuji, H. Hamabe, Y. Hayashi, H. Sekiya, A. Mori, Y. Nishimura, *J. Chem. Phys.* **1999**, *110*, 966.

52 F. A. Ensminger, J. Plassard, T. S. Zwier, *J. Phys. Chem.* **1993**, *97*, 4344.

53 K. Nishi, H. Sekiya, H. Kawakami, A. Mori, Y. Nishimura, *J. Chem. Phys.*1998*, *109*, 1589.

54 K. Nishi, H. Sekiya, H. Kawakami, A. Mori, Y. Nishimura, *J. Chem. Phys.* **1999**, *111*, 3961.

55 H. Sekiya, H. Takesue, Y. Nishimura, Z-H. Li, A. Mori, H. Takeshita, *J. Chem. Phys.* **1990**, *92*, 2790.

56 T. Tsuji, Y. Hayashi, H. Sekiya, H. Hamabe, Y. Nishimura, H. Kawakami, A. Mori, *Chem. Phys. Lett.* **1997**, *278*, 49.

57 T. Tsuji, Y. Hayashi, H. Hamabe, H. Kawakami, A. Mori, Y. Nishimura, H. Sekiya, *J. Chem. Phys.* **1999**, *110*, 8485.

58 P. R. Bunker, P. Jensen, *Molecular Symmetry and Spectroscopy*, 2nd Edn., NRC Research Press, Ottawa 1998.

59 S. F. Tayyari, F. Milani-Nejad, *Spectrochim. Acta, Part A*, **1998**, *54*, 255.

60 A. Alparone, S. Millefiori, *Chem. Phys.* **2003**, *290*, 15.

61 V. A. Benderskii, E. V. Vetoshkin, I. S. Irgibaeva, H. P. Trommsdorff, *Chem. Phys.* **2000**, *262*, 393.

62 T. D. Sewell, Y. Guo, D. L. Thompson, *J. Chem. Phys.* **1995**, *103*, 8557.

63 R. Meyer, T.-K Ha, *Mol. Phys.* **2003**, *101*, 3263.

64 D. P. Tew, N. C. Handy, S. Carter, *Mol. Phys.* **2004**, *102*, 2217.

65 T. Carrington, W. H. Miller, *J. Phys. Chem.* **1986**, *84*, 4364.

66 N. Shida, P. F. Barbara, J. E. Almlöf, *J. Chem. Phys.* **1989**, *91*, 4061.

67 K. Yagi, G. V. Mil'nikov, T. Taketsugu, K. Hirao, H. Nakamura, *Chem. Phys. Lett.* **2004**, *397*, 435.

68 Z. Smedarchina, W. Siebrand, M. Z. Zgierski, *J. Chem. Phys.* **1995**, *103*, 5326.

69 K. Giese, O. Kühn, *J. Chem. Phys.* **2004**, *120*, 4107.

70 N. Došlic, O. Kühn, *Z. Phys. Chem.* **2003**, *217*, 1507.

71 C. S. Tautermann, A. F. Voegele, T. Loerting, K. R. Liedl, *J. Chem. Phys.* **2002**, *117*, 1962.

72 V. Barone, C. Adamo, *J. Chem. Phys.* **1996**, *105*, 11007.

73 S. Sadhukhan, D. Muñoz, C. Adamo, G. E. Scuseria, *Chem. Phys. Lett.* **1999**, *306*, 83.

74 K. Yagi, T. Taketsugu, K. Hirao, *J. Chem. Phys.* **2001**, *115*, 10647.

75 N. Makri, W. H. Miller, *J. Chem. Phys.* **1989**, *91*, 4026.

76 M. D. Coutinho-Neto, A. Viel, U. Manthe, *J. Chem. Phys.* **2004**, *121*, 9207.

77 G. V. Mil'nikov, H. Nakamura, *J. Chem. Phys.* **2001**, *115*, 6881.

78 G. V. Mil'nikov, K. Yagi, T. Taketsugu, H. Nakamura, K. Hirao, *J. Chem. Phys.* **2004**, *120*, 5036.

79 D. P. Tew, N. C. Handy, S. Carter, S. Irle, J. Bowman, *Mol. Phys.* **2003**, *101*, 2513.

80 L. M. Jackman, J. C. Trewella, R. C. Haddon, *J. Am. Chem. Soc.* **1980**, *102*, 2519.

81 N. Makri, W. H. Miller, *J. Chem. Phys.* **1987**, *86*, 1451.

82 R. L. Redington, T. E. Redington, *J. Chem. Phys.* **2005**, *122*, 124304.
83 M. J. Wójcik, H. Nakamura, S. Iwata, T. Wiktor, *J. Chem. Phys.* **2000**, *112*, 6322
84 H. Shimanouchi, Y. Sasada, *Acta. Crystallogr. Sect. B*, **1973**, *29*, 81.
85 H. F. Hameka, J. R. de la Vega, *J. Am. Chem. Soc.* **1984**, *106*, 7703.
86 R. L. Redington, *J. Chem. Phys.* **2000**, *113*, 2319.
87 M. V. Vener, S. Scheiner, N. D. Sokolov, *J. Chem. Phys.* **1994**, *101*, 9755.
88 Z. Smedarchina, W. Siebrand, M. Z. Zgierski, *J. Chem. Phys.* **1996**, *104*, 1203.
89 S. Takada, H. Nakamura, *J. Chem. Phys.* **1995**, *102*, 3977.
90 J. J. Paz, M. Moreno, J. M. Lluch, *J. Chem. Phys.* **1995**, *103*, 353.
91 Y. Guo, T. D. Sewell, D. L. Thompson, *J. Phys. Chem. A* **1998**, *102*, 5040.
92 K. Giese, O. Kühn, preprint.
93 C. S. Tautermann, A. F. Voegele, T. Loerting, K. R. Liedl, *J. Chem. Phys.* **2002**, *117*, 1967.
94 H. Rostkowska, L. Lapinski, M. J. Nowak, L. Adamowicz, *Int. J. Quant. Chem.* **2002**, *90*, 1163.
95 M. J. Wójcik, M. Boczar, M. Stoma, *Int. J. Quant. Chem.* **1999**, *73*, 275.
96 J. J. Paz, M. Moreno, J. M. Lluch, *Chem. Phys.* **1999**, *246*, 103.
97 J. J. Nash, T. S. Zwier, K. S. Jordan, *J. Chem. Phys.* **1995**, *102*, 5360.
98 R. Burcl, S. Carter, N. C. Handy, *Chem. Phys. Lett.* **2003**, *373*, 357.
99 A. Kohen, J. P. Klinman, *Chem. Biol.* **1999**, *6*, R191.

2
Coherent Proton Tunneling in Hydrogen Bonds of Isolated Molecules: Carboxylic Dimers

Martina Havenith

2.1
Introduction

Due to the exceptional importance of hydrogen bonds in biology and chemistry the detailed investigation of their structure and dynamics has attracted the attention of several experimental and theoretical groups. Laser spectroscopic measurements and quantum mechanical calculations have led, in recent years, to remarkable progress towards the detailed understanding of important prototype systems such as $(NH_3)_2$ [1], $(H_2O)_2$ [2], and DNA base pairs [3].

Double hydrogen bonded systems play a crucial role in that they serve as model systems for the understanding of DNA base pairs. Proton transfer is of fundamental interest for biology as well as being a fundamental reaction in chemistry [4, 5]. Moreover, multiple-proton transfer in hydrogen bonded systems is one of the most fundamental processes in biology and chemistry. It governs oxidation–reduction reactions in many chemical and biological reactions [6]. In proton pumping mechanisms of trans membrane proteins protons are transported across the membrane by subsequent proton transfer. These reactions may incorporate strong quantum effects due to the low mass of the proton. The observation of pronounced isotopic effects is taken as an indication of a strong tunneling contribution. In contrast, the classical transmission probability is zero as long as the energy of the particle is lower than the barrier height ($V^{\ddagger}$) and is equal to 1 if the energy (E) exceeds this. For an ensemble at temperature T this leads to the well known Arrhenius equation for the reaction rate which is proportional to exp $(-(E - V^{\ddagger})/kT)$. The transmission will, therefore, depend solely on the barrier height and not on the width or on the exact shape of the transition barrier. Isotopic substitution will shift the zero point energies relative to the transition barrier and will, therefore, lead to a change in the reaction constant.

Quantum mechanical tunneling is a result of the wavelike nature of particles which allows transmission through a reaction barrier. The quantum mechanical transmission probability for energies below $V^{\ddagger}$ is governed by tunneling and reflection at the barrier. The transmission is larger than zero even well below the barrier and will depend crucially on the barrier width. In

Hydrogen-Transfer Reactions. Edited by J. T. Hynes, J. P. Klinman, H. H. Limbach, and R. L. Schowen

ISBN: 978-3-527-30777-7

the WKB approximation for 1-dim potentials the quantum mechanical transmission coefficient can be approximated by

$$P(E) = \exp\left(-1/\hbar \int_{-a}^{a} \sqrt{2\mu(V_a(x) - E)}\mathrm{d}x\right) \qquad (1)$$

with a, V_a and E corresponding to the classical turning points, the adiabatic barrier and the energy, respectively. This yields an exponential dependence on the reduced mass characterizing the tunneling motion. The weighting function for the contribution of thermally populated states is described by a Boltzmann distribution ($\propto \exp(-E/kT)$). When raising the temperature, over barrier processes become more and more feasible, implying that tunneling becomes less important.

2.2 Quantum Tunneling versus Classical Over Barrier Reactions

Recently, a controversial debate has arisen about whether "the optimization of enzyme catalysis may entail the evolutionary implementation of chemical strategies that increase the probability of tunneling and thereby accelerate reaction rates" [7]. Kinetic isotope effect experiments have indicated that hydrogen tunneling plays an important role in many proton and hydride transfer reactions in enzymes [8, 9]. Enzyme catalysis of horse liver alcohol dehydrogenase may be understood by a model of vibrationally enhanced proton transfer tunneling [10]. Furthermore, the double proton transfer reaction in DNA base pairs has been studied in detail and even been hypothesized as a possible source of spontaneous mutation [11–13].

For identification and understanding of the hydrogen transfer mechanism the nature of transfer has been the subject of several studies either supporting [14, 15] or questioning [16] the above hypothesis. Several criteria were raised to test the dominant contribution for enzyme catalyzed reactions [17], such as:

1. The elevated primary isotope kinetic isotope effect (KIE) (with the ratio of the rate constants k_H/k_D being larger than 8.9 at 20 °C)
2. A large difference in the activation energy for the different isotopes
3. A small ratio of the pre-exponential Arrhenius factors ($A_H/A_D<0.7$)
4. The Swain Schaad relationship which yields boundaries for the ratios of the rate constants for the primary kinetic isotope effects for H, D and T with deviations being interpreted as evidence of tunneling.

However, one has to point out that in a very recent theoretical study by Tautermann et al. [17] all these criteria are found to be inadequate to predict the dominant contribution of tunneling for a small benchmark system such as formic acid dimer.

In order to distinguish between the different contributions a comparison with sophisticated theoretical calculations is mandatory. However, it is necessary to test these models on small prototype systems. High resolution gas phase spectroscopy can now provide exact experimental data on proton transfer tunneling and thereby allow a test of tunneling criteria which could be used for more complex systems.

Gas phase experiments are ideally suited for this since the perturbations induced by the local environment are completely eliminated, providing an ideal tool for exact measurements. Studies of coherent proton transfer of isolated molecules in the gas phase are able to yield absolute numbers on proton transfer tunneling without being influenced by solvent effects or the interaction of a crystal surrounding. Moreover, high resolution spectroscopy can yield state specific tunneling frequencies. These studies allow direct comparison with the results of *ab initio* studies and will therefore provide a sensitive test for recently developed theoretical methods.

In summary, direct measurements of proton transfer tunneling in double hydrogen bonded dimer complexes in the gas phase are therefore of great interest.

2.3 Carboxylic Dimers

Carboxylic dimer systems are well suited to a study of coherent proton tunneling processes since they provide two identical hydrogen bonds. For the most stable of these, such as formic acid dimer or benzoic acid dimmer, the displacement of the protons can be well described by a double well potential. The barriers are high compared to the zero point energy, since these species are bound by two cooperatively strengthened $OH \cdots O = C$ hydrogen bonds. Quantum mechanically, the two protons which are involved in the hydrogen bonds can tunnel, even if the transition barrier is much higher than their kinetic energy. It is a point of discussion whether the coherent proton transfer can be described as a synchronous concerted double proton transfer or whether an asynchronous movement such as a step by step process will give an adequate picture. Most theoretical calculations propose a synchronous concerted proton transfer process for the carboxylic dimers, as is displayed in Fig. 2.1, where the two monomers first approach each other before tunneling can occur.

The proton transfer gives rise to a splitting of each rotational–vibrational state into two states (E_l and E_u) which are separated by $\Delta E = h\nu_{\text{tunneling}}$ with $\nu_{\text{tunneling}}$ describing the tunneling frequency for double proton transfer. Correspondingly, the proton transfer time τ is related to the tunneling splitting $\nu_{\text{tunneling}}$ by: $\tau = 1/(2\nu_{\text{tunneling}})$ [18].

Spectroscopically, tunneling will manifest itself in a doubling of the observed vibrational–rotational transitions. In the case of finite barriers the wave function of the upper and lower tunneling states can be described by the following wave functions:

$$\psi_{\mathrm{l}} = 1/\sqrt{2}\,(\psi_{\mathrm{L}} + \psi_{\mathrm{R}}) \text{ and } \psi_{\mathrm{u}} = 1/\sqrt{2}\,(\psi_{\mathrm{L}} - \psi_{\mathrm{R}}).$$

The wave functions ψ_{L} and ψ_{R} describe the dimer in one of the double minima (with R and L standing for right or left, respectively). The tunneling splitting ΔE depends sensitively on the barrier height V_{barrier} and on the barrier width. For carboxylic acid dimers the tunneling process cannot be described by a 1-dimensional coordinate but has to be described on a multidimensional potential energy surface. Obviously, the proton is not localized, so the effective barrier and thus the tunneling splitting will depend crucially on the part of the potential surface which has been sampled by the wave function during the optimal tunneling path. In a publication by Tautermann et al. [19], a semiclassical method was used to predict the energy barriers and the tunneling splittings of the numerous carboxylic acid dimers, such as formic acid dimer, benzoic acid dimer, carbamic acid dimer, fluoro acid dimer, carbonic acid dimer, glyoxylic acid dimer, acrylic acid dimer and *N,N*-dimethyl carbamic acid dimer and their deuterated analogs. The calculated

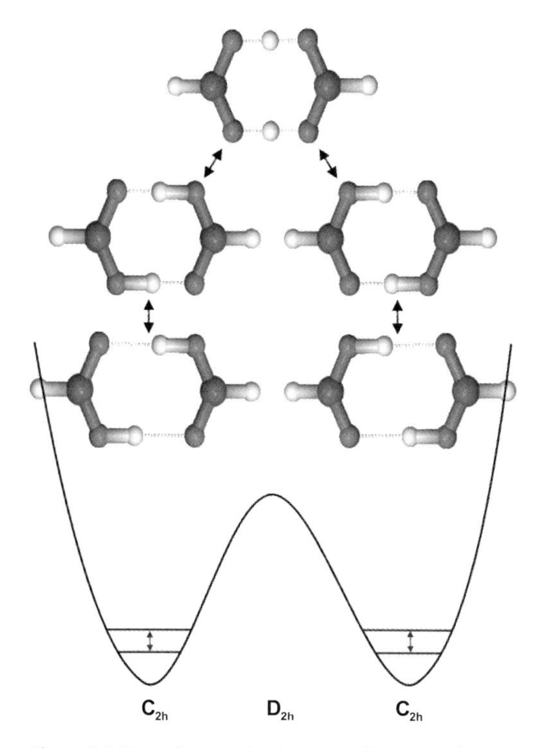

Figure 2.1 Tunneling motion in carboxylic dimer. The equilibrium structure corresponds to C_{2h} symmetry. In the transition state the dimer resembles D_{2h} symmetry.

barrier heights vary from 6.3 to 8.8 kcal mol^{-1}. The lowest barrier was found for carbamic acid dimer derivatives and the largest for the formic acid dimer, which was explained by a mesomeric stabilization of the transition state being completely missing for the formic acid dimer. For formic acid dimer, benzoic acid dimer, carbamic acid dimer, fluoro acid dimer and *N,N*-dimethyl carbamic acid dimer the proton transfer occurs from global minimum to global minimum and can be described by a double well potential, such as is displayed in Fig. 2.1. For glyoxylic acid dimer, acrylic acid and carbonic acid dimer the global minimum corresponds to a trans symmetry, as defined by the $C_2 = O$ groups for the first two. In the case of a double proton transfer the final configuration differs from the initial structure for these dimers and corresponds to an energetically higher state. Hence, no ground state tunneling splitting is expected. Only the energetically less stable cis conformers are expected to show a detectable tunneling splitting. The calculated tunneling splitting along the most probable tunneling path varies from 0.00053 cm^{-1} for fluoro acid dimer to 0.13 cm^{-1} for *N,N*-dimethyl carbamic acid dimer [19]. The variation was attributed to different barrier heights (smallest for *N,N*-dimethyl carbamic acid dimer) and, for equal barrier heights such as for formic acid dimer and fluoro acid dimer, to differences in the length of the most probable tunneling path.

The most recent calculations on the tunneling dynamics of double proton transfer in formic acid dimer and benzoic acid dimers have been reported by Smedarchina et al. [20]. They used direct dynamics calculations to predict the tunneling splittings of both carboxylic dimers.

The corresponding ground state tunneling splitting was predicted to be 441 MHz (0.015 cm^{-1}) for $(DCOOH)_2$ and 1920 MHz (0.064 cm^{-1}) for benzoic acid dimer. The increased splitting in the benzoic acid dimer is attributed to the stronger hydrogen bonding in the benzoic compared to the formic acid dimer.

With the tunneling splitting being very small, sophisticated high resolution spectroscopy techniques are required for an exact determination of this fundamental quantity. Whereas microwave spectroscopy has been proven to provide the ultimate frequency resolution, it is not suited to a study of cyclic carboxylic dimers since – due to their inherent symmetry – they have no permanent dipole moment. Therefore, the only spectroscopically accessible transitions are vibration–rotation–tunneling transitions in the infrared spectral region. For aromatic dimers such as benzoic acid dimer electronic–vibrational–rotational–tunneling transitions in the optical or ultraviolet spectral region are also accessible.

The exact determination of such a small splitting requires molecular beam techniques. The molecules are cooled in a supersonic expansion which leads to a population distribution among the rotational states corresponding to a Boltzmann distribution with 10–12 K. This leads to a considerable reduction of populated states and thereby to a simplification of the spectrum. In a supersonic expansion we find a considerable decrease in the line width due to the decrease in translational motion, especially when the laser is perpendicular to the expansion direction. The corresponding reduction in Doppler width allows an improved frequency resolution, which will later be shown to be crucial for the success of the experiment.

In the following we will present the results for two benchmark systems, the benzoic acid dimer and the formic acid dimer, which have been extensively studied in the past, both experimentally and theoretically. The study of these dimers in the gas phase allows a direct comparison between them.

2.4 Benzoic Acid Dimer

2.4.1 Introduction

Benzoic acid dimer was the first carboxylic dimer to be investigated in detail spectroscopically. The first studies involved the measurements of temperature dependent relaxation times via NMR. Due to the presence of two inherent chromophores it can also be investigated by means of optical spectroscopy.

The benzoic acid dimer is stabilized by two hydrogen bonds, with the two monomers being coplanar in the electronic ground state [21]. The proton transfer is shown to strongly couple to the heavy nuclei, implying that the whole benzoic acid frame rearranges itself slightly as the protons tunnel.

2.4.2 Determination of the Structure

X-ray structural studies showed that benzoic acid dimers crystallize in a monoclinic cell with the hydrogens aligned close to the crystallographic *c*-axis [22]. Time of flight neutron diffraction measurements were carried out by Brougham et al. [23] and these yielded the crystal structure of a deuterated benzoic acid. The structure of the monomer units is assumed to be unchanged upon complexation.

Remmers et al. measured the high resolution spectrum of benzoic dimer in the UV excitation spectrum of the $S_1 \leftarrow S_0$ transition and were able to deduce rotational constants for the ground and electronically excited states [24]. When assuming that the monomer structure remains unchanged upon complexation, the value of the intermolecular distance r and the geometry of the two monomers with respect to each other could be deduced. The value of $r = 3.622 \pm 0.147$ Å (r describing the distance from the center of mass of the monomer unit to the center of rotation) was found to be in good agreement with the previously determined O···O distance of 3.58 Å, as was deduced by Brougham et al. [23]. For the ground state a planar linear structure was anticipated. For the excited state a more bent structure was proposed, with the best agreement being for a bending angle of $\alpha = 3.4°$ (with α describing the angle between the longitudinal axis of the dimer and that of the monomer unit) and a reduced intermolecular distance of $r = 3.6$ Å. This structural change can be explained by the fact that the electronic excitation reduces the acidity of the monomer so that the two hydrogen bonds will no longer be equivalent [20].

2.4.3
Barriers and Splittings

All theoretical studies on benzoic acid dimer underlined the need for a multidimensional potential surface. These studies have investigated the temperature dependence of the transfer process: They included a density matrix model for hydrogen transfer in the benzoic acid dimer, where bath induced vibrational relaxation and dephasing processes are taken into account [25]. Sakun et al. [26] have calculated the temperature dependence of the spin–lattice relaxation time in powdered benzoic acid dimer and shown that low frequency modes assist the proton transfer. At high temperatures the activation energy was found to be 2.3 kcal mol^{-1}. The same assisted proton transfer was found in calculations of the synchronous proton transfer in benzoic acid crystal by Antoniou and Schwarz [27]. It was shown, that for benzoic acid dimer a coupling to the low frequency intramolecular modes will lower the effectively required activation energy.

In all these calculations it was found that although the activation energy seems to be rather low, which might support the occurrence of classical over barrier reactions, the actual process is, however, a pure quantum mechanical tunneling process.

The benzoic acid dimer has been extensively studied in various condensed phase environments where it experiences an asymmetric local environment. By doping the crystal the inherent asymmetry in the crystal was reduced. However, in general, the crystal structure and the dopant can influence the observed tunneling matrix element.

From these experiments the tunneling matrix element of (dye-doped) benzoic acid was determined and was found to be relatively independent of the dopant used [28, 29]. The measured values lie between 8.4 GHz (0.28 cm^{-1}) [28] and 6.56 GHz (0.22 cm^{-1}) [29].

The level structure of a benzoic acid dimer tautomer state of a pair of dimers sandwiching the dye molecule will depend on three parameters: The energy difference between the two tautomer forms in the crystal, the coupling of the two dimers stabilizing the degenerate configuration, and the tunneling matrix element J. In general, all parameters will depend on the specific electronic state. It should be noted that for a free dimer and the splitting will depend only on the tunneling matrix element. In dye- (seleno indigo) doped benzoic acid crystals the energy level structure of a pair of benzoic acid dimers was measured by hole burning spectroscopy [29]. Under the assumption that the tunneling matrix element is independent of the electronic state of the guest a tunneling matrix element of 6.5 ± 1.56 GHz for the benzoic acid dimer was deduced, yielding information on the proton transfer tunneling.

In 2000 Remmers et al. [24] used the combination of molecular beam techniques and high resolution laser spectroscopy (inherent line width 3 MHz) to obtain the first information on proton transfer tunneling in any isolated gas phase carboxylic acid dimer. They reported the measurement of a high resolution ultraviolet spectrum of benzoic acid dimer in the gas phase. The spectra showed two complete electronic rovibrational spectra with a constant frequency separation of (1107 ± 7 MHz) between both sets.

The transitions are labeled as a-type ($\Delta K_a = 0$, $\Delta K_c = 1$) transitions, b-type ($\Delta K_a = 1$, $\Delta K_c = 1$) or c-type transitions depending on whether absorption of a photon involved a change in the projection of the angular momentum along the main axes of inertia which are called the *a*- *b*- or *c*-axes. By definition, the moments of inertia are labeled as I_A, I_B and I_C with $I_A < I_B < I_C$. For carboxylic acid dimers the *a*-axis corresponds to the intermolecular axis connecting the two monomers. The *c*-axis corresponds to the out of plane axis. The transition dipole moment lies predominantly along the *b*-axis, which implies that in the spectrum we find preferentially b-type transitions involving a change in the projection of the angular momentum along the *a*-axis: $\Delta K_a = 1$.

After electronic excitation of one of the monomers to the electronic excited S_1 state the symmetry is broken and one expects two asymmetric potential energy surfaces where the electronic excitation takes place in one of the monomers. However, since both monomers are indistinguishable, coupling by an interaction term allows the electronic excitation to be exchanged between the monomers, which leads to two symmetrical potential curves, each consisting of a double well potential. In each double well potential tunneling causes an additional tunneling splitting. As a result, we obtain a potential energy scheme as is indicated in Fig. 2.2 [24]. Allowed transitions require a change of parity. Altogether, we expect to see four transitions; two of them were assigned to correspond to the experimental data. The observed splitting was attributed to the difference between the tunneling splitting in the excited and ground states.

In their analysis, Remmers et al. assumed that the splitting in the ground state amounts to 0.22 cm^{-1} (6500 MHz). This value corresponds to the previously determined tunneling matrix element [29]. The measured value ($\Delta = 1107 \pm 7$ MHz), which corresponds to the difference between the tunneling splitting in the ground and excited states, would imply that the ground state tunneling splitting is larger than the splitting in the excited state

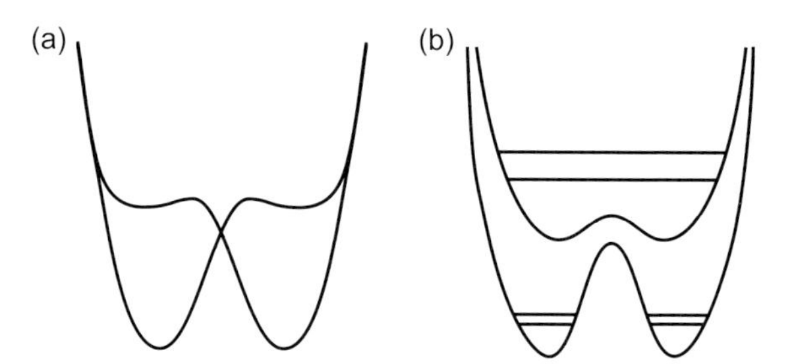

Figure 2.2 Potential energy surface of benzoic acid dimer. The ground state and electronically excited state are displayed. The potential for excitation of a single monomer (a) and after adiabatic coupling of the potential energy surface (b) are shown. This figure is adapted from Ref. [24].

However, in general, it is expected that the tunneling splittings in the electronically excited state will be much larger than the splittings in the ground states. (E.g., in dye-doped crystals splitting in the electronic excited state exceeds the splittings in the ground state by far (35 cm^{-1} versus < 1 cm^{-1}) [28].) In agreement with this expectation, Tautermann et al. predicted a tunneling splitting of 0.0022 cm^{-1} (6.6 MHz) for benzoic acid dimer [19]. If we assume this prediction to yield reliable predictions for the ground state of benzoic acid dimer as well (e.g. assume that the tunneling splitting of the ground state of benzoic acid dimer is of the order of 0.0022 cm^{-1}, which is well below experimental resolution in this rather congested spectrum) the measured splitting would have to be mainly attributed to the tunneling splitting in the excited state, which might far exceed the splitting in the ground state. The measured difference of tunneling splitting of 1107 MHz for the gas phase benzoic acid dimer would then correspond to the splitting in the excited state but would yield no further information on the tunneling splittings in the ground state.

However, based upon a direct dynamical calculation, which used a potential surface for benzoic acid dimer at a B3LYP/6-31 + G(d) level [20], Smedarchina et al. suggest that the hydrogen bonding is weaker in the excited state, in agreement with the experimentally observed increase in the intermolecular distance *r* [24]. This weakening of the bond is expected to yield a decrease in the tunneling splitting. They conclude, therefore, that the measurements are consistent with a splitting dominated by the ground state with a small or negligible contribution from the excited state. In summary, one can state that in order to come to an unambiguous assignment, further high resolution gas phase studies which determine the ground state tunneling splitting independently would be required .

2.4.4 Infrared Vibrational Spectroscopy

Following the study of Remmers et al. a fluorescence dip spectrum of benzoic acid dimer in the gas phase has been reported by Florio et al. [30]. Using this technique, the IR spectra of three isotopomers (the undeuterated, the ring-deuterated and the mixed OH/OD ring deuterated dimer) have been recorded showing a broad substructure in the region of the O–H stretching fundamental. Using a theoretical model they were able to explain the observed features in terms of large anharmonic effects which couple the O–H stretch with combination bands of appropriate symmetry. The resulting band structure is responsible for the wide spread and the observed substructure of the O–H in the region between 2600 and 3100 cm^{-1}. However, the influence of the intermolecular hydrogen bond vibrations in the electronic ground state, which were discussed extensively in the early theoretical papers, were found to result in negligible coupling to the O–H stretch.

In a newer study the fluorescence spectrum has been reinvestigated with improved spectral resolution. Accompanying DFT calculations suggested that in the electronically excited state the low frequency modes are mixed extensively with the carboxyl as well as the aromatic ring vibrations [31]. The experimentally observed geometry distortion was explained as a change in the acidic character of benzoic acid in the excited state.

2.5 Formic Acid Dimer

2.5.1 Introduction

The smallest organic complex serving as the prototype for multiple-proton transfer is the formic acid dimer. Formic acid dimerizes like all carboxylic acids with two hydrogen bonds forming a planar eight-membered ring between the two monomers (Fig. 2.1). The study of the double proton transfer in the hydrogen bonded dimer of formic acid has therefore attracted the interest of theoreticians for more than 50 years. As discussed before, the tunneling frequency will depend very sensitively on the height and the width of the barrier, which implies that high quality *ab initio* calculations are required for an exact description of the complete potential surface.

2.5.2 Determination of the Structure

Early theoretical work on FAD was concerned with the dimer equilibrium geometry and the electronic structure (see for example Refs. [32–36]). *Ab initio* molecular orbital studies on the structure of formic acid dimer in 1984 agreed very well with the experimental structures as determined by electron diffraction [37]. Due to the importance of the double proton transfer of FAD as a key prototype for multiple proton transfer reactions several theoretical studies have been reviewed in the literature [38]. Rotational constants for formic acid dimer were obtained by high resolution spectroscopy of $(DCOOH)_2$ [39] and by femtosecond degenerate four wave mixing experiments in the gas cell at room temperature and under supersonic jet experiments by Matylisky et al. [40].

The best agreement is obtained with the pure *ab initio* predictions by Neuheusser et al. [33] using MP2 with counterpointwise correction for basis set superposition error and a TZ2P basis set. These calculations predict an $O \cdots O$ distance of 2.672 Å. DFT calculations [41] yield a smaller $O \cdots O$ distance of 2.645 Å with a hence increased B-value. The MP2 calculations by Kim [42] and the extended molecular mechanics calculation slightly overestimated $O \cdots O$ (2.707 Å) thereby yielding a slightly too small B-value. The same argument holds for the results of the electron diffraction measurements which obtained an $O \cdots O$ distance of 2.696 Å. These values are in reasonable agreement with the rotational constants as predicted by the potential of Yokoyama et al. [43]. They postulated a semiempirical model potential which was set up to reproduce the experimental vibrational frequencies and the bond lengths in the monomers.

If we compare the structure in the MP2-study of Neuheusser with the structure as predicted by DFT [41] we see that they also differ in the position of the inner hydrogen atom (for example the angle COH). The relative changes in the rotational constant upon deuteration agree within (0.1–0.6)% when taking the structure as proposed by Hobza et al. [44]. More specifically : A comparison of the rotational constants for $(HCOOH)_2$ by Matylitsky et al. [40] with the rotational constants from Madeja and Havenith [39] $(DCOOH)_2$ via model structures based upon the results of *ab initio* calculations [44] yield an agreement within (0.1–0.6)%. However, both sets of rotational parameters disagree with the structure as deduced by electron diffraction studies [37].

Due to the inherent experimental limitations of the electron diffraction measurements the position of the inner hydrogen is expected to be less well determined. A displacement of the inner H-atom will affect the A rotational constant, which may explain the deviation between the rotational A-constant of the gas phase measurement [39, 40] and the constants as deduced from the diffraction measurements. We suggest that the deviation between the measured rotational constants and the rotational constants as would be predicted from the structure as deduced by electron diffraction can mainly be attributed to a difference in the intermolecular distance (for example O · · · O or O · · · H).

More specifically, the experimental changes in the rotational constants upon deuteration do not agree with the changes expected when the structure of the electron diffraction study holds. This confirms that this structure, which was obtained very early, although providing an excellent reference for the gas phase dimer structure at that time, probably suffers from the inherent problem of the exact position of the hydrogen atoms and from averaging effects due to zero point motions.

2.5.3
Tunneling Path

In 1995 Kim reported a direct dynamics calculation [42]. Their actual tunneling path is visualized in Fig. 2.1. During the proton transfer the heavy atoms move first to bring two formic acid monomers closer together, thereby decreasing the barrier for proton transfer. At the transition state the structure of formic acid dimer changes from C_{2h} to D_{2h} symmetry. Proton transfer gives rise to a splitting of each rotational–vibrational state separated by $\Delta E = h\nu$. The tunneling splitting ΔE depends sensitively on the barrier height $V_{barrier}$ and on the barrier width.

In the work by Shida et al. [45] a synchronous concerted double proton transfer is found to be the major mode of the reaction, which is confirmed in a recent molecular dynamics calculation by Wolf et al. [46]. In his molecular dynamics calculation the time evolution of the potential energy was additionally taken into account. The double proton transfer has also been investigated in a Car-Parinello *ab initio* molecular mechanics study by Miura et al. [41]. Quantum fluctuations are shown to cause significant deviations from the minimum energy path. While an asynchronous movement of the two protons close to the equilibrium structure was stated, synchronous movement becomes relevant on approaching the reaction barrier. However, in the study of Ushiyama and Takatsuka [47] the tunneling was described as a successive step by step process.

In a very recent calculation by Markwick et al. targeted molecular dynamics methods were implemented in the framework of Car-Parinello molecular dynamics to study the nature of the double proton transfer [48]. They predict a concerted proton transfer reaction. In the very early stages of this reaction the system enters a vibrationally excited pretransitional state. Whereas in the global minima large amplitude fluctuations have been found in the pretransitional region, the frequency of these fluctuations is found to increase dramatically while the amplitude of the oscillation decreases when approaching the transition state.

2.5.4
Barriers and Tunneling Splittings

The reported tunneling splitting for the proton transfer in the literature covers more than four orders of magnitudes. In 1981 a paper reported double proton transfer times in formic acid dimer between 95 min and 0.3 ns [49]. More recent calculations of transit times for the proton transfer still cover more than four orders of magnitude: Chang et al. predicted a proton transfer tunneling splitting of 0.3 cm^{-1} corresponding to a proton transfer time of $\tau = 55$ ps [36]. In 1991 Shida, Barbara and Almlöf used a reaction surface Hamiltonian to predict the proton transfer barrier for FAD [45]. Using MCPF (modified couple pair functionals) they predicted a tunneling splitting of 0.004 cm^{-1} ($\tau = 4.2$ ns). Molecular dynamics calculations by Ushiyama and Takatsuka [47] yielded transfer times of less than 150 fs which is 3–4 orders of magnitude faster than any of the previous calculations. It should be noted that in a paper by Vener et al. [50], using a similar approach to that of Shida et al., a tunneling splitting of 0.3 cm^{-1} instead of 0.004 cm^{-1} was calculated. Tautermann et al. [19] predict a ground state tunneling splitting of 0.0022 cm^{-1} for formic acid dimer (being equal to the predicted ground state tunneling splitting of benzoic acid dimer), when they apply a semiclassical method to determine the ground state tunneling splitting of various carboxylic acid derivative dimers. The energy barriers were calculated using hybrid density functional theory (B3LYP) and compared to the result of a high level method Gaussian theory (MP2).

Correspondingly, the effective barrier heights and the proposed tunneling path have also varied extensively in the literature of the last decade: The reported barrier heights include 14 kcal mol^{-1} [32]; 7.1–9.1 kcal mol^{-1} [32]; 15.6 kcal mol^{-1} [36]; 11.8 kcal mol^{-1} [45]; 8.9 kcal mol^{-1} calculated [46], 11.5 kcal mol^{-1} [50] and 8.39 kcal mol^{-1} [19].

2.5.5 Infrared Vibrational Spectroscopy

Experimentally, a very detailed spectroscopic study of carboxylic acids in the gas phase including various isotopic forms of FAD has been published by Maréchal [51]. He recorded FTIR spectra in the region 500–4000 cm^{-1}. IR band centers and relative intensities are reported. One vibrational band at 1740 cm^{-1} was reported for the C = O asymmetric stretch of FAD. Later, a high resolution study yielded two separate vibrational bands centered at 1738.5 and 1741.5 cm^{-1} [52]. However, the high density of lines did not allow a definitive assignment in this frequency range. Anomalous isotope effects in the intensities of the ($\nu_{C=O}$) band of FAD [51] and Fourier transform (FT) studies of DCOOH [53] indicate a resonance between the C = O asymmetric stretch and combination bands, which considerably complicates the assignment of the gas phase spectrum. For $(DCOOH)_2$ a Raman study including the IR band heads [54] is reported in the literature. There are no spectroscopic studies in the MW region due to the lack of a permanent dipole moment.

The first high resolution spectroscopic measurements of FAD with fully resolved rotational–vibrational–tunneling transitions were reported by Madeja and Havenith in 2002 [39]. They measured the C–O vibrational band of $(DCOOH)_2$ which could be analyzed in terms of an asymmetric top rigid rotor Hamiltonian. The vibrational frequency of the C–O stretch in $(DCOOH)_2$ was determined to be 1244.8461(2) cm^{-1}. This deviated considerably from values for the band center as given by Wachs et al. [55] (1231.85 cm^{-1}). Previous measurements include the Raman transition as reported by Bertie et al. (1230 $\pm$ 2 cm^{-1}) [54] and the value obtained by Millikan and Pitzer (1239 cm^{-1}) [56].

Very recently high resolution studies of $(HCOOH)_2$ in the gas phase have also been carried out by Ortlieb and Havenith, yielding a vibrational frequency of ν_0 = 1225.35 cm^{-1}. This also deviates significantly from the values of the vibrational band center (1215 cm^{-1}) as given in the paper by Maréchal [51]. The discrepancy can be explained by the fact that the vibrational band centers were determined at room temperature. In this case hot bands will contribute considerably to the observed low resolution spectrum and combination bands can also overlap with the fundamental. In consequence, both effects will shift the observed band center compared to the fundamental band.

In a paper by Florio et al. [57] a theoretical modeling of the O–H stretch in carboxylic dimers was reported. In general, large anharmonic effects and Fermi coupling result in a complicated band substructure in which the O–H stretch oscillator strength is spread over hundreds of wavenumbers. This can be the reason for the observed substructure of the first excited O–H stretch of 140 cm^{-1} in the paper by Robertson and Lawrence. Whereas good agreement with the experimentally observed vibrational substructure of the O–H stretching infrared spectrum of benzoic acid dimer [30] was found, the agreement with the experimental cavity ring down spectrum of Ito and Nakanaga [58] was rather disappointing. A complete FTIR spectrum of formic acid in the gas phase recorded in a supersonic jet expansion with a limited experimental resolution of 0.5 cm^{-1} was reported by Georges et al. [59].

We further want to mention an infrared matrix isolation study of FAD in argon matrices [60]. The study of Gantenberg et al. indicates the presence of a second structure, besides the cyclic double hydrogen bonded structure, which was identified as an acyclic dimer. The special dimerization process of FAD in ultracold matrices is found to be a more general phenomenon: In 2004 the observation of the same polar acyclic single hydrogen bonded structure was reported in suprafluid liquid helium droplets, establishing the unique cluster growth process at ultracold temperatures [61]. Based upon these experimental results a new alternative dimerization mechanism, in which two formic acid monomers proceed through an acyclic dimer to the cyclic dimer in a stepwise process, has been investigated theoretically and the corresponding barrier and dissociation energy predicted [62].

2.5.6
Coherent Proton Transfer in Formic Acid Dimer

We will now focus in this subsection on the experimental study providing accurate information on the proton transfer tunneling in isolated carboxylic acid dimers in the gas phase.

In an early paper by Robertson and Lawrence [63] the tunneling splitting of the ground state was estimated to be about 6 cm^{-1}. They assumed that the observed fine structure in the first excited O–H stretch of 140 cm^{-1} is attributed to the transfer tunneling splitting. However, in view of new theoretical studies and the recent experimental results [39], their conclusions were probably based on a misassignment.

In order to determine exact quantitative data it is necessary to obtain high resolution spectra of the dimer. Since a resolution of ≤ 0.001 cm^{-1} is beyond the scope of FTIR spectrometers, laser spectroscopic techniques were required. This could be achieved in an IR study using sensitive absorption measurements by a lead salt diode laser.

In the paper of Madeja and Havenith the observation of 409 lines in the frequency range from 1241.7 to 1250.7 cm^{-1} was reported. All lines could be assigned to the C–O stretch vibrational band. The splitting due to proton transfer tunneling was fully resolved (see Fig. 2.3). The frequency range from 1241.6 to 1242.6 cm^{-1} is displayed; it can clearly be seen that each rotational vibrational line is split into two fully resolved tunneling components.

For the selection rules a planar structure showing C_{2h} symmetry was anticipated. When assuming a synchronous tunneling motion, the molecular symmetry group corresponds to the permutation-inversion group G_8 which is isomorphic to D_{2h} (see Ref. [58] for further details). A detailed characterization of the different symmetries and their spin statistical weights can be found in Refs. [39, 64].

The proton transfer tunneling splittings in the ground and vibrationally excited states can be determined separately by measuring simultaneously two types of

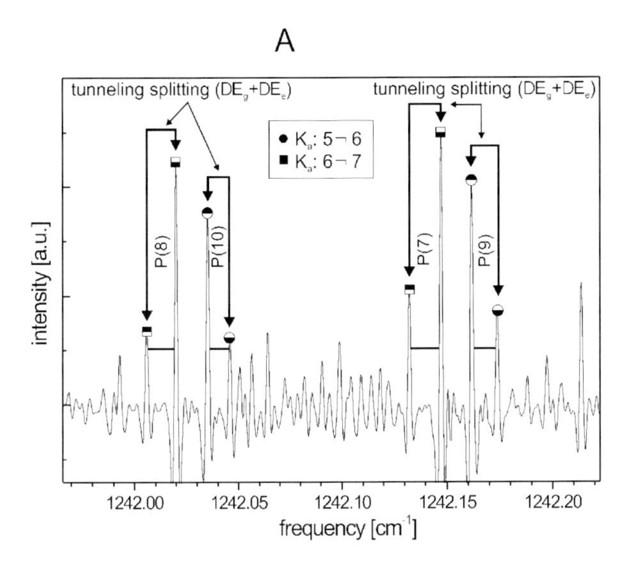

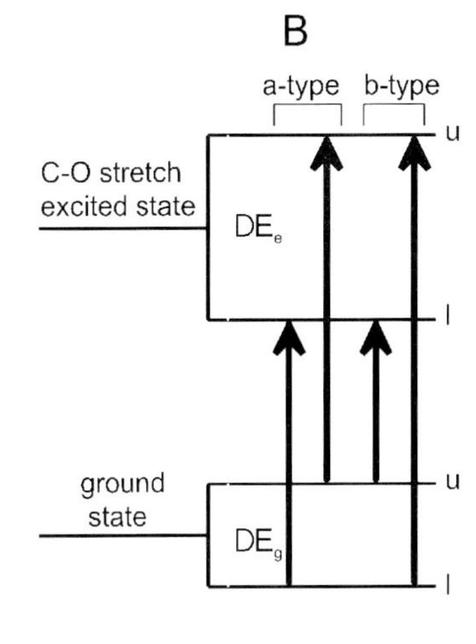

Figure 2.3 Zoomed part of the high resolution infrared spectrum of formic acid dimer showing the experimentally resolved proton transfer splitting. Each rovibrational transition is split into two, originating from different tunneling components. The expected intensity ratio is 3:1 due to different spin statistical weights. This figure is adapted from Ref. [39].

transitions (a- and b-type transitions) corresponding to distinct relative orientations of the change in angular momentum with respect to the axes of inertia of the molecule. For b-type transitions the selection rules require a change of tunneling state upon vibrational excitation; for a-type transitions upper to upper and lower to lower selection rules hold. By measuring *both* a- and b-type transitions, the sum and the difference of the tunneling splitting in the ground and vibrationally excited states can obtained simultaneously (see Fig. 2.3).

The direction transition dipole moment for the infrared spectrum is given by the dipole derivative of the C–O stretch relative to the different axis of inertia. A quantitative estimation would involve high level *ab initio* calculations. However, qualitatively the change in the dipole moment during the antisymmetric C–O stretch is expected also to have a projection along the intermolecular axis (*a*-axis) as well as perpendicular to this (*b*-axis). No out of plane motions are expected.

Both tunneling components with an approximate intensity ratio of 1 / 3 (in agreement with the corresponding spin statistics) were observed in the measurement [39]. Most lines (179) were assigned to b-type transitions from the upper tunneling state in the vibrational ground state to the lower tunneling state in the C–O stretch. Altogether 66 a-type transitions were observed. The a-type transitions were found to be weaker and are centered at the band origin, a frequency region which was experimentally difficult to access. Due to a lack of precise intensity measurements for the weaker a-type transitions it was difficult to assign whether they have to be attributed to transitions starting in the lower or upper tunneling component. In Ref. [39] two possible assignments were discussed which yielded the following tunneling splittings in the ground and vibrationally excited state, corresponding to different assignments: For the ground state tunneling splitting either 0.00286(25) or 0.0123(3) cm^{-1} can be obtained. For the excited state tunneling splitting 0.00999(21) cm^{-1} or alternatively 0.0031(3) cm^{-1} is determined. As long as the vibrational coordinate can be completely separated from the tunneling coordinate, the tunneling splitting should be unaffected by the vibrational excitation.

Initially [39], it was argued that an acceleration of the proton transfer upon vibrational excitation seems more likely, which favors the value of 0.00286(25) cm^{-1} for the ground state. This was based upon the fact that the vibrational analysis involves a mean decrease in the distance between the opposite O atoms of the two monomeric units. The reaction barrier height to double proton transfer is lowered, which is expected to decrease the proton transfer time and increase the tunneling splitting. Correspondingly, an enhancement of the tunneling splitting was observed for malonaldehyde when the O–O stretching vibration was excited [48, 65].

Whereas Tautermann et al., with a value of 0.0022 cm^{-1}, confirm the initial assignment, new IR studies in the carbonyl stretching region unambiguously show the second assignment to be correct with a tunneling splitting of 0.0125(3) cm^{-1} and 0.0031(3) cm^{-1} for the ground state and vibrationally excited state, respectively [64].

In 2004 and 2005 direct dynamics calculation based on instanton techniques are reported for tunneling splittings in the formic acid dimer and its deuterated isotopomers for the zero-point level and the lowest vibrationally excited level of a non-totally symmetric C–O stretch vibration [20, 66]. The calculated splittings of Smerdarchina et al. are in good agreement with the experimental observations for $(DCOOH)_2$ with the now confirmed reversed assignments. The deceleration upon vibrational excitation is attributed to the anti-symmetric nature of the vibration which handicaps a symmetrical proton transfer. The zero-point tunneling splitting of formic acid dimer is calculated to be 510 MHz, which is a factor of 1.4 larger than the observed splitting. This discrepancy is explained by the tendency of the B3LYP potential to underestimate the barrier height and width. This overestimation of the tunneling splitting would be consistent with the result for benzoic acid dimer: For the free dimer Smedarchina predicts a tunneling splitting of 1920 MHz which is a factor of 1.7 larger than the experimentally observed splitting. The significantly increased tunneling splitting in dye-doped crystals could also be explained. When inducing three symmetric lattice vibrations in the calculations they predict a tunneling splitting of 17 GHz for zero asymmetry, which can be compared with the tunneling matrix element of 8.4 ±1 GHz for the indigo-doped crystals [66].

2.6 Conclusion

In conclusion, high resolution spectroscopy allows an exact quantitative measurement of important dynamical processes. We have summarized the results for two benchmark systems: benzoic acid dimer and formic acid dimer.

The tunneling motion in carboxylic dimers is shown to be slow compared to the overall rotation, thus confirming the existence of deep local minima in a C_{2h} structure.

The proton transfer time of formic acid dimer in the electronic ground state is of the order of ns and can be compared with recent theoretical predictions which covered more than four orders of magnitude.

It is interesting to note that, in 2004, one additional example of proton transfer dynamics has been reported which was obtained via high resolution spectroscopy measurements in the gas phase [67]. Roscioli et al. reported the $S_1 \leftarrow S_0$ fluorescence excitation spectrum of 2-hydroxypyridine/2-pyridine dimer, which showed two subbands separated by 527 MHz. The experimentally determined observed splitting is of the same order of magnitude as the corresponding values for benzoic acid dimer (1107 MHz). Similar to the $S_1 \leftarrow S_0$ transition of benzoic acid dimer, these spectra only yield information on the difference between the tunneling splittings in the ground and electronically excited state; the values could not be determined independently. The proton transfer between the two bridging hydrogen atoms is estimated to occur in the order of ns, which is relatively long compared to the predicted values but very similar to the experimental values for formic acid dimer (nsec).

High resolution gas phase spectroscopy has now been able to nail down the size of the tunneling splitting for proton transfer in double hydrogen bonded systems.

If we come back to the question of whether the current tunneling criteria with enzyme supported reactions show the nature of the proton transfer process, formic acid dimer serves as a crucial test candidate. In the most recent theoretical study of Tautermann et al. it was shown that the tunneling criteria based on the calculated reaction rates for various isotopomers fail to predict tunneling behavior correctly: Instead of unraveling the pure quantum mechanical tunneling behavior of the proton transfer process in formic acid dimer the criteria indicate a classical over barrier process. It would be important to see whether the experimental data confirm these findings. In general, this shows that the measurement of small aggregates in the gas phase can add another aspect to the ongoing "Tunneling Enhancement by Enzyme" discussion.

Exact knowledge of the proton transfer in carboxylic dimers constitutes an important step towards the development of theoretical methods. While we have not tested the reliability of theoretical models on simple benchmark systems we probably will not be able to simulate quantitatively proton transfer tunneling in biological systems in the future.

References

1 D. D. Nelson Jr., G. T. Fraser, W. Klemperer, *Science* **1988**, *238*, 1670.
2 K. Liu, J. D. Gruzan, R. S. Saykally, *Science* **1996**, *271*, 929.
3 E. Nir, K. Kleinermanns, M. S. de Vries, *Nature* **2000**, *408*, 949.
4 G. A. Jeffrey, W. Saenger, *Hydrogen Bonding in Biological Structures*, Springer Verlag, Berlin, 1991.
5 H. H. Limbach, J. Manz, *Ber. Bunsenges. Phys. Chem.* **1998**, *102*, 289.
6 S. Hammes-Schiffer, *J. Chem. Phys.* **1996**, *105*, 2236.
7 B. J. Bahnson, J. P. Klinmann, *Methods Enzymol.* **1995**, *249*, 373.
8 A. Kohen, J. P. Klinman, *Chem. Biol.* **1999**, *6*, 191.
9 St. Scheiner, *Bioenergetics* **2000**, *1458*, 28.
10 B. J. Bahnson, T. D. Colby, J. K. Chin, B. M. Goldstein, J. P. Klinman, *Proc. Natl. Acad. Sci. USA*, **1997**, *94*, 12797.
11 P.-O. Lödin, *Rev. Mod. Phys.* **1963**, *35*, 724.
12 J. Florian, V. Hrouda, P. Hobza, *J. Am. Chem. Soc.* **1997**, *116*, 1457.
13 J. Maranon, A. Fantoni, J. R. Grigera, *J. Theor. Biol.* **1999**, *201*, 93.
14 A. Kohen, J. P. Klinmann, *Acc. Chem. Res.* **1998**, *31*, 397.
15 S. R. Billeter, S. P. Webb, P. K. Agarwal, T. Iordanov, S. Hammes-Schiffer, *J. Am. Chem. Soc.* **2001**, *123*, 11262.
16 K. M. Doll, B. R. Bender, R. G. Finke, *J. Am. Chem. Soc.* **2003**, *125*, 10877.
17 Ch. S. Tautermann, M. J. Loferer, A. F. Voegele, K. R. Liedl, *J. Chem. Phys.* **2004**, *120*, 11650.
18 J. M. Hollas, *High Resolution Spectroscopy*, Wiley & Sons, Sussex 1998.
19 Ch. S. Tautermann, A. F. Voegele, K. R. Liedl, *J. Chem. Phys.* **2004**, *120*, 631.
20 Z. Smedarchina, A. Fernández-Ramos, W. Siebrand, *J. Chem. Phys.* **2005**, *122*, 134309.
21 S. Nakaoka, N. Hirota, T. Matsuhita, K. Nishimoto, *Chem. Phys. Lett.* **1982**, *92*, 498.
22 R. Feld, M. S. Lehmann, K. W. Muir, J. C. Speakmann, *Z. Kristallogr.* **1981**, *157*, 215.
23 D. F. Brougham, A. J. Horsewill, A. Ikram, R. M. Ibberson, P. J. MacDonald, M. Pinter-Krainer, *J. Chem. Phys.* **1996**, *105*, 979.
24 K. Remmers, W. L. Meerts, I. Ozier, *J. Chem. Phys.* **2000**, *112*, 10890.
25 Ch. Scheurer, P. Saalfrank, *Chem. Phys. Lett.* **1995**, *245*, 201.
26 V. P. Sakun, M. V. Vener, N. D. Sokolov, *J. Chem. Phys.* **1996**, *105*, 379.
27 D. Antoniou, St. D. Schwarz, *J. Chem. Phys.* **1998**, *109*, 2287.
28 A. Oppenländer, Ch. Rambaud, H. P. Trommsdorff, J. Vial, *Chem. Phys.*, **1989**, *136*, 335.
29 Ch. Rambaud, H. P. Trommsdorf, *Chem. Phys. Lett.* **1999**, *306*, 124.
30 G. M. Florio, E. L. Sibert III, T. S. Zwier, *Faraday Discuss.* **2001**, *118*, 315.
31 Ch. K. Nandi, T. Chakraborty, *J. Chem. Phys.* **2004**, *120*, 8521.
32 S. Hayashi, J. Umemura, S. Kato, K. Morokuma, *J. Phys. Chem.* **1984**, *88*, 1330.
33 Th. Neuheusser, B. A. Hess, Ch. Reutel, E. Weber, *J. Phys. Chem.* **1994**, *98*, 6459.
34 C. Mijoule, M. Allavena, J. M. Leclerco, Y. Bouteiller, *Chem. Phys.* **1986**, *109*, 207.
35 W. Quian, S. Krimm, *J. Phys. Chem.* **1998**, *102*, 659.
36 Y.-T. Chang, Y. Yamaguchi, W. H. Miller, H. F. Schaeffer III, *J. Am. Chem. Soc.* **1987**, *109*, 7245.
37 A. Almeningen, O. Bastiansen, T. Motzfeld, *Acta Chem. Scand.* **1969**, *23*, 2848.
38 B. S. Jursic, *J. Mol. Struct.* **1997**, *417*, 89.
39 F. Madeja, M. Havenith, *J. Chem. Phys.* **2002**, *117, 7162.*
40 V. V. Matylitsky, C. Riehn, M. F. Gelin, B. Brutschy, *J. Chem. Phys.* **2003**, *119*, 10553.
41 Sh. Miura, M. E. Tuckermann, M. L. Klein, *J. Chem. Phys.* **1998**, *109*, 5290.
42 Y. Kim, *J. Am. Chem. Soc.* **1995**, *118*, 1522.
43 I. Yokoyama, Y. Miva, K. Madida, *J. Am. Chem. Soc.* **1991**, *113*, 6458.
44 J. Chocholoušová, J. Vacek, P. Hobza, *Phys. Chem. Chem. Phys.* **2004**, *4*, 2119.

45 N. Shida, P. F. Barbara, J. Almlöf, *J. Chem. Phys.* **1991**, *94*, 3633.
46 K. Wolf, A. Simperler, W. Mikenda, *Monatsh. Chem.* **1999**, *130*, 1031.
47 H. Ushiyama, K. Takatsuka, *J. Chem. Phys.* **2001**, *115*, 5903.
48 Ph. R. C. Markwick, N. L. Doltsinis, D. Marx, *J. Chem. Phys.* **2005**, *122*, 54112.
49 A. Agresti, M. Bacci, A. Ranfagni, *Chem. Phys. Lett.* **1981**, *79*, 100.
50 M. V. Vener, O. Kühn, J. M. Bowman, *Chem. Phys. Lett.* **2001**, *349*, 562.
51 Y. Maréchal, *J. Chem. Phys.* **1987**, *87*, 6344.
52 U. Merker, P. Engels, F. Madeja, M. Havenith, W. Urban, *Rev. Sci. Instrum.* **1999**, *70*, 1933.
53 K. L. Goh, P. P. Ong, T. L. Tan, *Spectrochim. Acta, Part A* **1999**, *55*, 2601.
54 J. E. Bertie, K. H. Michaelian, H. H. Eysel, D. J. Hager, *J. Chem. Phys.* **1986**, *85*, 4479.
55 T. Wachs, D. Borchardt, S. H. Bauer, *Spectrochim. Acta, Part A* **1987**, *43*, 965.
56 R. C. Millikan, K. S. Pitzer, *J. Am. Chem. Soc.* **1958**, *80*, 3515.
57 G. M. Florio, T. Zwier, E. M. Myshakin, K. D. Jordan, E. L. Sibert III, *J. Chem. Phys.* **2003**, *118*, 1735.
58 F. Ito, T. Nakanaga, *Chem. Phys.* **2002**, *277*, 163.
59 R. Georges, M. Freytes, D. Hurtmans, I. Kleiner, J. VanderAuwera, M. Herman, *Chem. Phys.* **2004**, *305*, 187.
60 M. Gantenberg, M. Halupka, W. Sander, *Chem. Eur. J.* **2000**, *6*, 1865.
61 F. Madeja, M. Havenith, K. Nauta, R. E. Miller, J. Chocholoušová, P. Hobza, *J. Chem. Phys.* **2004**, *120*, 10554.
62 N. R. Brinkmann, G. S. Tschumper, G Yau, H. F. Schaefer III, *J. Phys. Chem. A* **2003**, *107*, 10208.
63 R. Robertson, M. C. Lawrence, *Chem. Phys.* **1981**, *62*, 131.
64 M. Ortlieb, M. Havenith, in preparation.
65 N. Sato, S. Iwata, *J. Chem. Phys.* **1988**, *89*, 2932.
66 Z. Smedarchina, A. Fernández-Ramos, W. Siebrand, *Chem. Phys. Lett.* **2004**, *395*, 339
67 J. R. Roscioli, D. W. Pratt, Z. Smedarchina, W. Siebrand, A. Fernández-Ramos, *J. Chem. Phys.* **2004**, *120*, 11351.

3
Gas Phase Vibrational Spectroscopy of Strong Hydrogen Bonds

Knut R. Asmis, Daniel M. Neumark, and Joel M. Bowman

3.1
Introduction

The hydrogen bond interaction is key to understanding the structure and properties of water, biomolecules, self-assembled nanostructures and molecular crystals. However, much confusion remains about its electronic nature, a combination of van der Waals, electrostatic and covalent contributions, leading to a wide variety of hydrogen bonds with bond strengths ranging from 2 to 40 kcal mol^{-1}. In particular, our understanding of strong, low-barrier hydrogen bonds and their central role in enzyme catalysis [1], biomolecular recognition [2], proton transfer across biomembranes [3] and proton transport in aqueous media [4] remains incomplete. The central aim of this chapter is to outline some recent advances in the research on strongly hydrogen bonded model systems in the gas phase with emphasis on the work from our research groups.

Strong hydrogen bonds (A···H···B) are often classified based on their hydrogen bond energy; a typically cited lower limit is >15 kcal mol^{-1} [5]. Their most prominent physical properties are large NMR downfield chemical shifts and considerably red-shifted hydrogen stretch frequencies. Moreover, the H-atom transfer barrier, a characteristic feature of weak hydrogen bonds A–H···B, is either absent or very small in these systems (at their minimum energy geometry). Consequently, the H-atom in homoconjugated (A = B) strong hydrogen bonds is equally shared by the two heavy atoms forming two identical strong hydrogen bonds. This symmetry is lost in heteroconjugated (A ≠ B) systems, but the H-atom remains in a more centered position, i.e., the distance between the heavy atoms is smaller than in weaker hydrogen bonded systems. Strong hydrogen bonds can either be low-barrier, as in $(HO\cdots H\cdots OH)^-$, or single-well, as in $(Br\cdots H\cdots I)^-$, depending on the form of the potential curve along the H-atom exchange coordinate (see Fig. 3.1 and below).

Hydrogen bonds are very sensitive to perturbation, due to an intimate interdependence between the heavy atom separation, the H-atom exchange barrier and the position of the light H-atom leading to unusually high proton polarizabilities. Therefore it can prove advantageous to study strong hydrogen bonds in the gas

Hydrogen-Transfer Reactions. Edited by J. T. Hynes, J. P. Klinman, H. H. Limbach, and R. L. Schowen

ISBN: 978-3-527-30777-7

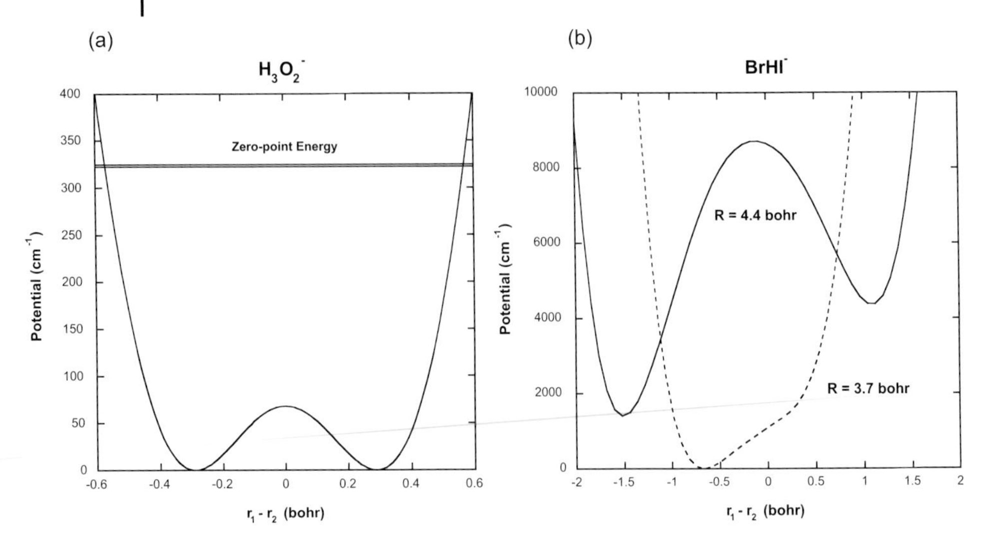

Figure 3.1 Typical potential energy curves for strong, low-barrier hydrogen bonds. The homoconjugated $H_3O_2^-$ (a) exhibits a (relaxed) symmetric, double-well potential as a function of the difference of the bridging hydrogen distance with the two oxygen atoms. The heteroconjugated $BrHI^-$ (b) is characterized by an asymmetric single-well potential at equilibrium and an asymmetric double well at a larger Br–I⁻ distance.

phase, in the absence of any perturbations from surrounding solvent or host molecules. Standard experimental techniques to study strong hydrogen bonds, including NMR, as well as X-ray and neutron diffraction, are currently limited to condensed phase probes. Gas phase experiments are hindered by low number densities and only vibrational spectroscopy exhibits the required sensitivity and selectivity to perform such studies.

Recently, advances in laser technology as well as in computational approaches have allowed significant progress in the study of strongly hydrogen bonded model systems. We first describe these improved experimental and theoretical methods and then discuss experiments and calculations on three prototypical systems containing strong hydrogen bonds: $BrHBr^-$, $BrHI^-$ and $H_5O_2^+$. These results demonstrate that the vibrational spectroscopy of triatomic systems involving strong hydrogen bonds has now been successfully solved, even when heavy atoms like iodine are involved. However, the study of slightly larger systems, like the protonated water dimer, remains challenging.

3.2 Methods

3.2.1 Vibrational Spectroscopy of Gas Phase Ions

Vibrational spectroscopy paired with quantum chemistry currently offers the most direct and generally applicable experimental approach to structural investigation of neutral and charged clusters in the gas phase [6]. Direct absorption measurements based on ion discharge modulation methods [7] can yield high resolution spectra of small and light molecular ions. Problems associated with high discharge temperatures can nowadays be overcome by using pulsed-slit supersonic expansions [8]. However, these types of experiments become increasingly difficult for larger and heavier molecular ions, particularly ion clusters, owing to spectral congestion, lower gas phase number densities and the presence of other absorbing species. Therefore alternative techniques have been developed in which the absorption of photons can be measured indirectly, by way of resonance enhanced photodissociation (or action) spectroscopy. Photodissociation techniques have the advantage that fragment ions can be detected background-free and with nearly 100% detection efficiency. A high selectivity can be achieved through mass selection of parent and fragment ions using appropriate mass filters.

An infrared photodissociation (IR-PD) spectrum is measured by irradiating ions with infrared radiation and monitoring the yield of fragment ions as a function of the irradiation wavelength. In order to induce fragmentation the parent ion AB^+ (the same line of argumentation holds for negative ions) is required to absorb sufficient energy to overcome the (lowest) dissociation threshold. Once a metastable ro-vibronic state is reached, intramolecular energy redistribution will eventually lead to dissociation, producing a charged and a neutral fragment:

$$AB^+ \xrightarrow{n \cdot h\nu} A^+ + B \tag{3.1}$$

As noted before, the dissociation energy of strong hydrogen bonds is roughly 5000 cm^{-1} or higher, while the fundamentals of the shared H-atom modes are found well below this limit. Process (3.1) therefore requires the absorption of multiple infrared photons. The coherent stepwise multiphoton excitation, where all photons are absorbed in one vibrational ladder, is unfavorable and becomes unrealistic for higher dissociation thresholds, because the laser falls out of resonance owing to the anharmonicity of vibrational potentials. The multiphoton process is better viewed as a sequential absorption of photons enhanced by rapid intramolecular vibrational energy redistribution at higher excitation [9]. Only the first few photons are absorbed in the "discrete" regime, in which a particular mode is excited resonantly. Higher excitation accesses the "quasi-continuum", in which the density of states is so high that the vibrational energy is rapidly randomized among all vibrational modes of the molecular ion. The ion continues to absorb photons until it has enough energy to dissociate. The transition between the two regimes depends on the vibrational density of states and the strengths of the interactions between vibrational modes. For larger molecular systems, the IR-PD spectrum often resembles the linear absorption spectrum [10]. For smaller systems with less internal degrees of freedom, the relative intensities may be different. However, if the laser fluence is kept at a moderate level, signal is only detected if the laser wavelength is resonant with a fundamental transition, that is,

$$AB^{+}(\nu = 0) \xrightarrow{h\nu} AB^{+}(\nu = 1) \xrightarrow{n \cdot h\nu} A^{+} + B \tag{3.2}$$

At high laser fluence the probability of directly exciting overtones is enhanced, complicating the interpretation of the IR-PD spectrum [11].

A useful method to avoid multiphoton excitation and measure IR-PD spectra in the linear regime is the messenger atom technique [12]:

$$AB^{+} \cdot Rg \xrightarrow{h\nu} (AB^{+} \cdot Rg)^{*} \longrightarrow AB^{+} + Rg \tag{3.3}$$

By forming ion–rare gas atom (Rg) complexes, the dissociation threshold of the system is lowered, generally below the photon energy and these predissociation spectra directly reflect the linear absorption spectrum. This technique has also been used to great effect in anion spectroscopy experiments [13]. The multiphoton dissociation approach remains attractive for systems in which the perturbation of the messenger atom cannot be neglected, or in instruments where rare gas attachment is difficult.

IR-PD experiments generally require intense and tunable radiation sources which, for many years, were only commercially available in the wavelength region up to ~4 μm (>2500 cm^{-1}), that is the region of X–H (X = C, N, O, halogen atom) stretches and overtones. The spectral signature of strong hydrogen bonds, however, is found at longer wavelengths. The application of free electron lasers (FELs) to molecular spectroscopy by Meijer and coworkers has bridged this gap [14]. More recently, several groups [15, 16] were able to access the region below 2000 cm^{-1} with higher energy, narrow bandwidth table-top laser systems, which make use of difference frequency mixing in an $AgGaSe_2$ crystal [17]. Even though the pulse energy in these table-top systems is roughly three orders of magnitude smaller than from FEL sources and generally not sufficient to perform multiphoton absorption experiments, it is enough to photodissociate messenger atom–ion complexes [16].

3.2.2
Experimental Setup

The experiments described here were performed on a guided ion beam tandem mass spectrometer [18] that was temporarily installed at the free electron laser facility FELIX (free electron laser for infrared experiments, FOM Institute for Plasma Physics, Nieuwegein, The Netherlands) [19]. A schematic of the experimental setup is shown in Fig. 3.2. Ions are generated in the ion source region (not

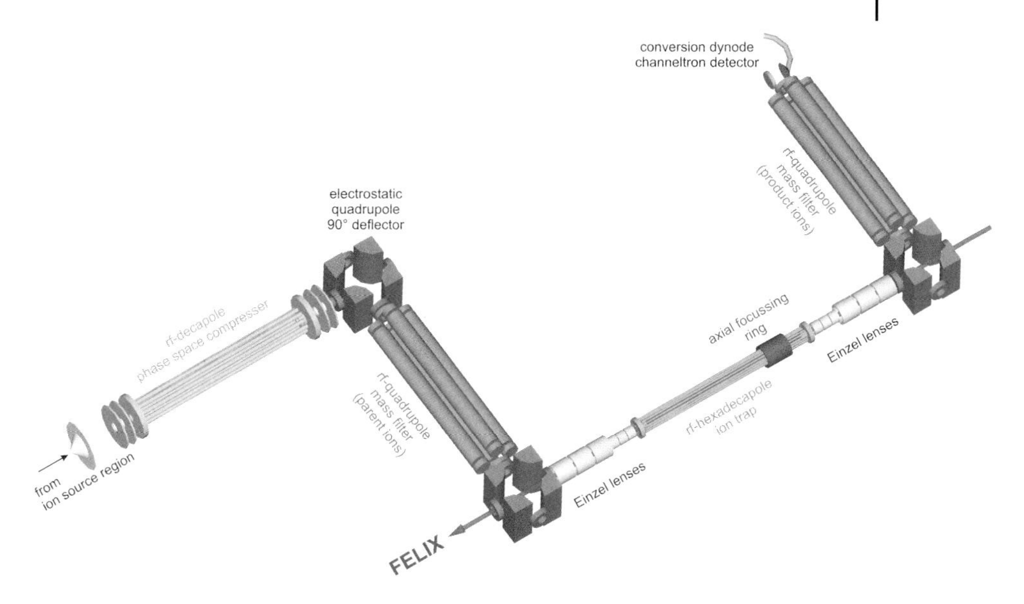

Figure 3.2 Schematic of the guided-ion-beam tandem mass spectrometer used in the present studies [60]. The instrument is housed in a five-stage differentially pumped vacuum chamber. The FEL radiation is applied collinearly to the axis of the ion trap.

shown) using either a pulsed supersonic jet crossed by a 1 keV electron beam for the bihalide anions or an ion spray source for the protonated water clusters. The ion beam, comprising a distribution of cluster ions of different size, is collimated and compressed in phase space in a gas-filled radio frequency (RF) ion guide and directed into the first quadrupole mass filter. Mass-selected ions are then guided into a temperature-adjustable RF ion trap. The trap consists of a linear RF ion guide and two electrostatic ion lenses contained in a cylindrical housing, which is connected to the cold head of a closed-cycle He cryostat. The cylinder is continuously filled with He (~0.2 mbar). The use of a buffer gas has several advantages: (i) The trap can be operated in a continuous ion-fill-mode. (ii) Trapped ions are collisionally cooled to the ambient temperature (approximately within a few milliseconds). Experiments can currently be performed at temperatures between 14 and 350 K.

IR-PD spectra are obtained by photoexcitation of the trapped ions with the pulsed FEL radiation and subsequent monitoring of the fragment ion signal. The FEL generates 5 μs macropulses at a repetition rate of 5 Hz. Each macropulse contains a series of several thousand ~1 ps micropulses with about 1 ns between micropulses. A new measurement cycle is triggered by the previous FEL macropulse. First, the ion trap is filled with mass-selected ions; typically for 150 ms. The trap is then closed and the ions are allowed to thermalize. Directly after FELIX fires, all ions are extracted and the mass-selected ion yield is monitored. This cycle is repeated multiple times, the signal is summed and then the FEL is set to the next wavelength. Overview spectra are measured first, in the region from 5.5 to

18 μm with a step size of 0.1 μm. Spectra with smaller step sizes and longer accumulation times are then measured in those spectral regions where signal is observed. The accuracy of the determined vibrational frequencies is generally within 1% of the central wavelength. The accuracy of the relative depletion intensities is less well defined, mainly due to the non-monotonic variation of the FELIX beam intensity, bandwidth, and waist size with wavelength. We try to minimize these variations and do not correct for them in our spectra.

3.2.3 Potential Energy Surfaces

Rigorous theoretical modeling of IR spectra of hydrogen bonded complexes consists of two parts. The first is the calculation of the potential energy surface and the dipole moment. The second is appropriate dynamics calculations using the potential and dipole moment surfaces. For simulations of the IR spectrum quantum, semi-classical or purely classical dynamics calculations are straightforward to do, in principle, in the weak-field limit. Practically, quantum dynamics calculations can be quite computationally demanding, depending on the number of degrees of freedom. Both aspects of this theoretical modeling are briefly reviewed next, followed by a review of recent applications to $BrHI^-$ and $H_5O_2^+$.

For triatomic molecules, even with heavy atoms, such as $BrHI^-$, it is now straightforward to obtain global potential energy surfaces, based on highly accurate *ab initio* calculations. One fairly common approach is to obtain the *ab initio* energies (and dipole moment) on a regular grid of roughly ten points per degree of freedom and to use interpolation, e.g., splines, to obtain the potential between grid points. This approach will be described below for $BrHI^-$. For larger systems, such as $H_5O_2^+$, the proton bound water dimer, and $H_3O_2^-$, mono-hydrated hydroxyl, the grid approach is impractical. For these systems a "scattered" approach has proved successful. In this approach tens of thousands of *ab initio* calculations are done at configurations of relevance to the dynamics, including stationary points, etc. The *ab initio* data are then fit using standard least-squares procedures with, however, some important new features of the basis functions. In the least-squares approach the data are represented by the compact expression

$$V(x_1, \cdots x_d) = \sum_{n=0} C_n g_n(x_1, \cdots x_d) \tag{3.4}$$

where x_i are the variables of the fit, g_n are a known set of linearly independent basis functions, and the C_n are coefficients that are determined by the least-squares procedure. The key to a successful fit to the data is the choice of basis functions and the variables of the fit. Somewhat surprisingly, fitting approaches done prior to our work did not explicitly incorporate permutational symmetry of like atoms into the basis functions. This was done recently in applications to CH_5^+ [20], $H_3O_2^-$ [21], and $H_5O_2^+$ [22]. The approach is to use all internuclear distances, r_i, as the basic variables and x_i is a suitable function of r_i. These variables form a closed set under permutation of atoms. Then, symbolically, V is given by

$$V(x_1, \cdots x_d) = \sum_{i_1 \cdots i_d} C_{i_1 \cdots i_d} S\{x_1^{i_1} \cdots x_d^{i_d}\} \tag{3.5}$$

where the symbol S indicates a symmetrization operator so that the symmetrized monomial $S\{x_1^{i_1} \ldots x_d^{i_d}\}$ is invariant with respect to any permutation of like nuclei. The approach we actually use is more involved and is described elsewhere [22].

With this new approach we have been able to obtain full-dimensional potentials using high quality *ab initio* calculations of electronic energies and dipole moments. Some of these results for $H_5O_2^+$ will be presented below.

3.2.4 Vibrational Calculations

As with potential surfaces, exact vibrational calculations for triatomics are essentially a "solved" problem. There are several numerically equivalent exact approaches that are currently in use. These basically differ in the choice of coordinates. Our recent calculations on $BrHI^-$ made use of two such approaches; one used so-called Jacobi coordinates and the other used internal valence coordinates, i.e., the BrH, HI^- bond lengths and the $BrHI^-$ bond angle. The kinetic energy operator in these coordinates is complex and so we refer the reader to the original literature [23] instead of giving it here. In Jacobi coordinates this operator is much simpler, and the full Hamiltonian for an "ABC" triatomic for a given value of the total nuclear angular momentum $\boldsymbol{J}$ (in a rotating frame) is given by

$$H = T_R + T_r + \frac{(\boldsymbol{J} - \boldsymbol{j})^2}{2\mu R^2} + V(R, r, \gamma) \tag{3.6}$$

where R is the distance of one atom, say A, to the center of mass of the diatom, say BC, r is the diatom internuclear distance, γ is the angle between the vectors $\boldsymbol{R}$ and $\boldsymbol{r}$, $\boldsymbol{J}$ is the total nuclear angular momentum operator, $\boldsymbol{j}$ is the diatom angular momentum operator, and V is the full potential.

For the $BrHI^-$ calculations, reviewed below, codes based on Jacobi coordinates and valence coordinates were used to obtain vibrational energies and wave functions. IR transition intensities were obtained for $J = 0$ to $J' = 1$ transitions using the exact wave functions and the *ab initio* dipole moment.

For larger hydrogen-bonded systems, rigorous calculations are far more difficult to carry out, both from the point of view of obtaining full-dimensional potentials and the subsequent quantum vibrational calculations. Reduced dimensionality approaches are therefore often necessary and several chapters in this volume illustrate this approach. With increasing computational power, coupled with some new approaches, it is possible to treat modest sized H-bonded systems in full dimensionality. We have already briefly reviewed the approach we have developed for potentials; for the vibrations we have primarily used the code Multimode (MM). The methods used in MM have been reviewed recently [24 and references therein, 25], and so we only give a very brief overview of the method here.

There are two versions of MM. One, that we refer to as "single-reference" MM is based on the exact Watson Hamiltonian, which is the Hamiltonian in rectilinear mass-weighted normal coordinates. The normal coordinates, as usual, are referenced to a single stationary point, which does not have to be a minimum. For calculations we review below the reference geometry was chosen as a saddle point, not a minimum. The other version of MM, which is much better suited for highly floppy systems, is based on the reaction path Hamiltonian [25]. A key element of both versions is the n-mode representation of the potential. In the single-reference version of MM, the potential in N normal modes is represented as

$$\begin{aligned} V(Q_1, \cdots Q_N) = & \sum_i V^{(i)}(Q_i) + \sum_{i \neq j} V^{(2)}(Q_i, Q_j) + \sum_{i \neq j \neq k} V^{(3)}(Q_i, Q_j, Q_k) + \\ & \sum_{i \neq j \neq k \neq l} V^{(4)}(Q_i, Q_j, Q_k, Q_l) + \cdots \end{aligned} \tag{3.7a}$$

For the reaction-path version of MM the potential is given by

$$\begin{aligned} V(\tau, Q_1, Q_2, \ldots) = & V^{(0)}(\tau) + \sum_i V_i^{(1)}(\tau, Q_i) + \sum_{ij} V_{ij}^{(2)}(\tau, Q_i, Q_j) + \\ & \sum_{ijk} V_{ijk}^{(3)}(\tau, Q_i, Q_j, Q_k) + \sum_{ijkl} V_{ijkl}^{(4)}(\tau, Q_i, Q_j, Q_k, Q_l) + \ldots \end{aligned} \tag{3.7b}$$

where the Q_i are rectilinear normal modes and τ is the large amplitude coordinate. In these representations n is less than N but the sums run over all sets of normal modes. This representation of the potential makes it feasible to perform the high-dimensional numerical quadratures, etc. needed to set up the Hamiltonian matrix for diagonalization. In recent applications n is typically 4 or 5 in Eq. (3.7a) and 3 or 4 in Eq. (3.7b). The basis used to construct the Hamiltonian matrix is the set of virtual excitations of an exact vibrational self-consistent field Hamiltonian, typically for the zero-point state. This virtual "CI" approach is denoted "VCI".

Relevant calculations using both versions of MM have been reported for $(OH^-)H_2O$ [21, 26] and $H_5O_2^+$ [27, 28] and some very recent results will be presented below. Diffusion Monte Carlo calculations, done by and in collaboration with Anne McCoy have also been done on these systems, however, these are not reviewed in detail here.

3.3 Selected Systems

3.3.1 Bihalide Anions

The bihalide anions XHY^-, where X and Y are halogen atoms, are among the simplest systems containing strong hydrogen bonds. In fact, these triatomic systems exhibit some of the strongest hydrogen bonds known, reaching 1.93 eV (44.5 kcal mol^{-1}) in FHF^- [29]. All bihalide anions are linear and have $D_{\infty h}$ (X=Y) or $C_{\infty h}$ (X≠Y) symmetry. As a result of the strong three-center bonding, the interatomic distance between the heavy atoms is small, leading to a pronounced red-shift of the antisymmetric stretch frequency ν_3 compared to the vibrational frequency of the diatomic X–H. Unusually low frequencies of the shared proton modes are a characteristic trademark for strong hydrogen bonds in general. The extensive sharing of the H-atom between the two halogen atoms makes these bonds highly susceptible to solvent perturbation [30]. In larger $Br^- \cdot (HBr)_n$ clusters (n>1), for example, the additional HBr molecules destroy the symmetry of the H-bonds and the H-atom is localized [31]. XHY^- anions are also of interest as transition state precursors in negative ion photoelectron spectroscopy experiments [32] and are isoelectronic with the rare gas compounds $RgHRg^+$, which exhibit very similar vibrational spectra [33].

The vibrational spectroscopy of bihalide anions has been studied extensively in cryogenic matrices [34]. This work showed that the hydrogenic stretching frequencies were very low, ranging from 1330 cm^{-1} in FHF^- to 645 cm^{-1} in IHI^-, in contrast to the uncomplexed HX frequencies of 4138 cm^{-1} in HF and 2309 cm^{-1} in HI [35]. These studies also indicated that the effects of the matrix on the vibrational frequencies could be significant, providing strong motivation to measure the unperturbed gas phase IR spectra of these bihalide anions. High resolution gas phase IR spectra over narrow frequency ranges have been measured for FHF^- and $ClHCl^-$ using diode laser absorption [36]. We recently applied a different and more general approach to gas phase ion vibrational spectroscopy. The result was the first broadband infrared spectra in the range from 600 to 1675 cm^{-1} of a homoconjugated ($BrHBr^-$) and a heteroconjugated ($BrHI^-$) gas phase bihalide anion measured by vibrational predissociation of the corresponding anion–Ar complex [11, 37].

The results of the $BrHBr^-$ study [37], together with recent data on $BrDBr^-$, are shown in Fig. 3.3. The spectra were measured by IR-PD of the anion–Ar complex. The addition of a single Ar atom is expected to result in only a minor shift (<5 cm^{-1}) of the ν_3 frequency compared to bare $BrHBr^-$. The fundamental of the antisymmetric stretch mode ν_3 is observed at 733 cm^{-1} and peaks at higher energies are attributed to $\nu_3+n\nu_1$ combination bands. As expected, the antisymmetric stretch frequency is red-shifted upon deuteration, while the peak spacing, related to the symmetric stretch mode ν_1, remains nearly unchanged. The ratio of $\nu_3(H)/\nu_3(D)$ is 1.45, slightly larger than the harmonic value of $\sqrt{2}$ and considerably larger than 1.2–1.3, the value usually quoted for strong hydrogen bonds [38]. The measured peak positions are in excellent agreement with the results from the anharmonic calculations of Del Bene and Jordan [39], indicated by the black bars in Fig. 3.3. They calculated two-dimensional vibrational eigenfunctions and eigenvalues on a high level *ab initio* potential energy surface for a collinear XHX^- system. The relative intensities in our gas phase spectra differ from the previously measured matrix spectra, presumably due to the underlying dissociation dynamics. Nonetheless, the strong relative intensity of the combination bands in both experiments indicate a complete breakdown of the (uncoupled) harmonic oscillator model in these prototypical systems and underline that explicit consid-

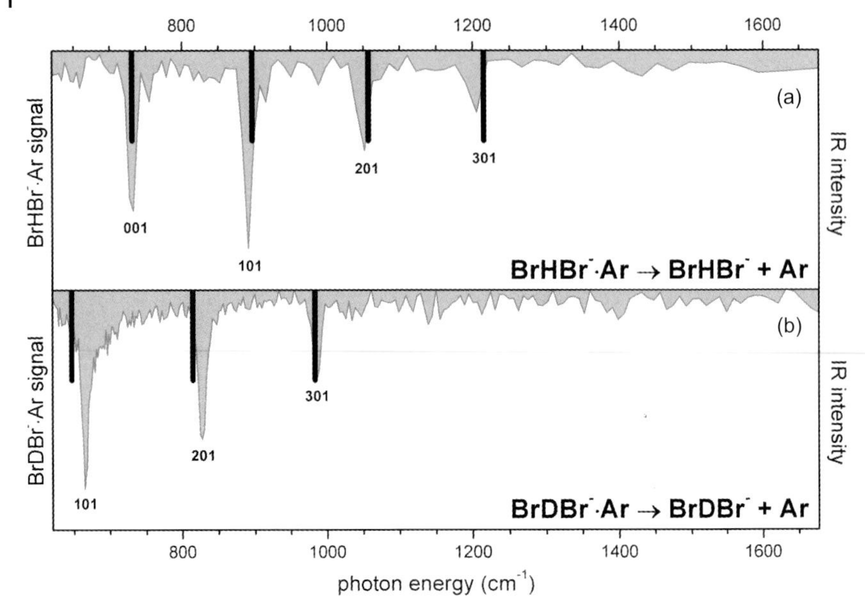

Figure 3.3 IR-PD spectra of $BrHBr^-\cdot Ar$ (a) and $BrDBr^-\cdot Ar$ (b) from 620 to 1680 cm^{-1}. The depletion of the parent ion signal is shown (gray trace and gray shaded-area) and compared to the eigenvalues (black bars) from a variational calculation on a two-dimensional CCSD(T)/aug-cc-pVTZ-based potential energy surface for bare $BrHBr^-$ and $BrDBr^-$ anions, respectively. The peak assignment is indicated by three numbers which give the vibrational quantum number for each mode (symmetric stretch ν_1, bend ν_2, antisymmetric stretch ν_3).

eration of intermode coupling is needed when calculating vibrational properties of strongly hydrogen bonded systems.

The vibrational spectra of the asymmetric bihalides are more complex and their assignment was only recently unraveled [11]. Earlier matrix IR work [40] on $BrHI^-$ indicated the presence of two absorbing species, a strongly asymmetric form, referred to as type I, and a more symmetric form (type II), the latter exhibiting a considerably lower H-atom stretch frequency owing to greater delocalization of the hydrogen atom, similar to the symmetric bihalides discussed above. To investigate this possible dual structure, spectra of $BrHI^-$ in the gas phase in the region of the bridging hydrogen stretch were recorded.

Parent ion depletion spectra for $BrHI^-\cdot Ar$ and $BrDI^-\cdot Ar$ are shown in Fig. 3.4. Compared to the IR spectra of the symmetric bihalides, these spectra are considerably more complex and their assignment, based on previous calculations [41], was problematic. The two most intense peaks in both spectra correspond closely to the peaks seen in the previous matrix works. Their frequencies exhibit red-shifts when going from the gas phase to the matrix from 12 to 64 cm^{-1}. All these peaks were assigned to the type I structure in the matrix. No peaks assigned to type II anions are observed in the gas phase, but additional peaks are observed in the gas phase spectra. In order to reach a satisfactory assignment of all the peaks observed

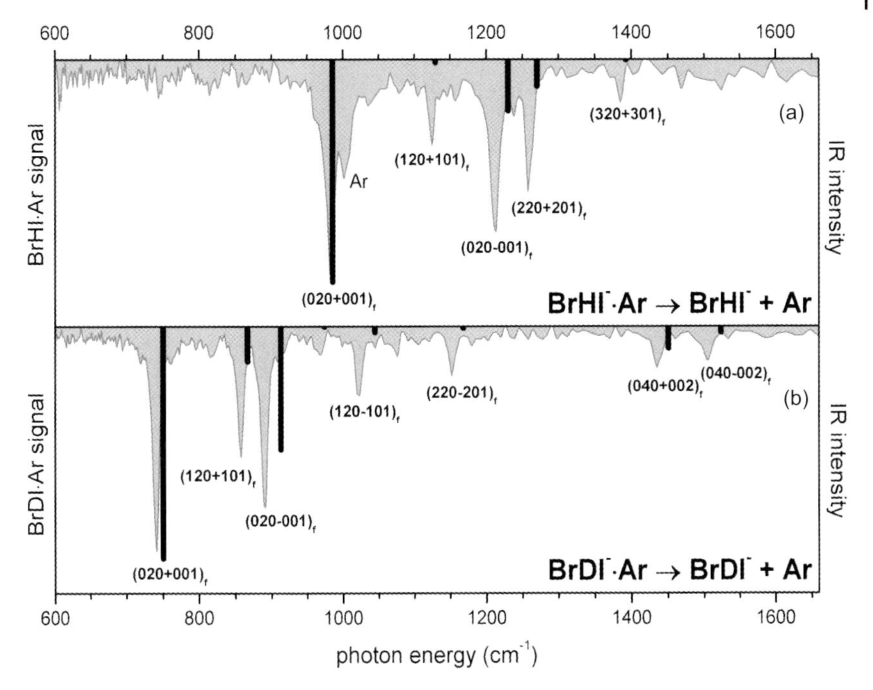

Figure 3.4 IR-PD spectra of $BrHI^-$·Ar (a) and $BrDI^-$·Ar (b). The depletion of the parent ion signal is shown (gray trace and gray shaded-area) and compared to the calculated frequencies and absorption intensities (black bars). Calculations were performed on the bare $BrHI^-$ and $BRDI^-$ anions and involved construction of a three-dimensional MRCI/AVQZ(cpp) potential energy surface, followed by ro-vibrational bound-state variational calculations (see text).

in the gas phase spectra new calculations on an improved potential energy surface had to be performed.

Such calculations were recently done. These included a new potential for this anion using multi-reference CI *ab initio* methods with large basis sets; the details are given elsewhere [11]. *Ab initio* energies and dipole moment values were obtained at roughly 2300 configurations taken from regular grids. Both the energy and dipole moment were subsequently interpolated using 3d splines. The potential along two cuts of this potential for the collinear geometry are shown in Fig. 3.1. In these plots, the distance between the Br and I^-, denoted R, is fixed while the potential is plotted as a function of the difference coordinate $r_1 - r_2$ where r_1 and r_2 are the distances between the H atom and Br and I^-, respectively. As seen, for large R the potential has a double-minimum, however, near the global minimum, where $R = 3.7\ a_0$, the potential is a single minimum. This surface and the corresponding *ab initio* dipole moment were used to calculate the IR spectra of $BrHI^-$ and $BrDI^-$ using two variational codes, one using valence coordinates and one using Jacobi coordinates. The results from the two calculations agree very well, as they should. The comparisons with experiment, given in Fig. 3.4, show very good agreement. This is especially satisfying because there was uncertainty about the assignment of the bands at 920 and 1171 cm^{-1}. The difficulty in making experimental assignments was illuminated and resolved by the calculations, which indicated a strong 2:1 Fermi mixing between the bend and the "asymmetric-stretch" modes. This mixing is depicted in Fig. 3.5 where plots of the corresponding wave functions are given in the upper two panels and plots of "deperturbed" wave functions are given in the lower two panels. The plots are in the Jacobi coordinates r and γ described above (and in the figure caption). The ener-

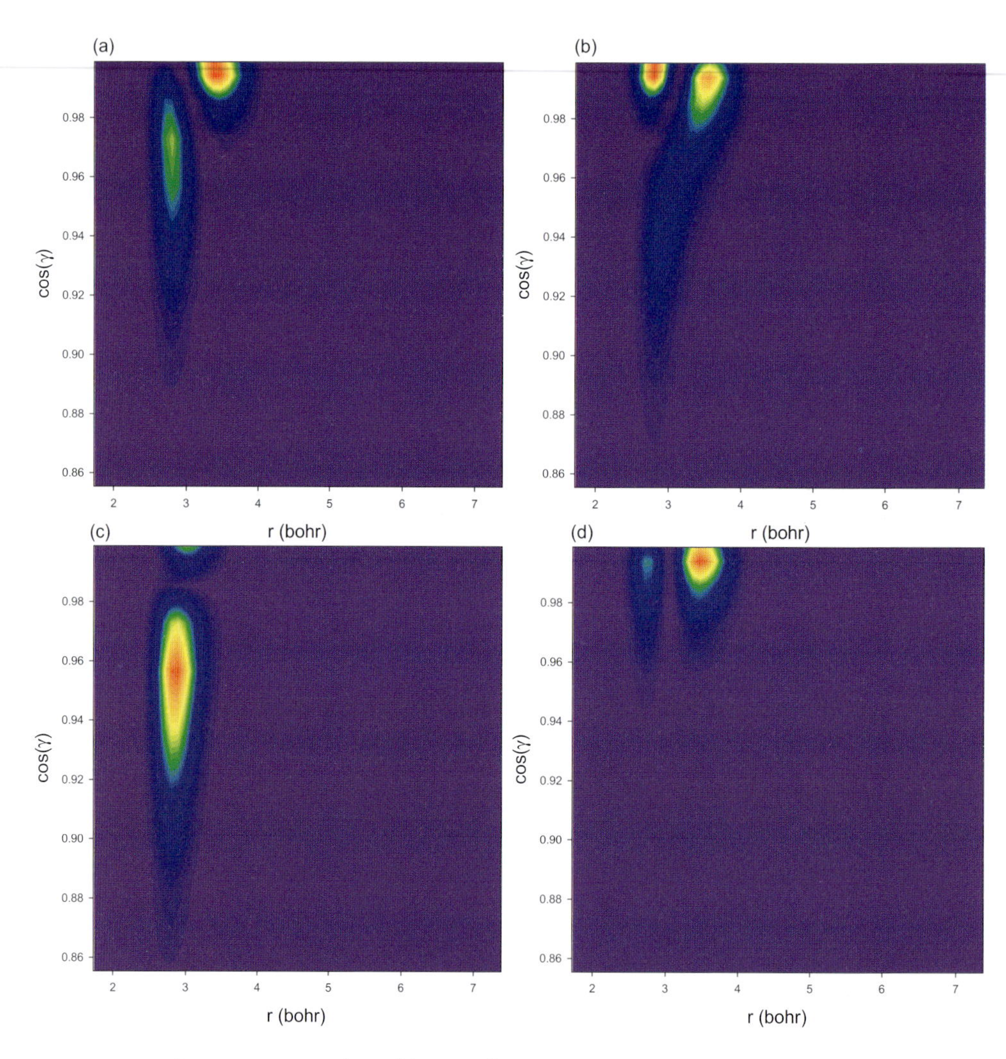

Figure 3.5 Contour plots of the eigen functions at 983 cm^{-1} (a) and 1127 cm^{-1} (b) and sum (d) and difference of the two (c), representing the pure states of the bending overtone and the H-atom stretch modes, respectively, in $BrHI^-$. The plots are two-dimensional slices in the plane defined by r_{HBr} and the Jacobi bending angle γ.

gies of these two perturbed states are 983 and 1127 cm^{-1}, in very good agreement with experiment. This analysis presents a satisfactory resolution of the assignment issue for these two lines. (A corresponding and completely analogous situation pertains to $BrDI^-$ for the two lines at 742 and 980 cm^{-1}.) Also, it was shown that these 2:1 resonances pervade the entire spectrum for both $BrHI^-$ and $BrDI^-$.

In summary, the gas phase vibrational frequencies observed for $BrHI^-\cdot Ar$ and $BrDI^-\cdot Ar$ are close to those seen in earlier matrix-isolation experiments and attributed to the type I form of the anion. The calculated vibrational energy levels show remarkable mode mixing between the H-atom stretch and bend modes, resulting in numerous Fermi resonance states, and thus more complex spectra than for the symmetric $XH(D)X^-$ bihalide anions. The calculations also suggest a reassignment of the matrix spectra. The matrix features originally attributed to the symmetric type II anion may actually arise from the bend fundamental and/or the associated combination band with the heavy atom stretch, rather than from a second form of the ion (type II) not observed in the gas phase. These results underline the breakdown of the harmonic approximation; mechanical anharmonicities lead to extended mixing of states, evidenced in the IR spectra by pronounced intensity borrowing effects in the form of Fermi resonances.

3.3.2 The Protonated Water Dimer $(H_2O\cdots H\cdots OH_2)^+$

The protonated water dimer is one of the most prominent ions containing strong hydrogen bonds. It plays a crucial role in the explanation of the unusually high proton mobility in water [42]. Protonated water clusters and $H_5O_2^+$ in particular have been suggested as proton release groups in bimolecular proton pumps like bacteriorhodopsin [3, 43]. The interpretation of the spectroscopic signature of hydrated protons in liquid water, characterized by a quasi-continuous absorption observed in the IR, has been a long running controversy [44], in particular the attribution of specific bands to the hydrated proton structures $H_5O_2^+$ and $H_9O_4^+$ [45]. More recently, *ab initio* molecular dynamics simulations have shown that the hydrated proton actually forms a fluxional defect in the hydrogen-bonded network with both $H_5O_2^+$ and $H_9O_4^+$ occurring only in the sense of limiting structures [4].

3.3.2.1 Experiments

For more than 30 years, starting with X-ray diffraction experiments of hydrated crystals [46], $H_5O_2^+$ has been the subject of extensive theoretical and experimental studies and its spectroscopy remains challenging until today [47 and references therein]. The vibrational spectroscopy of $H_5O_2^+$ in the gas phase was first studied in the region above 3500 cm^{-1}, the region of the free O–H stretch modes. The groundbreaking IR-PD experiments of Lee and coworkers [48] (see Fig. 3.6) in 1989 support a hydrated-crystal-like, C_2 symmetry $H_2O\cdots H^+\cdots OH_2$ structure, in which the proton is located symmetrically between the two water ligands. These experiments made use of a two-color excitation scheme; the first, tunable laser

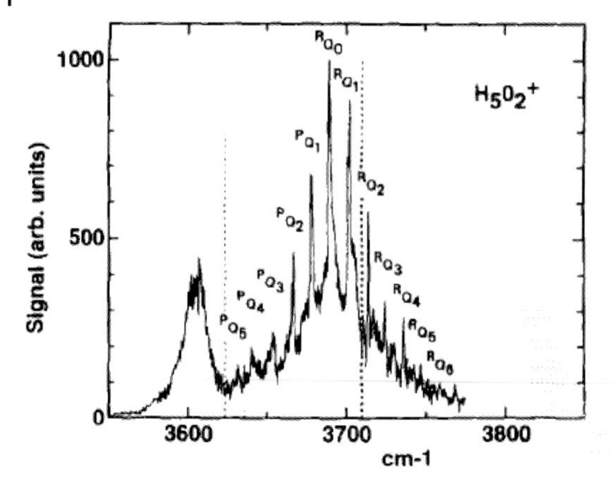

Figure 3.6 IR-PD spectrum of $H_5O_2^+$ measured in the region of the monomer O–H stretches. Figure taken from Ref. [48].

was used to probe the linear absorption regime and the second, fixed-frequency laser to overcome the $H_5O_2^+$ dissociation energy of 31.6 kcal mol^{-1} (11055 cm^{-1}) via multiphoton absorption [49]. The two bands in Fig. 3.6 were assigned to the symmetric and antisymmetric O–H stretches (either in phase or out of phase) of the two H_2O groups. The rotational progression is assigned to a series of Q branches. The observed peak spacing of 11.6 cm^{-1} is in good agreement with the calculated C_2, near symmetric top structure. In a later study [50], performed at considerably higher spectral resolution, more evidence for a very floppy molecule was found. The number of observed rovibrational lines was considerably more than one would expect for a rigid molecule and was speculatively attributed to tunneling splittings caused by large amplitude motion [51].

Interestingly, four (compared to two in Fig. 3.6) O–H stretching bands were observed in the IR-PD spectrum of the $H_5O_2^+ \cdot H_2$ complex [49]. Based on a comparison with computed vibrational frequencies, it was concluded that the perturbation exerted by H_2, which is bound by less than 4 kcal mol^{-1}, is already sufficient to break the symmetry and stabilizes an "asymmetric" $H_2O \cdots H_3O^+$ structure, for which four O–H bands are predicted.

The spectral region containing the shared proton modes, key to characterizing the potential energy surface for proton transfer, lies much lower in energy (<1600 cm^{-1}) and remained experimentally unexplored until 2003. Advances in free electron laser technology can be credited for making the spectral region of strong hydrogen bonds finally accessible to gas phase spectroscopic techniques. These experiments make use of a simpler one-color excitation scheme, in contrast to the two-color approach applied in the Lee experiments. The vibrational spectra of $H_5O_2^+$ and $D_5O_2^+$ measured in the region from 620 to 1900 cm^{-1} using radiation from the free electron laser for infrared experiments FELIX are plotted in Fig. 3.7 [52]. The spectrum of $H_5O_2^+$ exhibits four main bands, labeled B to E. Upon H–D substitution four bands (B′–E′) of similar width but different relative intensities appear red shifted with isotope shifts varying between 1.3 and 1.4. Based on previous theoretical studies, band E was assigned to the bending motion of the terminal water molecules and bands B to D to vibrations predominantly involving the shared proton. The assignment of the shared proton modes to specific vibrational states remains somewhat controversial and will be discussed in more detail together with the most recent calculations below.

The observed bands are more than twenty times broader than the bandwidth of the laser radiation (~5–15 cm^{-1}). The increased width could be caused by vibrational hot-bands, but these should be largely eliminated in the spectra because the ions are collisionally cooled prior to interaction with the laser pulse. However, the zero-point motion of the central proton on the extremely flat PES likely extends over an unusually large area, leading to the exploration of a much larger part of

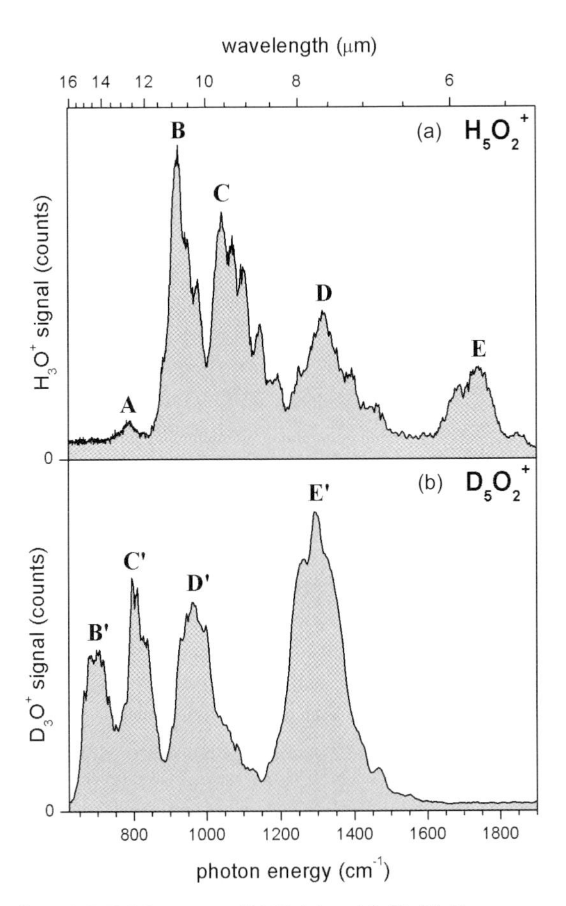

Figure 3.7 IR-PD spectra of $H_5O_2^+$ (a) and $D_5O_2^+$ (b). The spectra were measured by monitoring the formation of H_3O^+ and D_3O^+ as a function of FELIX wavelength (top axis). The spectra are plotted linearly against the photon energy (bottom axis). See Ref. [52] for experimental details.

the configurational space than for a typical rigid molecule, even at 0 K. Additionally, the anharmonicity of the shared proton modes is pronounced, leading to strong anharmonic coupling terms and closely related intensity borrowing effects. The result is an increase in the intensity of bands, which are normally weak or forbidden, like combination bands with low-frequency modes in the present case. Alternatively, multiphoton transitions in which two or more photons are resonant with a vibrational transition may contribute [11]. These transitions appear at shifted photon energies (compared to the single photon transition) due to the anharmonicity of the vibrational potential. In this way, a negative anharmonicity may lead to multiphoton transitions that contribute to the high energy tails of bands B and C. Information on the underlying potential energy surface may also be hidden in the structure observed for bands B to D, which we can currently not assign; its spacing is too large to be attributed to rotational transitions of a near symmetric top.

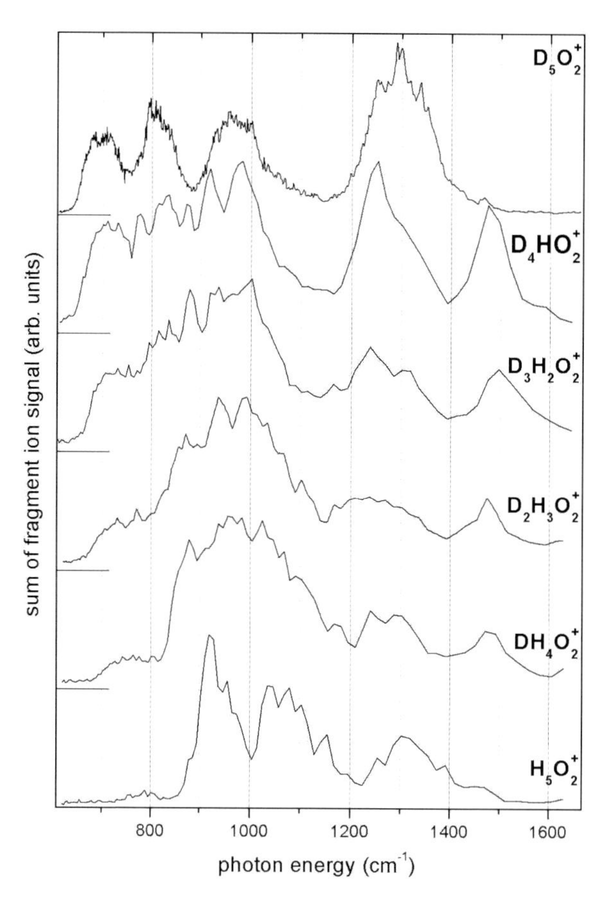

Figure 3.8 IR-PD spectra of $H_xD_{5-x}O_2^+$ with $x = 0$ to 5 (top to bottom) in the region of the shared proton (deuteron) modes (see text).

Additional information can be gained from the IR-PD spectra of the partially deuterated isomers $D_xH_{5-x}O_2^+$ (x=0,1,...,5) and these are shown in Fig. 3.8. Fragment ions of different mass are produced upon photodissociation of partially deuterated species and we measured IR-PD spectra for each possible fragment ion mass. We find nearly identical spectra for all fragment ions of a particular parent ion mass, indicating that efficient H/D scrambling occurs, and therefore only representative sum spectra over all fragment ions are shown for each parent ion mass in Fig. 3.8. Efficient intramolecular H/D exchange is made possible by the underlying sequential absorption mechanism in which the cluster ion can absorb photons during the complete duration of a FEL macropulse (~ 5 ms). If the barrier to internal rotation in $D_xH_{5-x}O_2^+$ ions is smaller than the dissociation limit, efficient intramolecular energy redistribution leads to complete H/D scrambling after sufficient energy is absorbed but before dissociation can occur. The mixed H/D spectra are more congested than the pure spectra, because multiple isomers are present. Nonetheless, some interesting features can be observed. The band at 1500 cm^{-1}, only present in the mixed H/D spectra is assigned to the bending motion of the terminal HDO monomers. The band centered at 770 cm^{-1} in $DH_4O_2^+$ is attributed to a shared proton mode of the symmetric $H_2O\cdots D^+\cdots OH_2$ isomer, red-shifted by 118 cm^{-1} compared to band B in Fig. 3.7. This band grows in intensity as the degree of deuteration is increased. Interestingly, it also continues to red-shift, from 770 cm^{-1} in $H_2O\cdots D^+\cdots OH_2$ to 700 cm^{-1} in $D_2O\cdots D^+\cdots OD_2$, indicating significant coupling between this shared proton mode and the modes of the terminal monomers. The intermediate region is more complex and its assignment requires a simulation of the vibrational spectra of the partially deuterated species.

Predissociation spectroscopy of argon-tagged $H_5O_2^+$ [16], which probes the complex in the linear absorption regime, yields a similar but simpler spectrum compared to the multiphoton dissociation spectrum of bare $H_5O_2^+$ discussed above (see Fig. 3.9). Headrick et al. argue that, unlike for H_2 [48], the extent of the Ar perturbation is not sufficient to stabilize the "asymmetric" $H_2O\cdots H_3O^+$ complex and change its chemical nature, but only leads to a small shift in the vibrational frequencies. Assuming an additive argon perturbation, they show that the 1080 cm^{-1} band of the $H_5O_2^+\cdot Ar$ complex, found at 1140 cm^{-1} in $H_5O_2^+\cdot Ar_2$, is placed close to the maximum of band C (in the multiphoton spectrum) in the bare ion. No feature corresponding to band D is found in the Headrick et al. spectrum. However a peak in this region is found in the recent (multiphoton) IR-PD by Fridgen et al. [53] (see Fig. 3.9) measured at the French free electron laser facilty CLIO [54]. VCI multimode calculations (see Fig. 3.11, later) also predict transitions with significant intensity in this spectral region. Below 1300 cm^{-1}, the spectra measured at FELIX and CLIO do not agree very well; the band maxima at 921 and 1043 cm^{-1} in the FELIX spectrum coincide with regions of low signal in the CLIO spectrum. Such discrepancies are surprising and can only in part be rationalized by different ion internal temperature distributions, $<$100 K in the FELIX study vs. room temperature in the CLIO study. More recent Ar- and Ne-

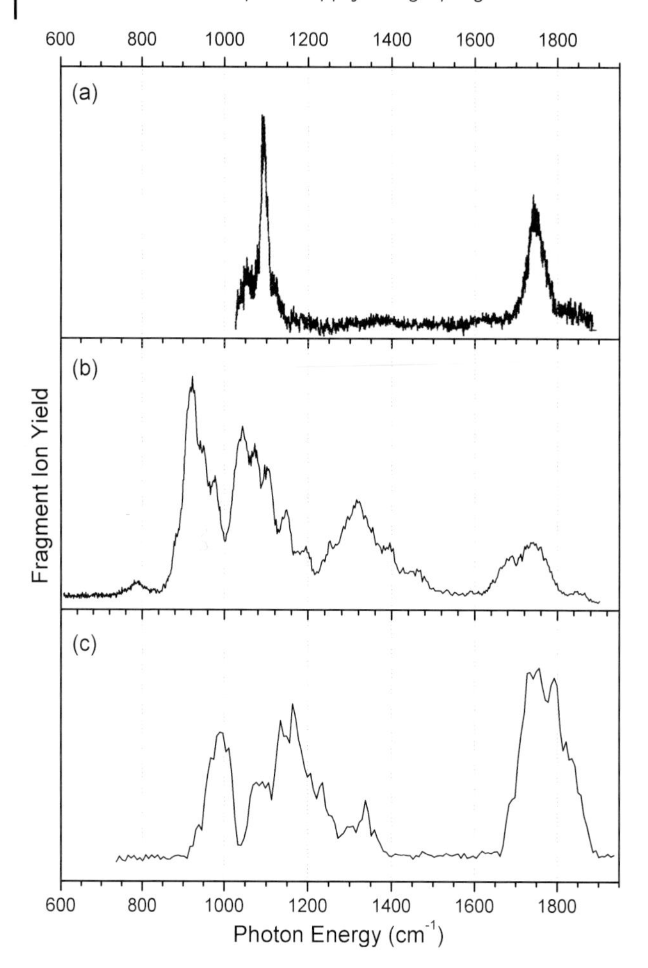

Figure 3.9 IR-PD spectra of (a) $H_5O_2^+$·Ar [16], (b) $H_5O_2^+$ measured at FELIX [52], and (c) $H_5O_2^+$ measured at CLIO [53].

tagged predissociation spectra extending below 1000 cm^{-1} [28] are in better agreement overall with the FELIX experiments than with the CLIO experiments.

3.3.2.2 **Calculations**

There have been a number of calculations recently on the vibrations of $H_5O_2^+$ beyond the harmonic approximation. As noted already, such calculations are essential due to the highly anharmonic nature of these vibrations. Attempts to go beyond the harmonic approximation have been done in reduced dimensionality [27, 56] and also with the additional vibrational adiabatic approximation [56]. These calculations selected the three proton degrees of freedom and the OO-stretch as the reduced dimensionality space. While such approaches are better in some respects than harmonic calculations in full dimensionality, they can miss important coupling with the monomer degrees of freedom. This was demonstrated in comparisons of fully coupled 4d and full dimensionality virtual configuration interaction (VCI) calculations [27]. The results of the comparison showed that for some states the reduced dimensionality treatment was quite realistic, but not for the proton anti-symmetric stretch. For that important mode the 4d estimate for the fundamental is nearly 300 cm^{-1} higher than the full dimensional calculation. The reason for this large difference was traced to the neglect of coupling of that mode to the monomer modes in the 4d calculations. This coupling is large and can be understood simply by noting that as the proton moves towards one monomer and away from the other, the closer monomer takes on H_3O^+ character while the other monomer takes on H_2O character and this clearly introduces monomer coupling into the proton anti-symmetric stretch.

The full-dimensional VCI calculations, mentioned above, were done with the single-reference version of Multimode. The calculations used a full-dimensional potential due to Ojamae, Shavitt, and Singer [57], version OSS3(p), and considered all four-mode couplings among the 15 normal modes. Until just recently, this was the only full dimensional potential surface for $H_5O_2^+$. In many respects the OSS3(p) potential is quite realistic; however, it is not spectroscopically accurate.

A summary of these published vibrational energies of $H_5O_2^+$ is given in Table 3.1 along with full-dimensionality correlation-consistent vibrational self-consistent field calculations [58] (CC-VSCF), which were restricted to two-mode coupling. All the calculations find the OO-stretch in the range 569–599 cm^{-1}. However, only the 2d+2d calculations predict the fundamental energy of the OH^+O-x bend below the asymmetric stretch; they also are alone in predicting the OH^+O-y bend at around 1000 cm^{-1}. With regard to the highly interesting OH^+O-anti-symmetric stretch, only the VCI/OSS3(p) calculations predict the fundamental below 1000 cm^{-1}. (Based on this prediction, the intense band labeled "B" in Fig. 3.7 was re-assigned [27] as the OH^+O-anti-symmetric stretch.) However, for the OH-monomer stretches agreement between the VCI/OSS3(p) calculations and experiment is only fair to poor. This appears to be due primarily to a shortcoming of the OSS3(p) potential. The CC-VSCF/MP2 calculations [58] for these stretches are in somewhat better agreement with experiment; however, not close to "spectroscopic" accuracy.

In reviewing published calculations on $H_5O_2^+$, it should be noted that diffusion Monte Carlo (DMC) calculations of the zero-point state using the OSS potential [47, 59] and the more recent *ab initio* potential [22, 28] (see below) have been reported. These calculations clearly indicate the highly delocalized nature of the zero-point state. In recent work, correlation function DMC calculations of the IR spectrum were reported using the OSS3(p) potential [47]. These calculations indicate a quite complex spectrum with much mixing among low-frequency modes. Two states with substantial OH^+O-anti-symmetric stretch character were identified at 737 and 870 cm^{-1}. The latter number is fairly close to the 902 cm^{-1} result for this state obtained in VCI calculations on the same potential energy surface (and shown in Table 3.1).

Tab. 3.1 Selected vibrational energies (cm^{-1}) of $H_5O_2^+$ calculated by a variety of methods and experiment for the OH-monomer stretches. "2d+2d" are the adiabatic 4d calculations of Ref. [56], "4d" are the fully coupled 4d calculations of Ref. [27], "CC-VSCF" are the correlation-consistent vibrational self-consistent field calculations of Ref. [58] and "VCI" are the virtual configuration interaction calculations of Ref. [27]. The potential used in these calculations is indicated after the back-slash.

Mode	2d+2d/MP2	4d/OSS3(p)	CC-VSCF/MP2	VCI/OSS3(p)	Exp[a]
OO-stretch	587	571	599	569	
OH+O-asym st	1158	1185	1209	902	
OH+O-x bend	968	1328	1442	1354	
OH+O-y bend	1026	1344	1494	1388	
OH-monomer1	NA	NA	3579	3320	–
OH-monomer2	NA	NA	3577	3420	3609
OH-monomer3	NA	NA	3593	3470	–
OH-monomer4	NA	NA	3518	3470	3684

Clearly an accurate simulation of the vibrational energies and IR spectrum of $H_5O_2^+$ requires accurate full-dimensional potential energy and dipole moment surfaces and a full-dimensional treatment of the dynamics. Very recently we reported a new *ab initio*, full dimensional potential energy surface and dipole moment for $H_5O_2^+$ [22]. The *ab initio* calculations were done at the CCSD(T) level of theory with an aug-cc-pVTZ basis for roughly 50 000 configurations. The potential energy surface is a fit to these energies. It is permutationally invariant with respect to like atoms and dissociates properly to H_2O + H_3O^+. It was obtained using the methods briefly reviewed in Section 3.2.3. The variables of the fit are given by $x_{i,j} = \exp(-r_{i,j}/a)$ where $r_{i,j}$ are all the internuclear distances and a was fixed at 2.0 a_0.

The fit is highly precise and the RMS fitting error is shown in Fig. 3.10 as a function of the maximum energy of the data set used in determining the error. As seen, the RMS error is quite small and is roughly 20 cm^{-1} for total energies up to 40 000 cm^{-1} above the global minimum. This value of the RMS is not in the range of "spectroscopic" accuracy, i.e., within 1 cm^{-1}; however, it is close to the expected accuracy of the *ab initio* method used.

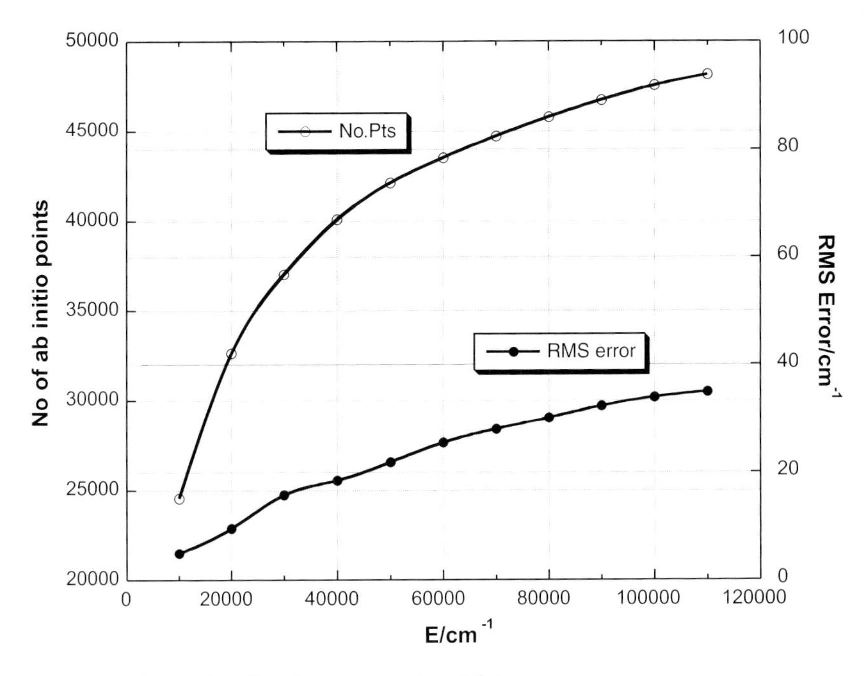

Figure 3.10 The number of configurations and RMS fitting error as a function of the energy cut-off, as explained in the text.

This new potential is highly anharmonic and fluxional with numerous low energy saddle points. Together with Anne McCoy, Stuart Carter and Xinchuan Huang, we have begun large-scale vibrational calculations using the new potential and dipole moment surfaces. These calculations are quite demanding and are being done with the single reference and reaction path versions of Multimode briefly described above and also with the DMC method. Some preliminary results have very recently been reported [28]; however, at present we only have a "rough" IR spectrum based on VCI calculations using the single-reference version of Multimode. This version of the code cannot describe internal torsional motion fully; however it can do crude IR simulations using the exact dipole moment. We present two plots of the calculated 0 K IR spectrum below. The first one, Fig. 3.11, is in the spectral region of the recent IR experiments, shown in Figs. 3.7–3.9. Based on unpublished comparisons of the VCI energies in this region with excited state DMC ones of McCoy we believe the VCI energies in this region are high by roughly 100 cm^{-1} Thus, we have shifted this VCI spectrum to the red by 100 cm^{-1}. As seen there are peaks in this calculated spectrum that correspond approximately to the experimental spectra shown in Figs. 3.7–3.9. (Note there is only rough agreement among these experimental spectra, as noted already.)

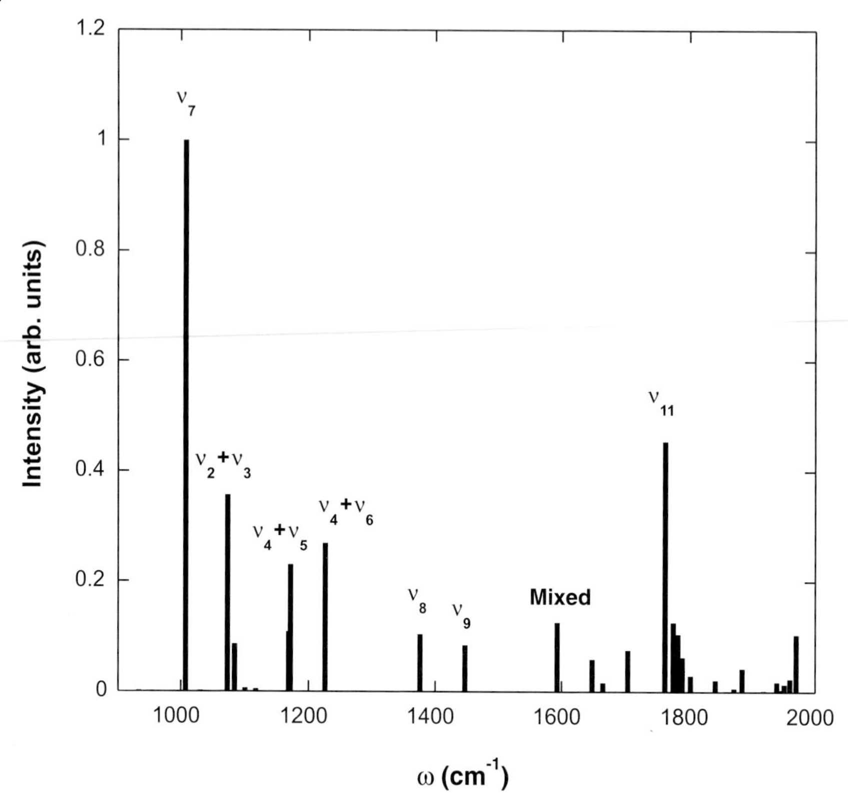

Figure 3.11 0 K IR spectrum obtained using "single-reference" VCI Multimode calculations using the new potential of Ref. [22]. Modes 2–5 are various low-frequency rock, wag and torsion modes, mode 6 is the OO-stretch, mode 7 is the OH^+O-asymmetric stretch, modes 8 and 9 are the OH+O x and y bends, and mode 11 is the asymmetric monomer bends.

The 0 K IR spectrum in the range 3600–3800 cm^{-1} is shown in Fig. 3.12 and is presented with no energy shift. As seen, the spectrum is fairly complex but does show intensities in qualitative accord with experiment, shown in Fig. 3.6. Our analysis indicates that the relatively intense lines at 3620 cm^{-1} are the monomer OH-symmetric stretch and those at 3710 cm^{-1} are the asymmetric stretch. We stress that these are preliminary results and likely not well converged. More accurate DMC and Multimode-reaction path calculations for the four monomer stretches indicate that the single reference energies are 40–100 cm^{-1} high [28].

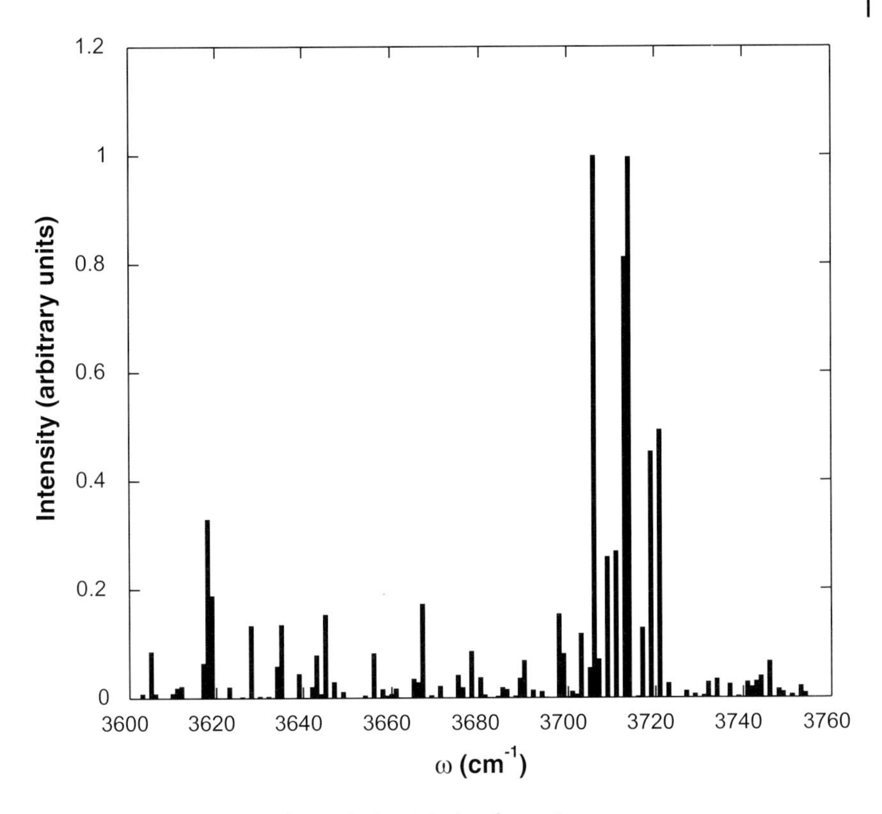

Figure 3.12 0 K IR spectrum obtained using "single-reference" VCI Multimode calculations using the new potential of Ref. [22].

3.4 Outlook

The vibrational signature of strong hydrogen bonds is found in a spectral region which can be accessed directly with novel radiation sources operating in the mid-IR. The photodissociation experiments described above all yield spectra without rotational resolution, mainly due to the inherent broad bandwidth of the FELIX radiation. Recent advances in the technology of table top tunable IR lasers are promising and these narrower bandwidth systems have the potential to go beyond this barrier and elucidate the structures of the clusters with considerably more detail based on high resolution spectra. The experiments using FELs should remain interesting, due to the considerably higher pulse energies, different temporal pulse structure and greater tunability available. Multiphoton absorption photodissociation spectra measured at high pulse energies, above the "quasi-linear" regime, contain additional information on the underlying PES that, in principle, could be extracted, if theoretical models could directly simulate the more complex multiphoton absorption processes.

The experimental vibrational spectra of strongly hydrogen bonded systems, especially $H_5O_2^+$ are complex, owing to strong mode-mixing. Their interpretation requires theoretical approaches beyond the harmonic approximation. The calculated spectra for $H_5O_2^+$ presented here look promising, but more work is needed using better converged calculations and also for non-zero temperature. Calculations for the various isotopomers of $H_5O_2^+$ are also needed and this will be forthcoming. More challenging will be calculations that directly address the experimental spectra that use either Ar-messenger or multiphoton dissociation techniques.

Acknowledgments

JMB thanks the National Science Foundation and the Office of Naval Research for financial support. He also gratefully acknowledges collaborators Bastiaan Braams, Stuart Carter, Xinchuan Huang and Anne McCoy. KRA and DMN wish to acknowledge contributions from their collaborators Nicholas L. Pivonka, Matthew J. Nee and Andreas Osterwalder at the University of California Berkeley; Ludger Wöste, Cristina Kaposta, Mathias Brümmer, Gabriele Santambrogio and Carlos Cibrián Uhalte at the Freie Universität Berlin and Gerard Meijer and Gert von Helden at the Fritz-Haber-Institut der Max-Planck-Gesellschaft in Berlin. KRA gratefully acknowledges financial support by the German Research Foundation DFG as part of the Collaborative Research Center 546 and the Research Training Group 788, as well as the "Stichting voor Fundamenteel Onderzoek der Materie (FOM)" in providing the required beam time on FELIX and the skillful assistance of the complete FELIX staff. He also thanks Travis Fridgen for supplying a copy of the CLIO spectrum. DMN acknowledges support from the Air Force Office of Scientific Research.

References

1 J. Trylska, P. Grochowski, J. A. McCammon, *Protein Sci.* **2004**, *13*, 513–528.

2 C. A. Hunter, *Angew. Chem., Int. Ed.* **2004**, *43*, 5310–5324.

3 V. Z. Spassov, H. Luecke, K. Gerwert, D. Bashford, *J. Mol. Biol.* **2001**, *312*, 203–219.

4 D. Marx, M. E. Tuckerman, J. Hutter, M. Parinello, *Nature* **1999**, *397*, 601.

5 T. Steiner, *Angew. Chem., Int. Ed.* **2002**, *41*, 48–76.

6 M. A. Duncan, *Int. J. Mass Spectrom.* **2000**, *200*, 545–569; E. J. Bieske, O. Dopfer, *Chem. Rev.* **2000**, *100*, 3963–3998.

7 N. H. Rosenbaum, J. C. Owrutsky, L. M. Tack, R. J. Saykally, *J. Chem. Phys.* **1986**, *84*, 5308–5313.

8 S. Davis, M. Fárník, D. Uy, D. J. Nesbitt, *Chem. Phys. Lett.* **2001**, *344*, 23–30.

9 S. Mukamel, J. Jortner, *J. Chem. Phys.* **1976**, *65*, 5204.

10 J. Oomens, A. G. G. M. Tielens, B. G. Sartakov, G. von Helden, G. Meijer, *Astrophysical Journal* **2003**, *591*, 968.

11 M. Nee, C. Kaposta, A. Osterwalder, C. Cibrian Uhalte, T. Xie, A. Kaledin, S. Carter, J. M. Bowman, G. Meijer, D. M. Neumark, K. R. Asmis, *J. Chem. Phys.* **2004**, *121*, 7259.

12 M. Okumura, L. I. Yeh, Y. T. Lee, *J. Chem. Phys.* **1985**, *83*, 3705–3706.

13 W. H. Robertson, M. A. Johnson, *Annu. Rev. Phys. Chem.* **2003**, *54*, 173–213.

14 G. von Helden, I. Holleman, G. M. H. Knippels, A. F. G. van der Meer, G. Meijer, *Phys. Rev. Lett.* **1997**, *79*, 5234–5237.

15 M. Gerhards, C. Unterberg, A. Gerlach, *Phys. Chem. Chem. Phys.* **2002**, *4*, 5563–5565; M. Gerhards, *Opt. Commun.* **2004**, *241*, 493–497; J. A. Stearns, A. Das, T. S. Zwier, *Phys. Chem. Chem. Phys.* **2004**, *6*, 2605–2610.

16 J. M. Headrick, J. C. Bopp, M. A. Johnson, *J. Chem. Phys.* **2004**, *121*, 11523–11526.

17 W. R. Bosenberg, D. R. Guyer, *J. Opt. Soc. Am. B-Opt. Phys.* **1993**, *10*, 1716.

18 K. R. Asmis, M. Brümmer, C. Kaposta, G. Santambrogio, G. von Helden, G. Meijer, K. Rademann, L. Wöste, *Phys. Chem. Chem. Phys.* **2002**, *4*, 1101–1104.

19 G. M. H. Knippels, G. H. C. Vanwerkhoven, E. H. Haselhoff, B. Faatz, D. Oepts, P. W. Vanamersfoort, *Nucl. Instrum. Methods A* **1995**, *358*, 308–310; D. Oepts, A. F. G. van der Meer, P. W. van Amersfoort, *Infrared Phys. Technol.* **1995**, *36*, 297–308.

20 A. Brown, B. J. Braams, K. Christoffel, Z. Jin, J. M. Bowman, *J. Chem. Phys.* **2003**, *119*, 8790–8793; A. B. McCoy, B. J. Braams, A. Brown, X. C. Huang, Z. Jin, J. M. Bowman, *J. Phys. Chem. A* **2004**, *108*, 4991–4994.

21 X. C. Huang, B. J. Braams, S. Carter, J. M. Bowman, *J. Am. Chem. Soc.* **2004**, *126*, 5042–5043.

22 X. Huang, B. J. Braams, J. M. Bowman, *J. Chem. Phys.* **2005**, *122*, 044308.

23 S. Carter, N. C. Handy, C. Puzzarini, R. Tarroni, P. Palmieri, *Mol. Phys.* **2000**, *98*, 1697–1712.

24 J. M. Bowman, S. Carter, X. C. Huang, *Int. Rev. Phys. Chem.* **2003**, *22*, 533–549; S. Carter, N. C. Handy, *J. Chem. Phys.* **2000**, *113*, 987–993; J. M. Bowman, S. Carter, N. C. Handy, in *Theory and Applications of Computational Chemistry: The First 40 Years*, C. Dykstra, G. Frenking, K, S. Kim, G. E. Scuseria (Eds.).

25 W. H. Miller, N. C. Handy, J. E. Adams, *J. Chem. Phys.* **1980**, *72*, 99–112; J. T. Hougen, P. R. Bunker, J. W. C. Johns, *J. Mol. Spectrosc.* **1970**, *34*, 136–172.

26 E. G. Diken, J. M. Headrick, J. R. Roscioli, C. Bopp, M. A. Johnson, A. B. McCoy, X. Huang, S. Carter, J. M. Bowman, *J. Phys. Chem. A* **2005**, *109*, 571.

27 J. Dai, Z. Bacic, X. Huang, S. Carter, J. M. Bowman, *J. Chem. Phys.* **2003**, *119*, 6571–6580.

28 A. B. McCoy, X. Huang, S. Carter, M. Landeweer, J. M. Bowman, *J. Chem. Phys.* **2005**, *122*, 061101; N. J. Hammer, E. G. Diken, J. R. Roscioli, E. M. Myshakin, K. D. Jordan, A. B. McCoy, X. Huang, S. Carter,

J. M. Bowman, M. A. Johnson, *J. Chem. Phys.* **2005**, *122*, 244301.
29 G. Caldwell, P. Kebarle, *Can. J. Chem.* **1985**, *63*, 1399–1406.
30 H. Gomez, G. Meloni, J. Madrid, D. M. Neumark, *J. Chem. Phys.* **2003**, *119*, 872–879.
31 N. L. Pivonka, C. Kaposta, G. von Helden, G. Meijer, L. Wöste, D. M. Neumark, K. R. Asmis, *J. Chem. Phys.* **2002**, *117*, 6493–6499.
32 D. M. Neumark, *Acc. Chem. Res.* **1993**, *26*, 33–39.
33 H. Kunttu, J. Seetula, M. Räsänen, V. A. Apkarian, *J. Chem. Phys.* **1992**, *96*, 5630–5635.
34 B. S. Ault, *Acc. Chem. Res.* **1982**, *15*, 103– 109.
35 D. U. Webb, K. N. Rao, *J. Mol. Spectrosc.* **1968**, *28*, 121; D. R. J. Boyd, H. W. Thompson, *Spectrochim. Acta* **1952**, *5*, 308.
36 K. Kawaguchi, *J. Chem. Phys.* **1988**, *88*, 4186–4189; K. Kawaguchi, E. Hirota, *J. Chem. Phys.* **1986**, *84*, 2953–2960; K. Kawaguchi, E. Hirota, *J. Chem. Phys.* **1987**, *87*, 6838–6841; K. Kawaguchi, E. Hirota, *J. Mol. Struct.* **1995**, *352*, 389–394.
37 N. L. Pivonka, C. Kaposta, M. Brümmer, G. von Helden, G. Meijer, L. Wöste, D. M. Neumark, K. R. Asmis, *J. Chem. Phys.* **2003**, *118*, 5275–5278.
38 P. A. Frey, *J. Phys. Org. Chem.* **2004**, *17*, 511–520.
39 J. E. Del Bene, M. J. T. Jordan, *Spectrochim Acta, Part A* **1999**, *55*, 719–729.
40 C. M. Ellison, B. S. Ault, *J. Phys. Chem.* **1979**, *83*, 832–837.
41 A. Kaledin, S. Skokov, J. M. Bowman, K. Morokuma, *J. Chem. Phys.* **2000**, *113*, 9479–9487.
42 N. Agmon, *Chem. Phys. Lett.* **1995**, *244*, 456–462.
43 R. Rousseau, V. Kleinschmidt, U. W. Schmitt, D. Marx, *Phys. Chem. Chem. Phys.* **2004**, *6*, 1848–1859.
44 G. Zundel, *Adv. Chem. Phys.* **2000**, *111*, 1–217.
45 N. B. Librovich, V. P. Sakun, N. D. Sokolov, *Chem. Phys.* **1979**, *39*, 351; J. Kim, U. W. Schmitt, J. A. Gruetzmacher, G. A. Voth, N. E. Scherer, *J. Chem. Phys.* **2002**, *116*, 737–746.
46 I. Olovsson, *J. Chem. Phys.* **1968**, *49*, 1063–1067.
47 X. Huang, H. M. Cho, S. Carter, J. Ojamae, J. M. Bowman, S. J. Singer, *J. Phys. Chem. A* **2003**, *107*, 7142–7151; H. M. Choi, S. J. Singer, *J. Phys. Chem. A* **2004**, *108*, **8691**.
48 L. I. Yeh, J. D. Myers, J. M. Price, Y. T. Lee, *J. Chem. Phys.* **1989**, *91*, 7319–7330.
49 Y. K. Lau, S. Ikuta, P. Kebarle, *J. Am. Chem. Soc.* **1982**, *104*, 1462–1469.
50 L. I. Yeh, Y. T. Lee, J. T. Hougen, *J. Mol. Spectrosc.* **1994**, *164*, 473–488.
51 D. J. Wales, *J. Chem. Phys.* **1999**, *110*, 10403–10409.
52 K. R. Asmis, N. L. Pivonka, G. Santambrogio, M. Brümmer, C. Kaposta, D. M. Neumark, L. Wöste, *Science* **2003**, *299*, 1375.
53 T. D. Fridgen, T. B. McMahon, L. MacAleese, J. Lemaire, P. Maitre, *J. Phys. Chem. A* **2004**, *108*, 9008–9010.
54 R. Prazeres, F. Glotin, C. Insa, D. A. Jaroszynski, J. M. Ortega, *Eur. Phys. J. D* **1998**, *3*, 87.
55 M.A. Johnson, personal communication.
56 M. V. Vener, O. Kühn, J. Sauer, *J. Chem. Phys.* **2001**, *114*, 240–249.
57 L. Ojamäe, I. Shavitt, S. J. Singer, *J. Chem. Phys.* **1998**, *109*, 5547–5564.
58 G. M. Chaban, J. O. Jung, R. B. Gerber, *J. Phys. Chem. A* **2000**, *104*, 2772–2779.
59 M. Mella, D. C. Clary, *J. Chem. Phys.* **2003**, *119*, 10048–10062.
60 K. R. Asmis, M. Brummer, C. Kaposta, G. Santambrogio, G. von Helden, G. Meijer, K. Rademann, L. Woste, *Phys. Chem. Chem. Phys.* **2002**, *4*, 1101–1104.

4
Laser-driven Ultrafast Hydrogen Transfer Dynamics

Oliver Kühn and Leticia González

4.1
Introduction

The ground state tunnel splitting in a symmetric double well potential, Δ_0, is associated with a tunneling time of $\tau_0 = h/2\Delta_0$. If we take the hydrogen transfer (HT) in tropolone [1] as an example we have $\Delta_0 \approx 1$ cm^{-1} and therefore $\tau_0 \approx 16.7$ ps. This is a rather long time as compared, e.g. with the period of an OH-stretch vibration or with the T_1 relaxation time of this vibration in solution [2]. Consequently, to study *ultrafast* HT in the electronic ground state the dynamics has to be initiated closer to or even above the reaction barrier what could be achieved by an interaction with an IR laser pulse. Of course, this draws on the analogy with electron-vibrational dynamics triggered after almost instantaneous optical excitation and thus switching of the electronic state. Here the driving force for the nuclear wave packet motion after excitation is due to the difference in the electron density which leads, for instance, to the rather rapid reactions observed for excited state HT [3–5].

Building further on this analogy one could think of adiabatically separating the fast OH-stretch dynamics from the slower motions of the molecular frame. This would result in adiabatic potential energy surfaces (PES) for the slow modes in a given state of the OH-stretch mode [6–8]. The HT can then be viewed as a slow mode wave packet moving on the adiabatic potential whose character changes gradually from an OH-stretch excitation localized in the reactant well to an excitation localized in the product well.

The first question, which is raised by this simple analogy, concerns the very possibility of exciting slow mode wave packets in a hydrogen bond at all. Taking a different perspective it touches the very issue of interpretation of the notoriously complex IR spectra [9]. In the condensed phase much of this complexity is hidden under bands broadened by the solvent interaction. Hence it was only recently that coherent wave packet motion of a 100 cm^{-1} hydrogen bond mode could be observed after OH-stretch excitation, although in a system which has only a single minimum potential [10]. Meanwhile coherent low-frequency dynamics has also been observed in a double minimum system (acetic acid dimer) [11]. With this

Hydrogen-Transfer Reactions. Edited by J. T. Hynes, J. P. Klinman, H. H. Limbach, and R. L. Schowen

ISBN: 978-3-527-30777-7

proof-of-principle at hand it is not far-fetched to tackle the goal of making use of the coherence contained in the hydrogen bond wave packet motion to control it by means of tailored IR laser pulses.

Influencing reaction pathways by means of designed laser fields is emerging as a powerful tool in femtochemistry and femtobiology [12–14]. In principle both frequency- and time-domain control schemes are being explored. Only the time-domain ultrashort laser pulse control will be discussed in the following since it gives direct access to the dynamics, particularly in the condensed phase where stationary spectra are usually very broad. In view of the importance of H-bonding and HT reactions it would be desirable to have at hand concepts for guiding the H-bond dynamics with laser pulses.

Two scenarios for H-bond dynamics will be addressed below: First, IR laser control of HT between different tautomers in the electronic ground state; second, H-bond breaking by combining IR pre-excitation with a second UV pulse which switches the electronic state. In both cases short few-cycle IR pulses will play an important role.

We start by outlining the theoretical methods putting emphasis on the determination of potential energy surfaces in Section 4.2. In Section 4.3.1 we demonstrate ultrafast laser driven HT in low-dimensional gas and condensed phase model systems using IR pulses encompassing a few-cycles only. However, turning to realistic multidimensional PES it is further shown that the anharmonicity of the PES for HT severely challenges any straightforward control attempt. In Section 4.3.2 we present results where the IR driven motion is restricted to the vicinity of the PES minimum, but of an anionic species of the type AHB^-. Here a second UV pulse can trigger a transition to a neutral PES. Provided that IR and UV pulses are properly timed the H-bond breaks selectively by virtue of the large forces coming along with the removal of an electron. Finally, a summary is given together with some perspectives in Section 4.4.

4.2 Theory

The theoretical description of HT Dynamics requires having at hand a PES and a method for solving the appropriate equations of motion. For few atom systems such as AHB^- one can express the PES in terms of bond distances and angles, or if only the bonded motion shall be considered, vibrational normal mode coordinates can be chosen (see, e.g., Ref. [15]). In the latter approach one relies on the fact that any desired geometry can be expressed in terms of displacements along normal mode coordinates. Since these coordinates are defined and thereby biased with respect to a reference geometry, it is to be expected that the representation of the total PES may not be very compact. Thus, the method is not very economical which hampers its application to large systems. Here one could, in principle, resort to the bond distance/bond angle approach with an additional assumption concerning the majority of degrees of freedom (DOF), which are not explicitly treated. They could be either frozen at some reference geometry or relaxed to the minimum subject to the constraints imposed by the explicit coordinates [16–18]. In both cases one makes an assumption concerning the time scales of the explicit and the neglected coordinates. It should be noted, however, that neither approximation can be fully justified. There will always be faster *and* slower coordinates as compared with the explicit ones.

A more rigorous formulation of this concept, which allows the selection of important DOF according to their actual coupling, is provided by reaction path and reaction surface approaches. Here one makes a distinction between one or a few large amplitude reactive coordinate(s) and many orthogonal harmonic vibrational coordinates [19]. In the case of a single reaction coordinate this is usually selected to be the minimum energy intrinsic reaction path [20]. HT reactions, however, are of the heavy–light–heavy type and the one-dimensional (1D) reaction path is known to be sharply curved. Accordingly, the coupling between the reaction coordinate and the orthogonal modes becomes rather strong which merely indicates that this choice of coordinates is not well suited for HT problems. To cope with this issue Miller et al. [21] and later Shida et al. [22] have proposed to use two and three, respectively, large amplitude coordinates to span a reaction surface. Still the majority of coordinates are treated in the harmonic approximation. Specifically, they have considered malonaldehyde and taken into account the two OH distances and the O–O distance as large amplitude coordinates.

If one aims to explore the PES this approach is rather illuminating. For dynamics calculations, however, one faces the challenge that due to the curvilinear nature of the coordinates, the kinetic energy operator takes a rather complicated form. This necessitates further approximations such as the adiabatic decoupling between the reaction coordinates and the orthogonal oscillator modes. It is probably because of this difficulty that the approach has not been fully explored for the use in HT reaction dynamics for many years. It has been only recently that interest has been revived and a number of studies focused on this approach [23–26].

As an alternative that solves the kinetic coupling problem, Miller and co-workers suggested an all-Cartesian reaction surface Hamiltonian [27, 28]. Originally this approach partitioned the DOF into atomic coordinates of the reactive particle, such as the H-atom, and orthogonal anharmonic modes of what was called the substrate. If there are N atoms and we have selected N_R reactive coordinates there will be $N_H = 3N - 6 - N_R$ harmonic oscillator coordinates and the reaction surface Hamiltonian reads

$$H = \sum_{i=1}^{N_R} \frac{p_i^2}{2} + V(x_1, \ldots, x_{N_R}) + \sum_{n=1}^{N_H} \frac{P_n^2}{2} - f_n(\mathbf{x})Q_n + \frac{1}{2} \sum_{m,n=1}^{N_H} Q_m K_{mn}(\mathbf{x}) Q_n \tag{4.1}$$

Here $V(x_1, \ldots, x_{N_R})$ is the potential for the reactive coordinates when the substrate coordinates are either frozen at some reference geometry or relaxed to minimize $V(x_1, \ldots, x_{N_R})$. Since in general we are not dealing with a minimum energy configuration, forces $f_n(\mathbf{x})$ appear in Eq. (4.1), which act on the normal mode coordinates $\{Q_n\}$. Finally, the last term in Eq. (4.1) contains the force constant matrix $K_{mn}(\mathbf{x})$, which is diagonal at that configuration for which the normal modes have been defined. Everywhere else there may be a mode coupling due to the motion of the reaction coordinates. The important point here is to notice that Eq. (4.1) does not contain any kinetic energy coupling, only potential energy coupling. Of course, this comes only at the price of not being able to take into account rotations. In that sense “substrate” implies that this approach is well suited when a light particle moves in the field of a surrounding which is almost rigid. Apart from an application to H-atom diffusion in crystalline silicon [29] this approach went essentially unnoticed until we have used it to study the proton transfer in 8-hydroxyimidazo[1,2-*a*]pyridine [30]. Subsequently, it has been applied to study the H-bond dynamics in phthalic acid monomethylester [31], the HT in salicylaldimine [32], 3-chlorotropolone [33], and 3,7-dichlorotropolone [34]. Very recently we have modified the original reaction surface approach by choosing *collective* reaction coordinates instead of the H-atom position [35]. This does not alter the all-Cartesian form of Eq. (4.1) but gives a more compact representation of the PES, which contains the most relevant motions of the heavy atoms already in the reaction coordinates. For the specific application to the HT in tropolone we could show that the reaction plane spanned by these reaction coordinates contains the intrinsic reaction path almost completely, that is, the remaining couplings to the orthogonal harmonic modes are rather modest [35].

The Hamiltonian in Eq. (4.1) has an almost product-like form since the majority of coordinates are treated as harmonic oscillators. This makes it rather suitable for quantum dynamics simulations, either in the time-dependent Hartree approximation [31] or using the more general multi-configuration time-dependent Hartree approach [36, 37].

However, Eq. (4.1) has another advantage in that it directly connects to the system–bath models used in condensed phase dynamics [38]. Here the reactive coordinates and the substrate modes comprise the relevant system and the bath, respectively. Larger molecules may provide their own bath and Eq. (4.1) can be used to calculate an *ab initio* system–bath Hamiltonian and microscopic relaxation and dephasing rates [33].

Assuming that the frequencies of the bath modes do not depend on the system coordinates one arrives at the system–bath Hamiltonian

$$
\begin{aligned}
H &= H_{\mathrm{S}} + H_{\mathrm{B}} + H_{\mathrm{SB}} \\
&= \sum_{i=1}^{N_{\mathrm{R}}} \frac{p_i^2}{2} + V(\mathbf{x}) + \frac{1}{2}\sum_{n=1}^{N_{\mathrm{H}}} [P_n^2 + \omega_n^2 Q_n^2] - \sum_{i=1}^{N_{\mathrm{H}}} f_n(\mathbf{x}) Q_n
\end{aligned}
\tag{4.2}
$$

Supplementing this equation with an additional set of solvent oscillators one can incorporate a solvent environment. Notice that this does not necessarily imply harmonic solvent motions. In fact the full anharmonicity of the solvent can be accounted for in the context of linear response theory [39] where the interaction is described in terms of an effective harmonic oscillator bath. This allows calculation of relaxation rates from classical molecular dynamics simulations of the force $f_n(\mathbf{x})$ exerted by the solvent on the relevant system. This approach has found application, for instance, for HOD in D_2O [40, 41] and for the intramolecular hydrogen bond dynamics in phthalic acid monomethyl ester [42]. Equipped with microscopic relaxation rates the dissipative dynamics can be modeled, e.g., by using a quantum master equation [42, 43]. This restricts the applicability to situations where the interaction between relevant system and bath is weak and Markovian [38]. To go beyond this limitation, quantum-classical methods have been developed and applied, e.g., to enzymatic hydride transfer reactions [44].

4.3 Laser Control

4.3.1 Laser-driven Intramolecular Hydrogen Transfer

The conceptionally simplest approach for IR-driven HT, the *pump–dump scheme* [45], can be illustrated using a one-dimensional double minimum potential as shown in Fig. 4.1. First we notice that the potential is asymmetric which allows one to distinguish between the initial state Ψ_i and the final state Ψ_f of the HT reaction. Initially a pump-pulse excites the system from the localized ground state Ψ_i to a delocalized intermediate state Ψ_b which usually is energetically above the reaction barrier. From there a second pulse dumps the system into the product

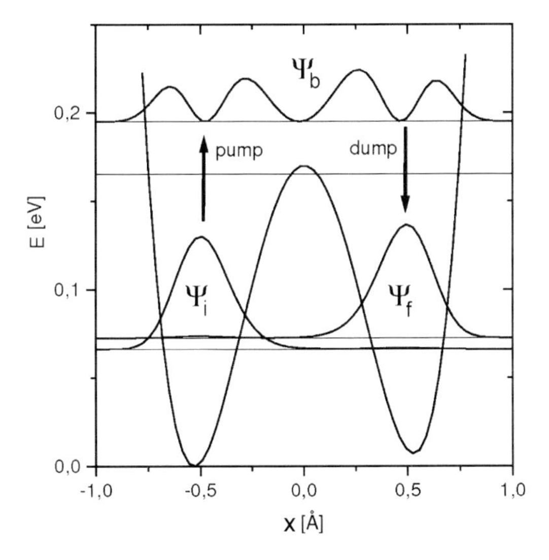

Figure 4.1 Illustration of the pump–dump scheme for a slightly asymmetric one-dimensional double minimum potential. In a first step a pump pulse excites the system from the initial state Ψ_i to an above-the-barrier state Ψ_b. From there a second pulse dumps the population into the target state Ψ_f.

well, thus populating state Ψ_f. The electric field for this above-the-barrier scheme takes the following form

$$E(t) = \sum_{i=1,2} E_i^0 \Theta(t - t_i)\Theta(\tau_i + t_i - t) \sin^2\left(\frac{\pi(t - t_i)}{\tau_i}\right) \cos(\omega_i t) \tag{4.3}$$

Here E_i^0 is the amplitude, τ_i the duration, and ω_i the frequency of the ith pulse. This scheme has been applied in Ref. [46] to a generic two-dimensional HT model which incorporated a H-atom reaction coordinate as well as a low-frequency H-bond mode. In a subsequent work [47] the approach has been specified to a simple model of HT in thioacetylacetone. The Hamiltonian was tailored to the form of Eq. (4.1) based on the information available for the stationary points, that is, the energetics as well as the normal modes of vibration. From these data an effective two-dimensional potential was constructed including the H-atom coordinate as well as a coupled harmonic oscillator, which describes the O–S H-bond motion. Although perhaps oversimplified, this model allowed the study of some principle aspects of laser-driven H-bond motion in an asymmetric low-barrier system.

A field like Eq. (4.3) with separate and overlapping pulses is capable of giving an almost 100% population transfer. This "above-the-barrier" mechanism is also obtained when using the more sophisticated optimal control theory. Here the laser pulse form is obtained from maximizing the functional (see, e.g., Ref. [48])

$$\begin{aligned} J = &|\langle \Psi(T)|\Psi_f\rangle|^2 - \kappa \int_0^T dt \frac{E^2(t)}{S(t)} \\ &- 2\mathrm{Re}\langle \Psi(t)|\Psi_f\rangle \int_0^T dt \langle \Phi(t)|\frac{\partial}{\partial t} + i[H - dE(t)]|\Psi(t)\rangle \end{aligned} \tag{4.4}$$

where T is the time at which the propagated wave function $\Psi(t)$ should coincide with the target state Ψ_f; $\Phi(t)$ is some auxiliary function and d is the dipole moment. The laser field is influenced by the shape function $S(t)$ as well as by the parameter κ, which sets a penalty for high field intensities [49]. Interestingly, upon increasing κ the dynamics of the system changes qualitatively. The above-the-barrier pathway becomes too expensive and a through-the-barrier tunneling pathway becomes operative. Because of its tunneling nature this pathway was coined "Hydrogen subway" in Ref. [50]. Needless to say both mechanisms are going beyond any perturbation theory with respect to the matter-field interaction.

Apart from requiring less intense fields the latter scheme has the advantage that during the time evolution highly excited vibrational states are not appreciably populated. Thus, the moving wave packet does not sample the full anharmonicity of the potential surface and might stay localized in a few DOF. The low degree of excitation is also preferable once considering condensed phase dynamics. Here the energy relaxation rates usually scale with the quantum number of the vibrational state [38]. In other words, exciting the system above the barrier would cause not only rapid energy relaxation but also phase relaxation and the means for controlling the dynamics might be lost. Indeed, using a simple system–bath model

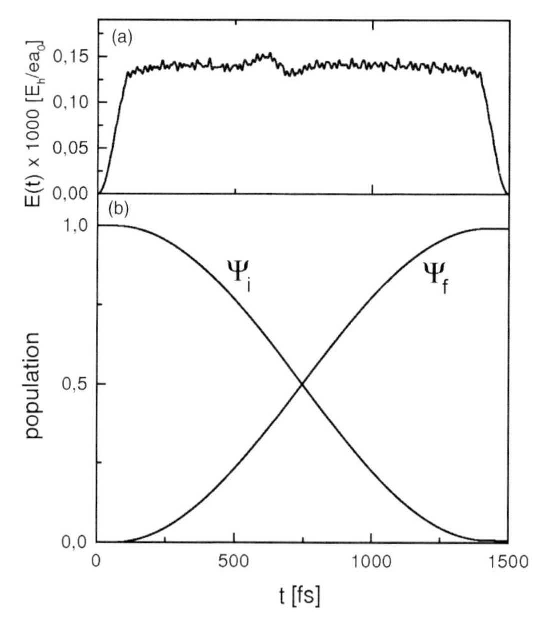

Figure 4.2 (a) Optimal control field for the system shown in Fig. 4.1 for a penalty parameter $\kappa = 400$ (see Eq. (4.4)). The pulse triggers quantum tunneling between the two lowest states whose populations are shown in (b); for details see also Ref. [50].

like Eq. (4.2) it was shown in Ref. [47] that the tunneling pathway is more effective in a dissipative situation.

The pulse forms in the simplest case where only the two lowest states are involved in the through-the-barrier tunneling dynamics are of plateau type; see Fig. 4.2. They serve to compensate for the potential asymmetry (initial rise), facilitate tunneling (plateau phase), and restore the asymmetry to stabilize the products (switch-off phase). It turns out that more realistic few-cycle pulses may realize the same net effect although not in the step-wise fashion as a plateau pulse [47, 51, 52]. In Fig. 4.3 we show such a few-cycle pulse together with the population dynamics of the two lowest states of a double minimum potential adapted to the situation in thioacetylacetone.

The time scale of the reaction is dictated by the tunneling time, i.e., the tunnel splitting for the degenerate levels. This implies that with increasing barrier height the ground state tunneling pair is no longer a good candidate for triggering an *ultrafast* isomerization. In Ref. [53] it was shown that in this case it is advantageous to combine an IR pulse causing vibrational excitation to a higher lying tunneling pair with a subsequent plateau type pulse which establishes the degeneracy of that pair.

The influence of the interaction with an environment on the control yield and pathways can be modeled using different strategies. In principle it is possible to determine optimized pulses in the presence of energy and phase relaxation [12,

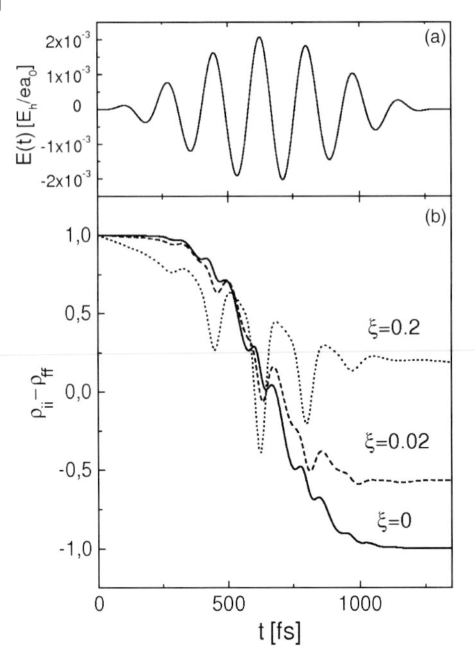

Figure 4.3 (a) Few-cycle control field which drives the HT in a one-dimensional model of thioacetylacetone. (b) Population difference between the two lowest states (initial and target state) for the coherent case and different strengths of the system–bath coupling as expressed by the Kondo parameter ξ. The dissipative propagation has been performed using a real-time path integral approach; for details see also Ref. [51].

13]. A simpler route studies the relaxation effects with reference to the optimized laser-driven dynamics of the isolated system. The interaction with the environment can be described using the quantum master equation approach (see Section 4.2) [38]. This requires that the coupling is weak and that a Markov approximation for the system–bath dynamics applies. Even if these two conditions are fulfilled, the presence of the rather strong external control field may lead to a modification of the system's energy spectrum and therefore to time-dependent relaxation rates. This has been investigated in Ref. [54] on the basis of a generalized Fokker–Planck theory. In Fig. 4.4 we show the time dependence of the low lying states of a 1D approximation to the double minimum potential of the HT in thioacetylacetone and of the relaxation rates for transitions between the two lowest levels. The latter clearly show the dependence on the time-dependent transition frequencies. While this extension of the quantum master equation approach is straightforwardly implemented, going beyond the restrictions imposed by perturbation theory and Markov approximation requires a considerable effort. One possibility is given by the real-time path integral approach [55]. This limits the applications to a few level systems coupled to a harmonic bath, but for this class of systems one can perform numerical exact quantum dynamics simulations. In Ref. [51] we have shown that under the conditions of strong system–bath coupling and strong-field driving the

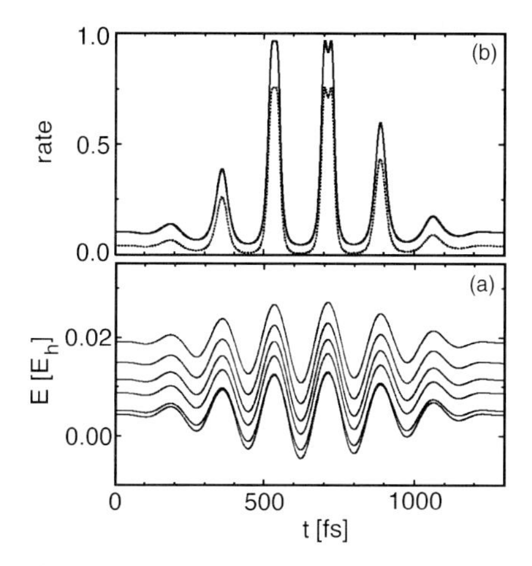

Figure 4.4 (a) Time-dependent energy levels for the one-dimensional model of thioacetylacetone for a few-cycle driving field similar to Fig. 4.3. In (b) we show the resulting rates (in units of the coupling strength) for upward (dotted) and downward (solid) population relaxation between the two lowest lying states; for details see Ref. [54].

dynamics is markedly different from what one observes when using the simple quantum master equation theory. Exemplary results are given in Fig. 4.3(b). Upon increasing the system–bath coupling the interplay between driving and relaxation according to the instantaneous spectrum – reflected in the oscillatory population dynamics – leads to a non-vanishing control yield, even for a relatively long laser pulse.

So far we have considered one- and two-dimensional model systems in the gas and condensed phases. Although the two-dimensional model of thioacetylacetone has been based on quantum chemistry data, the choice of a single reaction coordinate and a single coupled oscillator has been dictated by the desired simplicity of the model. The question arises whether the concepts, which have been developed for this model, can be transferred to more realistic situations. The requirements on a system have been spelled out already in Ref. [46]: (i) The PES has to be asymmetric to have a clear distinction between initial and final states. (ii) There should be a significant change in the dipole moment upon HT to provide directionality of the excitation. (iii) The barrier height should exceed the reorganization energy difference between the minimum and the transition state of the most strongly coupled oscillator. The reorganization energy of an oscillator in Eq. (4.1) is the energy, which is required to restore the oscillator's equilibrium position along the reaction coordinate. The last condition is of approximate character and has been motivated by the observation that upon increasing the coupling between the reaction coordinate and the coupled oscillator in a two-dimensional model, the eigenstates are energetically shifted above the barrier and the notion of localized reactant and product states becomes meaningless.

A systematic approach to the determination of multidimensional PES for HT reactions is given by the reaction surface Hamiltonian, Eq. (4.1). In Ref. [32] we have calculated a reaction surface for the tautomerism in salicylaldimine, a molecule which appeared to be suitable according to points (i) and (ii) above. The in-plane coordinates of the hydrogen atom were taken as large amplitude reaction coordinates and out of the remaining coordinates of the scaffold five harmonic oscillator DOF were identified as being strongly coupled. It turns out that the coupling of the OH-stretch fundamental excitation to the other coordinates is substantially reduced upon deuteration. While the time scale for intramolecular energy redistribution has been estimated to be ~700 fs, the OD-stretching vibration shows no appreciable decay but coherent oscillations during the first 5 ps after excitation [56]. In Fig. 4.5 we show the probability distributions of the ground state, as well as a state which is localized in the product well, for deuterated salicylaldimine. Here, the oscillator DOF have been fixed at their minimum position. Although the reorganization energy difference between reactant and product state is appreciable (3112 cm^{-1}) and larger than the barrier height (1940 cm^{-1}), a localized product state exists. In principle, triggering a transition between the two states in Fig. 4.5 using the methods discussed above appears to be feasible. However, we have not yet taken into account the dynamics of the oscillator coordinates. Their effect is most conveniently visualized after introduction of a diabatic basis, $\{\phi_a(\mathbf{x})\}$, describing the two reactive coordinates for fixed oscillator DOF, i.e. $\{Q_n = 0\}$:

$$[T_x + V(\mathbf{x})]\phi_a(\mathbf{x}) = E_a\phi_a(\mathbf{x}) \tag{4.5}$$

Two of these basis functions are shown in Fig. 4.5. The Hamiltonian Eq. (4.1) can be transformed into this representation which gives (adopting a vector/matrix notation) [32]

$$H_{\text{diab}} = \frac{\mathbf{P}^2}{2} + \sum_{a\beta} \left[U_a(\mathbf{Q})\delta_{a\beta} + V_{a\beta}(\mathbf{Q})(1-\delta_{a\beta}) \right] |a\rangle\langle\beta| \tag{4.6}$$

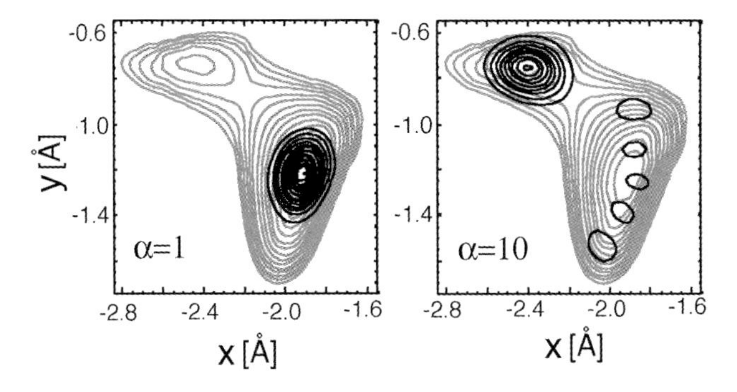

Figure 4.5 Probability densities for the diabatic ground state and the product state resulting from a reaction surface description of the deuterium transfer in salicylaldimine. Here *x* and *y* give the position of the H-atom and the substrate oscillators are frozen at their equilibrium value (cf. Eq. (4.1)); for details see Ref. [57].

with the diabatic PES: $U_a(\mathbf{Q}) = E_a - \mathbf{f}_{aa}\mathbf{Q} + 1/2\mathbf{Q}\mathbf{K}_{aa}\mathbf{Q}$ and the state couplings $V_{a\beta}(\mathbf{Q}) = \mathbf{f}_{a\beta}\mathbf{Q} + 1/2\mathbf{Q}\mathbf{K}_{a\beta}\mathbf{Q}$. A one-dimensional cut of the potential without state coupling along a HT promoting mode having a harmonic frequency of 861 cm^{-1} is shown in Fig. 4.6 [57]. The potentials corresponding to the two states in Fig. 4.5 are highlighted. Since the localized product state samples rather large forces, which result from distorting the structure from its reactant to the product configuration, the respective potential curve shows the largest shift as compared to the ground state. Consequently, it crosses with, and also couples to, several of the lower lying potential energy curves. Ideally one would like to know whether there is an eigenstate of this seven-dimensional model Hamiltonian that is localized on the product side similarly to Fig. 4.5. The determination of such an eigenstate, however, appeared to be not feasible, as can be appreciated by inspection of Fig. 4.6, which shows multiple curve crossings already for a one-dimensional cut of the potential.

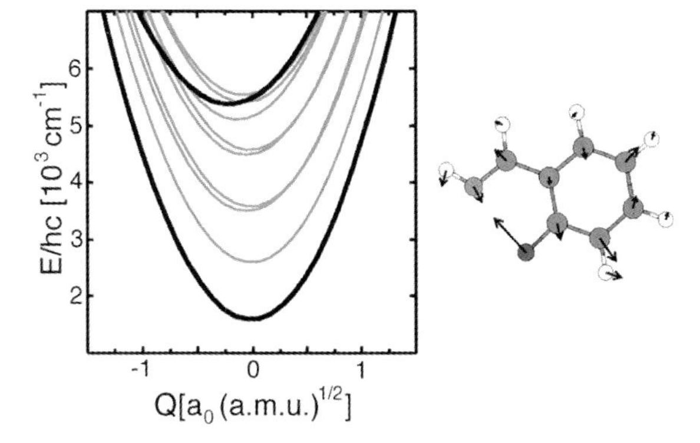

Figure 4.6 One-dimensional cuts through the diabatic PES for the deuterium transfer in salicylaldimine along the substrate mode shown in the right. The thick solid lines indicate the two diabatic states shown in Fig. 4.5; for details see Ref. [57].

Thus instead of aiming at populating an unknown eigenstate, e.g. by exploring different laser pulse parameters, the intention was to find out what the system does after being prepared in a state localized in the product well [57]. In Fig. 4.7 we show the population dynamics of the diabatic states starting with a wave packet, which corresponds to the oscillator ground state of the uncoupled PES ($V_{a\beta} = 0$ in Eq.(4.6)) in the diabatic state with $a = 10$ (see Fig. 4.5). The (field-free) propagation of this initial state yields a rapid decay during the first 50 fs, most notably into the OD-stretching dominated state ($a = 3$). Interestingly, about 5–10% of the population remains trapped in the initial state during the first picosecond. Recalling the coherent dynamics after excitation of the OD-stretching fundamental and "reversing" the field-free dynamics would now suggest a two-step mechanism for reaction control in this molecule with transitions like

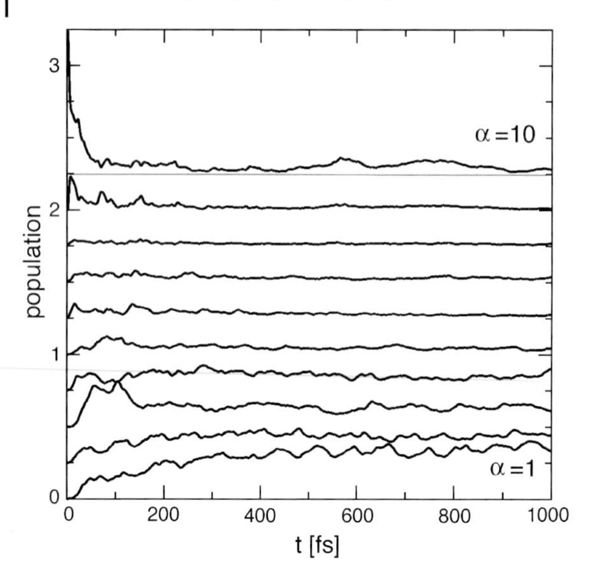

Figure 4.7 Population dynamics of the ten lowest diabatic states of a seven-dimensional model of the deuterium transfer in salicylaldimine after inital population of the product state shown in Fig. 4.5. (Each curve has been offset by $(a - 1)*0.25$). The vertical line shows the zero level for the initially populated state; for details see Ref. [57].

$a = 1 \xrightarrow{h\nu_1} 3 \xrightarrow{h\nu_2} 10$. Attempts to realize such a scheme have not yet been successful. In this respect one might also wonder whether the preferred relaxation pathway is, at the same time, the optimum excitation pathway. Clearly, answering this question would be a task for optimal control theory. However, the propagation of seven-dimensional wave packets in this strongly coupled system is rather demanding and an implementation of an approach like Eq. (4.4) is out of reach.

4.3.2
Laser-driven H-Bond Breaking

The already mentioned *pump–dump scheme* has also found application in the control of bond breaking. The pump pulse excites the system vertically from the electronic ground state to another suitable electronic state, from where the wave packet evolves freely following the steepest descent pathway. After some specific time delay, the wave packet arrives at a region of the excited PES appropriate to return the probability amplitude to the electronic ground state at a different nuclear configuration. Assuming that the electronic ground state possesses different channels for dissociation, the *dumped* wave packet is now biased to dissociate towards a particular target product, altering then the natural branching ratio of the system. Of course, the laser pulses used in this type of application are in the vis/UV frequency region in order to achieve resonant electronic transitions, rather than in the IR as in the previous applications (cf. Section 4.3.1). The interested reader can find numerous examples of *pump–dump* or more generally *pump–control* schemes intended for photodissociation, e.g. in Refs. [12, 13].

Triatomic systems like FHF^- or more generally AHB^- – where two different ligands are strongly bound through a H-bond – are good candidates to exercise control of selective bond breaking. In addition, bihalides ions like FHF^- and similar XHX^- compounds have been employed thoroughly in the field of transition state spectroscopy, a method that probes spectroscopically the transition state of the neutral system by electron photodetachment of the anion [58]. Note that, because these experiments aim at the characterization of the transition state vibrations, they employ nanosecond UV light with a frequency that goes beyond the resonant transition between the anion and neutral surfaces. As shown in Fig. 4.8(a), after photodetachment of the electron, the wave packet created in the neutral surface in the transition state region will divide into two equal branches along the channels XH+X and X+HX, giving rise to the fragments X–H and H–X, respectively. More generally speaking, the wave packet will split into two channels, AH+B and A+HB, with a given branching ratio that is dictated by the topology of the neutral PES or, in other words, determined by the position of the transition state with respect to the minimum of the anion potential.

An approach to achieving H-bond fragmentation in a selective manner is shown in Fig. 4.8(b). It consists of using a few-cycle IR pulse followed by a UV laser

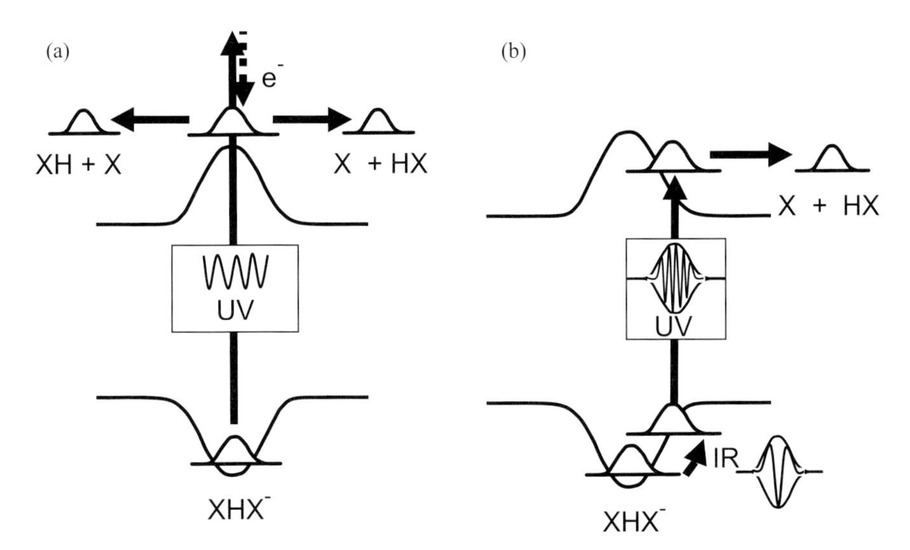

Figure 4.8 In (a) the principle of transition state spectroscopy is shown. A continuous wave UV laser pulse with a frequency exceeding the energy gap between anionic potential (XHX^-) and neutral potential (XHX) is employed. After photodetachment of the electron, dissociation will occur equally in the two possible channels: XH+X and X+HX. In (b) the principle of few-cycle IR + UV pulses is shown. First an ultrashort IR laser pulse creates a wave packet in the anionic state (XHX^-) which, at the appropriate time delay, is then excited to the neutral system (XHX). After photodetachment of the electron, dissociation is steered to occur along one single channel.

pulse. The sequence of these two pulses reminds one of the *pump–dump scheme* because the first laser pulse – a few-cycle IR laser pulse – creates a wave packet, in the electronic ground state however, and displaces it until a favored nuclear configuration, while the second pulse – the UV pulse – transfers the probability amplitude to the state where dissociation takes place. In contrast to the UV cw light, the combination of few-cycle IR + UV pulses applied to an XHX^- system can be optimized to enhance one of the exit channels, e.g. X +HX (compare Fig. 4.8(a) and (b)). The spirit of this control scheme is rather intuitive. A short ($\tau_i < 100$ fs), relatively intense IR laser pulse containing few-cycles is applied to drive the asymmetric stretching vibration of the anion or, equivalently, to force the oscillation of the H-atom between the two heavy end atoms. Each half cycle of the electric field is matched to half of the period of the asymmetric vibration of the H-bond of the system. With each half cycle of the pulse, the oscillating H-atom is pushed further away from its equilibrium position. When the displacement of the H is greatest at one of the turning points, photodetachment of the electron via the UV pulse will create a replica of the wave packet in a domain of the neutral surface far enough from the transition state window as to lead to the dissociation of the pre-excited bond. Needless to add, to make this scheme efficient the UV pulse must be shorter than the vibrational period of the oscillating H-atom. In this cartoon (cf. Fig. 4.8(b)), the reaction coordinate equals the hydrogen stretch and, therefore, the IR frequency equals the asymmetric H-bond stretching. In general, one could expect other degrees of freedom opening channels for competing processes via intramolecular vibration redistribution (IVR), reducing the efficiency of the proposed scheme. Yet, because the hydrogen vibration takes place very fast, we anticipate IVR, e.g. the bending mode, not to seriously influence the main conclusions obtained here. Encouraging examples of few-cycle IR + UV laser control are found in the theoretical work of Henriksen et al. [59, 60] for isotopically substituted ozone $^{16}O^{16}O^{18}O$.

A laser field (cf. Eq. (4.3)) consisting of one or several half cycle pulses takes the form

$$E(t) = \sum_i E_i^0 \sin\left(\frac{\pi(t-t_i)}{\tau_i} + \phi\right) \tag{4.7}$$

where the frequency of the IR laser pulse is given by $\omega_{IR}=\pi/\tau$ and the pulse duration τ is about half of the vibrational period of the asymmetric stretch of the H-bond. For convenience, the phase ϕ will be set to zero in the following applications. A one half-cycle pulse and a sequence of those are shown in Fig. 4.9(a). Their effect on FHF^- is illustrated using a 1D potential in Fig. 4.10. The frequency of the asymmetric stretching vibration in the 1D model of FHF^- is 1815 cm^{-1} and the pulse duration employed is τ=10 fs, which is approximately half of the period of the asymmetric stretch, $0.5\tau_{as} = 9$ fs. To measure the displacement of the initial wave function $\Psi_a(q,0) = \phi_{\upsilon=0}$ from its equilibrium position we calculate the modulus of the autocorrelation function in the anion potential V_a, $P_{anion} = |\langle\phi_{\upsilon=0}|\Psi_a(t>0)\rangle|^2$ as well as the mean position of the H-atom, q_{as}. While the field is switched on, population from υ=0 is moved to other vibrational states, creating a wave packet, and P_{anion} decreases accordingly from 1 to ca. 0.9 (cf. Fig. 4.9(b)). The wave packet begins oscillating and at $t = 9.5$ fs the H-atom has reached the turning point at $q_{as} \approx 0.07$ Å (cf. Fig. 4.9(c)). If three half-cycle pulses of similar duration are employed instead of one, the effect is similar to the classical situation of a driven oscillator: the mass is pushed after passing through each turning point and the amplitude of the resulting oscillation increases. Therefore, after three half-cycle pulses, the P_{anion} has decreased to ca. 0.5 and the H-atom has been maximally shifted from the equilibrium position until $q_{as} \approx 0.16$ Å at $t = 26.5$ fs (see also the corresponding wave functions in Fig. 4.10). One can then anticipate that the pre-stretched bond in the anion system will break selectively in the neutral system upon photodetachment. In order to visualize competing bond breaking we consider 2D PES. In view of the simplicity of XHX^- systems, PES can be constructed in terms of internal coordinates, and because we are interested in the competing

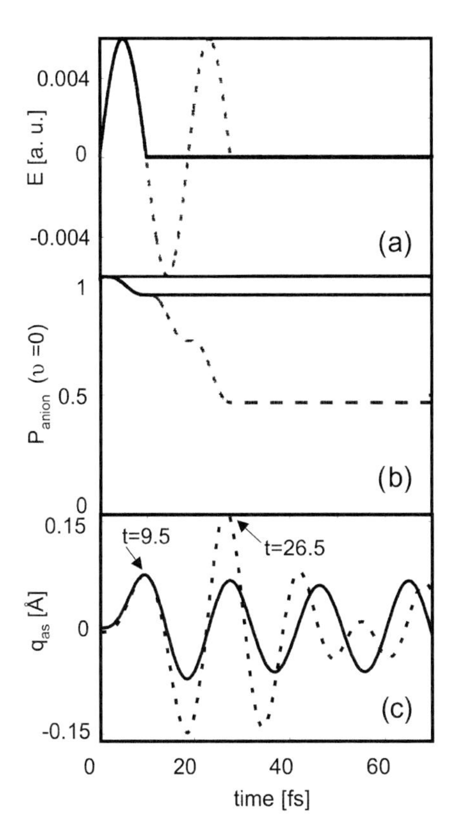

Figure 4.9 One (solid) and three half-cycle pulses (dashed) and their effects on the asymmetric stretching vibration of FHF^-. (a) Electric field (in atomic units) versus time (fs). The parameters are $E^0 = 5.8 \times 10^{-3}\ E_h/ea_0$ and $\tau_l = 10$. (b) Modulus of the autocorrelation function of the anionic wave function as defined in text. (c) Time evolution of the mean value of the H-atom driven by the laser field shown in (a). The time at which maximum displacements occur is indicated.

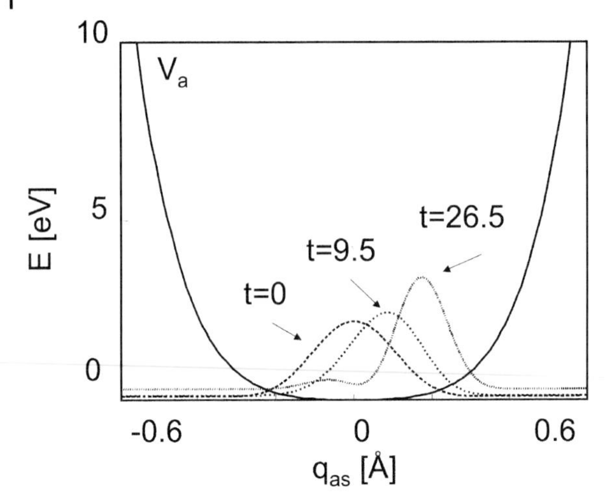

Figure 4.10 Potential energy curve of the anionic potential of FHF along the asymmetric stretching vibration given by $q_{as} = 1/2(R_1 - R_2)$, where R_1 and R_2 are the bond distances FH and HF. Embedded are the wave functions at $t = 0$ fs, after one half-cycle pulse at $t = 9.5$ fs, and after three half-cycle pulses at $t = 26.5$ fs, have been applied.

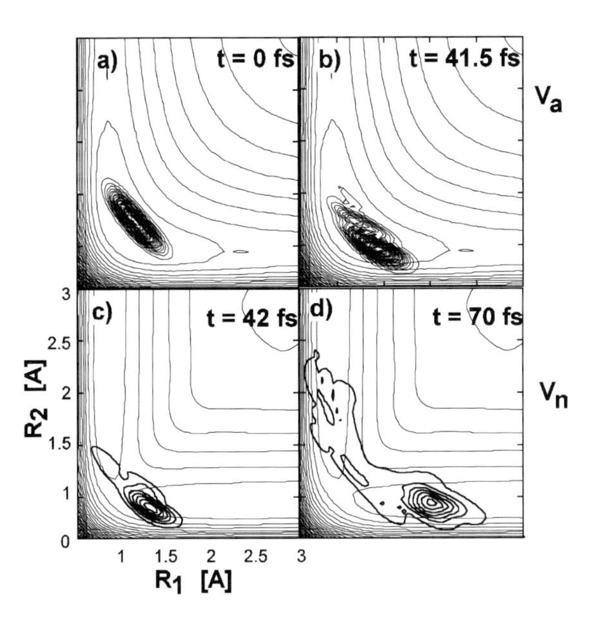

Figure 4.11 Selective H-bond breaking in FHF by photodetachment of the anion using few-cycle IR+UV laser pulses. (a)–(d) are sequential snapshots of the wave packet moving in the bound anionic and repulsive neutral V_a and V_n PES, respectively. The coordinates R_1 and R_2 are the bond distances FH and HF in Å. (a) $t = 0$ fs in V_a. (b) Maximum displacement of the H-atom at $t = 41.5$ fs in V_a. (c) $t = 42$ fs in V_n. (d) $t = 70$ fs in V_n.

ligand dissociation, the two internal bond distances X–H and H–X are considered in a collinear geometry, that is, the symmetric and asymmetric stretching normal modes are taken into account, whereas bending and rotations are excluded. Figure 4.11 shows the anion V_a and neutral V_n PES of FHF. Moreover, we replace the series of three half-cycle pulses by a smooth, switch-on–switch-off $\sin^2$-shaped laser field of the type given in Eq. (4.3) and shown in Fig. 4.12(b). The frequency ω_{IR} is chosen, as in the 1D model, to equal the H-bond oscillations, which in the 2D case correspond to longer vibrational periods of ca. 1500 cm^{-1}, therefore $\omega_{IR} = 1516$ cm^{-1} (0.188 eV). In Fig. 4.12(b) it can be seen how q_{as} encompasses the maximum amplitude of each half cycle. In passing we note that with increasing displacement of the wave packet its dispersion increases as well. Thus the optimum time delay between the IR and the UV pulse calls for a compromise between maximum displacement and minimum dispersion. As shown in Fig. 4.11(b), at $t = 41.5$ fs the wave packet remains relatively compact in the anion surface V_a and the mean value of q_{as} has reached 0.207 Å. When the IR pulse achieves maximum displacement, an ultrashort ($\tau = 5$ fs) resonant ($\omega_{UV} = 43548$ cm^{-1} = 5.5 eV) $\sin^2$-shaped UV pulse is applied to photodetach an electron, preparing the system on the PES of the neutral species. The maximum intensity needed for this selective preparation of the wave packet is ca. $I = 3.3$ TW cm^{-2}. For a laser pulse of duration $\tau = 5$ fs, mildly focused to a diameter of $s = 1$ mm, this corresponds to a pulse energy of approximately 500 μJ, which can be provided by state of the art laser systems. This few-cycle UV pulse excites the displaced wave packet to a downhill domain of V_n where it evolves predominantly along one dissociation

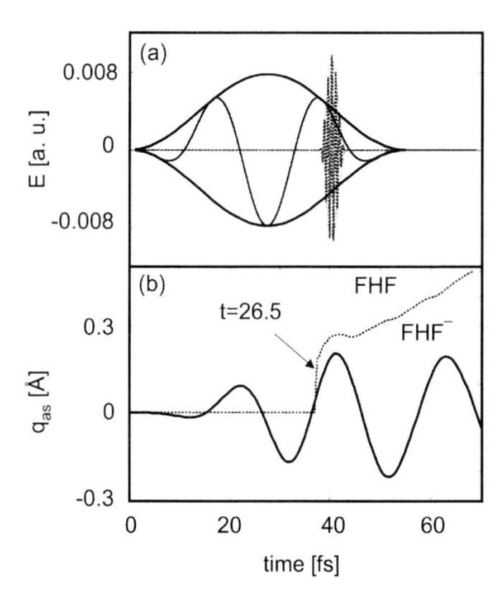

Figure 4.12 A few-cycle IR and UV pulse and its effect on the FHF system. (a) Electric field (atomic units) versus time (fs). The parameters are $E^0_{IR} = 7.8 \times 10^{-3}$ E_h/ea_0, $\omega_{IR} = 1516$ cm^{-1} and $\tau_{IR} = 55$ fs for the IR laser pulse and $E^0_{UV} = 9.7 \times 10^{-3}$ E_h/ea_0, $\omega_{UV} = 43548$ cm^{-1} and $\tau_{UV} = 5$ fs. (b) Time evolution of the mean value of the H-atom in the anion (solid) and neutral (dotted) driven by the laser field shown in (a).

channel. After the wave packet is created in V_n dispersion allows a small portion of the wave function to penetrate the non-desired domain (see Fig. 4.11(c, d)), and the branching ratio at $t = 42$ fs decreases slightly. Nevertheless, the branching ratio (F+HF:FH+F) is increased from its natural value of 50:50 to 75:25 [61].

The present IR+UV scheme then achieves symmetry breaking in the electronic ground state of the anion and subsequent selective bond breaking of the H-bond in the neutral system. The resulting atomic fragment is driven along the pre-selected bond, which has been stretched by the IR laser, whereas the molecular product is driven in the opposite direction. Although the resulting products, F+HF or FH+F, are chemically indistinguishable, the present laser strategy allows the choice of one or another, such that either the atomic or the molecular fragments are driven in a pre-selected direction while the counterparts are driven in the opposite one. As a consequence, this approach achieves spatial separation of the products [62], in contrast to cw UV lasers; see Fig. 4.13(a, b).

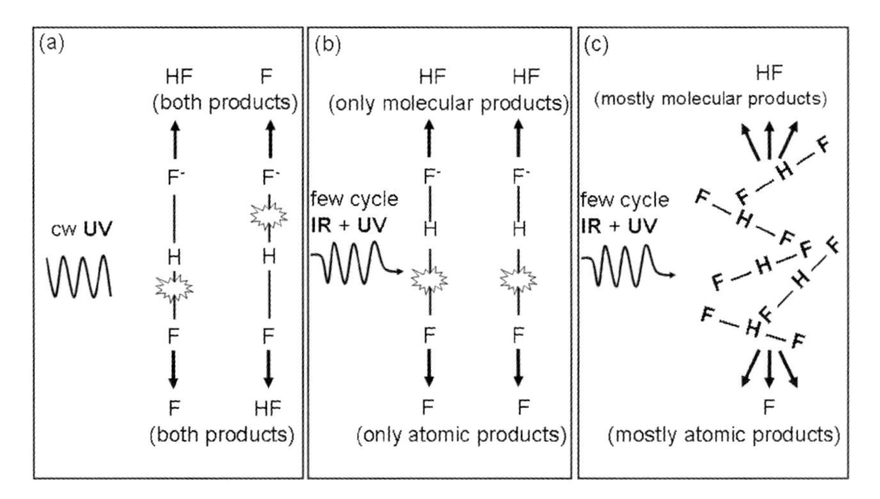

Figure 4.13 Comparison of traditional cw UV or IR+UV laser pulses (a), versus few-cycle IR+UV laser pulses in an oriented ensemble of FHF^- molecules (b) or a randomly oriented ensemble (c).

In these wave packet simulations, the molecular axis of the FHF^- system is assumed to be aligned along the space-fixed axis Z electric field vector. This assumption involves a maximum interaction of the IR and UV laser pulses with the system. Recalling that the time-dependent interaction potential is given by the scalar product of the electric field vector and the dipole vector, i.e. $E^0(t) \cdot \mu \cdot \cos\theta$, it is clear that for field polarizations perpendicular to the molecular axis ($\theta = 90°$) the interaction of the IR laser pulse with the anion vanishes, and for any molecular orientation different from $\theta = 0°$ or $180°$ the interaction is less efficient. Consider now an ensemble of randomly oriented FHF^- molecules, as in Fig. 4.13(c). Since the UV pulse is tuned to match the energy gap between anion and neutral surfaces at the geometry of maximum displacement, but for angles $\theta \neq 0°$ this geometry is never realized, the frequency of the UV pulse is increasingly off-resonant with angles from 0° to 90° and consequently the UV pulse no longer efficiently transfers population to the neutral surface. As a result, in going from 0° to 90° the probability of photodetachment decreases from 1 to 0, except for marginal probabilities of photodetachment due to the significant width of the ultrashort UV laser pulse. As documented in Ref. [62], the general trend for angle values increasing from 0° to 90° is the decrease in the probability of dissociating the molecular product to the direction $\Theta = \theta$, and the complementary increase in the probability to the direction $\Theta = \theta + 180°$. Both effects contribute to the dissociation of FH and F fragments in exactly opposite directions, thus spatially separating the products [62].

So far the few-cycle IR+UV laser scheme has been discussed in the FHF system for which the PES, anionic and neutral, are symmetric, and therefore the two dissociation channels are energetically equivalent. In passing we note that the isotopomer FDF has also been reviewed in Ref. [63]. Other systems found in the literature, like isotopically substituted ozone [59, 60] and HOD [64, 65] have different reduced masses for the diatomic fragments, hence allowing the observation of two different dissociation dynamics. Nonetheless, in these cases, the PES are also symmetric, implying that the transition state in the neutral PES can be found at the same nuclear configuration as the minimum energy configuration in the anionic PES and that the UV laser pulses are resonant for both dissociation channels. Much more challenging is the case of a chemically asymmetric system. This is the central part of the study performed in Ref. [66] using OHF^- as a model system.

OHF^- responds to the general pattern of a system possessing one H-bond much stronger than the other one. In the absence of pre-excitation photodetachment of OHF^- leads predominantly to the O+HF fragments, because the O···H bond is much weaker than the H···F bond. OHF has been the focus of a number of theoretical and experimental studies due to its environmental significance [67]. Neumark has measured the photodetachment spectrum of OHF^- (recall Fig. 4.8) [58], challenging theoretical groups to its simulation and comprehension [68–70]. The exoergicity of the reaction OH + F→O + HF has been well characterized [71]. However, controlling the branching ratio of the system was never previously tackled. In Ref. [66] we raised the question of whether the natural branching ratio can be enhanced to maximize the O + HF products or even reversed to maximize the amount of OH + F products.

In Fig. 4.14 we show the ground state vibrational wave function of OHF^- (Fig. 4.14(a)) on the neutral PES. Using a UV laser pulse of time duration $\tau_{UV} = 5$ fs one obtains a natural branching ratio of 76:24 for O+HF to OH+F products. Let us first consider now the use of few-cycle IR + UV pulses to enhance the breaking of the weaker hydrogen bond, i.e. O···H. As in the FHF^-/FHF case, the approach relies on designing an IR laser pulse which achieves maximum extension and compression of the bonds and thus maximum displacement of the wave function from its equilibrium position. At the minimum energy configuration the equilibrium geometry is calculated to be R_{OH} = 1.1 Å and R_{HF} =1.3 Å. The frequency of the IR pulse (ω_{IR} = 1565 cm^{-1}) is chosen to drive the asymmetric stretching vibration or the motion of the hydrogen between the O and F end atoms. As in previous cases, the time duration of the pulse is chosen τ_{IR}= 50 fs, so that the pulse contains a few half-cycle pulses. Each of them excites a superposition of several vibrational eigenstates. After a time delay of 19 fs (see Fig. 4.14(b)) the O···H has been stretched from 1.1 to 1.28 Å. The UV pulse prepares the wave packet on the neutral surface along the O+HF dissociation channel (see Fig. 4.14(c)) and exclusive O +HF products are obtained. At the end of the propagation the branching ratio is calculated as 97:3 for O+HF:OH+F, thus enhancing the breaking of the weak H-bond O···H (Fig. 4.14(d)). Notice that the wave packet after 60 fs (Fig. 4.14(c)) still remains compact and dispersion is small.

Let us now turn to the problem of breaking the stronger H-bond of the system, while leaving the weaker one intact. Since the previous IR laser pulse was designed to drive the asymmetric stretching vibration and create a dynamical wave packet that oscillates with the largest amplitude away from the equilibrium posi-

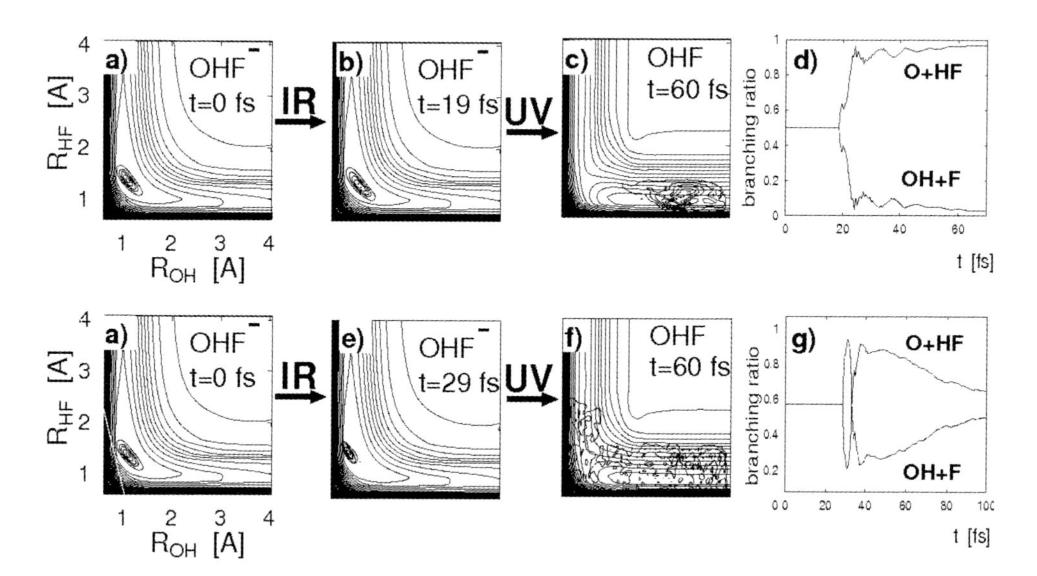

Figure 4.14 Selective H-bond breaking in OHF by photodetachment of the anion using few-cycle IR+UV laser pulses. The field parameters to maximize the products O+HF are $E^0_{IR} = 6.2 \times 10^{-3}\ E_h/ea_0$, ω_{IR} = 1565 cm^{-1}, τ_{IR} = 50 fs, $E^0_{UV} = 9.7 \times 10^{-3}\ E_h/ea_0$, ω_{UV} = 28228 cm^{-1} and τ_{UV} = 5 fs. The field parameters to maximize the products OH+F are $E^0_{IR} = 9.7 \times 10^{-3}\ E_h/ea_0$, ω_{IR} = 1565 cm^{-1}, τ_{IR} = 50 fs, $E^0_{UV} = 15.5 \times 10^{-3}\ E_h/ea_0$, ω_{UV} = 52423 cm^{-1} and τ_{UV} = 5 fs. (a) Initial wave function embedded in the anionic potential. (b) Wave packet in the anion potential after pre-excitation with the few-cycle IR laser pulse at t = 19 fs. (c) Wave packet in the neutral potential after the UV pulse in the neutral potential at t = 60 fs. (d) Branching ratio of the O+HF versus OH+F products. (e) Wave packet in the anion potential after pre-excitation with the few-cycle IR laser pulse at t = 29 fs. (f) Wave packet in the neutral potential after the UV pulse in the neutral potential at t = 60 fs. (g) Branching ratio of the O+HF versus OH+F products.

tion, the frequency and pulse duration of this pulse are maintained. Because of the asymmetry of the potential the field strength has to be intensified. The key point is that the UV laser pulse has to be fired at the time the bond of interest is maximally stretched. After 29 fs (see Fig. 4.14(e)) the wave packet has reached the opposite turning point, stretching the H···F bond from 1.3 to 1.41 Å. The new UV pulse not only has a different time delay in comparison with the previous case but also a different frequency, since it has to match the new gap between the anion and neutral PES. Despite the relatively compact form of the wave packet in the anion surface, because the transition state area is considerably flat, as soon as the wave packet is excited to the neutral PES, it spreads noticeably. As seen in Fig. 4.14(f) it begins traveling towards the center of the neutral potential instead of being confined along the dissociation channel OH+F. As a result, the wave packet disperses in both directions losing selectivity. The branching ratio shown in Fig. 4.14(g) shows that most of the 90% selectivity in the OH+F channel gained during the first 30 fs is lost in the next 10 fs. After 100 fs the wave packet density is stable and the branching ratio is steady. The calculated ratio is 57:43 for O+HF versus OH+F. This means that, after all, we have achieved a significant gain in breaking the strong H-bond H···F versus the weak O···H one.

The previous results demonstrate that it should be possible to exert control over the H-bond fragmentation using a sequence of few-cycle IR + UV laser pulses. The key steps for the success of such an approach are: (i) the control of the carrier envelope phase ϕ of the IR pulse, and (ii) the perfect timing of the UV pulse with respect to the IR one. Both aspects are related and their importance cannot be overemphasized. Depending on the carrier phase of the IR laser pulse the half-cycle pulses will have maximum amplitude at different times. Because it is the maxima of each of these half-cycle pulses which determine the maximum displacement of the H-atom, the phase inherently imposes the appropriate time delay at which the UV laser pulse should be fired. The successful implementation of these laser schemes in the laboratory requires therefore the reproducible generation of pulses with the same absolute phases. Though not straightforward, it is encouraging to see that in recent years control of the phase of attosecond X-ray [72, 73] as well as few-cycle femtosecond, near-IR (800–1700 nm) lasers (as the ones employed here!) is becoming a reality [74–76]. A phase of π does not shift the peaks of the half-cycle pulses, just changes the sign, or in other words, determines the direction in which the initial wave packet begins oscillating. For asymmetric systems, like OHF^-, this indeed plays again an important role in choosing the time delay of the UV laser pulse. Of course the frequency of the UV laser pulse has to match the energy gap between anion and neutral surfaces at the corresponding geometry achieved by the IR laser pulse. This means that, in asymmetric systems, the frequency of the UV pulse optimized to match one exit channel will not, in general, be efficient in achieving population inversion at the other exit channel.

The latter issue serves to bring up the role of molecular orientation (or alignment for symmetric systems) in the proposed laser scheme. In the case of FHF it has already been mentioned that even starting from a random ensemble of all possible orientations spatial separation of the products can be achieved due to (i) the inefficiency of the IR laser pulse in exciting molecules which are not in the optimum orientation angle ($\theta = 0°$, i.e. parallel to the molecular axis); and (ii) the off-resonant character of the UV laser pulse acting on those molecules which have not been optimally pre-excited with the IR laser pulse [62]. Similar behavior can be expected in OHF, greatly enhanced by the fact that the UV pulses in both channels have very different frequency.

Another way to overcome this problem is of course by achieving pre-orientation of the sample before starting the control experiment. The rotational periods of FHF^- and OHF^- are calculated as 48.6 and 49.6 ps, respectively [77]. Because the symmetry breaking and posterior bond selective photodissociation takes place in less than 100 fs, it is reasonable to assume that the molecule will stay oriented while the laser-driven control takes place.

4.4 Conclusions and Outlook

The manipulation of molecular systems by means of tailored laser pulses has triggered a host of experimental and theoretical efforts. Due to their importance for various processes in chemical and biological systems H- bonds are a particularly attractive target in this respect. However, despite the impressive progress in the experimental control in the optical wavelength range [12, 13], IR-driven reaction dynamics is just becoming reality [78]. In this chapter we have reviewed control schemes for HT reactions in the electronic ground state. It has been demonstrated that for low-dimensional model systems control can be achieved, e.g., by over-the-barrier and through-the-barrier driving. Stepping to more realistic multidimensional models it turned out that control schemes derived for simpler systems may not be transferable. Of course, this does not imply that laser control of H-bond dynamics is not feasible at all. We merely want to point out that most likely for many real (medium-sized organic) systems the anharmonic coupling coming along with the change in the electronic structure upon H-transfer is rather pronounced and therefore a serious challenge to any control attempt. Feedback control in the optical domain has been shown to be a very powerful tool [14] and similar ideas might provide useful when it comes to multidimensional H-bond dynamics. On the other hand, even if the control of a H-transfer reaction might prove too difficult for many systems due to rapid intramolecular vibrational energy redistribution, driving the ground state wave packet away from its minimum position in a controlled manner could give – if combined with nonlinear spectroscopy – valuable information about the anharmonic PES.

In the second part of this chapter we have shown that such displaced wave packets could be the first step in an alternative control scheme involving a properly timed UV excitation. In principle this might cause a transition to some electronically excited state and the subsequent excited state H-bond dynamics could be controlled. Here, however, we focused on the transition between an anionic and a neutral PES accompanied by electron photodetachment. For H-bonds of the type AHB^- it was demonstrated that selective bond breaking can be controlled. In view of the experimental progress [79, 80] such an IR/UV scheme does not seem to be too far-fetched.

Acknowledgments

The work related to the few-cycle IR plus UV control strategies on FHF and OHF is part of the PhD work of N. Elghobashi. L.G. wishes to thank her for providing the corresponding figures and her continuous enthusiasm. Further we wish to thank N. Došlić (Zagreb) for her contributions to the "Hydrogen Subway", M. Petković who carried out the calculations for salicylaldimine, and Prof. J. Manz (Berlin) for many stimulating discussions and his continuous support. The financial support from the "Graduiertenkolleg" 788, "Hydrogen Bonding and Hydrogen Transfer" and the Sfb450 is also gratefully acknowledged.

References

1 K. Tanaka, H. Honjo, T. Tanaka, H. Kohguchi, Y. Ohshima, Y. Endo, *J. Chem. Phys.* **1999**, *110*, 1969.
2 E. T. J. Nibbering, T. Elsaesser, *Chem. Rev.* **2004**, *104*, 1887.
3 T. Elsaesser, in *Femtosecond Chemistry*, J. Manz, L. Wöste (Eds.), Verlag Chemie, Weinheim, **1995**, p. 563.
4 A. Douhal, F. Lahmani, A. H. Zewail, *Chem. Phys.* **1996**, *207*, 477.
5 S. Lochbrunner, E. Riedle, *Rec. Res. Devel. Chem. Phys.* **2003**, *4*, 31.
6 B. I. Stepanov, *Nature* **1945**, *157*, 808.
7 N. Sheppard, in *Hydrogen Bonding* (Ed.: D. Hadži), Pergamon Press, New York, **1957**, p. 85.
8 M. V. Basielevsky, M. V. Vener, *Russ. Chem. Rev.* **2003**, *72*, 3.
9 S. Bratos, J. C. Leicknam, G. Gallot, H. Ratajcak, in *Ultrafast Hydrogen Bonding Dynamics and Proton Transfer Processes in the Condensed Phase*, T. Elsaesser, H. J. Bakker (Eds.), Kluwer Academic, Dordrecht, **2002**, p. 5.
10 J. Stenger, D. Madsen, J. Dreyer, E. T. J. Nibbering, P. Hamm, T. Elsaesser, *J. Phys. Chem. A* **2001**, *105*, 2929.
11 K. Heyne, N. Huse, J. Dreyer, E. T. J. Nibbering, T. Elsaesser, S. Mukamel, *J. Chem. Phys.* **2004**, *121*, 902.
12 S. Rice, M. Zhao, *Optical Control of Molecular Dynamcis*, John Wiley and Sons, Hoboken, **2000**.
13 M. Shapiro, P. Brumer, *Principles of Quantum Control of Molecular Processes*, John Wiley and Sons, Hoboken, **2003**.
14 T. Brixner, G. Gerber, *ChemPhysChem* **2003**, *4*, 418.
15 X. Huang, H. M. Cho, S. Carter, L. Ojamäe, J. Bowman, S. J. Singer, *J. Phys. Chem. A* **2003**, *107*, 7142.
16 M. V. Vener, O. Kühn, J. Bowman, *Chem. Phys. Lett.* **2001**, *340*, 562.
17 M. V. Vener, O. Kühn, J. Sauer, *J. Chem. Phys.* **2001**, *114*, 240.
18 N. Došlić, O. Kühn, *Z. Phys. Chem.* **2003**, *217*, 1507.
19 E. Kraka, in *Encyclopedia of Computational Chemistry*, P. v. Rague-Schleyer (Ed), Wiley, New York, **1998**, p. 2437.
20 W. H. Miller, N. C. Handy, J. E. Adams, *J. Chem. Phys.* **1980**, *72*, 99.
21 T. Carrington, W. H. Miller, *J. Chem. Phys.* **1986**, *84*, 4364.

22 N. Shida, P. F. Barbara, J. E. Almlöf, *J. Chem. Phys.* **1989**, *91*, 4061.
23 B. Fehrensen, D. Luckhaus, M. Quack, *Z. Phys. Chem.* **1999**, *209*, 1.
24 R. Meyer, T. K. Ha, *Mol. Phys.* **2003**, *101*, 3263.
25 X. Huang, B. J. Braams, S. Carter, J. Bowman, *J. Am. Chem. Soc.* **2004**, *126*, 5042.
26 D. Luckhaus, *Chem. Phys.* **2004**, *304*, 79.
27 R. Jaquet, W. H. Miller, *J. Phys. Chem.* **1985**, *89*, 2139.
28 B. A. Ruf, W. H. Miller, *J. Chem. Soc., Faraday Trans. 2* **1988**, *84*, 1523.
29 K. M. Forsythe, N. Makri, *J. Chem. Phys.* **1998**, 6819.
30 H. Naundorf, J. A. Organero, A. Douhal, O. Kühn, *J. Chem. Phys.* **1999**, *110*, 11286.
31 G. K. Paramonov, H. Naundorf, O. Kühn, *Eur. Phys. J. D* **2001**, *14*, 205.
32 M. Petković, O. Kühn, *J. Phys. Chem. A* **2003**, *107*, 8458.
33 R. Xu, Y. J. Yan, O. Kühn, *Eur. Phys. J. D* **2002**, *19*, 293.
34 K. Giese, D. Lahav, O. Kühn, *J. Theor. Comput. Chem.* **2004**, *3*, 567.
35 K. Giese, O. Kühn, *J. Chem. Phys.* **2005**, *123*, 054315.
36 H.-D. Meyer, G. A. Worth, *Theor. Chem. Acc.* **2003**, *109*, 251.
37 H. Naundorf, G. A. Worth, H.-D. Meyer, O. Kühn, *J. Phys. Chem. A* **2002**, *106*, 719.
38 V. May, O. Kühn, *Charge and Energy Transfer Dynamics in Molecular Systems*, 2nd edn., Wiley-VCH, Weinheim, **2004**.
39 N. Makri, *J. Phys. Chem. B* **1999**, *103*, 2823.
40 R. Rey, J. T. Hynes, *J. Chem. Phys.* **1996**, *104*, 2356.
41 C. P. Lawrence, J. L. Skinner, *J. Chem. Phys.* **2002**, *117*, 5827.
42 H. Naundorf, O. Kühn, *Phys. Chem. Chem. Phys.* **2003**, *5*, 79.
43 K. Heyne, E. T. J. Nibbering, T. Elsaesser, M. Petković, O. Kühn, *J. Phys. Chem. A* **2004**, *108*, 6083.
44 P. K. Agarwal, S. Billeter, S. Hammes-Schiffer, *J. Phys. Chem. B* **2002**, *106*, 3283.
45 M. V. Korolkov, J. Manz, G. K. Paramonov, *Adv. Chem. Phys.* **1997**, *101*, 327.
46 N. Došlić, O. Kühn, J. Manz, *Ber. Bunsenges. Phys. Chem.* **1998**, *102*, 292.
47 N. Došlić, K. Sundermann, L. González, O. Mó, J. Giraud-Girard, O. Kühn, *Phys. Chem. Chem. Phys.* **1999**, *1*, 1249.
48 W. Zhu, J. Botina, H. A. Rabitz, *J. Chem. Phys.* **1998**, *108*, 1953.
49 K. Sundermann, R. deVivie-Riedle, *J. Chem. Phys.* **1999**, *110*, 1896.
50 N. Došlić, O. Kühn, J. Manz, K. Sundermann, *J. Phys. Chem. A* **1998**, *102*, 9645.
51 O. Kühn, *Eur. Phys. J. D* **1999**, *6*, 49.
52 N. Došlić, O. Kühn, *Chem. Phys.* **2000**, *255*, 247.
53 H. Naundorf, K. Sundermann, O. Kühn, *Chem. Phys.* **1999**, *240*, 163.
54 O. Kühn, Y. Zhao, F. Shuang, Y. Yan, *J. Chem. Phys.* **2000**, *112*, 6104.
55 W. Sim, N. Makri, *Comput. Phys. Commun.* **1997**, *99*, 335.
56 M. Petković, O. Kühn, in *Ultrafast Molecular Events in Chemistry and Biology*, J. T. Hynes, M. M. Martin (Eds.), Elsevier, Amsterdam, **2004**, p. 181.
57 M. Petković, O. Kühn, *Chem. Phys.* **2004**, *304*, 91.
58 D. M. Neumark, *Acc. Chem. Res.* **1993**, *26*, 33.
59 B. Amstrup, N. E. Henriksen, *J. Chem. Phys.* **1992**, *97*, 8285.
60 B. Amstrup, N. E. Henriksen, *J. Chem. Phys.* **1996**, *105*, 9115.
61 N. Elghobashi, L. González, J. Manz, *J. Chem. Phys.* **2004**, *120*, 8002.
62 N. Elghobashi, J. Manz, *Isr. J. Chem.* **2003**, *43*, 293.
63 N. Elghobashi, L. González, J. Manz, *Z. Phys. Chem.* **2003**, *217*, 1577.
64 M. Machholm, N. E. Henriksen, *J. Chem. Phys.* **1996**, *105*, 9115.
65 N. Elghobashi, P. Krause, J. Manz, M. Oppel, *Phys. Chem. Chem. Phys.* **2003**, *5*, 4805.
66 N. Elghobashi, L. González, *Phys. Chem. Chem. Phys.* **2004**, *6*, 4071.
67 R. P. Wayne, *Chemistry of Atmospheres*, Oxford University Press, Oxford, **2000**.
68 R. N. Dixon, H. Tachikawa, *Mol. Phys.* **1999**, *97*, 195.
69 L. González-Sánchez, S. Gómez-Carrasco, A. Aguado, M. Paniaga, M. L. Hernández, O. Roncero,

J. M. Alvariño, *J. Chem. Phys.* **2004**, *121*, 309.

70 L. González-Sánchez, S. Gómez-Carrasco, A. Aguado, M. Paniaga, M. L. Hernández, O. Roncero, J. M. Alvariño, *J. Chem. Phys.* **2004**, *121*, 9865.

71 J. J. Sloan, D. G. Watson, J. M. Williamson, J. S. Wright, *J. Chem. Phys.* **1981**, *75*, 1190.

72 A. Apolonski, A. Poppe, G. Tempea, C. Spielmann, T. Udem, R. Holzwarth, T. W. Hänsch, F. Krausz, *Science* **2000**, *288*, 635.

73 D. J. Jones, S. A. Diddams, J. K. Ranka, A. Stenz, R. S. Windeler, J. L. Hall, S. T. Cundiff, *Science* **2001**, *293*, 825.

74 G. G. Paulus, F. Grasbon, H. Walther, P. Villoresi, M. Nisoli, S. Stagira, E. Priori, S. DeSilvestri, *Nature* **2001**, *414*, 182.

75 M. Nisoli, G. Sansonse, S. Stagira, S. DeSilvestri, C. Vozzi, M. Pascolini, L. Poletto, P. Villoresi, G. Tondello, *Phys. Rev. Lett.* **2003**, *91*, 213905.

76 G. Sansonse, C. Vozzi, S. Stagira, M. Pascolini, L. Poletto, P. Villoresi, G. Tondello, S. DeSilvestri, M. Nisoli, *Phys. Rev. Lett.* **2004**, *92*, 113904.

77 N. Elghobashi, PhD Thesis, Free University, Berlin, **2005**.

78 L. Windhorn, J. S. Yeston, T. Witte, W. Fuß, M. Motzkus, D. Proch, K. L. Kompa, C. B. Moore, *J. Chem. Phys.* **2003**, *119*, 641.

79 B. Schenkel, J. Biegert, U. Keller, C. Vozzi, M. Nisoli, G. Sansone, S. Stagira, S. De Silvestri, O. Svelto, *Opt. Lett.* **2003**, *28*, 1987.

80 T. M. Fortier, David J. Jones, S. T. Cundiff, *Opt. Lett.* **2003**, *28*, 2198.

Part II
Hydrogen Transfer in Condensed Phases

In Ch. 5 Ceulemans reviews the protonation of alkanes in solid matrices by radical cations created by γ-radiation, studied by EPR. The protonation does not take place at a particular atom but at a particular C–H or C–C bond, resulting in the formation of a three-center two-electron bond in a carbon-site specific way. Of great importance is the mobility of the matrix which greatly enhances the molecular reactivity.

In Ch. 6 Limbach describes the use of NMR for the study of single and multiple hydrogen/deuterium transfers in liquids and solids. Using suitably labeled molecules, rate constants can now be obtained in a large temperature range from the second to the picosecond timescale. Expressions for multiple kinetic isotope effects of double to quadruple hydrogen transfer reactions are derived by formal kinetics and used in combination with the Bell tunneling model modified by Limbach. Thus, Arrhenius curves of single to quadruple hydrogen transfers can be simulated. The detection of stepwise vs. concerted reaction pathways and the role of H-bond compression in H-transfer reactions is discussed. Criteria are developed which allow one to detect hidden pre-equilibria. A number of cases known in the literature are re-analyzed in this sense, and pre-equilibria are proposed as the origin for non-conventional Arrhenius curves which could not explained before with simple tunneling models.

In Ch. 7 Douhal examines ultrafast photoinduced proton transfers of dye molecules trapped in nanocavities. The main factors that determine the issue of a created wavepacket in the cage are the structure, the orientation of the guest, the docking and rigidity of the complex, and the polarity of the cage. Water molecules located inside and at both gates of cyclodextrins have special properties reminiscent of biological water. Their restricted dynamics is slower than that found in bulk water. This abnormal behavior plays a key role in many chemical and biological processes. Further nanospaces in membranes, micelles, polymers, lipid vesicles, liquid crystals, sol–gels, dendrimers, proteins, DNA, zeolites and nanotubes are explored.

In Ch. 8, Waluk reviews the spectroscopy of porphycenes embedded in supersonic jets, liquids and polymer matrices. Tunnel splittings caused by delocalization of two inner hydrogen atoms is observed for porphycenes in supersonic jets.

Hydrogen-Transfer Reactions. Edited by J. T. Hynes, J. P. Klinman, H. H. Limbach, and R. L. Schowen

ISBN: 978-3-527-30777-7

The barrier to this transfer is higher in the lowest excited singlet state than in the ground state. The transfer mechanism involves a concerted double hydrogen tunneling process, activated by excitation of a low-frequency mode, which modulates the NHN separation. This separation can also be strongly altered by peripheral substitution. In porphycenes with alkyl substituents on the ethylene bridges, the NHN distances become very small. It is shown that the tautomerism of porphycenes can also be monitored on a single-molecule level.

Finally, Ch. 9 by Vener is devoted to the proton dynamics in hydrogen bonded crystals. In particular, the effects on the vibrations of systems with low-barrier hydrogen OHO-hydrogen bonds are studied. DFT calculations with and without periodic boundary conditions indicate that the coupling of the proton to the heavy atom coordinate is stronger in crystals as compared to the isolated systems. This coupling accounts for major structural and spectroscopic properties observed experimentally for molecular crystals, e.g. geometric H/D isotope effects, low-frequency shifts of the asymmetric stretch of the O···H··O fragment and large variations for the isotopic frequency ratio.

5
Proton Transfer from Alkane Radical Cations to Alkanes

Jan Ceulemans

Alkane radical cations are very strong Brønsted acids capable of protonating alkanes. Radiation-chemical studies have proven highly successful for the investigation of such proton transfer from alkane radical cations to alkane molecules, using n-alkane nanoparticles embedded in cryogenic CCl_3F matrices for the study of symmetric proton transfer ($RH^{\bullet+} + RH \rightarrow R^{\bullet} + RH_2^{+}$) and mixed n-alkane crystals for the study of asymmetric proton transfer ($R_IH^{\bullet+} + R_{II}H \rightarrow R_I^{\bullet} + R_{II}H_2^{+}$). The mechanism of the radiolytic processes involved is discussed in considerable detail. Selectivity with respect to the site of both proton donation and proton acceptance has been studied, using EPR spectroscopy at 77 K for the study of proton donation and gas chromatographic analysis after melting for the study of proton acceptance. The experiments described allow one to conclude that the proton-donor site is related very strictly to the structure of the semi-occupied molecular orbital of the parent radical cation, with proton transfer taking place from those C–H bonds that carry appreciable unpaired-electron and positive-hole density. With respect to the site of proton acceptance, it is observed that chemical transformation due to protonation of n-alkanes by alkane radical cations in the systems studied is restricted to C–H bonds at secondary carbon atoms (no chemical transformation resulting from proton transfer to C–C bonds or to C–H bonds at primary carbon atoms). It is argued that the absence of chemical transformation due to C–C protonation has a complex origin in which thermochemical and structural (cage) effects are involved as well as the intrinsic stability of the carbonium ions, but that the absence with respect to protonation of C–H bonds at primary carbon atoms has (in all likelihood) a purely thermochemical origin. As to protonation of n-alkanes at secondary C–H bonds, extensive evidence is presented supporting a marked preference for the penultimate position, with considerably lower (and mutually equal) transfer to the interior sites (intrinsic acceptor-site selectivity). In mixed n-alkane crystals, additional selectivity with respect to the site of proton acceptance results from structural factors in combination with the donor-site selectivity (structurally determined acceptor-site selectivity).

Hydrogen-Transfer Reactions. Edited by J. T. Hynes, J. P. Klinman, H. H. Limbach, and R. L. Schowen

ISBN: 978-3-527-30777-7

5.1
Introduction

Three major types of cationic species that can be derived from saturated hydrocarbons are alkyl carbenium ions (R^+), alkane radical cations ($RH^{\bullet+}$) and alkyl carbonium ions (RH_2^+). The term “carbocations” is usually reserved to denote alkyl carbenium and carbonium ions only. Pentacoordinated alkyl carbonium ions (protonated alkanes) are the species that result from protonation of alkane molecules; they are of paramount importance as reactive intermediates/transition states in the initiation of (Brønsted) acid-catalyzed conversions of saturated hydrocarbons. Upon dissociation of alkyl carbonium ions, trivalent alkyl carbenium ions are formed and these are responsible for the further progression of acid-catalyzed conversions of alkanes. Alkyl carbenium ions may also be formed by ionization of neutral alkyl radicals and by proton addition to olefins. In both carbenium and carbonium ions, the positive charge is very much located on a particular part of the cation.

Alkane radical cations are the species that result from ionization (electron removal) of neutral alkane molecules. They have both paramagnetic and ionic properties, hence the term radical cations. In alkane radical cations, the unpaired-electron and positive-charge (hole) density is distributed over a large part of the molecular frame in a manner that depends on the carbon chain conformation. As will be amply demonstrated in the present chapter, alkane radical cations are very strong Brønsted acids that are capable of protonating neutral alkanes. As such, they offer an interesting and elegant alternative to superacids for the study of site selectivity in the protonation of alkanes. Most of the information on alkane radical cations and essentially all the information on their ability to protonate alkane molecules has been derived from radiation chemistry, a rather specialized field that involves the study of the chemical effects of ionizing radiation. The radiolytic processes that allow the study of (donor and acceptor) site selectivity in the proton transfer from alkane radical cations to alkane molecules will therefore be discussed in some detail.

5.2
Electronic Absorption of Alkane Radical Cations

Radical cations of saturated hydrocarbons have strong electronic absorptions in the visible and near-infrared region of the spectrum. The strongly colored nature of alkane radical cations is in striking contrast to neutral alkanes that absorb electronically only in the vacuum UV. The electronic absorption of alkane radical cations has been studied in the solid phase by matrix isolation using γ-irradiation [1–3] and in the gas phase by ion cyclotron resonance (ICR) photodissociation in either the steady-state or pulsed mode of operation [4]. Both methods have their specific merits and drawbacks. A major concern in matrix isolation spectroscopy is spectral purity (because of the possible presence of other absorbing species) and elaborate deconvolution techniques have been developed to overcome this problem. With gas phase dissociation spectra the effects of threshold truncation must be taken into account as photon energies must be sufficient for bond cleavage and truncation of the spectra at some wavelength in the red or near-infrared may be problematic.

The strong electronic absorption of alkane radical cations is readily understood in molecular orbital terms. Extending down from the highest occupied molecular orbital (HOMO) is a rather closely packed set of valence molecular orbitals, that are clearly displayed in the photoelectron spectra (PES) of neutral alkanes. The electronic absorption of alkane radical cations is due to transitions (induced by photon absorption) of electrons from such lower-lying molecular orbitals to the semi-occupied molecular orbital (SOMO), which is the highest-occupied molecular orbital in the ground-state ion. By illumination within the (broad and largely unstructured) absorption band of alkane radical cations, electronically excited states of alkane radical cations can thus be created in a quite convenient way.

5.3 Paramagnetic Properties of Alkane Radical Cations

Information on the paramagnetic properties of alkane radical cations has been obtained by conventional EPR spectroscopy using matrix isolation [5–8] and by fluorescence detected magnetic resonance (FDMR) spectroscopy [9]. With the former technique, chlorofluorocarbon matrices (such as CCl_3F, CCl_2FCF_2Cl and CCl_3CF_3) and perfluorocarbon matrices (such as perfluoromethylcyclohexane) as well as SF_6 have routinely been used in combination with X- or γ-irradiation. With the matrices employed, EPR signals from matrix radicals are anisotropically spread over a wide range (and are consequently very weak) and the paramagnetic absorption of alkane radical cations can be observed without serious interference. FDMR spectroscopy derives its special interest from the fact that it allows the observation of alkane radical cations in irradiated alkane systems.

The paramagnetic absorption of alkane radical cations is critically dependent on their conformation. In neat n-alkane crystals, alkane molecules are in the extended all-trans conformation (see Fig. 5.1) and FDMR spectroscopy unequivocally shows that alkane radical cations retain that conformation in such systems. The extended all-trans conformation is also the preferred conformation of many n-alkane radical cations in chlorofluorocarbon and perfluorocarbon matrices. In this conformation, the unpaired electron occupies the planar σ molecular orbital and delocalizes over the entire extended chain. Only two C–H bonds (both chain-end, one on each side) are in the planar σ molecular frame in the extended structure and high unpaired-electron and positive-hole density appears only on these in-plane protons. Alkane radical cations in the extended conformation (as well as in other conformations) are thus σ-delocalized paramagnetic species. The associated hyperfine interaction with the two (equivalent) in-plane chain-end protons results in a 1:2:1 three-line (triplet) EPR spectrum. The fact that the hyperfine in-

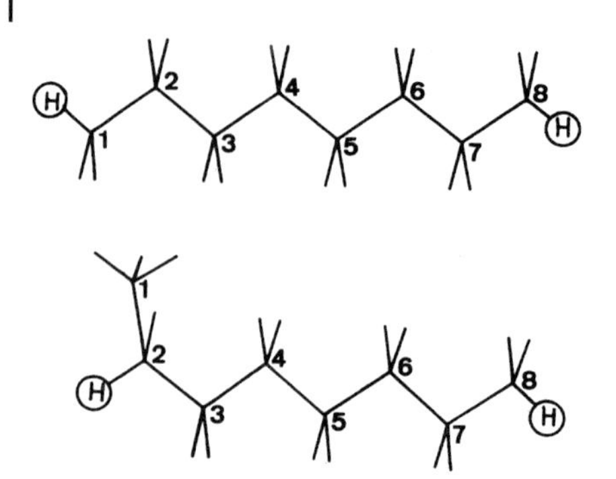

Figure 5.1 Extended all-trans and gauche-at-C_2 conformers of octane.

teraction is due to two *chain-end* C–H bonds has unequivocally been demonstrated by experiments using specific deuteration. With increasing chain length, EPR spectra of n-alkane radical cations in the extended conformation become increasingly contracted as the hyperfine interaction decreases due to increased delocalization of the unpaired electron. The major proton hyperfine coupling constants of n-alkane radical cations in the extended conformation show a steady decrease with increasing chain length.

A second conformation that is frequently encountered in chlorofluorocarbon matrices is the gauche-at-C_2 conformation, which is obtained from the extended conformer by 120° rotation around C_2–C_3 (see Fig. 5.1). In this conformation, the unpaired electron and positive hole delocalize over the planar part of the C–C skeleton as well as over one in-plane chain-end and penultimate C–H bond (on opposite sides of the radical cation). Other more twisted conformations (such as obtained from the extended conformer by 120° rotation around the penultimate C–C bond as well as around C_2–C_3, either both clockwise or one clockwise and the other counterclockwise) have also been observed, but they are less common in chlorofluorocarbon matrices. In n-alkanes in the liquid state, many conformers are present and the paramagnetic absorption reflects this by the fact that an unresolved spectrum is observed for the radical cations, resulting from hyperfine coupling with many different protons [10]. As a result of increased delocalization, the line width of such composite spectra decreases with increasing chain length.

5.4 The Brønsted Acidity of Alkane Radical Cations

Alkane radical cations are very strong Brønsted acids and form an acid–base pair with neutral alkyl radicals as conjugate base according to the *acid–base half-reaction*

$$RH^{\bullet+} \rightarrow R^{\bullet} + H^{+} \qquad (5.1)$$

For n-alkane radical cations, both primary and secondary alkyl radicals may act as conjugate base, depending on the conformation of the radical cation. This is a direct consequence of the fact that the site of proton donation from alkane radical cations is strictly dependent on their electronic structure, that is, a high unpaired-electron and positive-hole density in a particular C–H bond results in proton transfer from that bond (evidence for this is given in Sections 5.7 and 5.8), whereas the electronic structure of the radical cations in turn depends on their conformation. For n-alkane radical cations in the extended conformation, in which the unpaired-electron and positive-hole delocalize over the planar C–C skeleton and two in-plane chain-end C–H bonds (one on each side), proton transfer takes place exclusively from chain-end positions and the corresponding primary alkyl radicals act as conjugate base. For n-alkane radical cations in the gauche-at-

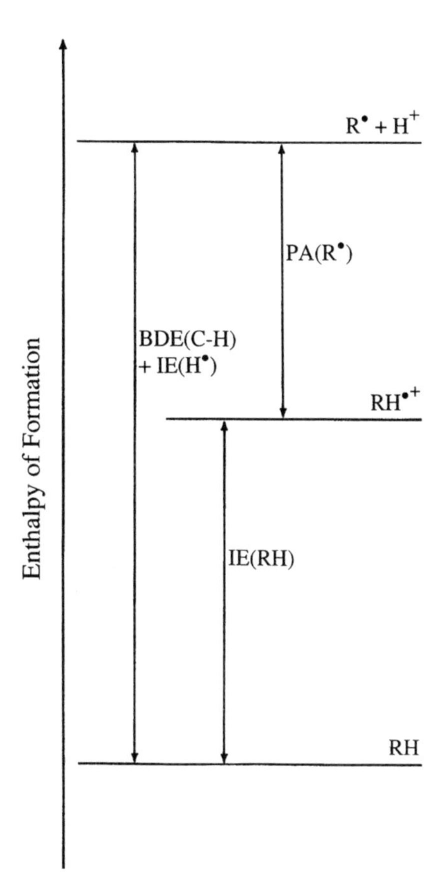

Figure 5.2 Energy diagram relating the proton affinity of neutral alkyl radicals, PA($R^{\bullet}$), to the ionization energy of the corresponding alkane, IE(RH), with the bond dissociation energy of the appropriate C–H bond, BDE(C–H), and the ionization energy of atomic hydrogen, IE($H^{\bullet}$), as additional parameters.

C_2 conformation, in which the unpaired electron and positive hole delocalize over the planar part of the C–C skeleton as well as over one planar chain-end and penultimate C–H bond, proton transfer can take place from both a terminal and a penultimate position (on opposite sides of the radical cation); in such conformers, primary as well as secondary alkyl radicals act as conjugate base.

As is evident from Fig. 5.2, the proton affinity of neutral alkyl radicals, $E_{pa}(R^\bullet)$, may be derived from the ionization energy of the corresponding alkane, $E_i(RH)$, the bond dissociation energy of the appropriate C–H bond, $E_{bd}(C{-}H)$, and the ionization energy of atomic hydrogen, $E_i(H^\bullet)$, according to Eq. (5.2).

$$E_{pa}(R^\bullet) = E_{bd}(C{-}H) + E_i(H^\bullet) - E_i(RH) \tag{5.2}$$

Protonation energies of linear primary and secondary alkyl radicals show a steady increase with increasing chain length. This increase is closely associated with the corresponding decrease in ionization energy of the related alkane molecule, which is due to the fact that alkane radical cations are σ-delocalized species for which the degree of delocalization of the unpaired-electron and positive-hole density (and the corresponding stabilization) increases with increasing chain length. As such, it is also related to the decrease with increasing chain length of the proton hyperfine coupling constants in the EPR spectra of n-alkane radical cations.

5.5 The σ-Basicity of Alkanes

Conventional organic bases (such as amines) are compounds containing heteroatoms on which lone-pair electrons reside. The basicity of organic compounds is not limited, however, to such nonbonding electron systems. It has long been recognized that unsaturated aliphatic as well as aromatic hydrocarbons have a tendency to act as bases in acid–base reactions. The reactivity of olefins, acetylenes and aromatic hydrocarbons toward electrophiles lies in the π-electron donor abilities of the unsaturated double and triple bonds and π-aromatic systems. The addition of a proton to olefinic double bonds results in the formation of alkyl carbenium ions. In the absence of lone-pair and π-electrons, i.e. in the case of saturated aliphatic hydrocarbons, C–H and C–C bonds (both σ-bonds) may also act as proton acceptors, as was evidenced by the work of Olah and coworkers on superacids [11–13]; very strong acids are required to protonate saturated hydrocarbons. As will amply be demonstrated in the present chapter, alkane radical cations offer an elegant alternative to superacids for the protonation of alkanes. Protonation of alkanes by alkane radical cations that are derived from the same alkane will be termed “symmetric”; the term “asymmetric proton transfer” will be applied if this condition is not fulfilled. It is to be remarked at this point that the reaction of an alkane with a *free* proton is a very exothermic process; the non-occurrence of protonation of alkanes by conventional acids is thus due to the fact that the tendency for proton retention of these acids is too strong and not to “lack of appetite” of the alkane. As a matter of fact, very strong acids may be characterized as "weak" in the sense that they easily give in to the demand for protons by other species.

In contrast to conventional organic bases that contain heteroatoms, in which protonation takes place at the heteroatom involving the lone-pair electrons, protonation in alkanes does not take place at a particular atom but at a particular C–H or C–C bond, resulting in the formation of a three-center two-electron bond. The σ-basicity of alkanes is based on the general electron-pair donor ability of *shared* electron pairs in single bonds, in contrast to the donor ability of *unshared* electron pairs (lone-pair electrons) in conventional organic bases. In C–H protonated alkanes, bonding is provided by a three-center two-electron bond resulting from the overlap of the σ-orbital of a hydrogen molecule with an empty sp^3 orbital of the appropriate carbon atom. In C–C protonated alkanes, the bonding is due to an overlap of two sp^3 orbitals belonging to adjacent carbon atoms and the 1s orbital of atomic hydrogen; this group also contains only two electrons. As a result of electron-density differentiation, protonation takes place on the main lobes of the C–H or C–C bonds (where the major part of the electron density resides) and not on the back lobes, that is, the carbon atoms themselves. It is to be remarked that with the formation of three-center two-electron bonds the octet rule for the carbon atoms involved is not violated; carbon atoms involved in three-center two-electron bonds in alkyl carbonium ions are pentacoordinated but they are *not* pentavalent.

Experimental data on the protonation energies of saturated hydrocarbons are very scarce. A useful experimental approach consists in the determination of the dissociation energy of the carbonium ions by ion-equilibrium measurements, as performed by Kebarle and Hiraoka using high-pressure ion source mass spectrometry [14]. The thermodynamics of the protonation of alkanes and the energetics of pentacoordinated alkyl carbonium ions are not easily accessible experimentally, but they are linked through the dissociation energy of the carbonium ions to the energetics of the corresponding carbenium ions that can more readily be studied experimentally. The approach only works properly for the lowest members of the alkyl carbonium ion series, however. Because of the shortage of direct experimental observations of carbonium ions and of experiments that provide numerical data on their energies, the structure and energy of these species have been extensively studied theoretically. Considerable progress on this has been made in recent years by high-level theoretical calculations allowing the establishment of a *σ-basicity scale* (viz., C–C > tertiary C–H > secondary C–H > primary C–H > CH_4) that reflects the thermodynamics of the protonation of the σ bonds [15–17]. However, theoretical studies are also largely limited to short-chain ($n_C \leq 4$) cations and reliable high-level theoretical calculations on long-chain alkyl carbonium ions still appear difficult to perform.

5.6 Powder EPR Spectra of Alkyl Radicals

Proton transfer from alkane radical cations to alkane molecules results in the transformation of these cations into neutral alkyl radicals (the conjugate bases). The nature of these radicals is determined by the site of proton donation in the alkane radical cation. Information on the site of proton donation in the proton transfer from alkane radical cations to alkane molecules can thus be derived from EPR spectral analysis of the neutral alkyl radicals formed. To aid the reader in appreciating the results that are presented on this matter below and in understanding related spectra from the literature, a section on the characterization of neutral alkyl radicals by EPR spectroscopy in solid systems is included at this point.

Neutral alkyl radicals derived from n-alkanes can be separated into three distinct classes with respect to their EPR spectrum, viz. chain-end ($^{\bullet}CH_2–CH_2–CH_2–\cdots$), penultimate ($CH_3–^{\bullet}CH–CH_2–\cdots$) and interior ($\cdots–CH_2–^{\bullet}CH–CH_2–\cdots$). The possibility of cancellation of the anisotropic interactions of α-hydrogens is highly relevant with respect to the powder EPR spectra of alkyl radicals. Such annulment is possible for chain-end radicals, but not for penultimate and interior radicals. In alkyl radicals both the isotropic and anisotropic hyperfine interactions with the α-protons are "extensive", that is, they affect the paramagnetic absorption to an extent that can clearly be discerned in the powder EPR spectra; in contrast, there is only a large isotropic interaction with the β-protons (anisotropic interactions with the β-protons are much smaller in value). When in 1-alkyl radicals the two α-proton spins are antiparallel, the hyperfine anisotropy largely cancels and relatively sharp intense hyperfine lines result, which are easily discernible in an EPR spectrum. In contrast, all EPR absorption bands of penultimate and interior n-alkyl radicals are strongly anisotropically broadened because cancellation of the anisotropic hyperfine interaction with the α-proton cannot take place. Another specific characteristic of the powder EPR spectrum of 1-alkyl radicals is the outermost transition, which is severely anisotropically broadened because the α-proton spins are parallel and manifests itself in a first-derivative spectrum as a slightly double-humped curve; (more centrally located anisotropic absorptions are not clearly observable due to mutual interference and spectral interference by the "isotropic" lines). Both features (the relatively narrow and largely isotropic lines and the double-humped curve) make the powder EPR spectrum of 1-alkyl radicals easily recognizable. The structure and overall appearance of the powder spectrum of 1-alkyl radicals are apparent from the simulated spectra shown in Fig. 5.3.

Experimental information on the powder spectra of authentic chain-end, penultimate and interior alkyl radicals can be obtained from γ-irradiated solid *cis*-decalin-d_{18} containing appropriate chloro- and bromoalkanes [18]. The EPR spectrum obtained after γ-irradiation of neat *cis*-decalin-d_{18} consists of a broad unresolved band, which extends over a relatively restricted spectral region as a result of spectral contraction due to deuteration (see Fig. 5.4(a)). The addition of chloro- and bromoalkanes before irradiation results in a considerable distortion of this unre-

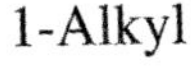

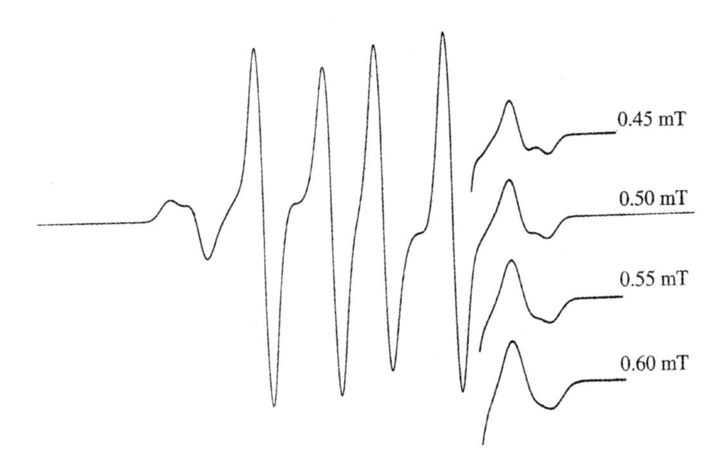

Figure 5.3 Spectrum of 1-alkyl radicals simulated on the basis of hyperfine coupling constants of Ref. [35]. The spectrum is simulated using the Gaussian type lineshape function, with $\Delta H_{ms} = 0.5$ mT; the effect of different linewidths on the outermost anisotropic transition is also shown.

solved absorption and in the appearance of very characteristic additional EPR absorptions in the lateral regions of the spectra. The additional EPR absorptions are solely due to very specific (solute-dependent) alkyl radicals, formed through dissociative electron attachment to the solute chloro- and bromoalkanes.

$$R\text{–}X + e^- \rightarrow R^\bullet + X^- \tag{5.3}$$

Hole trapping by these compounds is excluded because of the low ionization energy of decalin. Chromatographic analyses have shown that alkyl radicals formed by γ-irradiation of chloro- and bromoalkanes in *cis*- and *trans*-decalin are characteristic for the haloalkane solute and explicitly rule out the occurrence of radical isomerization [19]. The identity of the alkyl radicals observed is thus unambiguously determined by the choice of the chloro- and bromoalkane solutes. The lateral parts of the spectra obtained from γ-irradiated *cis*-decalin-d_{18} containing 1 mol% 1-, 2- and 3-bromooctane, shown in Fig. 5.4(b)–(d), can thus fully be attributed to chain-end ($R_I^\bullet$), penultimate ($R_{II}^\bullet$) and interior ($R_{III}^\bullet$) alkyl radicals, respectively. The relatively narrow (largely isotropic) lines and the double-humped curve, that make the powder EPR spectrum of 1-alkyl radicals easily recognizable, are clearly discernible in the lateral parts of the spectrum obtained from γ-irradiated *cis*-decalin-d_{18} containing 1 mol% 1-bromooctane (Fig. 5.4(b)). Unambiguous information on the band shape of (the lateral parts of) the EPR absorption of penultimate and interior alkyl radicals is provided by the spectra of γ-irradiated *cis*-decalin-d_{18} containing 2- and 3-bromooctane (Fig. 5.4(c, d)). As is evident from these spectra, all

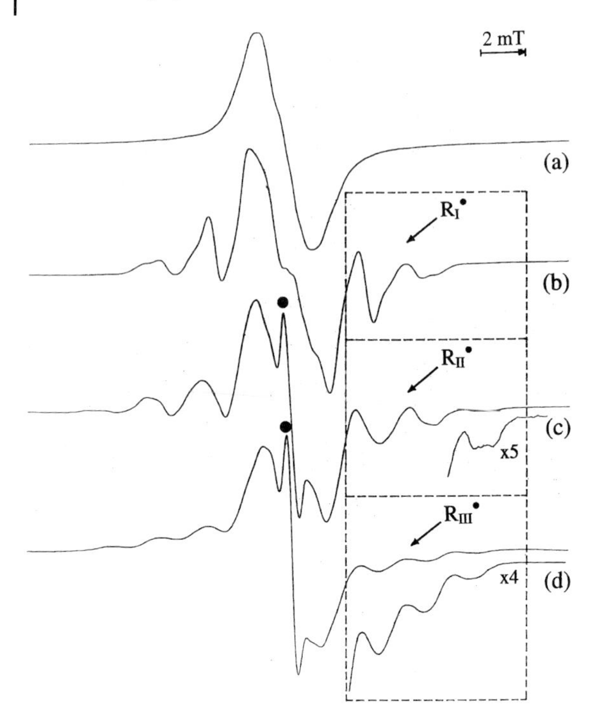

Figure 5.4 First-derivative EPR spectra obtained after γ-irradiation of neat *cis*-decalin-d_{18} (a) and of *cis*-decalin-d_{18} containing 1 mol% 1-bromooctane (b), 2-bromooctane (c) and 3-bromooctane (d). The dashed rectangles contain spectral features that are highly typical for chain-end ($R_I^•$), penultimate ($R_{II}^•$) and interior ($R_{III}^•$) alkyl radicals, respectively; ● indicates a background absorption.

absorption bands of these radicals are rather broad. The EPR absorption of interior alkyl radicals is less well resolved than that of penultimate alkyl radicals; as a result, equally weighted combinations of the experimental powder spectra of authentic penultimate and interior alkyl radicals rather tend to correspond to the spectrum of penultimate radicals and quite considerable fractions of interior radicals may be "hidden" under the spectrum of penultimate radicals without seriously affecting the composite spectrum. Most importantly, the spectrum of penultimate and interior alkyl radicals extends over a wider spectral region than that of chain-end radicals, making the former detectable in the presence of the latter by careful examination of the region next to the double-humped curve; the overall appearance of convoluted spectra may also provide an indication of the presence of secondary alkyl radicals.

5.7
Symmetric Proton Transfer from Alkane Radical Cations to Alkanes: An Experimental Study in γ-Irradiated n-Alkane Nanoparticles Embedded in a Cryogenic CCl_3F Matrix

5.7.1
Mechanism of the Radiolytic Process

Radiolysis of cryogenic trichlorofluoromethane containing a suitable n-alkane as solute has proven very suitable for the study of symmetric proton transfer from alkane radical cations to alkane molecules. At low concentration of the alkane solute (RH) in the binary CCl_3F/alkane system, absorption of ionizing radiation mainly occurs by trichlorofluoromethane resulting in its excitation and ionization,

$$CCl_3F \rightsquigarrow CCl_3F^* \tag{5.4a}$$

$$CCl_3F \rightsquigarrow CCl_3F^{\bullet+} + e^- \tag{5.4b}$$

leaving the solute alkane largely unaffected as far as direct interaction with the ionizing radiation is concerned. The solute alkane is thus little involved in the initial "violent" phase of energy deposition by the ionizing radiation, in which massive excess energy is often available that is not conducive to site-selective processes. As a result of long-range electron tunneling, the positive hole is transferred efficiently from trichlorofluoromethane radical cations to the alkane solute, however, resulting in the neat and selective formation of alkane radical cations (without concomitant neutral alkyl radical formation as would be the case in neat alkanes).

$$CCl_3F^{\bullet+} + RH \rightarrow CCl_3F + RH^{\bullet+} \tag{5.5}$$

Though gas phase ionization energies of CCl_3F and higher alkanes (such as undecane) are quite different ("evaluated" ionization energies of 11.68 eV and 9.56 eV have been reported [20] for CCl_3F and undecane, respectively), the excess energy imparted to solute alkane radical cations is much lower than this difference suggests because of dimer cation formation in irradiated trichlorofluoromethane. The stabilization energy due to $(CCl_3F)_2^{\bullet+}$ dimer cation formation is unknown at present, but is likely to be extensive. Similar stabilization of solute alkane radical cations by radical cation adduct formation with CCl_3F or by dimer cation formation can be ruled out on the basis of the relative inaccessibility of the *semi*-occupied molecular orbital (SOMO) in alkane radical cations and (for the adduct formation) by the great difference in ionization energy between CCl_3F and the solute alkane. EPR spectra of alkane radical cations in CCl_3F are incompatible with both adduct and dimer cation formation.

At cryogenic temperatures, alkane radical cations are stable when fully isolated in the CCl_3F matrix, because electrons formed in the ionization process react with trichlorofluoromethane by dissociative electron attachment.

$$CCl_3F + e^- \rightarrow CCl_2F^\bullet + Cl^- \qquad (5.6)$$

It is now well established, however, that alkanes form small aggregates in CCl_3F, aggregates to which positive-hole transfer still occurs efficiently [21–23]. The degree of aggregation increases with increasing alkane concentration and at a specific concentration increases quite strongly with increasing chain length of the alkane solute. As a result, symmetric proton transfer may take place from solute alkane radical cations to alkane molecules yielding alkyl radicals and protonated alkanes (alkyl carbonium ions).

$$RH^{\bullet+} + RH \rightarrow R^\bullet + RH_2^+ \qquad (5.7)$$

As became apparent from our experiments, only secondary C–H protonation leads to effective chemical transformation in γ-irradiated CCl_3F/alkanes (see below). As secondary C–H protonated alkanes are characterized by negative dissociation energies, protonation is followed immediately by dissociation of the C–H protonated alkanes into alkyl carbenium ions and molecular hydrogen.

$$RH_2^+ \rightarrow R^+ + H_2 \qquad (5.8)$$

Upon melting, alkyl carbenium ions react with chloride ions, both species being trapped in the solid system at 77 K. This leads to the formation of isomeric chloroalkanes, with the position of the chlorine atom being indicative for the original site of C–H protonation; the coulombic attraction involved and the high mobility of chloride ions make this neutralization process highly competitive with respect to ion–molecule reactions involving exclusively long-chain species.

5.7.2 Physical State of Alkane Aggregates in CCl_3F

Information on the physical state (crystalline versus amorphous) of alkane aggregates in CCl_3F can be derived from EPR experiments on γ-irradiated odd n-alkanes with $n_C \geq 11$ in this matrix, by virtue of the dependence of intermolecular radical-site transfer on the molecular alignment in n-alkane crystals [24]. Odd n-alkanes with $n_C \geq 11$ have an orthorhombic crystalline structure and with such a mode of packing the nearest neighboring C–H bond (of adjacent molecules) to a primary radical site belongs to a chain-end methyl group, as can be seen in Fig. 5.5. As a result, transformation of primary into secondary alkyl radicals by intermolecular radical-site transfer does not take place in orthorhombic crystals. Alternatively and viewed from a different perspective, if extensive radical transformation in γ-irradiated odd n-alkanes with $n_C \geq 11$ in CCl_3F occurs, then this constitutes evidence that the alkane aggregates do not pack according to the orthorhombic crystalline structure. In view of general considerations on the crystallization of n-alkanes, it can be concluded that in such a case they must be amorphous with no specific structure at all.

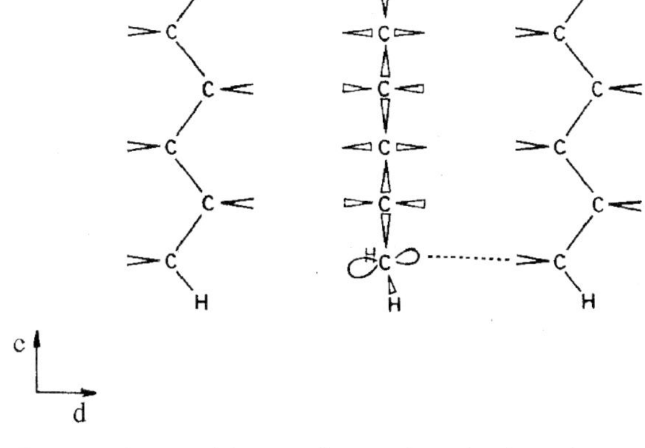

Figure 5.5 Structural diagram showing the molecular packing in odd n-alkanes ($n_C \geq 11$) with an orthorhombic crystalline structure as a projection on a diametrical plane, dc, that forms an angle of 45° with the *ac* and *bc* planes of the crystal.

EPR spectra obtained after γ-irradiation of undecane at various concentrations in CCl_3F are shown in Fig. 5.6. With increasing undecane concentration, the disappearance of undecane radical cations and transformation of primary into secondary undecyl radicals (by intermolecular radical-site transfer)

$$1\text{-}C_{11}H_{23}^{\bullet} + n\text{-}C_{11}H_{24} \rightarrow n\text{-}C_{11}H_{24} + sec\text{-}C_{11}H_{23}^{\bullet} \quad (5.9)$$

are obvious. The most notable changes in the EPR spectra with increasing undecane concentration which indicate this are (i) the appearance of a paramagnetic absorption outside the spectral region of 1-undecyl radicals; (ii) a gradual shift of the 1-undecyl radical spectrum away from the baseline, i.e., it becomes superimposed on a broad and (in first-derivative spectral terms and at the right-hand side of the spectra) consistently negative paramagnetic absorption; (iii) the gradual decrease and disappearance of the spectral absorption attributed to undecane radical cations, i.e., the pseudo-singlet (contracted triplet) denoted $RH^{\bullet+}$ and (iv) the gradual decrease and disappearance of the spectral absorption due to primary undecyl radicals, i.e., the relatively narrow (largely isotropic) lines and the double-humped curve denoted $R_I^{\bullet}$. The first two spectral changes listed point to the appearance and gradual increase in importance of secondary undecyl radicals. At 4 mol %, it is hard to see even a trace of the spectral characteristics of 1-undecyl radicals that were so prominent at lower concentrations.

The transformation of primary into secondary undecyl radicals and the dominance of secondary undecyl radicals at much higher concentrations unequivocally show that, as a rule, undecane aggregates in CCl_3F are amorphous and this conclusion can reasonably be extended to all n-alkanes with $n_C \leq 11$ for concentrations below about 20 mol%. In the EPR spectra of γ-irradiated CCl_3F/undecane primary radical features reappear from about 7 mol%, however, and these features

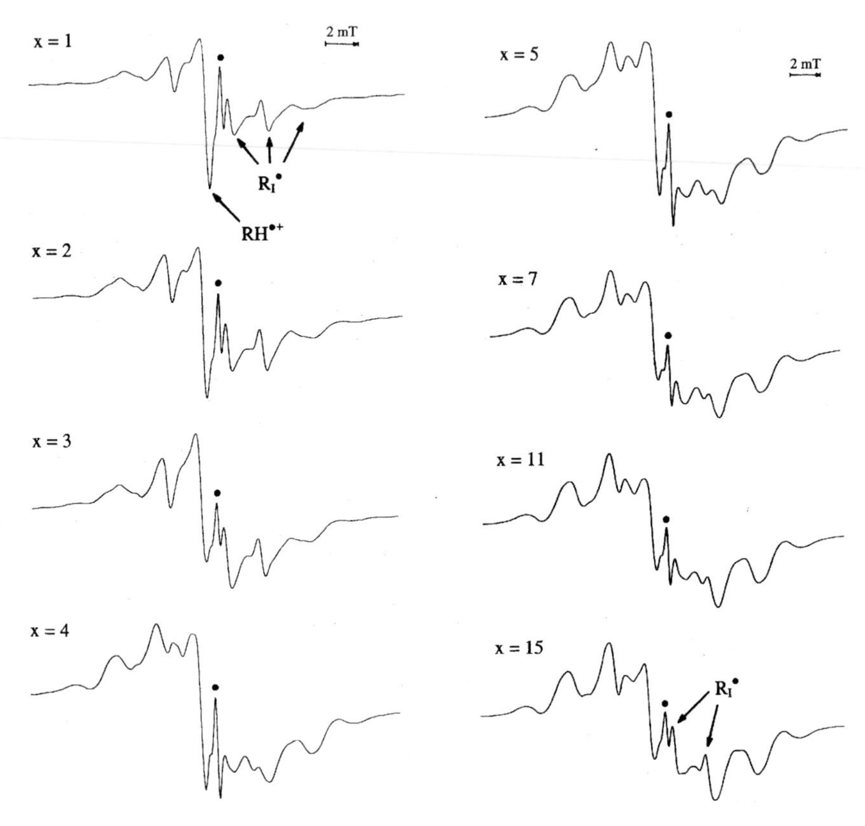

Figure 5.6 First-derivative EPR spectra obtained after γ-irradiation of undecane at various concentrations in CCl_3F. Spectral features due to undecane radical cations and primary undecyl radicals are indicated, respectively, by $RH^{\bullet+}$ and $R_I^{\bullet}$; ● indicates a background absorption.

become more prominent at higher undecane concentration (see Fig. 5.6) pointing to the onset and increasing prominence of crystallization. With undecane this is still very restricted, even at high concentration, but with longer n-alkanes (such as tridecane) crystallization is observed at much lower concentrations and becomes quite prominent at high alkane concentration.

5.7.3
Evidence for Proton-donor and Proton-acceptor Site Selectivity in the Symmetric Proton Transfer from Alkane Radical Cations to Alkane Molecules

5.7.3.1 Proton-donor Site Selectivity

Information on the site of proton donation in the symmetric proton transfer from alkane radical cations to alkane molecules can be derived from the nature of the alkyl radicals formed in γ-irradiated CCl_3F/n-alkanes at 77 K but only at relatively low alkane concentration, viz. at or around the onset of alkane aggregation and alkyl radical formation [18, 25]. Working around the onset of alkyl radical formation is essential because otherwise all radicals are transformed into the thermodynamically most stable ones by intermolecular radical-site transfer. The most notable result, the EPR spectrum obtained after γ-irradiation of CCl_3F containing 1.75 mol heptane, is shown in Fig. 5.7. The spectrum mainly consists of a (distorted) triplet due to heptane radical cations in the extended all-trans conformation. In addition to the triplet spectrum, a weak resonance signal that can be attributed to 1-heptyl radicals is observed in the lateral region of the spectrum. The intensity of the lateral heptyl radical absorption increases with increasing heptane concentration, but this increase is accompanied by gradual transformation of primary into secondary heptyl radicals by intermolecular radical-site transfer. The result on γ-irradiated CCl_3F/heptane characterizes the reaction of (higher) alkane radical cations with alkane molecules unequivocally as proton transfer (hydrogen abstraction would lead to preferential formation of secondary heptyl radicals) and shows that the site of proton donation is related very strictly

Figure 5.7 First-derivative EPR spectrum obtained after γ-irradiation at 77 K of 1.75 mol% heptane in CCl_3F; ● indicates a background absorption.

to the structure of the semi-occupied molecular orbital in the radical cation. The high unpaired-electron and positive-hole density in the *in-plane chain-end C–H bonds* leads to proton transfer from those sites, giving rise to the selective formation of 1-heptyl radicals. Experiments with n-alkane radical cations in different conformations confirm the relation between electronic structure and site of proton donation. Octane radical cations in CCl_3F, for instance, are largely in a gauche-at-C_2 conformation obtained by one 120° rotation around C_2–C_3 in the extended conformer (see Fig. 5.1). Proton transfer results in the formation of secondary *as well as* primary octyl radicals in this case, as evidenced by the fact that both are present from the very first appearance of octyl radicals with increasing octane concentration in γ-irradiated CCl_3F/octane. This is again in accordance with the electronic structure, as in the gauche-at-C_2 conformation there is large unpaired-electron and positive-hole density in the planar part of the C–C skeleton as well as in *one chain-end and one penultimate C–H bond* (both in-plane) at opposite sides.

5.7.3.2 **Proton-acceptor Site Selectivity**

Information on the site of proton acceptance in the symmetric proton transfer from alkane radical cations to alkane molecules can be derived from chromatographic analysis of chloroalkanes formed in CCl_3F/alkanes after γ-irradiation at 77 K and subsequent melting [26, 27]. Because of the amorphous state of alkane aggregates in CCl_3F, any acceptor site selectivity observed must be intrinsic in nature. A manifest effect observed upon analysis of chloroalkanes with the same carbon number as the parent alkane is the gradual reduction in the contribution of the chain-end isomer with increasing alkane concentration; data for CCl_3F/decane are shown in Fig. 5.8. The effect is related to the fact that the protonation energy for primary C–H protonation is substantially lower than that for C–C pro-

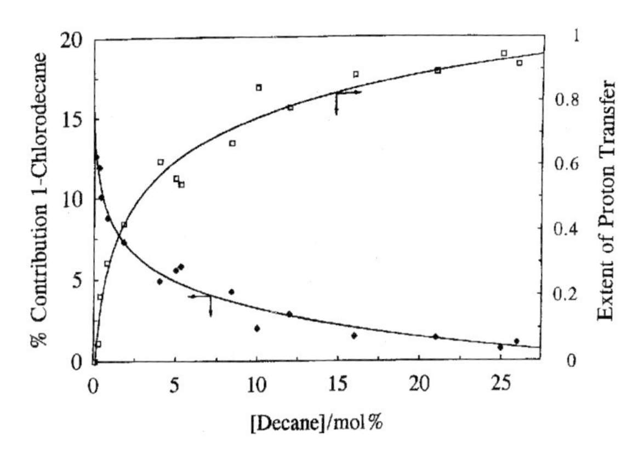

Figure 5.8 Contribution percentage of 1-chlorodecane to chlorodecane formation and extent of proton transfer from decane radical cations to decane molecules as a function of decane concentration in γ-irradiated CCl_3F/decane.

tonation and secondary C–H protonation and provides further evidence for a gradual increase in proton transfer from alkane radical cations to alkane molecules with increasing alkane concentration in γ-irradiated CCl_3F/alkanes. The extent of proton transfer as a function of decane concentration in CCl_3F/decane calculated from these data is shown in Fig. 5.8. The isomeric composition of chloroalkanes at high alkane concentration (extensive proton transfer) unequivocally shows protonation to take place preferentially at the penultimate (rather than at the interior) C–H bonds in n-alkanes, an observation that is supported by thermochemical data. The weaker protonation at the different inner (non-penultimate) C–H bonds in n-alkanes apparently occurs to about the same extent at each bond.

A search for 1-chloroalkanes with smaller carbon number than the parent alkane revealed the absence of such products in CCl_3F/decane after γ-irradiation at 77 K and subsequent melting, providing strong evidence for the absence of chemical transformation due to C–C protonation of n-alkanes by proton transfer from alkane radical cations. Indeed, pentacoordinated alkyl carbonium ions that have the C–C protonated structure can be formed from and dissociate into alkyl carbenium ions and neutral alkanes (a process characterized by positive dissociation energies) and it is quite logical to expect that neutralization with chloride ions will take place accordingly, i.e., with formation of shorter-chain 1-chloroalkanes and alkanes. The absence of chemical transformation due to C–C protonation was originally considered to be purely thermochemical in nature, but now appears to have a more complex origin in which thermochemical and structural (cage) effects are involved as well as the intrinsic stability of the carbonium ions. This relates specifically to (i) the (modest) endothermicity of the proton transfer from alkane radical cations to alkane molecules for both C–H and C–C protonation, (ii) the sign of the dissociation energy with respect to the normal fragmentation of alkyl carbonium ions and (iii) the size of the corresponding fragmentation products. The dissociation energy of C–C protonated alkanes for fragmentation into smaller-sized alkyl carbenium ions and neutral alkanes is positive and the fragmentation products of such a dissociation are held together by the cage effect, effectively preventing the dissociation of C–C protonated alkanes in condensed phases. This, in association with the endothermicity of the proton transfer from alkane radical cations to alkanes, results in the transfer of the proton from C–C protonated alkanes *back* to the associated alkyl radicals and thus in the absence of chemical transformation due to C–C protonation. In contrast, the dissociation energy of secondary C–H protonated alkanes for fragmentation into alkyl carbenium ions and molecular hydrogen is negative, i.e., the fragments are more stable than the carbonium ion itself, and molecular hydrogen is small enough to escape from the cage in which it is formed. As a result, back transfer of the proton does not occur and chemical transformation due to secondary C–H protonation is observed. The absence of chemical transformation due to primary C–H protonation is, in all likelihood, due to the very absence of such protonation and is, as such, purely thermochemical in nature; (the positive sign of the dissociation energy of primary C–H protonated alkanes for fragmentation into alkyl carbenium ions and molecular hydrogen may be an additional factor in the absence of chemical transformation). It is to be noted that the results imply that no trapped alkyl carbonium ions are present after irradiation of n-alkanes under cryogenic conditions, either neat or in CCl_3F matrices in the solid state.

The negative sign of the dissociation energy of secondary (and tertiary) C–H protonated alkanes for fragmentation into alkyl carbenium ions and molecular hydrogen follows from high-level theoretical calculations on the energetics of pentacoordinated carbonium ions and simply represents the fact that the enthalpy of formation of the dissociation products is lower than that of the corresponding carbonium ions. Negative dissociation energies are not easily conceived by experimental chemists and cannot readily be determined by ion-equilibrium measurements on the stability of cationic species. When a particular cation is intrinsically unstable, its dissociation energy is intuitively assumed to be (near) zero. This misconception is at the origin of various erroneous representations of the relative order of the energies of C–C and secondary C–H protonation [14, 27].

5.7.4 Comparison with Results on Proton Transfer and "Deprotonation" in Other Systems

Information on selectivity with respect to the site of proton donation (but not of proton acceptance) in the symmetric proton transfer from alkane radical cations to alkane molecules has been derived from systems other than γ-irradiated CCl_3F/alkanes. Two approaches have basically been taken to bring alkane radical cations into contact with alkane molecules in solid systems in order to study their reaction: (i) increasing the temperature of the irradiated system and bringing it to a point where, as a result of diffusion and migration, alkane radical cations disappear and neutral alkyl radicals are formed and (ii) increasing the concentration of the alkane solute to a point where the alkane molecules are not fully isolated and the appearance of neutral alkyl radicals is observed experimentally. Thermal conversion of alkane radical cations into neutral alkyl radicals has been conducted in SF_6 and CCl_2FCF_2Cl and conflicting results have been obtained in these matrices. n-Alkane radical cations (C_4–C_7) radiolytically produced in SF_6 at 77 K exhibit the planar extended structure with no detectable gauche conformers; upon warming above 100 K transformation into 1-alkyl radicals is observed exclusively, as is expected from the unpaired-electron and positive-hole distribution [28]. In sharp contrast, 2-alkyl radicals are selectively formed by thermal conversion in CCl_2FCF_2Cl, regardless of the conformation of the alkane radical cations [5–8, 29]. This has been assumed to come from prior thermal conversion of extended conformers into gauche-at-C_2 conformers, but an alternative explanation based on charge neutralization in CCl_2FCF_2Cl is far from inconceivable. The thermal conversion of alkane radical cations into neutral alkyl radicals can phenomenologically be termed "deprotonation" and is usually denoted as such in the literature. Different reaction processes may be at the origin of such thermal "deprotonation". Increasing the temperature of the irradiated system, to a point where diffusion and migration of species takes place, not only brings alkane radical cations into contact with alkane molecules but also with the counter anions trapped in the system; these have a relatively low concentration, but are attracted to the radical cations by coulombic attraction and (in the case of chloride ions) are much more mobile species than long-chain alkane molecules. The mechanism of thermal conversion is thus considerably less certain than that of concentration studies, in which reaction of the alkane radical cations with alkane molecules is implicated by the increasing alkane concentration. Concentration studies have been conducted in the synthetic zeolites ZSM-5 and Linde-5A [30]. Radical cations of C_6 and C_8 n-alkanes prepared by irradiation at 4 K in ZSM-5 at low alkane concentration are in the extended all-trans conformation. At high alkane concentration, alkyl radicals were observed with dominant (but not exclusive) formation of the chain-end isomer; the minor formation of secondary alkyl radicals has been attributed to intermolecular radical-site transfer. In Linde-5A only 1-octyl radicals were observed, but no spectrum of octane radical cations was reported in this zeolite.

5.8 Asymmetric Proton Transfer from Alkane Radical Cations to Alkanes: An Experimental Study in γ-Irradiated Mixed Alkane Crystals

5.8.1 Mechanism of the Radiolytic Process

Radiolysis of mixed alkane crystals consisting of a lower alkane (R_IH), such as pentane, to which relatively low concentrations (about 1 mol%) of a higher alkane ($R_{II}H$) and a suitable electron acceptor (for instance CO_2 or a chloroalkane) have been added, has proven very suitable for the study of asymmetric proton transfer from alkane radical cations to alkane molecules. Radiation-chemical mechanisms of alkanes are inherently very complex and the discussion will be limited to processes that have some bearing on the concepts discussed.

The primary interaction of γ-irradiation with systems composed of a lower alkane with 1 mol% of a higher alkane and with a specific chloroalkane or CO_2 as solute consists mainly in excitation and ionization of the main component, i.e., the lower alkane.

$$R_IH \rightsquigarrow R_IH^* \tag{5.10a}$$

$$R_IH \rightsquigarrow R_IH^{\bullet+} + e^- \tag{5.10b}$$

Excited alkane molecules may dissociate into alkyl radicals and hydrogen atoms,

$$R_IH^* \rightarrow R_I^\bullet + H^\bullet \tag{5.11}$$

or, by molecular dissociation, into the corresponding alkene and molecular hydrogen. Alkane radical cations formed in the ionization process may also carry quite considerable amounts of excitation energy. Indeed, removal by irradiation of an electron in alkanes from lower-lying molecular orbitals results directly in the formation of *electronically excited* radical cations. These may react by dissociation into alkyl carbenium ions and hydrogen atoms

$$R_IH^{\bullet+*} \rightarrow R_I^+ + H^\bullet \qquad (5.12)$$

and by proton transfer to adjacent alkane molecules yielding alkyl carbonium ions and neutral alkyl radicals.

$$R_IH^{\bullet+*} + R_IH \rightarrow R_I^\bullet + R_IH_2^+ \qquad (5.13)$$

The alkyl carbonium ions formed may dissociate into alkyl carbenium ions and molecular hydrogen or alternatively transfer the proton back to the associated alkyl radicals. Overall, little or no site selectivity is expected in such reactions (5.12) and (5.13). Many of the lower-lying molecular orbitals in neutral alkanes undoubtedly contribute to C–H bonding at various sites in the molecule and electron removal will result in unpaired-electron and positive-hole density at those sites allowing dissociation at and proton transfer from such positions. Because of the high amount of excess energy available, subtle differences in the tendency of (secondary) C–H bonds to act as proton acceptor are also not likely to play a major role in this initial "violent" phase of energy deposition by the ionizing radiation.

Neutralization by electrons remains an important process in the radiolysis of alkanes containing chloroalkanes or CO_2 as solute. Many of the electrons formed in the ionization process have insufficient energy to escape the Coulomb field of their associated cation and, on returning, neutralize the corresponding radical cations, carbenium ions or carbonium ions (geminate recombination). Part of the electrons formed do not return, however, but react with the chloroalkane solute by dissociative electron attachment,

$$R_{EA}\text{–Cl} + e^- \rightarrow R_{EA}^\bullet + Cl^- \qquad (5.14)$$

or by electron attachment to CO_2.

$$CO_2 + e^- \rightarrow CO_2^{\bullet-} \qquad (5.15)$$

By this process, the negative charge becomes trapped in the solid system as a chloride ion or as a CO_2 anion. For the study of proton-donor site selectivities, which are conducted by EPR spectroscopy after irradiation at 77 K, the use of CO_2 as electron acceptor is to be preferred as the CO_2 anion has only a rather narrow central absorption that does not hinder the analysis. Chloroalkanes can be used in such studies if the respective deuterated products are employed. For the study of proton-acceptor site selectivities, which are conducted by gas-chromatographic analysis after warming and melting the sample, chloroalkanes are highly preferred over CO_2 as electron acceptor because of the nature of the respective neutralization products. Upon warming and melting, alkyl carbenium ions that are also trapped in the solid system will be neutralized by chloride ions resulting in the formation of chloroalkanes.

$$R_I^+ + Cl^- \rightarrow R_ICl \quad (5.16)$$

Despite their low concentration, a non-negligible fraction of molecules of the higher alkane solute will be located in regions where the initial "violent" phase of energy deposition by the ionizing radiation takes place. This will result in the formation and trapping of alkyl carbenium ions of the higher alkane solute. As with the lower alkane matrix, little or no site selectivity is to be expected overall from this process and the corresponding neutralization reaction

$$R_{II}^+ + Cl^- \rightarrow R_{II}Cl \quad (5.17)$$

is at the origin of a "random" (non-site-selective) formation of chloroalkanes of the higher alkane solute, upon which the site-selective formation (that yields information on proton-acceptor selectivities) is superimposed.

Information on proton-donor and proton-acceptor site selectivities in γ-irradiated mixed crystals is yielded by processes that are separated from the initial energy deposition in both space and time. A sizable fraction of the lower-alkane radical cations formed in the ionization process are not overly excited or become sufficiently deactivated before reacting by fragmentation or proton transfer. At this point, fragmentation is excluded and proton transfer from radical cations that are in their electronic ground state is not at all efficient in neat n-alkane crystals (and microscopic neat parts in n-alkane mixed crystals). Also, as a result of electron scavenging, a number of such cations escape "immediate" neutralization by electrons. The "unreactive" lower-alkane radical cations are not trapped in the solid system, not because they are mobile themselves, but because of positive-hole transfer to (electron transfer from) adjacent matrix molecules, resulting in positive-hole migration.

$$R_IH^{\bullet+} + R_IH \rightarrow R_IH + R_IH^{\bullet+} \quad (5.18)$$

In the inhomogeneous coulombic field generated by the trapped cations and anions formed by irradiation, positive holes will migrate in the direction of trapped anions. When, as a result of this process, a matrix (R_IH) radical cation becomes adjacent to a solute ($R_{II}H$) molecule, positive-hole transfer to the solute may occur.

$$R_IH^{\bullet+} + R_{II}H \rightarrow R_IH + R_{II}H^{\bullet+} \quad (5.19)$$

In addition, proton-transfer reactions take place. In irradiated mixed alkanes the higher-alkane radical cations are trapped next to matrix molecules and part of them react by proton transfer.

$$R_{II}H^{\bullet+} + R_IH \rightarrow R_{II}^{\bullet} + R_IH_2^+ \quad (5.20)$$

Also, proton transfer from matrix radical cations to solute molecules takes place in competition with positive-hole transfer (reaction (5.19) in the same direction.

$$R_IH^{\bullet+} + R_{II}H \rightarrow R_I^{\bullet} + R_{II}H_2^+ \quad (5.21)$$

Proton transfer from alkane radical cations that are in their electronic ground state is greatly facilitated by the fact that in mixed n-alkane crystals planar chain-end C–H bonds, from which proton donation takes place, come into close contact with secondary C–H bonds in adjacent molecules (see the structural diagrams below). In neat n-alkane crystals, there is only close contact with primary C–H bonds that have much lower protonation energies.

5.8.2
Evidence for Proton-donor and Proton-acceptor Site Selectivity in the Asymmetric Proton Transfer from Alkane Radical Cations to Alkanes

Evidence for the occurrence of asymmetric proton transfer in mixed n-alkane crystals has been obtained in close connection with the demonstration of site selectivity in the donor and acceptor processes. Such evidence can only be gathered properly from species that are related to the solute (higher alkane) molecule, because any site selectivity from matrix species is completely wiped out by an overwhelming amount of nonselective processes (see above). With respect to the site of proton donation, proton transfer from solute radical cations to matrix molecules was therefore studied, whereas, with respect to the site of proton acceptance, proton transfer from matrix radical cations to solute molecules was investigated.

The site of proton donation in the proton transfer from solute radical cations to matrix molecules has been investigated using γ-irradiated pentane-d_{12} containing 0.5 mol% octane as well as trapped CO_2 [31]. Octane radical cations are formed by positive-hole transfer from matrix cations in this system and part of them react by proton transfer to matrix molecules,

$$\text{n-}C_8H_{18}^{\bullet+} + \text{n-}C_5D_{12} \rightarrow \text{1-}C_8H_{17}^{\bullet} + \text{n-}C_5D_{12}H^+ \tag{5.22}$$

while others simply remain trapped in the solid system. The first-derivative EPR spectrum of the system is shown in Fig. 5.9. The EPR spectrum is largely due to deuterated pentyl radicals (this spectrum is narrowed by the deuteration); the very intense and relatively sharp asymmetric feature near the center of the spectrum, with a pseudodoublet structure at the high-field side, is due to $CO_2^{\bullet-}$. The (weak) lateral features can be attributed with certainty to 1-octyl radicals; no secondary octyl radicals are observable in the system. In association with the results of Ichikawa et al., that show that octane radical cations in pentane matrices are in the extended all-trans conformation [32], these data thus confirm the relation between the electronic structure of alkane radical cations and the site of proton donation.

With respect to the site of proton acceptance in the proton transfer from matrix radical cations to solute molecules, two systems have been investigated exhaustively, viz. γ-irradiated heptane/octane/1-chlorohexane [33] and heptane/decane/1-chloroheptane [34]. A third system, viz. γ-irradiated pentane/decane/1-chloropentane has also been studied but in considerably less detail. In the first system, proton transfer takes place from heptane radical cations to octane molecules

$$\text{n-}C_7H_{16}^{\bullet+} + \text{n-}C_8H_{18} \rightarrow \text{1-}C_7H_{15}^{\bullet} + \text{2-}C_8H_{19}^{+} \tag{5.23}$$

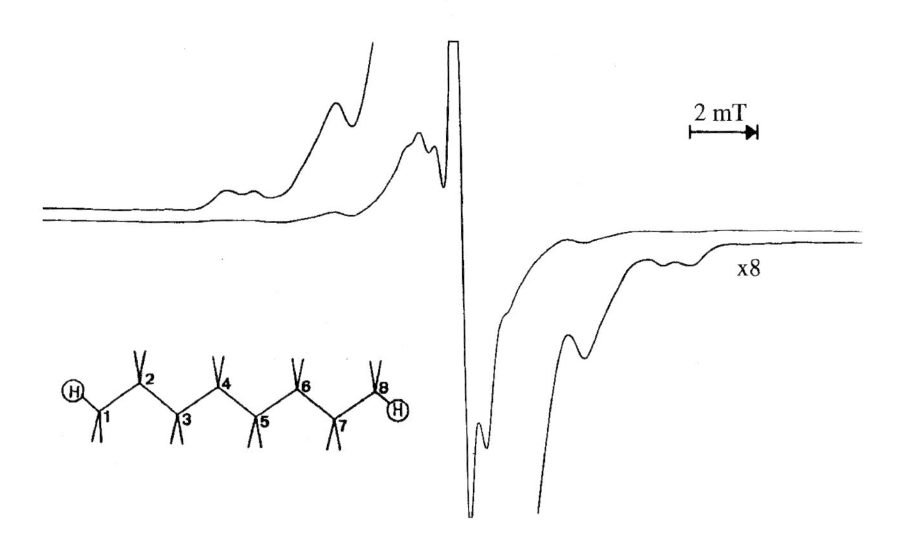

Figure 5.9 First-derivative EPR spectrum obtained after γ-irradiation at 77 K of 0.5 mol% octane in perdeuterated pentane, containing CO_2 as electron acceptor.

followed by dissociation and subsequent neutralization of the octyl carbenium ions by chloride ions upon melting.

$$2\text{-}C_8H_{17}^+ + Cl^- \rightarrow 2\text{-}C_8H_{17}Cl \tag{5.24}$$

From Fig. 5.10 it is evident that penultimate C–H bonds in octane, which for reasons of energetics have the greatest propensity to act as proton acceptor, are also structurally favored over the interior C–H bonds with respect to proton acceptance from planar chain-end C–H bonds in the heptane radical cations. Both structural and thermodynamic factors thus favor the penultimate position and this translates into a very high selectivity with respect to this site (C_2/C_3: 5.3, C_2/C_4: 13.4, at 3 mol%). As a matter of fact, the minor formation of 3- and 4-chlorooctane can (largely) be attributed to random processes related to reaction (5.17).

In the heptane/decane/1-chloroheptane system protonation of decane by heptane radical cations

$$n\text{-}C_7H_{16}^{\bullet+} + n\text{-}C_{10}H_{22} \rightarrow 1\text{-}C_7H_{15}^{\bullet} + sec\text{-}C_{10}H_{23}^+ \tag{5.25}$$

results in the formation of decyl carbonium ions. Selectivity with respect to the site of proton acceptance is studied by analyzing the chlorodecanes formed by dissociation of these cations and subsequent neutralization by chloride ions upon melting of the carbenium ions produced.

$$sec\text{-}C_{10}H_{21}^+ + Cl^- \rightarrow sec\text{-}C_{10}H_{21}Cl \tag{5.26}$$

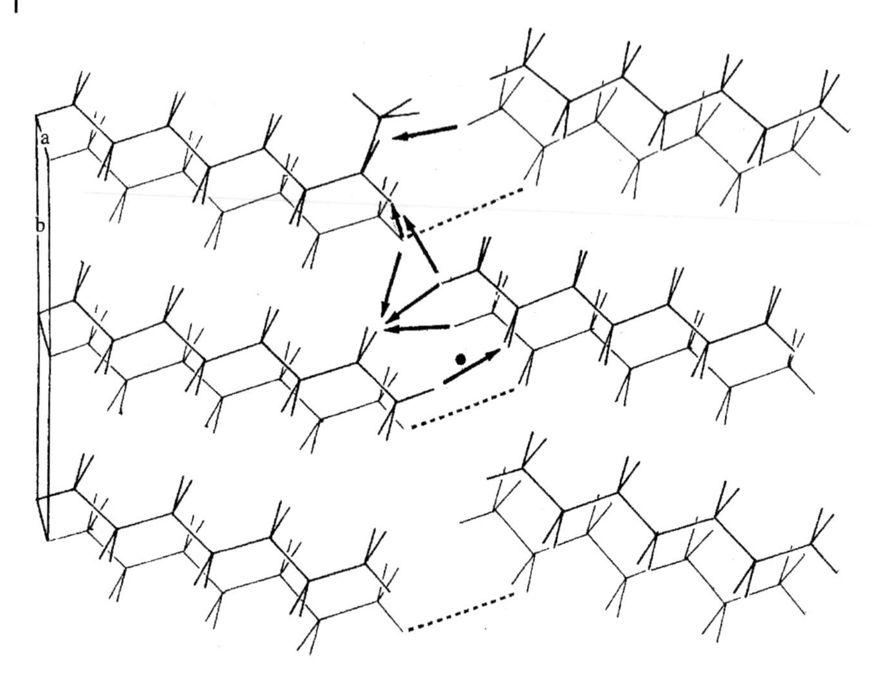

Figure 5.10 Structural diagram depicting the packing in heptane crystals containing octane molecules in the extended all-trans and gauche-at-C_2 conformation. The arrows indicate potential sites for proton transfer from planar chain-end C–H bonds in heptane radical cations to penultimate C–H bonds in octane molecules (and vice versa heptane versus octane, indicated by ●). The nearest approach of planar chain-end C–H bonds in heptane radical cations to penultimate C–H bonds in heptane molecules is indicated by dashed lines.

Information on the accessibility of the different C–H bonds in decane for planar chain-end C–H bonds in heptane radical cations may be derived from the structural diagrams shown in Fig. 5.11 and 5.12. Some of the proton transfers indicated therein may take place from heptane radical cations *above as well as below* the decane molecule, whereas other transfers have no such alternative, either due to steric hindrance by the gauche methyl group or due to the transfer occurring in the *b*-plane of the crystal. Competition between different sites for the same planar chain-end C–H bond in heptane radical cations must also be taken into account. From the structural diagrams it can be seen that the C–H bonds at the C_4 position in decane are directly accessible to planar chain-end C–H bonds in heptane radical cations, but to a much lesser extent than C–H bonds at the C_3 position; C–H bonds at the C_5 position in decane are only accessible to a very minor extent. The formation of 5-chlorodecane can (largely) be attributed to random processes associated with reaction (5.17). C–H bonds at the C_2 and C_3 position in decane are both easily accessible with a clear positive bias towards C_3 for the gauche-at-C_2 conformer. In stark contrast, the selectivity for the proton transfer to the penultimate position is higher than for the transfer to the C_3 position, as can be derived from the isomeric composition of the chlorodecanes. The site selectivity factor (relative to C_3) obtained after correction for a yield due to random processes (as defined by the yield of 5-chlorodecane) amounts to 3.29, *confirming that the proton transfer in n-alkanes preferentially occurs to the penultimate C–H bonds.* In contrast to the situation in CCl_3F/decane, the inner (C_3 to C_5) positions are *not* protonated to about the same extent. The disparity between the site selectivity for the protonation of the inner positions of decane in heptane versus in CCl_3F, as illustrated by the site selectivity factors (heptane/CCl_3F; C_3:1/1, C_4:0.49/1.02, C_5:0/0.96), can be attributed to *structurally-determined acceptor site selectivity* in heptane crystals, i.e., acceptor site selectivity resulting from the donor site selectivity in combination with accessibility differences due to structural factors.

In the pentane/decane/1-chloropentane system, proton transfer takes place from pentane radical cations to decane molecules,

$$\text{n-}C_5H_{12}^{\bullet+} + \text{n-}C_{10}H_{22} \rightarrow \text{1-}C_5H_{11}^{\bullet} + \text{5-}C_{10}H_{23}^{+} \tag{5.27}$$

followed by dissociation and subsequent neutralization of the decyl carbenium ions by chloride ions upon melting.

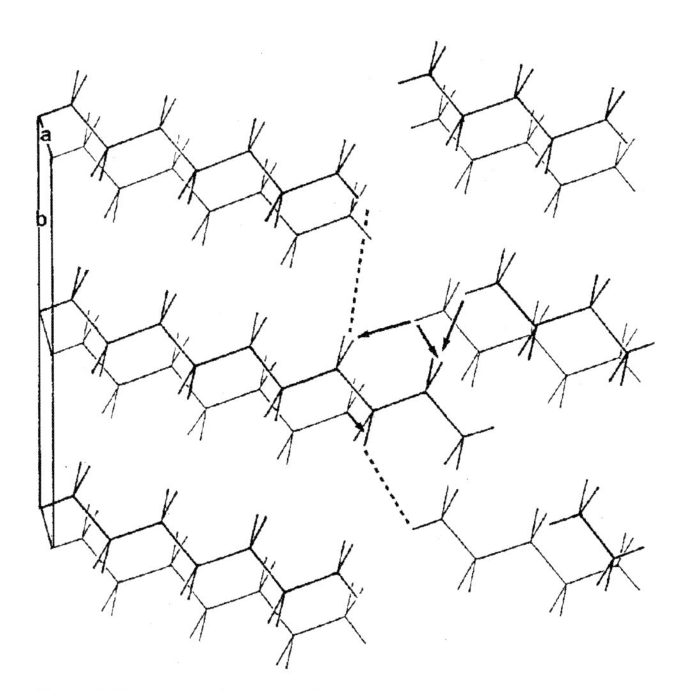

Figure 5.11 Structural diagram depicting the packing in heptane crystals containing decane molecules in the extended all-trans conformation. The arrows indicate potential sites for proton transfer from planar chain-end C–H bonds in heptane radical cations to C_2, C_3 and C_4 carbon–hydrogen bonds in decane molecules; points of approach that are deemed irrelevant or of minor importance to the proton-transfer process for structural reasons are indicated by dashed lines.

$$5\text{-}C_{10}H_{21}^{+} + Cl^{-} \rightarrow 5\text{-}C_{10}H_{21}Cl \qquad (5.28)$$

In this system, planar chain-end C–H bonds in pentane radical cations from which proton donation takes place only come into close contact with secondary C–H bonds at the inner (C_5) position in decane, as well as with primary C–H bonds; the latter have a much lower protonation energy than secondary C–H bonds, however, and thus cannot compete effectively as acceptor in the protonation process. Experiments show a marked predominance of 5-chlorodecane over more lateral secondary chlorodecanes, in accordance with the restricted accessibility of secondary C–H bonds in decane to planar chain-end C–H bonds in pentane radical cations. Perhaps even more importantly, no substantial preference is observed for the penultimate position relative to the C_3 and C_4 positions. This shows unequivocally that the preference for the penultimate position in the experiments with other systems described above is *not* due to the transformation of alkyl carbenium ions by hydride transfer, i.e., reactions such as

$$5\text{-}C_{10}H_{21}^{+} + n\text{-}C_{10}H_{22} \rightarrow n\text{-}C_{10}H_{22} + 2\text{-}C_{10}H_{21}^{+} \qquad (5.29)$$

but instead can be attributed unambiguously to a greater propensity of penultimate than of more interior secondary C–H bonds for proton acceptance.

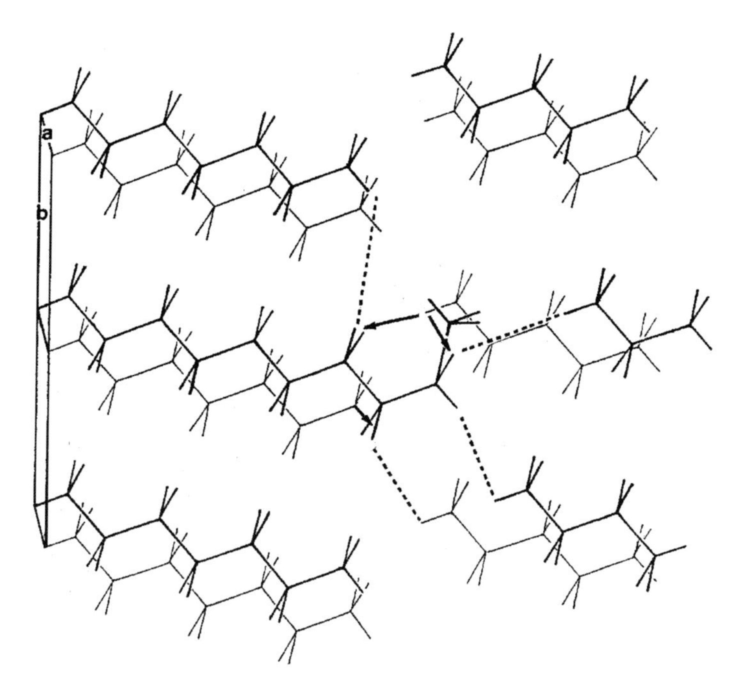

Figure 5.12 Structural diagram depicting the packing in heptane crystals containing decane molecules in the gauche-at-C_2 conformation. The arrows indicate potential sites for proton transfer from planar chain-end C–H bonds in heptane radical cations to C_2, C_3 and C_4 carbon–hydrogen bonds in decane molecules; points of approach that are deemed irrelevant or of minor importance to the proton-transfer process for structural reasons are indicated by dashed lines.

References

1 P. W. F. Louwrier, W. H. Hamill, *J. Phys. Chem.* **1970**, *74*, 1418–1421.

2 G. Wolput, M. Neyens, M. Strobbe, J. Ceulemans, *Radiat. Phys. Chem.* **1984**, *23*, 413–418.

3 A. Van den Bosch, M. Strobbe, J. Ceulemans, in *Photophysics and Photochemistry above 6 eV*, F. Lahmani (Ed.), Elsevier, Amsterdam, **1985**, pp. 179–189.

4 R. C. Benz, R. C. Dunbar, *J. Am. Chem. Soc.* **1979**, *101*, 6363–6366.

5 K. Toriyama, K. Nunome, M. Iwasaki, *J. Phys. Chem.* **1981**, *85*, 2149–2152.

6 K. Toriyama, K. Nunome, M. Iwasaki, *J. Chem. Phys.* **1982**, *77*, 5891–5912.

7 M. Lindgren, A. Lund, G. Dolivo, *Chem. Phys.* **1985**, *99*, 103–110.

8 G. Dolivo, A. Lund, *Z. Naturforsch.a* **1985**, *40*, 52–65.

9 D. W. Werst, M. G. Bakker, A. D. Trifunac, *J. Am. Chem. Soc.* **1990**, *112*, 40–50.

10 V. I. Borovkov, V. A. Bagryansky, I. V. Yeletskikh, Y. N. Molin, *Mol. Phys.* **2002**, *100*, 1379–1384.

11 G. A. Olah, Y. Halpern, J. Shen, Y. K. Mo, *J. Am. Chem. Soc.* **1971**, *93*, 1251–1256.

12 G. A. Olah, Y. Halpern, J. Shen, Y. K. Mo, *J. Am. Chem. Soc.* **1973**, *95*, 4960–4970.

13 G. A. Olah, *Angew. Chem., Int. Ed. Engl.* **1973**, *12*, 173–254.

14 K. Hiraoka, P. Kebarle, *J. Am. Chem. Soc.* **1976**, *98*, 6119–6125.

15 P. M. Esteves, C. J. A. Mota, A. Ramírez-Solís, R. Hernández-Lamoneda, *J. Am. Chem. Soc.* **1998**, *120*, 3213–3219.

16 C. J. A. Mota, P. M. Esteves, A. Ramírez-Solís, R. Hernández-Lamoneda, *J. Am. Chem. Soc.* **1997**, *119*, 5193–5199.

17 P. M. Esteves, G. G. P. Alberto, A. Ramírez-Solís, C. J. A. Mota, *J. Phys. Chem.* A **2000**, *104*, 6233–6240.

18 D. Stienlet, J. Ceulemans, *J. Phys. Chem.* **1992**, *96*, 8751–8756.

19 D. Stienlet, A. Vervloessem, J. Ceulemans, *J. Chromatogr.* **1989**, *475*, 247–260.

20 P. J. Linstrom, W. G. Mallard, Eds., *NIST Chemistry WebBook, NIST Standard Reference Database Number 69*, National Institute of Standards and Technology, Gaithersburg MD, **2003**. (Accessible at URL: http://webbook.nist.gov/chemistry).

21 G. Luyckx, J. Ceulemans, *J. Chem. Soc., Chem. Commun.* **1991**, 988–989.

22 G. Luyckx, J. Ceulemans, *J. Chem. Soc., Faraday Trans.* **1991**, 87, 3499–3504.

23 G. Luyckx, J. Ceulemans, *Radiat. Phys. Chem.* **1993**, *41*, 567–573.

24 D. Stienlet, G. Luyckx, J. Ceulemans, *J. Phys. Chem. B* **2002**, *106*, 10873–10883.

25 D. Stienlet, J. Ceulemans, *J. Chem. Soc., Perkin Trans. 2* **1992**, 1449–1453.

26 A. Demeyer, J. Ceulemans, *J. Phys. Chem. A* **1997**, *101*, 3537–3541.

27 A. Demeyer, J. Ceulemans, *J. Phys. Chem. A* **2000**, *104*, 4004–4010.

28 K. Toriyama, K. Nunome, M. Iwasaki, *J. Phys. Chem.* **1986**, *90*, 6836–6842.

29 G. Dolivo, A. Lund, *J. Phys. Chem.* **1985**, *89*, 3977–3984.

30 K. Toriyama, K. Nunome, M. Iwasaki, *J. Am. Chem. Soc.* **1987**, *109*, 4496–4500.

31 D. Stienlet, J. Ceulemans, *J. Phys. Chem.* **1993**, *97*, 8595–8601.

32 T. Ichikawa, M. Shiotani, N. Ohta, S. Katsumata, *J. Phys. Chem.* **1989**, *93*, 3826–3831.

33 A. Demeyer, D. Stienlet, J. Ceulemans, *J. Phys. Chem.* **1994**, *98*, 5830–5843.

34 L. Slabbinck, A. Demeyer, J. Ceulemans, *J. Chem. Soc., Perkin Trans. 2* **2000**, 2241–2247.

35 K. Toriyama, M. Iwasaki, M. Fukaya, *J. Chem. Soc., Chem. Commun.* **1982**, 1293–1295.

6
Single and Multiple Hydrogen/Deuterium Transfer Reactions in Liquids and Solids

Hans-Heinrich Limbach

Overview

In this chapter, Arrhenius curves of selected single and multiple hydrogen transfer reactions for which kinetic data are available over a large temperature range are reviewed. The curves are described by a combination of formal kinetics of reaction networks and the one-dimensional Bell–Limbach tunneling model for each reaction step. The main parameters of this model are the barrier heights and barrier widths of the isotopic reactions, the tunneling masses, the pre-exponential factor and a minimum energy for tunneling to occur. This approach allows one to compare efficiently very different reactions studied in different environments and to prepare the kinetic data for higher-dimensional quantum-mechanical treatments. The first type of reactions discussed is concerned with those where the hydrogen bond geometries of the reacting molecules are well established and where kinetic data of the isotopic reactions are available over a large temperature range. Here, it is possible to study the relation between kinetic isotope effects and chemical structure. Examples are the tautomerism of porphyrin, of the porphyrin anion and related compounds exhibiting intramolecular hydrogen bonds of medium strength, and the solid state tautomerism of pyrazoles and of benzoic acid in cyclic associates. One main result is the finding of pre-exponential factors of the order of $kT/h \cong 10^{13}\ s^{-1}$, as expected by transition state theory for vanishing activation entropies. The barriers of multiple H-transfers are found to be larger than those of single H-transfers.

The second type of reactions discussed refers mostly to liquid state solutions and involves major heavy atom reorganization. Here, equilibria between reactive and non- or less reactive molecular configurations may play a role. Several cases are discussed where the less reactive forms dominate at low or at high temperature, leading to unusual Arrhenius curves. These cases include examples from small molecule solution chemistry like the base-catalyzed intramolecular H-transfer in diaryltriazene, 2-(2′-hydroxyphenyl)-benzoxazole, 2-hydroxy-phenoxyl radicals as well as an enzymatic system, thermophilic alcohol dehydrogenase. In the latter case, temperature dependent kinetic isotope effects are interpreted in terms of a transition between two regimes with different temperature independent kinetic isotope effects.

Hydrogen-Transfer Reactions. Edited by J. T. Hynes, J. P. Klinman, H. H. Limbach, and R. L. Schowen

ISBN: 978-3-527-30777-7

6.1 Introduction

Since the introduction of modern spectroscopic and kinetic techniques, H-transfer reactions constitute an active field of research because of their importance in chemistry and physics [1]. As in electron transfer reactions, tunneling plays an important role. However, in contrast to the Marcus theory of electron transfer [2], there is no widely accepted theory of H-transfer available to date. This is because H-transfer can occur in a variety of forms. Proton and hydride transfer are coupled to charge transfer and are strongly affected by solvent phenomena [3]. In contrast, hydrogen atom transfer and multiple proton transfers often take place without net charge transfer. As the electron/heavy atom mass ratios are smaller than 10^{-4} the tunneling motion of the electron is well separated from heavy atom reorganization. This may not be the case in H-transfers, where heavy atom tunneling may play a role. In contrast to electron transfer, hydrogen bond formation strongly affects the way H is transferred. Only H-transfer in weak and moderately strong hydrogen bonds can be described in terms of rate processes, as for electron transfer; in strong hydrogen bonds protons may no longer experience a barrier of transfer but are localized in the hydrogen bond center [3]. Finally, the three isotopes of hydrogen add to the complexity of H-transfer but also constitute important mechanistic tools.

Many different spectroscopic and kinetic techniques have been applied to the study of H-transfer reactions. Among these are dynamic nuclear magnetic resonance techniques. In the past decades it has been shown that NMR is an especially powerful tool for the study of degenerate single and multiple hydrogen transfer reactions in model hydrogen bonded systems embedded in liquid and solids [4]. Traditionally, the dynamic range of NMR was limited to the millisecond timescale but, for H-transfers in the solid state, it has been possible to extend this scale to the micro- to nanosecond timescale. In addition, it has been possible to elucidate multiple kinetic hydrogen isotope effects over large temperature ranges which enables one to detect tunnel effects at low temperatures from the observation of concave Arrhenius curves of the isotopic reactions. These curves can serve as benchmarks in order to check quantum-mechanical theories of rate processes in condensed matter.

Kinetic isotope effects of single H-transfers in organic liquids have often been interpreted in terms of a combination of Eyring's transition state theory [5] and isotope fractionation theory as proposed by Bigeleisen [6]. In this theory, kinetic isotope effects arise mainly from the difference in zero-point energies between the transition and the initial state. However, as has been shown by Bell [7], in hydrogen transfer reactions one has to take into account tunneling through the barrier, as has been mentioned above. His one-dimensional semiclassical "Bell tunneling model" has been very successful for the interpretation of Arrhenius curves of single hydrogen transfers using empirical parameters. The model was developed in times when computers were not available, i.e. was designed for the case of slow proton transfer, mainly hydrogen abstraction from carbon. Typically, reactions in solution were studied around room temperature over a limited range of temperatures. For these cases, the so-called "Bell tunneling correction" to the Arrhenius curves, elucidated in terms of classical transition state theory, was sufficient and has, therefore, often been taken as synonymous with the "Bell tunneling model". Full semiclassical tunneling calculations employing modified barriers have been performed by various other authors e.g. Ingold et al. [8], Limbach et al. [9], and Sutcliffe et al. [10]. Other semiclassical models of single proton abstractions have been proposed by Kuznetsov and Ulstrup [11, 12] and modified by Knapp et al. [13] for use in enzyme reactions. Siebrand et al. [14] have proposed a golden rule treatment of H-transfer between the eigenstates of the reactants and products where low-frequency vibrations play an important role varying the heavy atom distances.

Various quantum-mechanical theories have been proposed which allow one to calculate isotopic Arrhenius curves from first principles, where tunneling is included. These theories generally start with an *ab initio* calculation of the reaction surface and use either quantum or statistical rate theories in order to calculate rate constants and kinetic isotope effects. Among these are the "variational transition state theory" of Truhlar [15], the "instanton" approach of Smedarchina et al. [16], or a Redfield-relaxation-type theory as proposed by Meyer et al. [17]. However, these methods require extensive theoretical work and are, generally, not available for the experimentalist in the stage where he needs to simulate his Arrhenius curves. For this stage, empirical tunneling models are important.

This is especially true for the case of multiple hydrogen transfer reactions. In a number of papers, Limbach et al. have proposed to use formal kinetics in order to describe multiple kinetic isotope effects of intramolecular [18–23] and intermolecular HH-transfers [24]. This method has been extended to triple [25] and quadruple transfer reactions [26, 27]. In order to solve the problem of multiple particle transfer two limiting cases were considered. The "concerted" transfer refers to the case where several hydrogen and heavy atoms are transferred at the same time in such a way that they can be treated as a single particle. The "stepwise" transfer refers to the case where intermediates are involved. Here, the overall rate constants of the isotopic reactions studied are related using formal kinetics, to those of the individual reaction steps. Each reaction step, consisting of a single or concerted multiple H-transfer, is treated in terms of a tunneling model based on a modified Bell model [9]. In order to avoid possible confusion, this model will be denoted as the "Bell–Limbach" model. We note that the formal kinetic treatment of the stepwise transfer does not need to be combined with this model, but can also be combined with any theory treating a single step. For example, Smedarchina et al. [16] have used this approach in connection with their instanton approach.

The Bell–Limbach model is not designed to give definite interpretations of Arrhenius curves of hydrogen transfer reactions which have to come from more sophisticated methods. However, it provides an opportunity to check whether the number of parameters describing a given set of Arrhenius curves matches or exceeds the number of parameters necessary to describe the same set in terms of sums of single Arrhenius exponentials. This check also tells whether it is useful to apply a more sophisticated tunnel model containing a larger number of parameters. Finally, the Bell–Limbach model provides a platform which allows one to compare different reactions with each other and to derive general trends important for the kineticist to guide future research.

The scope of this chapter is, therefore, (i) to review the Bell–Limbach tunneling model in comparison with other models and its use for describing single steps of multiple hydrogen transfer networks and (ii) to review applications of this approach in a number of cases which have been studied mainly by NMR. A description of the techniques used for the determination of rate constants of H-transfer will not be included in this chapter; readers interested in this problem are referred to a recent review [4].

6.2 Theoretical

In this section first different models of single H-transfer will be reviewed, including primary kinetic H/D isotope effects, where the focus is on the Bell–Limbach model. Then formal kinetics will be used to describe multiple hydrogen transfers and their kinetic isotope effects.

6.2.1 Coherent vs. Incoherent Tunneling

The simplest model proposed to accommodate a degenerate hydrogen transfer process has been derived [28] from the theory of the one-dimensional symmetric double oscillator [29]. As illustrated in Fig. 6.1(a), this model assumes a symmetric double well with delocalized vibrational hydrogen states which are given in approximation as the positive and the negative linear combinations of the corresponding harmonic oscillator states. The energy splitting is hJ, where J can be interpreted as the frequency of coherent hydrogen tunneling of a wave packet created at $t = 0$, oscillating between the two wells. This type of coherent H tunneling has been verified for small molecules such as malonaldehyde [30], tropolone [31] (Chapt. 1), or formic acid dimer [32] (Chapt. 2) in the gas phase. In contrast, when malonaldehyde is embedded in condensed matter, intermolecular interactions lift the gas phase symmetry of the double well, leading to localized protons [33]. The situation is depicted schematically in Fig. 6.1(b). Here, the one-dimensional double oscillator theory predicts vibrational states with more or less localized proton wavefunctions. In order to arrive at a rate process, more dimensions have to be taken into account. One way is to couple the double oscillator to a bath of harmonic oscillators which will result in vibrational relaxation (VR), taking place on the femtosecond timescale [34]. VR is responsible for the transfer of vibrational energy between the double oscillator and the bath. By VR the localized reactant ground state is converted to higher vibrational states producing a nuclear wave packet. The latter is transferred by tunneling to the product well and deactivated to the product ground state. Usually, if VR is fast enough, the H tunneling process can be described in terms of rate constants for the forward and the backward H-transfer.

In the liquid state, the effective symmetry of the potential is restored by solvent relaxation (Fig. 6.1(c)). In principle, H-transfer is not restricted to symmetric configurations but can also take place in asymmetric configurations. Hynes et al. [35, see also Chapter 10] have proposed an alternative view which is sketched in Fig. 6.1(d). Here, localized states exhibiting asymmetric single or double wells are first converted by solvent relaxation to a strong symmetric low-barrier hydrogen bond in which the transfer takes place adiabatically before it is completed again by solvent relaxation. In conclusion, in condensed matter coherent tunneling generally becomes incoherent and can then be described by a rate constant instead of a tunnel frequency.

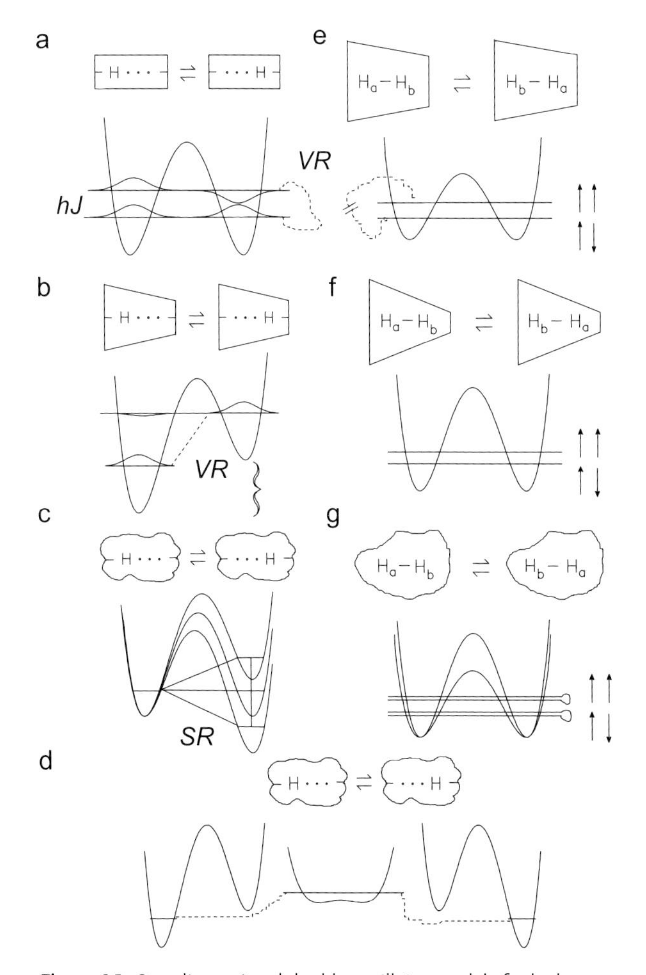

Figure 6.1 One-dimensional double oscillator models for hydrogen transfer and dihydrogen exchange under different conditions.

An exception to this rule is, however, the case of exchange of the nuclei of dihydrogen pairs bound to a transition metal center, where the exchange exhibits a barrier. This phenomenon will be sketched only briefly; for further information the reader is referred to Chapter 21 and to a recent minireview [36].

The situation is illustrated for the gas phase in Fig. 6.1(e). The nuclear spins – characterized by arrows – play a decisive role. The lower tunnel state is symmetric with respect to a permutation H_a–H_b $\leftrightarrow$ H_b–H_a and the upper state is antisymmetric. Since hydrogen has a nuclear spin 1/2 the lower state has to be coupled with the antisymmetric nuclear spin function $\uparrow\downarrow$ (antiparallel spins) in order to fulfill the Pauli exclusion principle. In contrast, the upper tunnel state is antisymmetric and has to be combined with the symmetric nuclear spin states $\uparrow\uparrow$ (parallel spins), as in the case of o-H_2 and p-H_2. Thus, interconversion of the delocalized dihydrogen states also involves a nuclear spin conversion which is much slower than VR. Thus, the tunnel splitting hJ can survive even in larger molecules, even on the NMR timescale. In the crystalline solid the barrier of interchange and the value of J may be altered but not the inherent symmetry of the process, Fig. 6.1(f). When placing the molecule in a multitude of different exchanging environments, i.e. in a liquid (Fig. 6.1(g)) solvent relaxation (SR) will lead to an average temperature dependent tunnel splitting as long as nuclear spin conversion is slow.

As this chapter focuses on hydrogen transfers in liquids and solids, it will be assumed that the transfer constitutes a rate process which can be described in terms of rate constants, for which the usual rate theories can be applied, in particular those derived from transition state theory.

6.2.2 The Bigeleisen Theory

In the theory of Bigeleisen [6], a combination of the theory of equilibrium isotope effects with Eyrings transition state theory [5], kinetic H/D isotope effects can be expressed by

$$\frac{k^{\mathrm{H}}}{k^{\mathrm{D}}} = \frac{Q^{\mathrm{D}} Q^{\mathrm{H}\ddagger}}{Q^{\mathrm{H}} Q^{\mathrm{D}\ddagger}} \tag{6.1}$$

Here, Q^{H} and Q^{D} represent the partition functions of the protonated and deuterated initial state and $Q^{\mathrm{H}\ddagger}$ and $Q^{\mathrm{D}\ddagger}$ those of the transition state. In order to evaluate Eq. (6.1) the vibrational frequencies of the initial and the transition states are needed. Generally, they are calculated using quantum-mechanical *ab initio* methods in harmonic approximation. The main source of kinetic isotope effects arises then from zero-point energy changes of the protons and deuterons in flight, as has been discussed by Bell [7, 37]. These effects are illustrated for a triatomic model system in Fig. 6.2. There is only a single stretching vibration of AH in the initial state. The transition state exhibits three normal vibrations, an imaginary antisymmetric stretch, a real symmetric stretch (not illustrated), and a real bending vibration. The antisymmetric stretch does not involve any zero-point energy, neither for H nor for D. The symmetric stretch is not isotope sensitive. However,

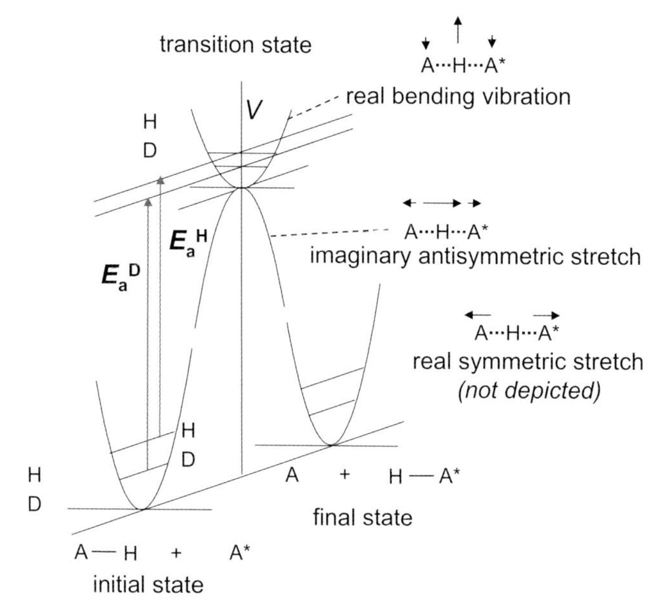

Figure 6.2 Triatomic model of H-transfer illustrating changes in zero-point energies of normal vibrations between the initial and transition states.

the bending vibration contains different zero-point energies for H and for D. Overall, the model predicts a different barrier for H and D transfer, and a kinetic H/D isotope effect of about 6 at room temperature.

6.2.3
Hydrogen Bond Compression Assisted H-transfer

Hydrogen bonding dominates H-transfers from and to oxygen, nitrogen, and fluorine. The barrier of the transfer depends strongly on the hydrogen bond geometry. Unfortunately, a general relation between both is not available to date. In fact, H-transfers in hydrogen bonds constitute a multidimensional problem where many different modes can contribute to the reaction coordinate. Experimentally, it is not easy to identify these modes and to take them into account in simple tunneling models. There is, however, one exception: from an empirical standpoint, hydrogen bond compression has been identified as one important mode which can be taken into account using empirical hydrogen bond correlations which will be described in this section.

With any hydrogen bond A–H···B one can normally associate two distances, the A–H distance $r_1 \equiv r_{AH}$ for the diatomic unit AH, and the H···B distance $r_2 \equiv r_{HB}$ for the diatomic unit HB. According to Pauling [38], one can associate with these distances so-called valence bond orders or bond valences, which correspond to the "exponential distances"

$$p_1 = \exp\{-(r_1 - r_1^\circ)/b_1\},\ p_2 = \exp\{-(r_2 - r_2^\circ)/b_2\},\ \text{with}\ p_1 + p_2 = 1 \tag{6.2}$$

b_1 and b_2 are parameters describing the decrease of the bond valences of the AH and the HB units with the corresponding distances. r_1° and r_2° are the equilibrium distances of the fictive non-hydrogen bonded diatomic molecules AH and HB. If one assumes that the total valence for hydrogen is unity, it follows that the two distances depend on each other, leading to an ensemble of allowed r_1 and r_2 values representing the "geometric hydrogen bond correlation". The hydrogen bond angle does not appear in Eq. (6.2). This correlation may be transformed into a correlation between the natural hydrogen bond coordinates $q_1 = {}^1/_2(r_1 - r_2)$ and $q_2 = r_1 + r_2$. For a linear hydrogen bond, q_1 represents the distance of H from the hydrogen bond center and q_2 the distance between atoms A and B. Experimentally, hydrogen bond correlations have been established using X-ray and neutron diffraction crystallography [39], as well as by NMR [40]. Note, however, that correlations of the type of Eq. (6.2) were also used a long time ago in the context of describing the "bond energy bond order conservation" reaction pathway of the H_2 + H reaction [41].

A typical geometric hydrogen bond correlation according to Eq. (6.2) derived for NHN-hydrogen bonded systems [42] is depicted in Fig. 6.3. When H is transferred from one heavy atom to the other, q_1 increases from negative values to positive values, and q_2 goes through a minimum which is located at $q_1 = 0$ for hydrogen bonded systems of the AHA-type and near 0 for those of the AHB-type. This correlation implies that, as an approximation, both proton transfer and hydrogen bonding coordinates can be combined into a single coordinate.

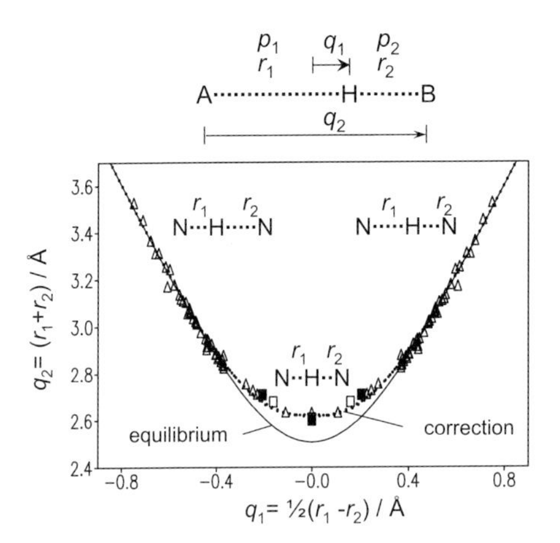

Figure 6.3 Correlation of the hydrogen bond length $q_2 = r_1 + r_2$ with the proton transfer coordinate $q_1 = {}^1/_2(r_1 - r_2)$. Solid line: correlation for equilibrium distances calculated with $b_1 = b_2 =$ 0.404 Å and $r_1^\circ = r_2^\circ = 0.992$ Å. Dotted line: empirical correction for zero-point vibrations. Adapted from Ref. [42].

The solid correlation line in Fig. 6.3 is calculated by adapting the parameters of Eq. (6.2) to experimental hydrogen bond geometries established by low-temperature neutron diffraction and NMR data. A deviation has been observed between the experimental data and the solid correlation line in the region of strong hydrogen bonds around $q_1 = 0$. This effect has been associated with zero-point energy vibrations which are not taken into account in Eq. (6.2), which is valid only for equilibrium geometries. An empirical correction leading to the dotted line in Fig. 6.3 has been proposed [36]. A similar empirical correction has also been proposed for OHN-hydrogen bonds [43].

The shortest possible equilibrium heavy atom distance is given by [39b]

$$q_{2min} = 2(r_o - b \ln {}^1/_2) \tag{6.3}$$

which leads to the values for symmetric hydrogen bonds listed in Table 6.1. These distances provide interesting references for characterizing transition states of H-transfers obtained by quantum-mechanical calculations. For example, hydride transfer distances between two carbon atoms at the transition state were calculated to be in the range 2.69–2.75 Å for various enzyme reactions [44].

Table 6.1 Shortest possible heavy atom distances of symmetric H-bonds predicted by the valence bond order model.

	r_o/Å	b/Å	q_{2min}/Å
OHO	0.95	0.37	2.41
NHN	0.99	0.404	2.53
CHC	1.1	~0.4	~2.75

6.2.4
Reduction of a Two-dimensional to a One-dimensional Tunneling Model

How do the H-bond geometries change during a typical H-transfer process? It is clear that at the minimum value of the heavy atom coordinate q_2 only a single geometry is realized, which is consistent with a single well potential for the H-motion. In contrast, at other geometries, the correlation curve indicates the possibility of double well situations, where the barrier height E_d increases with increasing value of the heavy atom coordinate q_2.

The situation is illustrated schematically in Fig. 6.4(a) for the case of degenerate H-transfers. One-dimensional cuts $V(q_1)$ at different values of q_2 through a two-dimensional potential surface of a degenerate H-transfer are displayed. The barrier height E_d of the double well describing the H-transfer decreases when q_2 is

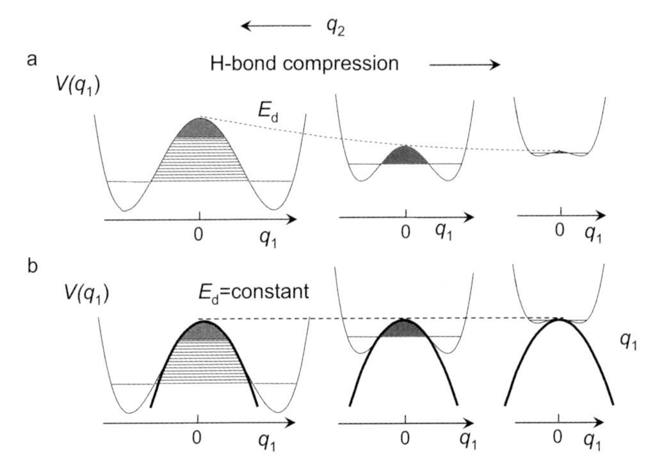

Figure 6.4 (a) One-dimensional cuts $V(q_1)$ through a two-dimensional potential energy surface of a degenerate H-transfer at different values of q_2. (b) Reduction of the two-dimensional double-well potential problem to a one-dimensional Bell model. Adapted from Refs. [9] and [26].

decreased, and eventually a single well configuration is reached. There are only a small number of AH vibrational states available; here, only the vibrational ground states are depicted.

As has been pointed out in Ref. [9], such a two-dimensional model can be reduced to a one-dimensional model by setting E_d = constant, as indicated in Fig. 6.4(b) and by assuming a continuous distribution of rapidly interconverting configurations with different values of q_2. Such a situation can be reached practically by excitation of low-frequency H-bond vibrations or phonons. The situation of Fig. 6.4(b) can practically be replaced by an inverted parabola as a barrier, with a continuous distribution of vibrational levels on both sides of the barrier.

A similar argument holds for non-degenerate H-transfers, as illustrated schematically in Fig. 6.5. Here, the asymmetry of the potential curve, i.e. the difference in the energy between the two wells will disappear in the region of the strongest H-bond compression.

In order to evaluate qualitatively the expected kinetic isotope effects one has to discuss (i) zero-point energy (ZPE) changes of H in the transition state as compared to the initial state and (ii) tunnel effects.

The expected changes in the zero-point energies of the H transferred are illustrated schematically in Fig. 6.6 for the degenerate case, as proposed by Westheimer [45]. The antisymmetric stretch in the initial state exhibits quite different ZPEs for H and for D as the force constants are large. This vibration becomes imaginary in the transition state, which is assumed here to be located in the minimum of q_2 , i.e. the ZPE of the antisymmetric stretch is lost in the transition state. The ZPE of the symmetric stretch in the transition state is small and exhibits little isotope dependence. We note that ZPE is built up in the bending vibration in the

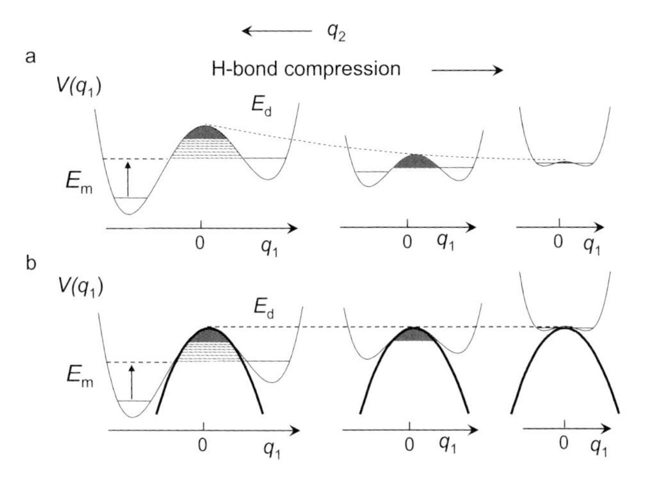

Figure 6.5 (a) One-dimensional cuts $V(q_1)$ through a two-dimensional potential energy surface of a *non-degenerate* H-transfer at different values of q_2. (b) Reduction of the two-dimensional double-well potential problem to a one-dimensional Bell model. E_m refers to a minimum energy for tunneling to occur.

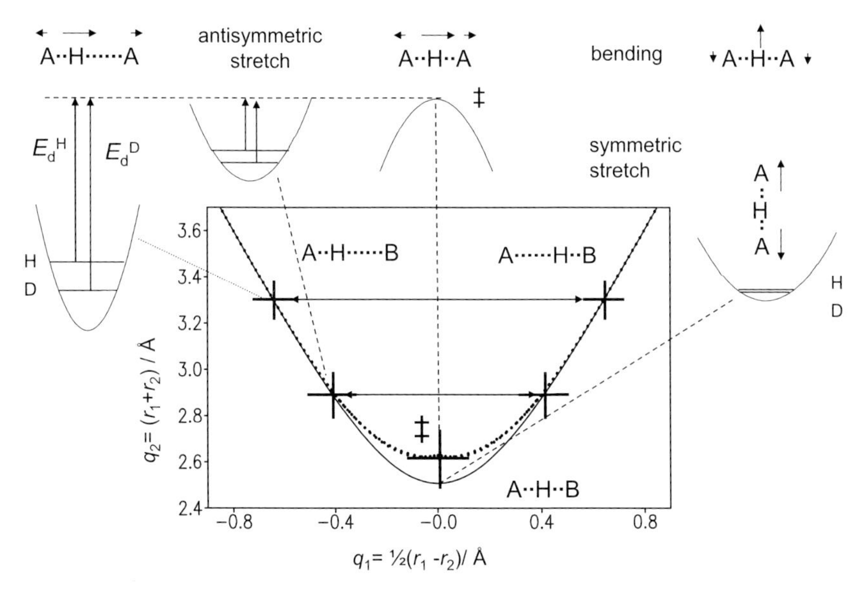

Figure 6.6 Hydrogen bond correlation and zero-point vibrations for a degenerate H-transfer.

transition state. Overall, a substantial difference in the effective barriers for the H-transfer and for the D-transfer is expected. Tunneling pathways can occur at larger values of q_2, which is expected to remain constant during the tunnel process. As only hydrogen isotopes move at constant q_2 the tunneling masses are then 1 for H and 2 for D.

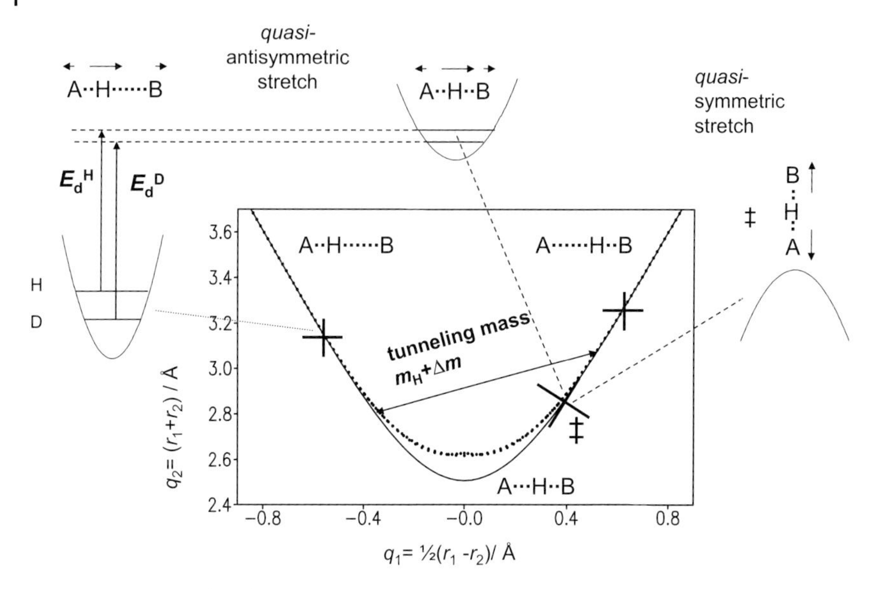

Figure 6.7 Hydrogen bond correlation and zero-point vibrations for a non-degenerate H-transfer.

In contrast, if the transfer is non-degenerate, a situation may occur as illustrated in Fig. 6.7. At the transition state there is remaining ZPE in the antisymmetric stretch. This will lead to a decrease in the difference between the effective barriers for H and for D, as has been proposed by Westheimer [45]. This decrease has also been called the "Westheimer-effect" [46]. Tunneling pathways may no longer involve only changes in q_1, but also a substantial heavy atom motion. This means that the effective tunneling masses will be increased by an additional mass Δm as illustrated schematically.

6.2.5 The Bell–Limbach Tunneling Model

The simplest tunnel model which allows one to calculate Arrhenius curves of H-transfer reactions is the Bell tunneling model [7] which has been modified in our laboratory [9]. The model has been reviewed recently by Limbach et al. [26]. It is visualized in Fig. 6.8 which will be explained in the following.

According to Bell, the probability of a particle passing through or crossing a barrier is given by [7]

$$G(W) = \frac{1}{1 + D(W)^{-1}} \tag{6.4}$$

where W represents the energy of the particle and $D(W)$ the transmission coefficient, given according to the Wentzel–Kramers–Brillouin approximation [47] by

$$D(W) = \exp\left(-\frac{2}{\hbar}\int_{-a'}^{a'} p\mathrm{d}x\right) = \exp\left(-\frac{2}{\hbar}\int_{-a'}^{a'} \sqrt{2m(V(x) - W)}\mathrm{d}x\right) \tag{6.5}$$

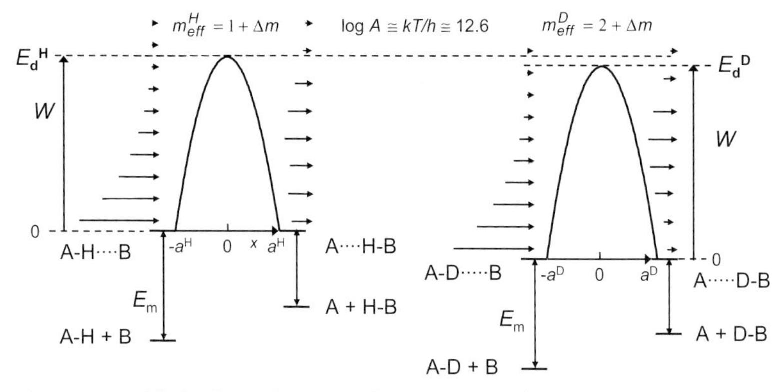

Figure 6.8 Modified Bell tunneling model for H and D transfer. Adapted from Refs. [9] and [26].

p represents the momentum and m the mass of the particle moving in the $q_1 \equiv x$-direction, and $V(x)$ the potential energy experienced by the particle. $|\alpha'|$ represents the position of the particle when it enters or leaves the barrier region at energy W. $V(0) = E_d$ represents the energy of the barrier i.e. the "barrier height" and $2a$ the width of the barrier at the lowest energy where tunneling can occur. Classically $G(W)= 0$ for $W < E_d$ and $G(W)= 1$ for $W > E_d$, but quantum-mechanically $G(W) > 0$ for $W \leq E_d$ and $G(W) < 1$ for $W \geq E_d$. Note here that D is related to the energy splitting of a symmetric double oscillator by [28]

$$\Delta E = \frac{h\nu}{\pi} D^{1/2} \tag{6.6}$$

Assuming that the barrier region can be approximated by an inverted parabola it follows that

$$V(x) = E_d\left(1 - \frac{x^2}{a^2}\right) \text{ and that } W = E_d\left(1 - \frac{a'^2}{a^2}\right) \tag{6.7}$$

It has been shown by Bell [48] that

$$D(W) = \exp\left(-\frac{2\pi(E_d - W)}{h\nu_t}\right), \quad \nu_t = \frac{1}{\pi a}\sqrt{\frac{E_d}{2m}} \tag{6.8}$$

where ν_t represents a "tunnel frequency". The fraction of particles in the energy interval dW is given by the Boltzmann law

$$\frac{dN}{N} = \frac{\exp(-W/kT)dW}{\int_0^\infty \exp(-W/kT)dW} = \frac{1}{kT}\exp(-W/kT)dW \tag{6.9}$$

The classical integrated reaction probability is then given by

$$\left(\frac{\Delta N}{N}\right)_{\text{class}} = \frac{1}{kT}\int_{E_d}^{\infty} \exp(-W/kT)dW = \exp(-E_d/kT) \tag{6.10}$$

the quantum mechanical integrated reaction probability by

$$\left(\frac{\Delta N}{N}\right)_{QM} = \frac{1}{kT}\int_0^\infty G(W)\exp(-W/kT)\mathrm{d}W \tag{6.11}$$

and the ratio of both quantities by

$$Q_t = \frac{\left(\frac{\Delta N}{N}\right)_{QM}}{\left(\frac{\Delta N}{N}\right)_{class}}$$

$$= \frac{\frac{1}{kT}\int_0^\infty G(W)\exp(-W/kT)\mathrm{d}W}{\exp(-E_d/kT)} = \int_0^\infty \frac{G(W)}{kT}\exp((E_d - W)/kT)\mathrm{d}W \tag{6.12}$$

Using an Arrhenius law for the classical temperature dependence it follows that

$$k = k_{class}Q_t = A\exp(-E_d/kT)\int_0^\infty \frac{G(W)}{kT}\exp((E_d - W)/kT)\mathrm{d}W \tag{6.13}$$

Replacing the Boltzmann constant by the gas constant, and introducing a superscript as label for the isotope L = H, D it follows that

$$k^L = A^L\exp(-E_d^L/RT)\int_0^\infty \frac{G^L(W)}{RT}\exp\left(\left(E_d^L - W\right)/RT\right)\mathrm{d}W, \mathrm{L} = \mathrm{H, D} \tag{6.14}$$

At very high temperatures, the integral becomes unity and one obtains the classical expression

$$k^L = A^L\exp(-E_d^L/RT), \mathrm{L} = \mathrm{H, D} \tag{6.15}$$

From Eq. (6.14) one obtains the following expression for the "primary" kinetic isotope effect for the H-transfer as a function of temperature

$$P = \frac{k^H}{k^D} = \frac{A^H Q^H\exp(-E_d^H/RT)}{A^D Q^D\exp(-E_d^D/RT)} \tag{6.16}$$

where Q^L has been called the "tunnel correction". This equation has been introduced by Bell [7]. The energy difference $\Delta\varepsilon = E_d^D - E_d^H$ describing the losses of zero-point energy between the reactant and the transition state can be calculated using Bigeleisen theory [6].

In the low temperature regime for $W = 0$ it follows from Eq. (6.8) that

$$G(0) \cong D(0) = \exp\left(-\frac{2\pi^2 a}{h}\sqrt{2mE_d}\right) \tag{6.17}$$

Therefore,

$$k_o^L = A^L D^L(0) = A^L\exp\left(-\frac{2\pi^2 a^L}{h}\sqrt{2m^L E_d^L}\right) \tag{6.18}$$

The width of the barrier for the D-transfer can be calculated from Eq. (6.7) and is given by

$$2a^{\mathrm{D}} = 2a^{\mathrm{H}}\sqrt{\frac{E_{\mathrm{d}}^{\mathrm{D}}}{E_{\mathrm{d}}^{\mathrm{H}}}} \tag{6.19}$$

With $m^{\mathrm{H}} = 1$ and $m^{\mathrm{D}} = 2$, the low-temperature rate constant k_0^H is then determined mainly by a^{H} for a given value of E_{d}^H. The low-temperature and temperature independent kinetic isotope effect k_0^H/k_0^D is, therefore, determined by E_{d}^D which is obtained experimentally at high temperatures. In other words, k_0^H/k_0^D and the high-temperature kinetic isotope effects cannot be varied independently of each other, which is not in agreement with experimental data. This effect can be associated with heavy atom tunneling during the H-transfer. The tunneling mass is increased and the low-temperature H/D isotope effect decreased.

In order to take heavy atom tunneling into account, the expansion

$$(a\sqrt{m})^{\mathrm{L}} = \sqrt{\sum_i a_i^2 m_i} = \left((a^{\mathrm{L}})^2 m_{\mathrm{L}} + \sum_k a_k^2 m_k\right)^{1/2} = a^{\mathrm{L}}\sqrt{m^{\mathrm{L}} + \Delta m} \tag{6.20}$$

is used [9], where

$$\Delta m = \sum_k \left(\frac{a_k}{a^{\mathrm{L}}}\right)^2 m_k,\ \mathrm{L} = \mathrm{H, D}\ \text{and}\ k = \text{heavy atoms} \tag{6.21}$$

The heavy atom contribution reduces generally the value of k_0^H/k_0^D. For example, if during H-tunneling in an OHO-hydrogen bond over $2a^{\mathrm{H}} = 0.5$ Å both oxygen atoms are displaced, each by $2a^{\circ} = 0.05$ Å, it follows that $\Delta m = 0.32$, and the total tunneling mass is 1.32 instead of 1.

Equation (6.14) is visualized in Fig. 6.8. Particles of different kinetic energies W, given by a Boltzmann distribution, hit the barrier from the left side, where the probabilities of finding given energies are symbolized by arrows of different length. The arrows on the right side represent the particles which came through the barrier by tunneling. As the tunneling mass of H is smaller than that of D, at a given temperature, the energy for the maximum number of H tunneling through the barrier is smaller than for D.

As Limbach et al. have proposed, Eq. (6.14) needs to be modified in a minor way for application in multiple proton transfer reactions [9, 18, 21, 25, 49]. The most important change is to replace the lower integration limit in Eq. (6.13) by a minimum energy E_{m} for tunneling to occur as indicated in Fig. 6.8, i.e.

$$k^{\mathrm{L}} = A^{\mathrm{L}}\exp(-E_{\mathrm{d}}^{\mathrm{L}}/RT)\int_{E_m}^{\infty}\frac{G^{\mathrm{L}}(W)}{RT}\exp((E_{\mathrm{d}}^{\mathrm{L}} - W)/RT)\mathrm{d}W,\ \mathrm{L} = \mathrm{H, D} \tag{6.22}$$

This modification is necessary for example, when the reaction pathway involves an intermediate. Tunneling can then take place only at an energy which corresponds to the energy of the intermediate. Then, one can identify E_{m} with the energy E_{i} of this intermediate. However, E_{m} may also represent a reorganization energy E_{r} necessary for a heavy atom rearrangement preceding the tunneling process. Thus, E_{m} includes the "work term" in Marcus theory of electron transfer [2]. In addition, E_{m} may include the reaction enthalpy ΔH of a pre-equilibrium to H-transfer as discussed above.

A set of Arrhenius curves calculated using Eq. (6.22) depends then on the following parameters:

1. A single pre-exponential factor A in s^{-1} is used for all isotopic reactions, i.e. a possible mass dependence [7] is neglected within the margin of error. If solvent reorganization and pre-equilibria are absent, A is expected to be about 10^{13} s^{-1}. According to transition state theory, pre-exponential factors are given by $kT/h = 10^{12.6}$ s^{-1} for $T = 298$ K.
2. $E_m = \Delta H + E_r + E_i$ represents the minimum energy for tunneling to occur as described above and is assumed to be isotope independent. Note that a similar effect on the Arrhenius curves may be obtained by using more complex barrier shapes [10].
3. E_d^H is the barrier height for the H-transfer step of interest. Therefore, the sum $E_m + E_d^H$ represents the total barrier height for the H-transfer.
4. $2a^H$ is the barrier width of the inverted parabola used to describe the barrier of the H-transfer at the energy E_m. This parameter indicates the tunnel distance of H. $2a^D$ can be approximated by Eq.(6.19).
5. $\Delta\varepsilon = E_d^D - E_d^H$ represents the increase in the barrier height when H is replaced by D.
6. The tunneling masses are given by $m_{eff}^L = m^L + \Delta m$, L = H, D with $m^H = 1$ and $m^D = 2$.

Δm corresponds to the contribution of heavy atom displacements during the tunneling process.

In order to illustrate the formalism typical Arrhenius curves of H and D transfer using arbitrary parameters are plotted in Fig. 6.9. From the slope of the curves at high temperature one can obtain the quantities $E_m + E_d^H$ and $E_m + E_d^D$ which were the same in all graphs of Fig. 6.9. Because of the different slopes of the H and the D curve, temperature dependent kinetic H/D isotope effects occur in the high-temperature range, as illustrated by the dotted lines. At low temperatures, parallel Arrhenius curves are expected, exhibiting a slope given by E_m. The low-temperature branches are also illustrated by dotted lines. By extrapolation of the low-temperature branches to high temperatures, the values of k_0^H and of k_0^D are obtained. According to Eq. (6.18), they provide information about the barrier width $2a^H$ and the heavy atom tunneling extra mass Δm.

Let us discuss the effects of $2a^H$ and Δm on the Arrhenius curves. In Fig. 6.9(a) and (b) the H curves are identical, as the product $2a^H(1+\Delta m)^{1/2}$ is the same. However, the introduction of an extra tunneling mass of 3 in Fig. 6.9(b) reduces the kinetic isotope effects in the low-temperature regime but not in the high-temperature regime. On the other hand, by comparison of Fig. 6.9(a) with Fig. 6.9(c), or of Fig. 6.9(b) with Fig. 6.9(d), it follows that when a barrier of constant height becomes narrower, the regime of temperature independent kinetic isotope effects is reached at higher temperatures.

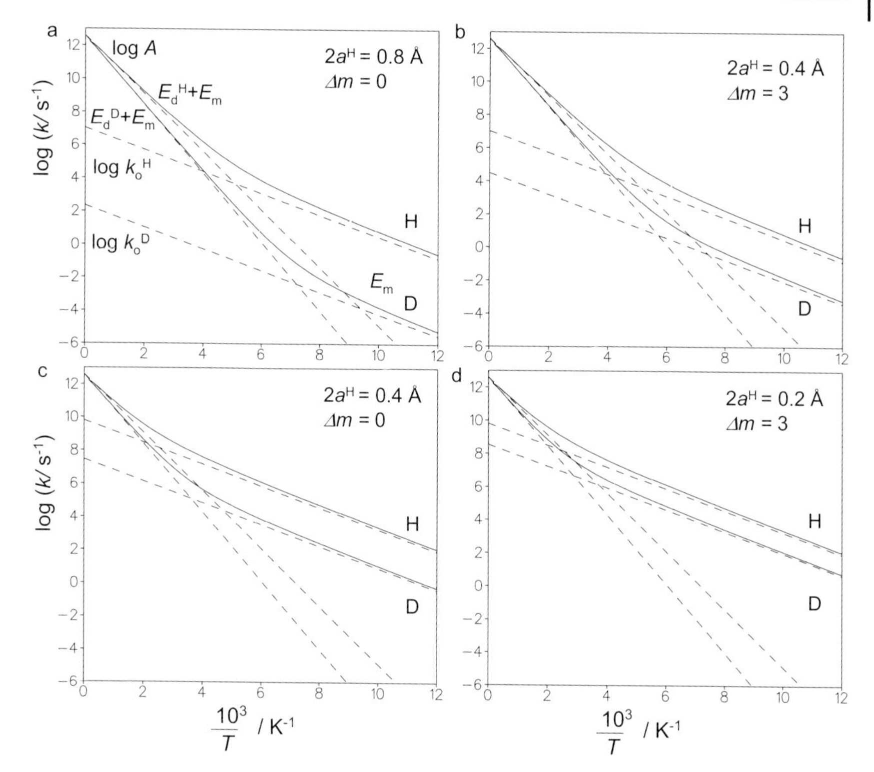

Figure 6.9 Arrhenius curves of H and D transfer calculated according to the Bell–Limbach tunneling model. Minimum energy for tunneling to occur $E_d^H = 12.55$ kJ mol^{-1}, barrier heights $E_d^H = 20.9$ kJ mol^{-1}, $E_d^D = 27.2$ kJ mol^{-1}, tunneling masses $m^H = 1 + \Delta m$, $m^D = 2 + \Delta m$. Barrier width $2a^D = 2a^H(E_d^D / E_d^H)^{1/2}$ according to Eq. (6.19). k_o^H and k_o^D are the extrapolated ground state tunneling rates as defined in Eq. (6.18).

6.2.6
Concerted Multiple Hydrogen Transfer

Equation (6.22) may not only be used in the case of a single proton transfer but also in the case of concerted multiple proton transfers, according to Fig. 6.10, characterized by a single barrier. Only some minor changes are necessary.

1. $\Delta\varepsilon = E_d^D - E_d^H$ continues to represent the increase in the barrier height when a given H is replaced by D. For successive replacements of H by D, generally, the so-called "Rule of the Geometric Mean" (RGM) derived for equilibrium isotope effects [6b] is applied, i.e. it is assumed that replacement of each H by D leads to the same increase $\Delta\varepsilon$ of the barrier height

$$E_d^{HD} = E_d^{HH} + \Delta\varepsilon,\ E_d^{DD} = E_d^{HH} + 2\Delta\varepsilon, \ldots. E_d^{DDDD} = E_d^{HHHH} + 4\Delta\varepsilon \tag{6.23}$$

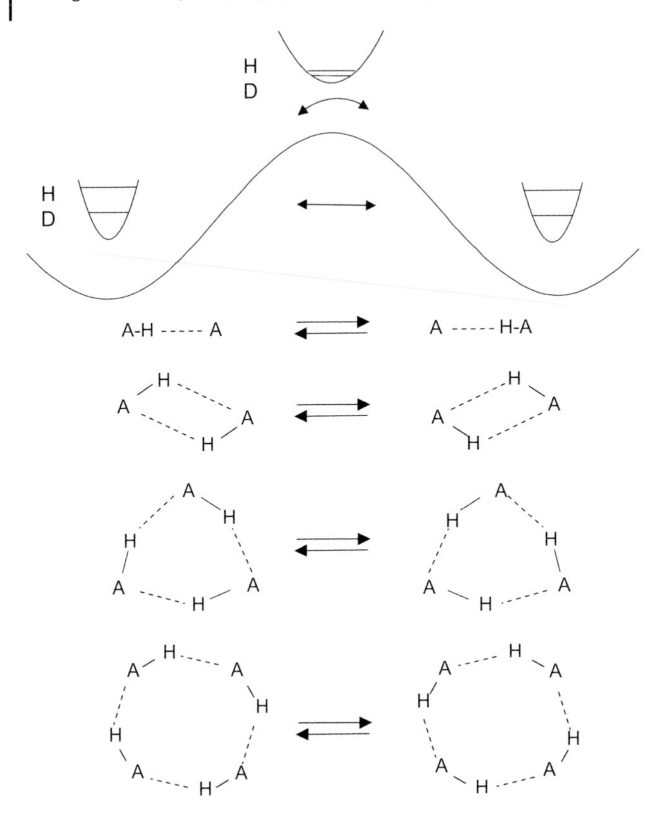

Figure 6.10 Schematic one-dimensional energy profile of degenerate single-barrier hydron (H, D , T) transfers in coupled networks of 1 to 4 cyclic hydrogen bonds. The hydron transfer can be an over-barrier process or a tunneling process, as indicated by the double arrows in the energy profile. The over-barrier process leads to kinetic isotope effects because of the loss of zero-point energy at the top of the barrier for each proton transferred, as indicated schematically. The tunneling process leads to kinetic isotope effects because of different tunneling masses for the hydrogen isotopes. Reproduced with permission from Ref. [26].

$\Delta\varepsilon$ is symbolized schematically in Fig. 6.10 by the different spacings of H and D levels in the ground state and the transition states.

2. The tunneling mass of a given isotopic reaction is written as $m_{\text{eff}} = \sum_i m_i^L + \Delta m$

6.2.7 Multiple Stepwise Hydrogen Transfer

In a number of papers, Limbach et al. have proposed to use formal kinetics in order to describe the case of stepwise HH-transfer [18, 24]. This method has been extended to the stepwise triple [25] and quadruple transfer cases [26]. In these papers the Bell–Limbach model was employed for the treatment of each reaction step. We note again that the same equations developed can also be used when each reaction step is described in terms of a first-principle theory. In this section the main results of this research are reviewed only briefly. For a more detailed description the reader is referred to Ref. [26].

6.2.7.1 HH-transfer

In Fig. 6.11 is depicted a general scheme of a stepwise HH-transfer reaction between the initial state A and the final state D. B and C are intermediates whose concentration is small. In each reaction step a single H is transferred, the other H is bound. Let us denote the formation of the intermediate as "dissociation" and the backward reaction as "neutralization". The corresponding free energy reaction profile is illustrated in Fig. 6.11(b).

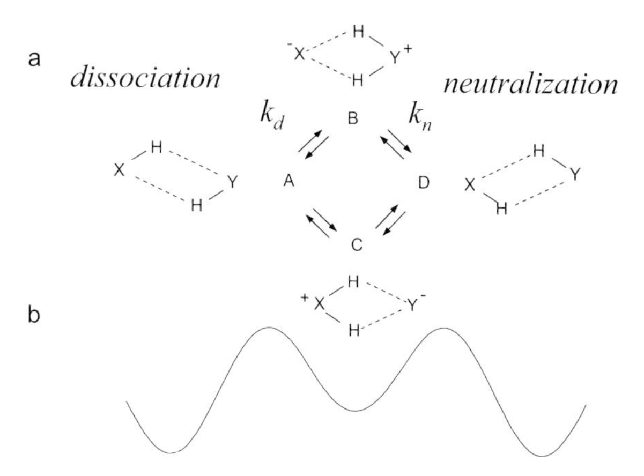

Figure 6.11 Degenerate stepwise HH-transfer involving metastable intermediates. (a) Chemical reaction network as base for a description in terms of formal kinetics. (b) Corresponding free energy diagram. Reproduced with permission from Ref. [26].

Using the steady-state approximation it can easily be shown that the rate constant of the interconversion between A and D is given by

$$k_{AD} = \frac{k_{AB}k_{BD}}{k_{BA} + k_{BD}} + \frac{k_{AC}k_{CD}}{k_{CA} + k_{CD}} \tag{6.24}$$

For the isotopic rate constants it follows that

$$k_{AD}^{HH} = \frac{k_{AB}^{HH} k_{BD}^{HH}}{k_{BA}^{HH} + k_{BD}^{HH}} + \frac{k_{AC}^{HH} k_{CD}^{HH}}{k_{CA}^{HH} + k_{CD}^{HH}}, \quad k_{AD}^{HD} = \frac{k_{AB}^{HD} k_{BD}^{DH}}{k_{BA}^{HD} + k_{BD}^{DH}} + \frac{k_{AC}^{DH} k_{CD}^{HD}}{k_{CA}^{DH} + k_{CD}^{HD}}$$

$$k_{AD}^{DH} = \frac{k_{AB}^{DH} k_{BD}^{HD}}{k_{BA}^{DH} + k_{BD}^{HD}} + \frac{k_{AC}^{HD} k_{CD}^{DH}}{k_{CA}^{HD} + k_{CD}^{DH}}, \quad k_{AD}^{DD} = \frac{k_{AB}^{DD} k_{BD}^{DD}}{k_{BA}^{DD} + k_{BD}^{DD}} + \frac{k_{AC}^{DD} k_{CD}^{DD}}{k_{CA}^{DD} + k_{CD}^{DD}} \tag{6.25}$$

The primary and secondary kinetic isotope effects of a given reaction step ij can be written in the form

$$P_{ij} = \frac{k_{ij}^{H(L)}}{k_{ij}^{D(L)}} \quad \text{and} \quad S_{ij} = \frac{k_{ij}^{L(H)}}{k_{ij}^{L(D)}} \tag{6.26}$$

where the brackets indicate the bound hydrogen. The corresponding isotopic fractionation factor is given by

$$\phi_{ij} = \frac{P_{ji}}{P_{ij}} \frac{S_{ji}}{S_{ij}} \tag{6.27}$$

For the case of degenerate HH-transfers, the following isotopic reaction rate constants have been derived [18a–c, 26]

$$\begin{aligned} k_{AD}^{HH} &= k_{d}^{HH} \\ k_{AD}^{HD} &= k^{DH} = k_{d}^{HH} P_{d}^{-1} \left[\frac{\phi_{d} S_{n}^{-1} + S_{d}^{-1}}{\phi_{d} S_{n}^{-1} + P_{d}^{-1}} \right] \\ k_{AD}^{DD} &= k_{d}^{DD} = k_{d}^{HH} P_{d}^{-1} S_{d}^{-1} \end{aligned} \tag{6.28}$$

The subscript d refers to the dissociation and the subscript n to the neutralization step in Fig. 6.11(a). In the absence of isotopic fractionation $\phi_{d} = 1, S_{n} = S_{d} = S$, $P_{n} = P_{d} = P$, and Eq. (6.28) simplifies to

$$\begin{aligned} k_{AD}^{HH} &= k_{d}^{HH} \\ k_{AD}^{HD} &= k_{AD}^{DH} = k_{AD}^{HH} \left[\frac{2}{S + P} \right] = k_{AD}^{DD} \left[\frac{2}{P^{-1} + S^{-1}} \right] \\ k_{AD}^{DD} &= k_{d}^{DD} = k_{AD}^{HH} P^{-1} S^{-1} = k_{d}^{HH} P^{-1} S^{-1} \end{aligned} \tag{6.29}$$

and in the absence of secondary isotope effects to

$$\begin{aligned} k_{AD}^{HH} &= k_{d}^{HH} \\ k_{AD}^{HD} &= k_{AD}^{DH} = k_{AD}^{HH} \left[\frac{2}{1 + P} \right] = k_{AD}^{DD} \left[\frac{2}{P^{-1} + 1} \right] = k_{d}^{DD} \left[\frac{2}{P^{-1} + 1} \right] \\ k_{AD}^{DD} &= k_{d}^{DD} = k_{AD}^{HH} P^{-1} = k_{d}^{HH} P^{-1} \end{aligned} \tag{6.30}$$

As the bound H does not contribute, the notation can be simplified by setting $k_{d}^{LL} = k_{d}^{L}$, by dropping the labels for the tautomeric states and by keeping in mind that according to Eq. (6.26)

$$P = k^{H}/k^{D} = k^{HH}/k^{DD}$$

Thus, it follows that

$$k^{HH} = k^{H}$$
$$k^{HD} = k^{DH} = \frac{2k^{H}}{1 + k^{H}/k^{D}} = \frac{2k^{D}}{k^{D}/k^{H} + 1} \tag{6.31}$$
$$k^{DD} = k^{D} = k^{HH}P^{-1} = k^{H}P^{-1}$$

This means that at low temperature where P is large the HD reaction is ca. twice as fast as the DD reaction. Equation (6.31) has been used in connection with the Bell–Limbach tunneling model to describe the stepwise double proton transfer in porphyrins, azophenine, and oxalamidines, as will be discussed in Section 6.3. Smedarchina et al. [16] used the same equations for their quantum-mechanical treatment of the porphyrin tautomerism.

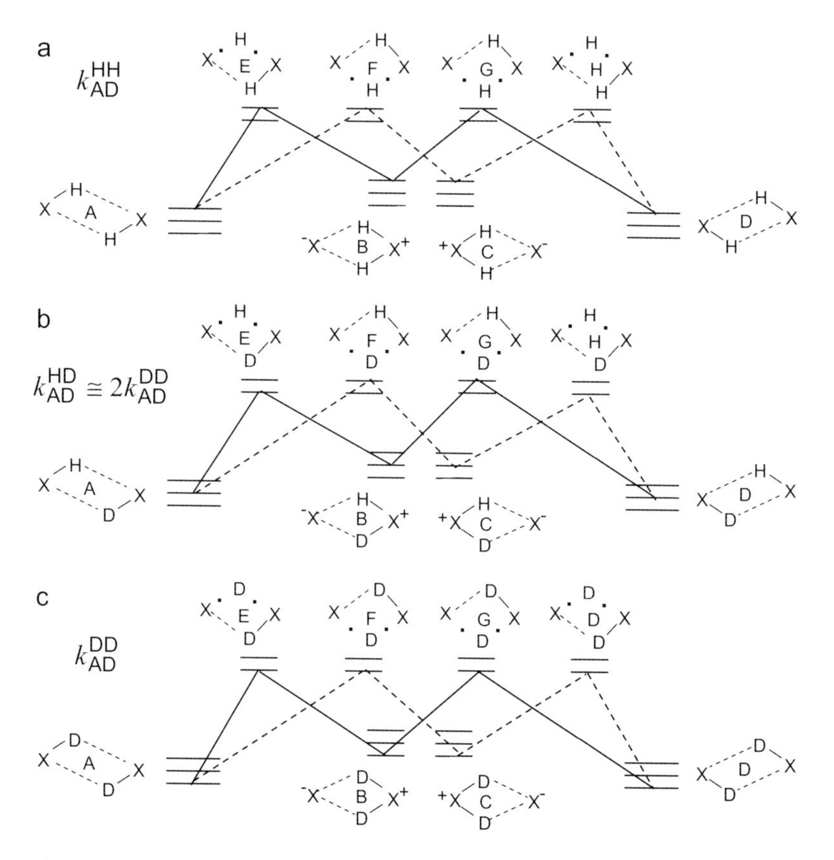

Figure 6.12 Free energy correlation (shown schematically) for the H and D zero-point vibrations for a degenerate stepwise double hydrogen transfer reaction according to Eq. (6.31), where secondary kinetic isotope effects and isotopic fractionation between the initial and the intermediate state were neglected. Adapted from Ref. [18c].

Equation (6.30) can be visualized in the free energy diagrams shown in Fig. 6.12. Different free energies for different isotopic reactions arise either from zero-point energy differences or from tunneling contributions. In all isotopic processes there are equivalent pathways via either intermediate B or C. The initial and final states A and D as well as the two intermediates B and C have two bound hydrogen isotopes, leading to three isotopic states of different free energy, containing either mobile HH, HD, or DD isotopes. In contrast, the states where one hydrogen isotope is in flight are characterized by only two isotopic states of different free energy because there is only one bound hydrogen isotope. Note that the term "state with a hydrogen isotope in flight" can be either a conventional transition state or a state where the isotope tunnels from a thermally activated state through the reaction energy barrier. Let us first compare the HH and the DD reaction profiles in Fig. 6.12(a) and (c). Both profiles are symmetric. Therefore, internal return causes the intermediate to react to the product D only with a probability of $1/2$; it returns with the same probability to A. The factor of $1/2$ entering the expression for k_{AD}^{HH} and k_{AD}^{DD} is, however, canceled in Eq. (6.30) because there are two equivalent pathways via B and via C for all isotopic reactions. The DD reaction is slower than the HH reaction because of the loss of zero point energy of the XH/XD stretching vibration in the transition states or because of tunneling. In contrast, the reaction profile of the HD process is asymmetric and the transition states E and G (and F and H) are now no longer equivalent. Therefore, the problem of internal return is absent. If one neglects secondary kinetic isotope effects, the energy necessary to reach the H transition states is similar to that in the HH process and the energy necessary to reach the D transition states is similar to that in the DD case. The D transfer step constitutes, therefore, the rate-limiting step of the reaction. The HD reaction has then the same free energy of activation as the DD reaction; however, since in the HD reaction all molecules that have passed the transition state react to products in contrast to the DD reaction, it follows that $k_{AD}^{HD} \cong 2k_{AD}^{DD}$ (Eq. (6.30)).

In contrast, for the case of a non-degenerate reaction where A represents the dominant form and B the dominant intermediate it has been shown [19b] that

$$k_{AD}^{HD} \cong k_{AD}^{DD} \quad \text{and} \quad k_{AD}^{DH} = k_{AD}^{HH} \tag{6.32}$$

These results can again be visualized in a free energy diagram as shown in Fig. 6.13. Since the transition states H are higher in energy than G, the pathways symbolized by the dashed lines do not contribute to the reaction rates. These pathways are, therefore, no longer discussed; only the favored pathways characterized by the solid lines that involve the transition states E and G are considered in the following. The true transition state of all isotopic reactions is G. In the latter, the loss of zero-point energy of XD is larger than of XH and hence the DD reaction is slower than the HH reaction. Thus, neglecting secondary kinetic isotope effects, it follows that the DH reaction is as fast as the HH reaction as H is transferred in the rate-limiting step. By contrast, the D isotope is transferred in the rate-limiting step of the HD reaction which exhibits, therefore, similar rate constants to those of the DD reaction.

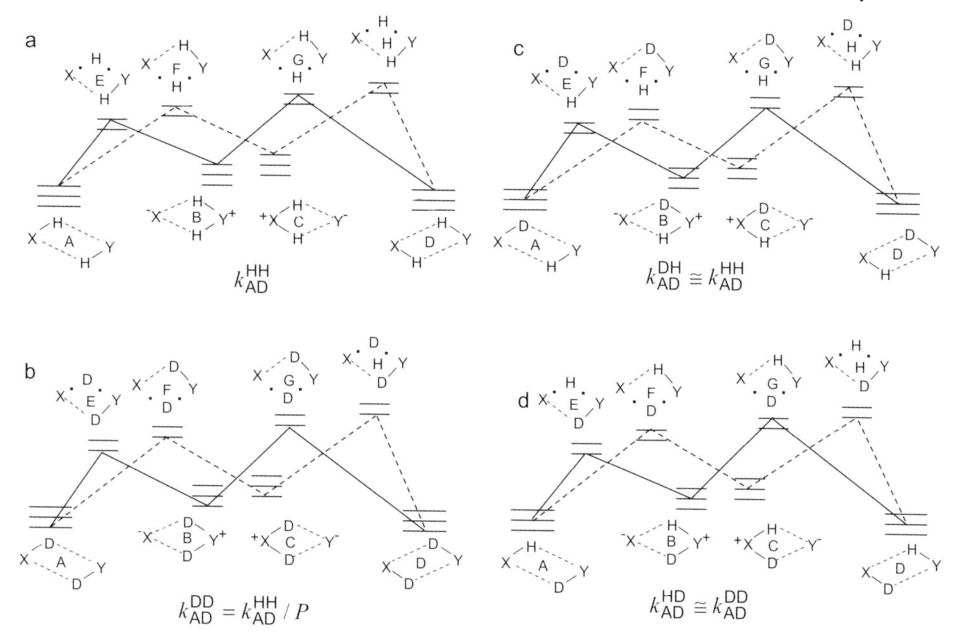

Figure 6.13 Free energy correlation (shown schematically) for the H and D zero-point vibrations for a non-degenerate stepwise HH-transfer reaction. The secondary kinetic isotope effect *S* was set to unity. Adapted from Ref. [18c].

Some calculated kinetic isotope effects are illustrated in the Arrhenius diagrams of Fig. 6.14. If the secondary isotope effects were equal to the primary ones, this would mean that both protons are in flight in the rate-limiting step and the double barrier case would reduce to the single barrier case. Then, one would obtain the rule of the geometric mean (RGM) with $k^{HH}/k^{HD} = k^{HD}/k^{DD} = P$, and the overall isotope effect is $k^{HH}/k^{DD} = P^2$. This result represents in fact a derivation of the RGM for the single barrier case. In the Arrhenius plot of Fig. 6.14(a) the validity of this rule was assumed. However, this rule is valid only in the absence of tunneling and if both proton sites are equivalent. In the presence of tunneling $k^{HH}/k^{HD} > k^{HD}/k^{DD}$ as has been verified previously.

In Fig. 6.14(b) the two-barrier or stepwise transfer Arrhenius diagrams are plotted, where it was assumed that the secondary isotope effects of dissociation and neutralization are small, i.e. equal to 1. In addition, absence of isotopic fractionation is assumed, i.e. $\phi_d = 1$. In this case, k^{DD}/k^{HH} is equal to the kinetic isotope effects $P_d = P_n$ of the dissociation and neutralization steps. When these isotope effects are large, which is the case at low temperatures, k^{HD}/k^{DD} is equal to 2. The statistical factor arises from the fact that in the DD reaction D is transferred in both steps. Therefore, when the intermediate is reached, return to the reactant as well as reaction to the product occurs with equal probability. By contrast, there is no internal return in the HD reaction which exhibits only a single rate-limiting step, i.e. the one in which D is transferred. Note that Eq. (6.30) is valid even in the presence of tunneling.

When isotopic fractionation takes place, for example through a strenghtening of the H-bond in the intermediate leading to reduced zero-point energies [50], the factor will be larger than 2, leading to an increase of k^{HD}/k^{DD} as illustrated in Fig. 6.14c.

The effect of an increasing reaction asymmetry is illustrated in Fig. 6.14(d)–(f). When the asymmetry is small, the HD and DH processes are almost equally fast, and again characterized by approximately twice the rate constant of the DD process. When the asymmetry becomes larger, the HD curve merges rapidly with the DD curve, and eventually the DH curve with the HH curve.

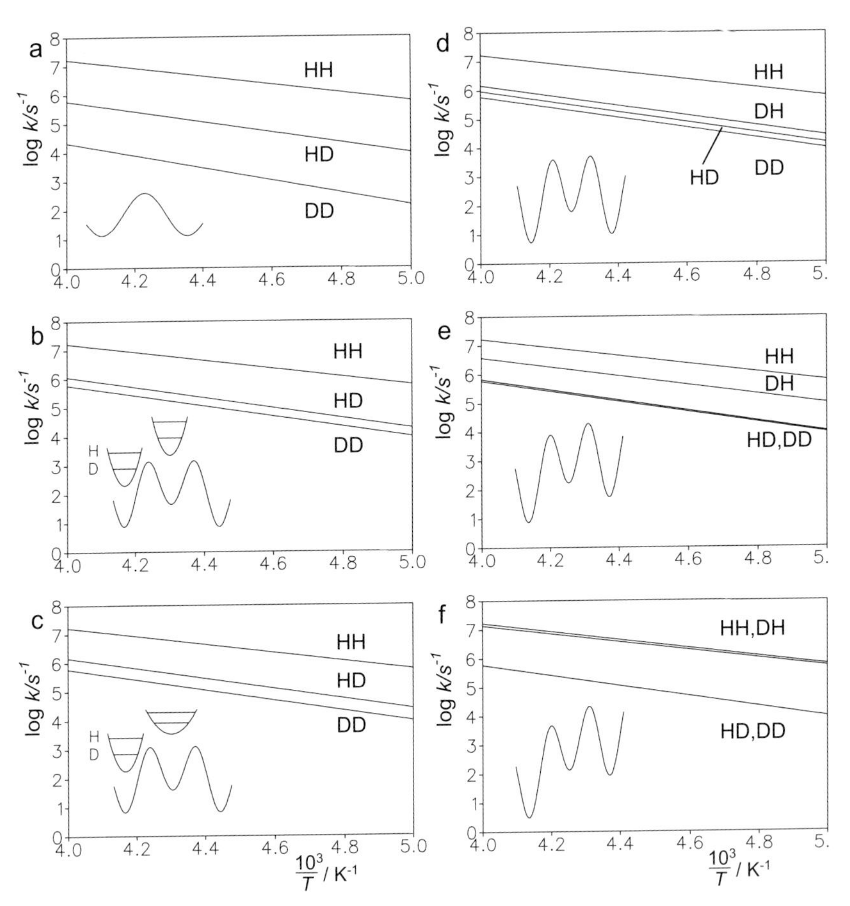

Figure 6.14 Arrhenius diagrams of a degenerate double hydron transfer using the following arbitrary parameters: $k^{HH} = 10^{13}\exp(-27.6\ \text{kJ mol}^{-1}/RT)$, $P = P_d = P_f = \exp(-7\ \text{kJ mol}^{-1}/RT)$. (a) Single barrier case. (b) Double barrier case with the H/D fractionation factor of dissociation $\phi_d = 1$. (c) As (b) but with an arbitrary value of $\phi_d = \phi_f = \exp(-0.92\ \text{kJ mol}^{-1} / RT)$;. Adapted from Ref. [26]. (d) to (f) Effect of increasing asymmetry on the kinetic HH/HD/DH/DD isotope effects.

Finally, note that the concerted and the stepwise HH-transfer constitute limiting cases and that various intermediate cases are possible. Rauhut et al. [51] have studied "plateau" reactions which are realized when the energy of the intermediate is raised, producing a very wide flat single barrier region which cannot be described in terms of an inverted parabola.

Meschede and Limbach [24c] have pointed out that compression of both hydrogen bonds of a double proton transfer system leads eventually to a single barrier situation, even if at large H-bond distances a stepwise reaction mechanism is realized.

6.2.7.2 Degenerate Stepwise HHH-transfer

The case of a stepwise triple proton transfer [25] is illustrated in Fig. 6.15. It can take place along different pathways via at least two metastable intermediates. For example, if hydrons {1}, {2} and {3} are transferred one after the other the reac-

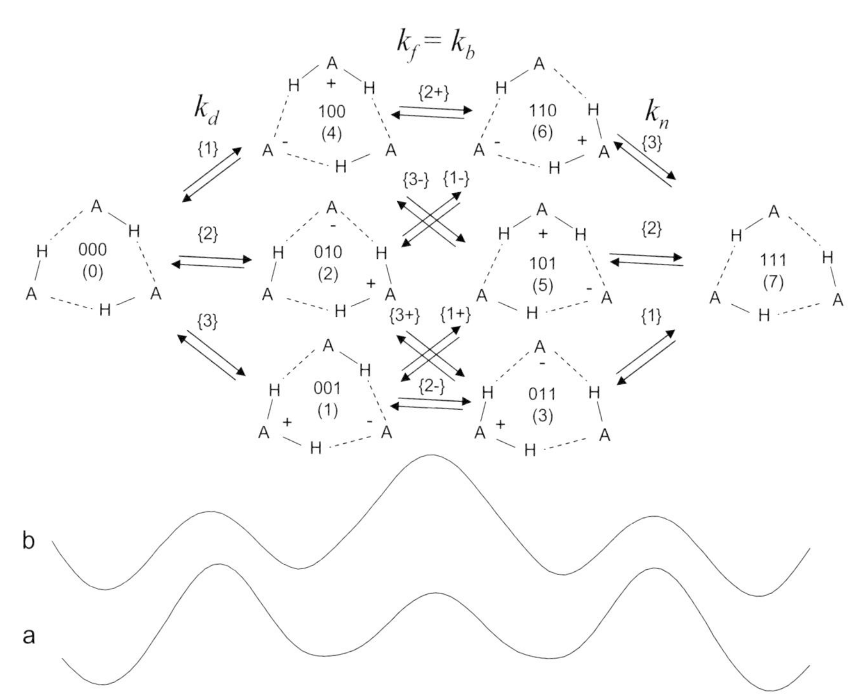

Figure 6.15 Degenerate triple-barrier triple hydron transfer involving two zwitterionic intermediates. A complete transfer consists of a dissociation step, one or more cation or anion "propagation" steps, characterized by the forward and backward rate constants $k_f = k_b$, and a neutralization step. These isotope dependent rate constants depend on whether a cation or an anion is propagated. Cation and anion propagation are included in the brackets indicating the hydron number transferred as a plus or minus sign in the brackets. The rate-limiting step may correspond to the propagation (a) or the dissociation (b). In each step only a single hydron loses zero-point energy at the configuration at the top of the barrier. Adapted from Ref. [26].

Table 6.2 Kinetic isotope effects of degenerate triple proton transfers according to Ref. [9].

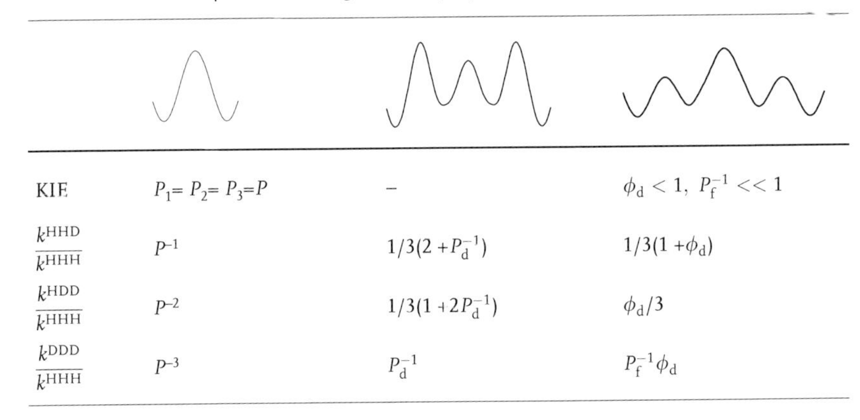

KIE	$P_1 = P_2 = P_3 = P$	–	$\phi_d < 1,\ P_f^{-1} << 1$
$\frac{k^{HHD}}{k^{HHH}}$	P^{-1}	$1/3(2 + P_d^{-1})$	$1/3(1 + \phi_d)$
$\frac{k^{HDD}}{k^{HHH}}$	P^{-2}	$1/3(1 + 2P_d^{-1})$	$\phi_d/3$
$\frac{k^{DDD}}{k^{HHH}}$	P^{-3}	P_d^{-1}	$P_f^{-1}\phi_d$

tion pathway is 000→100→110→111. As a degenerate reaction is considered the reaction energy profile must be symmetric, leaving only two cases: in the first case the central reaction step is rate-limiting and in the second case the first and third steps, as indicated schematically at the top of Fig. 6.15.

The derivation of expressions for the multiple kinetic isotope effects of the triple hydrogen transfer case is analogous to the HH-transfer but more tedious. Therefore, the reader is referrred to refs. [25] and [26]. The main results are included in Table 6.2. As in the case of the HH-transfer, the kinetic isotope effects derived for the stepwise transfers are valid in the presence of tunneling and are independent of the tunneling model used. By contrast, the kinetic isotope effects of the single barrier reaction are affected by tunneling.

The Arrhenius curves calculated for a small temperature range without tunneling contributions are illustrated in Fig. 6.16. For the single barrier case (Fig. 6.16(a)) Table 6.2 again predicts the rule of the geometric mean. Now, the overall isotope effect is larger than in the HH case, i.e. $k^{HHH}/k^{DDD} = P^3$, however the individual isotope effects are the same if P is kept constant.

In the stepwise triple barrier case with the central propagation as the rate-limiting step the overall isotope effect is given by $k^{HHH}/k^{DDD} = P_d$. It seems astonishing that the individual isotope effects $k^{HHH}/k^{HHD} = 3/2$ and $k^{HHH}/k^{HDD} = 3$ are given only by statistical values when P_d is large (Table 6.2). Thus, the Arrhenius curves of the HHH, HHD and HDD reaction are grouped closely together as illustrated in Fig. 6.16(b). This result can be explained as follows. When the central step is rate-limiting either H is transferred in this step with the probability 1/3 or D with the probability 2/3. The latter D-transfer does not contribute substantially to the overall rate constant, as H is transferred in the central step as fast as in the HHH reaction. Therefore, the HHH/HHD and the HHD/HDD isotope effects are given only by statistical factors.

When the propagation is the rate-limiting step (Fig. 6.16(c)) a similar phenomenon occurs. In the HDD reaction there are pathways where H is transferred both in the first and the final rate-limiting steps, e.g. the sequence 000 → 100 → 101

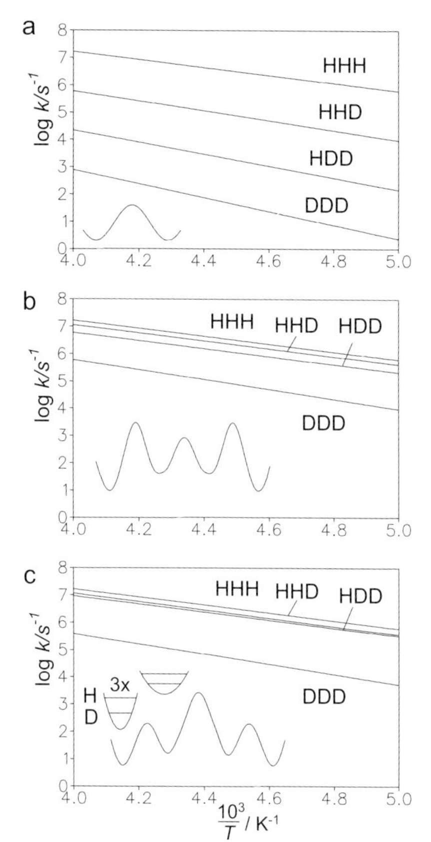

Figure 6.16 Arrhenius diagrams of a degenerate triple hydron transfer. For the HHH-transfer an arbitrary Arrhenius law $k^{HHH} = 10^{13}\exp(-27.6 \text{ kJ mol}^{-1}/RT)$ in s^{-1} was assumed, involving the single proton kinetic isotope effects $P = P_d = P_f = \exp(-7 \text{ kJ mol}^{-1}/RT)$. (a) Single barrier case. (b) Triple barrier case with the dissociation as rate-limiting step. The rate constants are independent of the fractionation factor ϕ_d of the dissociation. (c) Triple barrier case with $\phi_d = \exp(-0.92 \text{ kJ mol}^{-1}/RT)$ and equal rate constants $k_f = k_b$ for the cation and anion propagation, i.e. $a = 1$. Reproduced with permission from Ref. [26].

→ 001 → 011 → 111. Therefore, also in this case only replacement of the last proton by a deuteron exhibits a non-statistical kinetic isotope effect. However, the HDD process will experience isotopic fractionation of the dissociation process. Moreover, the DDD process is additionally affected by isotopic fractionation.

6.2.7.3 Degenerate Stepwise HHHH-transfer

In Ref. [26] three limiting cases were considered, i.e. a single barrier (Fig. 6.10), a double barrier (Fig. 6.17) and a quadruple-barrier reaction pathway (Fig. 6.18). The first process does not involve any intermediate. The second process consists essentially of consecutive double proton transfer steps, where each step involves a single barrier. There are two possibilities, either protons {1, 2} are transferred first, followed by protons {3, 4}, or vice versa, proceeding via the zwitterionic intermediates 1100 or 0011. It is again assumed that the intermediates can be treated as separate species, i.e. that there are no delocalized states involving different potential wells. This assumption will be realized when the barriers are large. Each reaction step is then characterized by an individual rate constant. The process con-

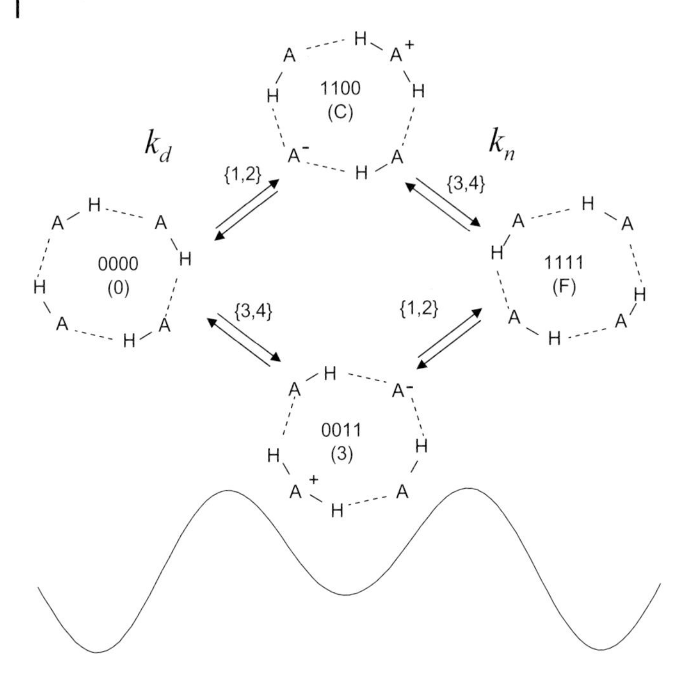

Figure 6.17 Degenerate two-barrier quadruple hydron transfer involving a single zwitterionic intermediate. Two hydrons lose zero-point energy in the configuration corresponding to the top of the barrier. Isotopic fractionation ϕ_d can occur for the dissociation. Reproduced with permission from Ref. [26].

sists of consecutive single proton transfer steps, corresponding to dissociation (k_d), two propagation steps (k_f and k_b), and a neutralization step (k_n). In the propagation steps one of the four protons is transferred and coupled to the transfer of either a positive or a negative charge (cation or anion propagation), indicated by + or – signs. Two possible intermediates, 0101 and 1010 are double zwitterions involving very high energies and are, therefore, neglected.

The expected kinetic isotope effects obtained by neglecting secondary kinetic isotope effects are summarized in Table 6.3. The results are visualized in Fig. 6.19, where again a simple Arrhenius law was assumed for the HHHH reaction with the same arbitrary parameters as for the HH and the HHH reactions in Fig. 6.14 and 6.16.

In the case of the single-barrier mechanism, all four hydrons are in flight in the transition state. Subsequent replacement of H by D involves similar primary kinetic isotope effects P, leading to equally spaced Arrhenius curves of the isotopic reactions in Fig. 6.19(a), with an overall kinetic isotope effect of $k^{HHHH}/k^{DDDD} \approx P^4$. This result is analogous to the single-barrier HH and the HHH-transfer cases discussed above. Note that, generally, the transfer of n hydrons is expected to give rise to an overall kinetic isotope effect of P^n [26].

For the two-barrier HHHH case it was assumed that the first and second primary kinetic isotope effects of the double proton transfer in the dissociation steps

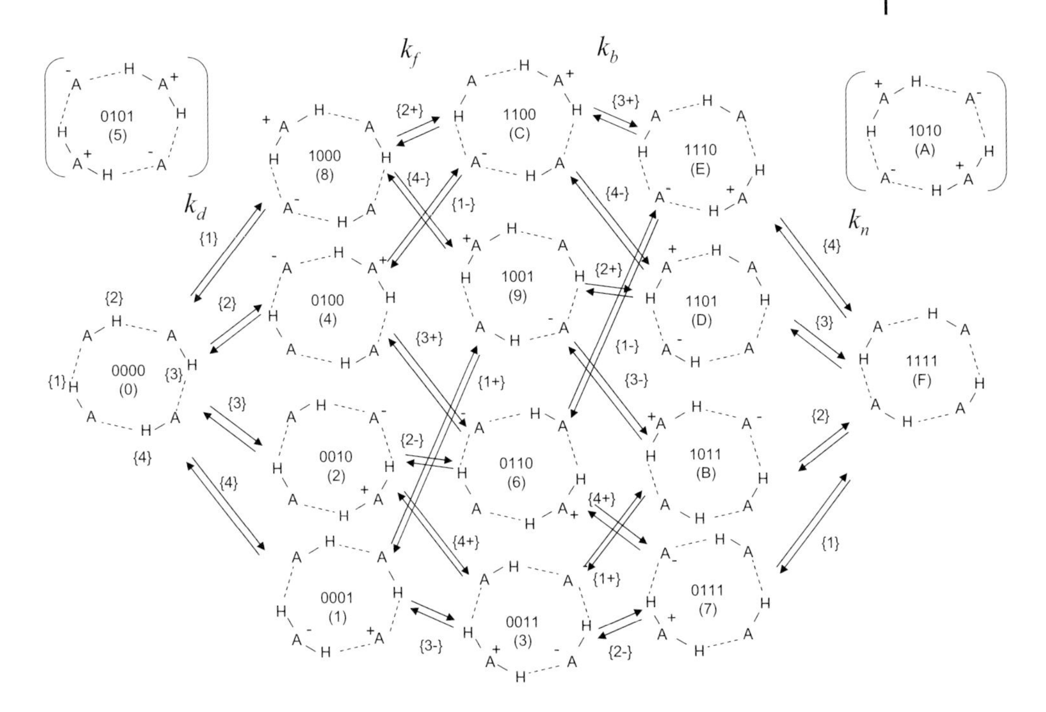

Figure 6.18 Degenerate quadruple-barrier quadruple hydron transfer. A complete transfer consists of a dissociation step, two or more propagation steps and a neutralization step. The double-cation–double-anion intermediates 0101 and 1010 are not further taken into account because of their high Couloumb energy. Reproduced with permission from Ref. [26].

P_{d1} and P_{d2} are equal. ϕ_d represents the single H/D fractionation factor of the dissociation step, corresponding to the equilibrium constant of the formal reaction

$$\text{initial state (D)} + \text{intermediate state (H)} \rightleftharpoons \text{initial state (H)} + \text{intermediate state (D)}. \tag{6.33}$$

The results are depicted in Fig. 6.19(b) and (c). Whereas in Fig. 6.19(b) $\phi_d=1$, in Fig. 6.19(c) it was assumed that $\phi_d = \exp(-0.92 \text{ kJ mol}^{-1}/RT) < 1$. This fractionation factor takes into account the fact that the zero-point energies of each proton in the intermediate may be reduced due to low-frequency shifts of the proton vibrations, as expected for an increase in the H-bond strength in the intermediate.

One observes three groups of Arrhenius curves, i.e. the HHHH curve, the group of the HHHD and the HDHD curves, and the group of the HHDD, HDDD and DDDD curves. Within each group the differences are small. They are further attenuated by isotope fractionation (Fig. 6.19c). The overall kinetic isotope effect, given by $k^{HHHH}/k^{DDDD} = P_d^2$, is typical for a concerted double proton transfer reaction. It is interesting to note that replacement of the first H by D already leads to a

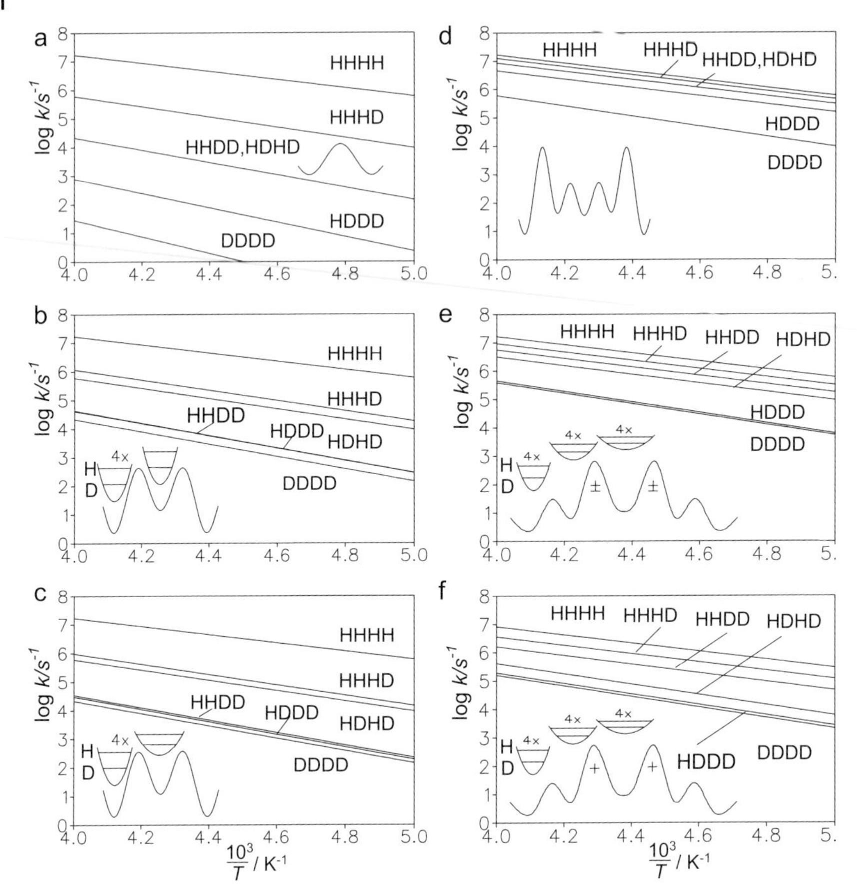

Figure 6.19 Simulated Arrhenius diagrams of a degenerate quadruple hydron transfer. Arrhenius laws are assumed for the HHHH-transfer. (a) Single-barrier case. (b) Double-barrier case with $\phi_d = 1$. (c) Double-barrier case with $\phi_d = \exp(-0.92 \text{ kJ mol}^{-1}/RT)$. (d) Quadruple-barrier case with dissociation as rate-limiting step. (e) Quadruple-barrier case with propagation as rate-limiting step, equal cation and anion propagation $a = 1$ (as indicated by the ± signs), $\phi_d = \phi_f = \exp(-0.92 \text{ kJ mol}^{-1}/RT)$. (f) Quadruple-barrier case with propagation as rate-limiting step, only cation propagation a=0 (as indicated by the + signs), $\phi_d = \phi_f = \exp(-0.92 \text{ kJ mol}^{-1}/RT)$. Adapted from Ref. [26].

substantial isotope effect of $k^{\mathrm{HHHH}}/k^{\mathrm{HHHD}} = P_d/2$ when isotopic fractionation is absent.

In the quadruple-barrier case one needs to distinguish whether dissociation/neutralization or propagation are the rate-limiting steps. Furthermore, a parameter a is needed describing the ratio of the forward reaction rate constants of the anion and the cation propagation, i.e.

$$a = \frac{k_f^{i-}}{k_f^{i+}} \cong \frac{k_b^{i-}}{k_b^{i+}} \tag{6.34}$$

Table 6.3 Kinetic isotope effects of degenerate quadruple proton transfers according to Ref. [26].

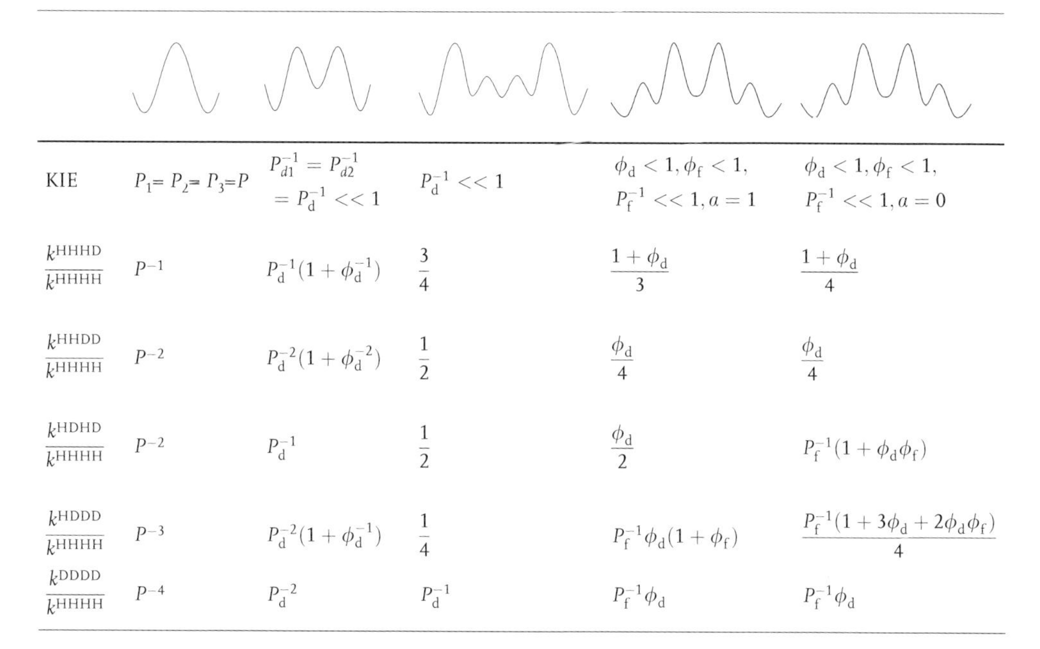

KIE	$P_1 = P_2 = P_3 = P$	$P_{d1}^{-1} = P_{d2}^{-1} = P_d^{-1} << 1$	$P_d^{-1} << 1$	$\phi_d < 1, \phi_f < 1, P_f^{-1} << 1, a = 1$	$\phi_d < 1, \phi_f < 1, P_f^{-1} << 1, a = 0$
$\frac{k^{HHHD}}{k^{HHHH}}$	P^{-1}	$P_d^{-1}(1 + \phi_d^{-1})$	$\frac{3}{4}$	$\frac{1 + \phi_d}{3}$	$\frac{1 + \phi_d}{4}$
$\frac{k^{HHDD}}{k^{HHHH}}$	P^{-2}	$P_d^{-2}(1 + \phi_d^{-2})$	$\frac{1}{2}$	$\frac{\phi_d}{4}$	$\frac{\phi_d}{4}$
$\frac{k^{HDHD}}{k^{HHHH}}$	P^{-2}	P_d^{-1}	$\frac{1}{2}$	$\frac{\phi_d}{2}$	$P_f^{-1}(1 + \phi_d \phi_f)$
$\frac{k^{HDDD}}{k^{HHHH}}$	P^{-3}	$P_d^{-2}(1 + \phi_d^{-1})$	$\frac{1}{4}$	$P_f^{-1} \phi_d (1 + \phi_f)$	$\frac{P_f^{-1}(1 + 3\phi_d + 2\phi_d \phi_f)}{4}$
$\frac{k^{DDDD}}{k^{HHHH}}$	P^{-4}	P_d^{-2}	P_d^{-1}	$P_f^{-1} \phi_d$	$P_f^{-1} \phi_d$

where i–and i+ indicate the transfer of hydron i coupled to the transfer of a negative or positive charge. In Table 6.3 P_f and P_b constitute the kinetic isotope effects of the two propagation steps; ϕ_d corresponds to the fractionation factor of the dissociation.

When dissociation and neutralization are the rate-limiting steps (Fig. 6.19(d)), an overall kinetic isotope effect of $k^{HHHH}/k^{DDDD} = P_d$ is expected, again typical for a single H-transfer. Surprisingly, only replacement of the last H by D is affected by P_d. As in the stepwise HHH case, this effect arises again from the possibility of reaction pathways involving transfer of H in the rate-limiting steps, even in the HDDD reaction.

In the case of the propagation as rate-limiting step, an overall isotope effect of $k^{HHHH}/k^{DDDD} = P_f \phi_d$ is expected. The rate constants of the other isotopic reactions are between the values of k^{HHHH} and k^{DDDD}. In Fig. 6.19(e) it is assumed that both cation and anion propagation exhibit the same rate constants, and in Fig. 6.19(f) it is assumed that anion propagation does not contribute fundamentally to the rate constants. This effect leads to a reduction in k^{HDHD}.

6.2.8 Hydrogen Transfers Involving Pre-equilibria

According to Eigen's scheme of H-transfer [52] in solution the reaction partners have first to meet in order to react. For an intermolecular H-transfer from AH to B one can write

$$\text{A–H} + \text{B} \overset{K}{\rightleftharpoons} \text{A–H}\cdots\text{B} \overset{k}{\rightleftharpoons} \text{A}\cdots\text{H–B} \rightleftharpoons \text{A} + \text{HB} \tag{6.35}$$

In Eq. (6.35) electrical charges are omitted. k represents the intrinsic forward rate constant. The pre-equilibrium constant is given by

$$K = \frac{c_{\text{AB}}}{c_{\text{A}} c_{\text{B}}} \tag{6.36}$$

where c_{A} and c_{B} represent the concentrations of the reactants AH and B, and c_{AB} the concentration of the reactive complex AHB. The total concentrations are then given by $C_{\text{A}} = c_{\text{A}} + c_{\text{AB}}$ and $C_{\text{B}} = c_{\text{B}} + c_{\text{AB}}$. The experimentally accessible forward pseudo-first order rate constant is then given by [53]

$$\begin{aligned} k_{\text{obs}} &= -\frac{\text{d}C_{\text{A}}}{C_{\text{A}}\text{d}t} = -\frac{d(c_{\text{A}} + c_{\text{AB}})}{C_{\text{A}}\text{d}t} = -\frac{dc_{\text{AB}}}{C_{\text{A}}\text{d}t} = \frac{kc_{\text{AB}}}{C_{\text{A}}} \\ &= \frac{k}{2KC_{\text{A}}}\left[-\sqrt{(C_{\text{A}} - C_{\text{B}})^2K^2 + 1 + 2K(C_{\text{A}} + C_{\text{B}})} + 1 + K(C_{\text{A}} + C_{\text{B}})\right] \end{aligned} \tag{6.37}$$

Equation (6.37) predicts changes in the apparent reaction order as a function of C_{A} and C_{B}. It can easily be shown that

$$k_{\text{obs}} = kKC_{\text{B}} \text{ for } K \ll 1 \text{ and } k_{\text{obs}} = k \text{ for } K \gg 1 \tag{6.38}$$

For the case of two proton donors AH of the same kind which exchange two protons in a cyclic dimer $(\text{AH})_2$ the following equation has been derived

$$k_{\text{AA}} = \frac{k}{4KC_{\text{A}}}\left[-\sqrt{1 + 8KC_{\text{A}}} + 1 + 4KC_{\text{A}}\right] \tag{6.39}$$

which can also be obtained from Eq. (6.37) by setting $C_{\text{A}} = C_{\text{B}}$ and replacing K by $2K$ [24]. Again,

$$k_{\text{obs}} = 2kKC_{\text{B}} \text{ for } K \ll 1 \text{ and } k_{\text{obs}} = k \text{ for } K \gg 1 \tag{6.40}$$

Finally, the intramolecular case is interesting, where AH and B are functional groups of the same molecule. The equilibrium constant between the non-reactive form AH + B and the reactive form AHB is given by

$$K = \frac{c_{\text{R}}}{c_{\text{NR}}} \tag{6.41}$$

With the total concentration $C = c_{\text{NR}} + c_{\text{R}}$ it follows that

$$c_{\text{R}} = \frac{KC}{1 + K} \tag{6.42}$$

Often, only the sum of the concentrations of NR and R is measured, i.e. the kinetics cannot distinguish between NR and R. As the interconversion between

NR and R is assumed to be fast, the observed first order or pseudo-first order rate constant is then given by

$$k_{obs} = -\frac{1}{C}\frac{dc_R}{dt} = \frac{kc_R}{C} = \frac{kK}{1+K} \tag{6.43}$$

In the case where $K \gg 1$ it follows that

$$k_{obs} = k \tag{6.44}$$

By contrast, if $K \ll 1$ it follows that

$$k_{obs} = kK \tag{6.45}$$

In other words, the observed rate constants depend in a similar way on the equilibrium constants in the three cases discussed. In order to have an impression of the effects on the Arrhenius curves let us discuss the intramolecular case given by Eq. (6.43).

The temperature dependence of the pre-equilibrium constant is given by

$$K = \exp(-\Delta H/RT + \Delta S/R) \tag{6.46}$$

where ΔH and ΔS represent the enthalpy and entropy of the pre-equilibrium. Let us assume a simple Arrhenius law for the intrinsic H-transfer step

$$k = A\exp(-E_a/RT) \tag{6.47}$$

with the arbitrary parameters $A = 10^{13}$ s^{-1} and $E_a = 30$ kJ mol^{-1}, represented by the dashed lines in the Arrhenius diagrams of Fig. 6.20. The effective Arrhenius curves calculated using Eq. (6.43) are represented by the solid lines.

Figure 6.20(a) depicts the case where the formation of the reactive state involves a negative enthalpy and a negative entropy, as expected for a hydrogen bond association between the reaction partners AH and B. Thus, the reacting state predominates at low temperatures where $k_{obs} = k$. The true Arrhenius curve is then measured, with normal pre-exponential factors. At high temperatures, however, non-reactive state dominates and $k_{obs} = kK$. As $K \ll 1$ in this region, the observed Arrhenius curves exhibits a convex curvature and unusually small pre-exponential factors. The effective activation energy is given by $E_a - |\Delta H|$.

Figure 6.20(b) depicts the case where the formation of the reactive state involves a positive entropy and enthalpy. Such a case could happen if the reaction partners AH and B are involved in strong interactions with other species. For example, AH could be hydrogen bonded to any proton acceptor, or B to any proton donor, which requires this interaction to be broken before the partners can react. Now, the reacting state predominates at high temperatures and the non-reactive state at low temperatures. Only at high temperatures is the true Arrhenius curve measured, exhibiting a normal pre-exponential factor of about 13. At low temperatures, the

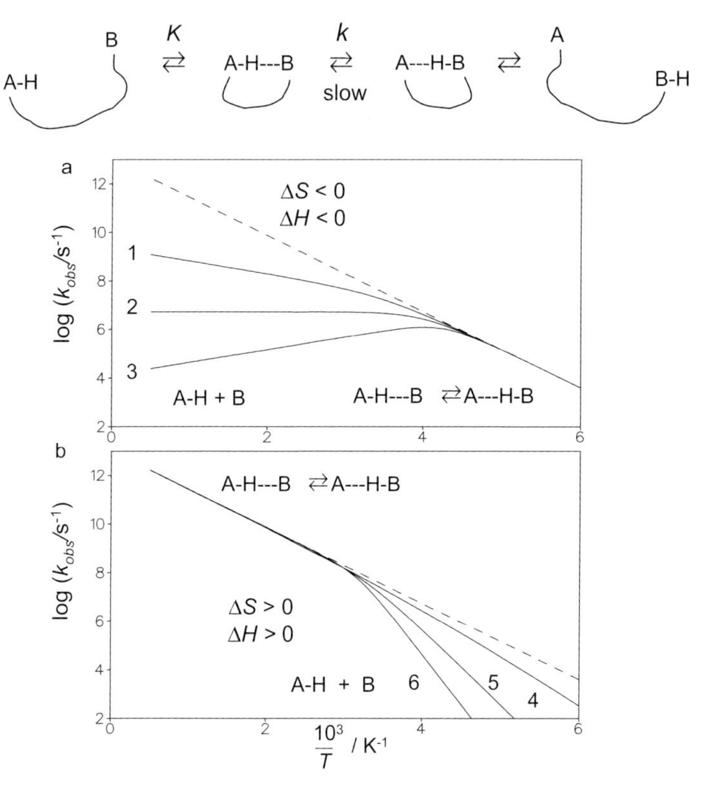

Figure 6.20 Arrhenius curves of a H-transfer in the presence of a pre-equilibrium. Arbitrary parameters of the Arrhenius curves in the reactive complex: log$A$ = 13, and E_a = 30 kJ mol^{-1}. Parameters of the formation of the active complex: (1) ΔH = – 20 kJ mol^{-1}, ΔS = – 70 J K^{-1} mol^{-1}; (2) ΔH = – 30 kJ mol^{-1}, ΔS = – 120 J K^{-1} mol^{-1}; (3) ΔH = – 40 kJ mol^{-1}, ΔS = – 170 J K^{-1} mol^{-1} ; (4) ΔH = 10 kJ mol^{-1}, ΔS = 40 J K^{-1} mol^{-1}; (5) ΔH = 30 kJ mol^{-1}, ΔS = 100 J K^{-1} mol^{-1}; (6) ΔH = 50 kJ mol^{-1}, ΔS = 160 J K^{-1} mol^{-1}. Adapted from Ref. [53].

observed rate constants are slower than the intrinsic ones, the effective activation energy is given by $E_a + |\Delta H|$. In addition, the observed pre-exponential factor is unusually large.

6.3 Applications

In this section various hydrogen transfer systems are reviewed for which Arrhenius curves of the different isotopic reactions are available over a large temperature range. Mainly the systems are discussed exhibiting degenerate hydrogen transfers which could be studied by dynamic NMR. The main question is how the reaction properties are related to the molecular structure.

Table 6.4 Bell–Limbach tunneling model parameters of various H-transfers[1])

System	Ref	k_{298K} / s^{-1}	KIE_{298K}	E_m / kJ mol^{-1}	ΔH / kJ mol^{-1}	ΔS / J K^{-1} mol^{-1}	log (A/s^{-1})	E_d / kJ mol^{-1}	Δm / a.m.u.	$2a$ / Å	$\Delta\varepsilon$ / kJ mol^{-1}
tetraphenylporphyrin organic solvents/solid state	18b	5600	HH/HD 16 HH/DH 4 HH/DD	26.8	–	–	12.9	29.3	1.5	0.48	HD 5.0
porphyrin organic solvents/solid state	18d, 18e	16000	HH/DD 11.5 HH/HD 6.5 HD/DD 1.9 H/D 11.4 D/T 3.4 H/T 39	22.7	–	–	12.6	28.7	2.5 [1.5]	0.48 [0.68]	HD 4.9 [4.95] DT 3.0 [2.2]
porphyrin anion organic solvent/solid phosphazene matrix	23b	10^5	H/D 16.5 D/T 3 H/T 49.6	10.0	–	–	12.6	34.3	0 [1.5]	0.87 [0.78]	HD 6.5 [7.74] DT 4.2 [3.8]
phthalocyanine α-form solid	66b	1.1×10^5	–	29.3	–	–	12.9	17.6	1.5	0.48	HD 5.9
phthalocyanine β-form solid	66b	4.3×10^5	–	24.7	–	–	12.6	15.5	1.5	0.48	HD 5.9

1) Square brackets indicate values published previously.

System	Ref	k_{298K} / s^{-1}	KIE_{298K}	E_m / kJ mol^{-1}	ΔH / kJ mol^{-1}	ΔS / J K^{-1} mol^{-1}	log (A/s^{-1})	E_d / kJ mol^{-1}	Δm / a.m.u.	$2a$ / Å	$\Delta\varepsilon$ / kJ mol^{-1}
acetyl-porphyrin organic solvents AC → AD	20b	2870	HH/HD 16 HH/DH 4 HD/DD 1.2 DH/DD 4.5	24.3	–	–	12.6	31.0	1.5	0.48	HD 5.9
trans-cis and cis-trans step BD → AD				23.0	–	–	12.6	22.6	1.5	0.48	HD 5.9
tetraphenyl-chlorin organic solvents	19b	15	HH/HD 2.6	33.5	–	–	12.6	36.8	1.5	0.48	HD 5.9
tetraphenyliso-bacterio-chlorin organic solvents k_1	19a	4×10^5	–	16.7	–	–	12.6	26.4	1.5	0.48	HD 5.9
tetraphenyliso-bacterio-chlorin organic solvents k_2	19a	350	–	29.3	–	–	12.6	29.3	1.5	0.6	HD 5.9
indigodiimin HH-transfer organic solvent	68	4×10^8	–	17.6	–	–	12.6	29.7	1.5	0.4	HD 5.9

System	Ref	k_{298K} / s^{-1}	KIE_{298K}	E_m / kJ mol^{-1}	ΔH / kJ mol^{-1}	ΔS / J K^{-1} mol^{-1}	log (A/s^{-1})	E_d / kJ mol^{-1}	Δm / a.m.u.	$2a$ / Å	$\Delta\varepsilon$ / kJ mol^{-1}
indigodiimin NH_2 rotation organic solvent	68	2.6×10^{9}	–	16.7	–	–	12.6	16.3	1.5	0.48	HD 5.9
porphycene solid state	70	5×10^{7}	–	5.9	–	–	12.6	24.7	1.5	0.48	HD 5.9
DTAA solid state		2.5×10^{6}	–	20.5	–	–	12.6	17.6	3	0.34	HD 1.05
TTAA solid state	49	3×10^{9}	1.8	3.4 [2.9]	–	–	12.6 [12.4]	15.1 [14.2]	3 [3]	0.17 [0.50]	1.05 [3.0]
$(PhCOOH)_2$ solid state	74	9×10^{10}		0.84			11.6	5.4	1.8	0.48	
(PhCOOH/ PhCOOD) solid state	74		HH/HD ≈2.4 (298 K) HH/HD 24 (15 K)	0.84			11.6	7.5	1.8	0.52	
$(PhCOOD)_2$ solid state	74		HH/HD ≈ 6 (298 K) HD/DD 21 (15 K)	1.0			11.6	12.1	1.8	0.44	
DPBrP crystal	27	6500	HH/HD 5 HD/DD 5 HH/DD 25	5.6	–	–	12.65	47.5	2.3	0.55	HH/HD HD/DD

System	Ref	k_{298K} / s^{-1}	KIE_{298K}	E_m / kJ mol^{-1}	ΔH / kJ mol^{-1}	ΔS / J K^{-1} mol^{-1}	log (A/s^{-1})	E_d / kJ mol^{-1}	Δm / a.m.u.	$2a$ / Å	$\Delta\varepsilon$ / kJ mol^{-1}
DMP crystal	25, 27	990	HHH/HHD 3.8 HHD/HDD 3.7 HDD/DDD 3.4 HHH/DDD 47	8.4	–	–	12.3	48.1	2.8	0.43	HHH/HHD 2.7 HHD/HDD 2.7 HDD/DDD 2.7
DPP crystal	27	10000	HHHH/HDLL 3, HDLL/DDLL 4 HDLL/DDLL 4 HHHH/DDDD 12	19	–	–	12.6	32.5	4	0.384	HHLL/HDLL 2.7 HDLL/ DDLL 2.7
azophenine in organic solvents	21	720	HH/HD 4.1 HD/DD 1.4	27.2	–	–	12.6	30.1	1.5	0.6	HD 3.8
tetraphenyl-oxalamidine in CD_2Cl_2	22a	1500	HH/HD 3	44.4	–	–	12.6	24	1.5	0.42	HD 2.5
oxalamidine OA7 in methyl-cyclohexane	22b	14	HH/HD 3.1 HD/DD 1.5	52.7	–	–	12.6	27.2	1.5	0.2	HD 2.9
oxalamidine OA7 in acetonitrile	22b	75	HH/HD 3.2 HD/DD 1.6	52.7	–	–	12.6	16.7	1.5	0.2	HD 3.8
F-amidine in THF	24	10^7	HH/HD 4.1 HD/DD 3.5 HH/DD 14.3	5.4			12.2	26.4	1.05	0.55	HH/HD 3.0 HD/DD 3.0

System	Ref	k_{298K} / s^{-1}	KIE_{298K}	E_m / kJ mol^{-1}	ΔH / kJ mol^{-1}	ΔS / J K^{-1} mol^{-1}	log (A/s^{-1})	E_d / kJ mol^{-1}	Δm / a.m.u.	$2a$ / Å	$\Delta\varepsilon$ / kJ mol^{-1}
Me-BO	83	2×10^9	H/D 14.5	0.293	-9	-60	12.6	19.7	2.1	0.29	5.44
Ingold radical	8	$\approx 10^4$	H/D 18	4.2	-19	-85	12.6	50.2	1.9	0.34	6.7
2-hydroxyphen-oxyl radical CCl_4+ dioxane	86	2×10^7	H/D 56	0.0	21	38	12.6	27.2	0	0.3	6.7
di-tertbutyl-2-hydroxyphenox-yl radical in heptane	85	4×10^9	H/D 9.8	1.26	–	–	12.6	23.9	1	0.17	3.3
CH_3COOH+ CH_3OH in THF	9b	750	HH/HD 5.1 HD/DD 3.1 HH/DD15.5	16.5	16.5	-42	12.6 (10.4)	36	0	0.44	HH/HD 0.64 HD/DD 0.64
2 CH_3COOH+ CH_3OH in THF	9b	3200	HHH/DDD 11.5 HHH/DHH 2.1	27.2	27.2	-34	12.6 (11)	33.5	0	0.2	0.25
Thermophilic Dehydrogenase	91	90	H/D 4.8	0.0	96	326	12.6	67.0	0	0.515	0.0
pure methanol and calix[4]ar-ene	93, 74	10^{11}	–	0.13	–	–	12.4	8.8	2.8	0.49	–

According to Eigen's scheme of H-transfer (Eq. (6.35)) it can be divided into two steps, i.e. a diffusion step and an intrinsic H-transfer step in a hydrogen bonded complex. This picture can be specified further, for example by introduction of heavy atom reorganization in the intrinsic H-transfer step, either before or during the actual H-transfer.

Therefore, in the first part of this section, intramolecular hydrogen transfers or intermolecular hydrogen transfers in preformed hydrogen bonded complexes in the solid state which are coupled only to minor heavy atom motions are discussed. H-transfers coupled to major heavy atom motions will then be treated in the second part; they include pre-equilibria, hydrogen bond switches, conformational changes, solvent motions etc.

For a better comparison, the Arrhenius curves of all hydrogen transfers discussed in this section have been recalculated for this review using the Bell–Limbach tunneling model described in the theoretical section. Some systems have already been presented recently in a mini-review [54]. The parameters used are assembled in Table 6.4. Finally, note that in all cases where hydrogen isotopes are transferred from and to nitrogen the compounds had to be enriched for NMR measurements with the ^{15}N isotope.

6.3.1 H-transfers Coupled to Minor Heavy Atom Motions

6.3.1.1 Symmetric Porphyrins and Porphyrin Analogs

The tautomerism of porphyrin and of its analogs is illustrated in Fig. 6.21. Experimental aspects have been reviewed recently by Elguero et al. [55] and theoretical aspects by Maity et al. [56].

The thermal tautomerism of *meso*-tetraphenylporphyrin (TPP) – which is more soluble than the parent compound porphyrin – was discovered by Storm and Teklu [57] using liquid state ^{1}H NMR. A large kinetic HH/DD isotope effect was observed which was interpreted in terms of a concerted double proton transfer. Hennig et al. [18a, 58] established in the late 70s and early 80s of the last century the intramolecular pathway of the reaction, measured the HH and DD reaction rates and later also the HD rates of TPP over a wide temperature range. The data were interpreted in terms of tunneling. As NMR methods to obtain rate constants were still developing, the low-temperature rate constants were overestimated by Hennig et al. [58] as criticized by Stilbs and Moseley [59]. As a consequence, the methods used were improved and led to a novel NMR pulse sequence based on "magnetization transfer in the rotating frame" [60], referred to later as "CAMELSPIN" [61] and then as "ROESY" [62], one of the most used NMR pulse sequences nowadays.

In the late 70s and early 80s, tunneling theories indicated a preference for tunneling in symmetric double wells vs. asymmetric wells, supporting the concerted double proton transfer [63]. Thus, it seemed at that time that tunneling observed experimentally was only compatible with a concerted reaction pathway, supported

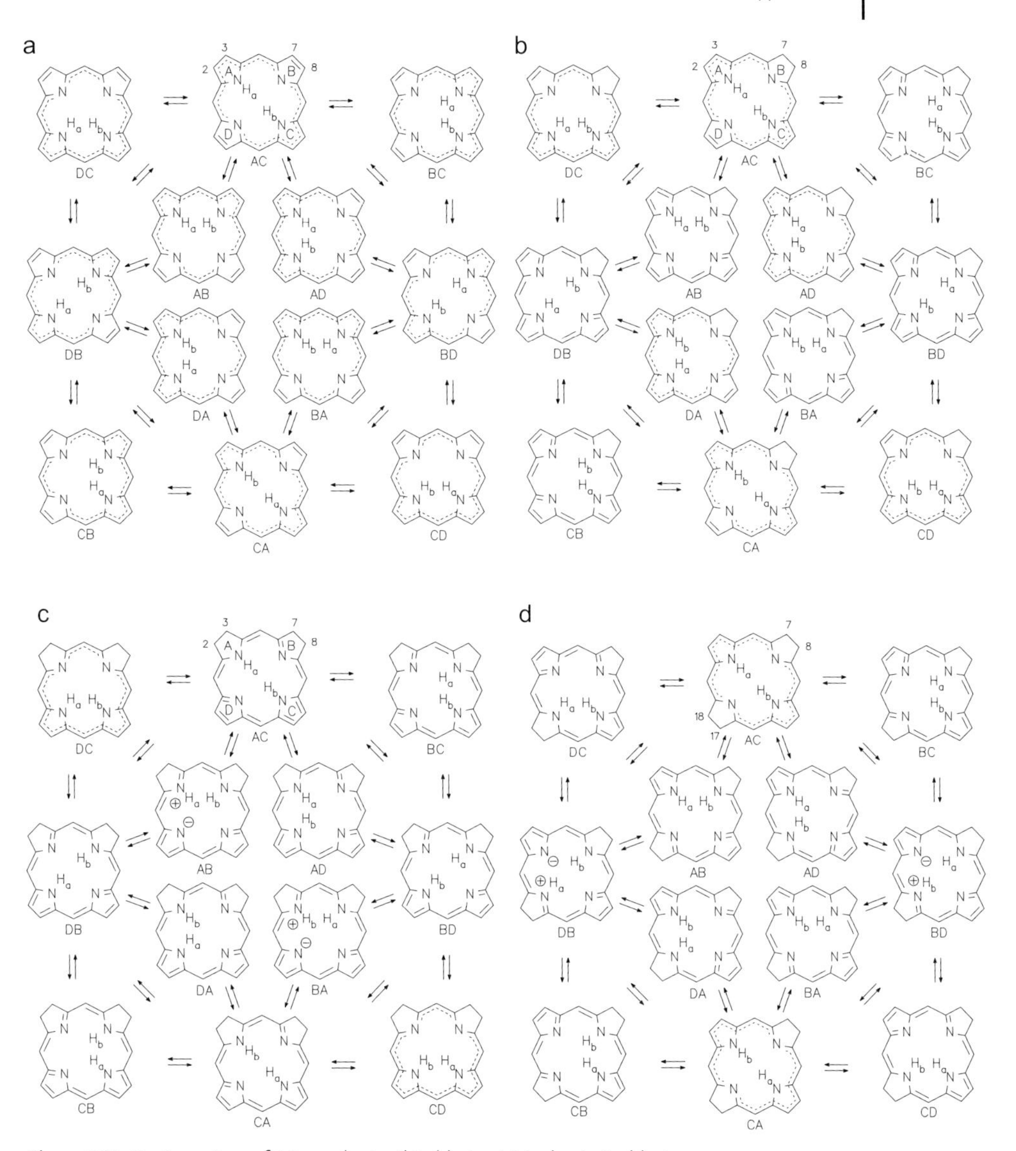

Figure 6.21 Tautomerism of (a) porphyrin, (b) chlorin, (c) isobacteriochlorin and (d) bacteriochlorin. Adapted from Ref. [19a].

by the finding that the two NH-stretches in the ground state of porphyrins are coupled [63]. In a number of papers, a stepwise transfer was proposed by Sarai [64], promoted by specific vibrations of the porphyrin skeleton which lower the N···N distance for H-transfer. In parallel, quantum-mechanical calculations indicated that cis-tautomers represent metastable intermediates, as indicated in Fig. 6.21(a). To our knowledge, the most recent *ab initio* calculation was published by Maity et al. [65]. Further theoretical progress will be discussed later.

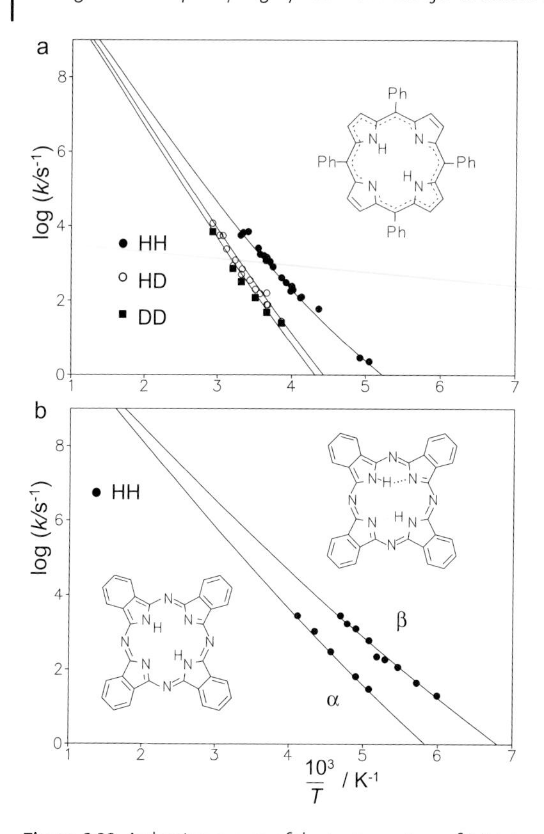

Figure 6.22 Arrhenius curves of the tautomerism of (a) tetraphenylporphyrin in the liquid and solid state [18b] and (b) of the α- and β-forms of solid phthalocyanin [66b]. The Arrhenius curves were calculated using the parameters listed in Table 6.4.

In the meantime, Limbach et al. [58b] showed, for *meso*-tetraarylporphyrins, that the transfer also takes place in the solid state [58c], but it was found that the degeneracy of the reaction can be lifted by solid state interactions. Thus, the tautomerism of *meso*-tetratolylporphyrin was found to be degenerate in the solid state, but the tautomerism of triclinic TPP was perturbed in the sense that the trans-tautomers BD and DB exhibited a larger energy than the trans-tautomers AC and CA (Fig. 6.21(a)). However, by co-crystallization with a small amount of Ni-TPP a tetragonal structure was obtained in which the degeneracy of the tautomerism was restored, as shown by Schlabach et al. [18b]. No difference between the rate constants of the degenerate reactions in the liquid and the solid state could then be observed. Schlabach et al. also published the final set of rate constants for the HH, HD, and the DD tautomerism of TPP depicted in Fig. 6.22(a). The kinetic isotope effects were interpreted in terms of Eq. (6.31) for degenerate stepwise HH-transfers in the absence of isotopic fractionation between initial and intermediate state and of secondary isotope effects, re-written as

$$k^{HH} = k^{H}, k^{HD} = \left[\frac{2k^{D}}{1 + k^{D}/k^{H}}\right], k^{DD} = k^{D} \quad (6.48)$$

Here k^H and k^D represent the single H-transfer rate constants of the formation of the individual intermediates in Fig. 6.21(a). Equations (6.31) and (6.48) had already been discussed by Limbach et al. [58] but used only after independent confirmation in the cases of azophenine and oxalamidines [21, 22], discussed below. Equation (6.48) was visualized in Fig. 6.12 and 6.14(b). The reaction energy profile of the HH-transfer involves two transition states of equal height. Thus, the product side is reached only with probability $1/2$ as the internal return to the initial state also exhibits the same probability. The same is true for the DD reaction, only the effective barriers are larger. However, the symmetry is destroyed in the HD reaction. The rate-limiting step is the D-transfer which involves the same barrier as the corresponding process of the DD reaction. But as there is only a single barrier of this type in contrast to the DD reaction the HD reaction is about 2 times faster than the DD reaction.

The fit of the experimental data to Eq. (6.48) is very satisfactory, as illustrated in Fig. 6.22(a), where the solid lines were recalculated here using the Bell–Limbach model, with the parameters included in Table 6.4. This result also means that there is no substantial decrease in the zero-point energies of the two protons in the cis-intermediate states as compared to the initial and final trans-states, as this would increase the HD/DD isotope effect beyond the value of 2 as was illustrated in Fig. 6.14(c).

The solid state tautomerism of solid phthalocyanine was discovered [66a] and studied by Limbach et al. [66b]. As illustrated in Fig. 6.22(b), there are two forms which differ in the arrangement of the central nitrogen atoms. They are arranged in a square in the α- form but in a rectangular way in the β-form [66b]. Thus, the latter contains two weak inner NHN-hydrogen bonds.

The reaction rates observed are substantially increased as compared to TPP; the increase is larger for the β-form as compared to the α- form (Fig. 6.22(b)). Kinetic H/D isotope effects have not yet been studied. The difference in the reaction kinetics of the two forms has been explained as follows. The observed tautomerism in the α- form was interpreted with a circular tautomerism as illustrated in Fig. 6.21(a), with similar transfer rates for the formation of all intermediates. However, the observed transfer in the β-form was assigned to a local HH-transfer within the two intramolecular hydrogen bonds which led to an extra increase in the rate constants.

The thermal tautomerism of the unsubstituted solid parent compound porphyrin was discovered by Wehrle et al. [18c]. Again, the degeneracy of the tautomerism was not lifted. The HH, HD, DD rate constants in the liquid and the solid state were determined by Braun et al. [18d] leading to the Arrhenius diagram of Fig. 6.23(b). Again, no kinetic liquid–solid state effects could be observed. The motivation of these studies was to elucidate the influence of substituents on the tautomerism and to facilitate the comparison with theoretical studies which are generally performed on the non-substituted parent compound. In fact, although the observed isotopic pattern is similar to that of TPP, it is found that the reaction in the parent compound is considerably faster than in TPP, but slower than in phthalocyanine (Table 6.4). These findings indicate that NHN-hydrogen bond compression is necessary for the HH-transfer in porphyrins and its analogs to occur, and that this compression is hindered by substituents in the *meso*-positions but facilitated by replacement of *meso*-carbon by *meso*-nitrogen atoms.

As matrix effects on the tautomerism of porphyrin are absent, it is justified to combine the data obtained by NMR with those obtained at low temperatures using optical methods for porphyrin embedded in solid hexane [67], leading to the full Arrhenius diagram depicted in Fig. 6.23(a) [18d]. In contrast, rate constants obtained for substituted and unsubstituted porphyrins should not be included in a single Arrhenius diagram.

Before the full Arrhenius diagram is discussed in detail, let us first include the results of a subsequent study of Braun et al. [18e] who measured also the rate constants k^{HT} and k^{TT} using liquid state ^{3}H NMR of tritiated porphyrin dissolved in toluene. In order to discuss the new data it is convenient to convert the rate constants k^{LL} into the rate constants k^L using Eq. (6.31) which is naturally valid also

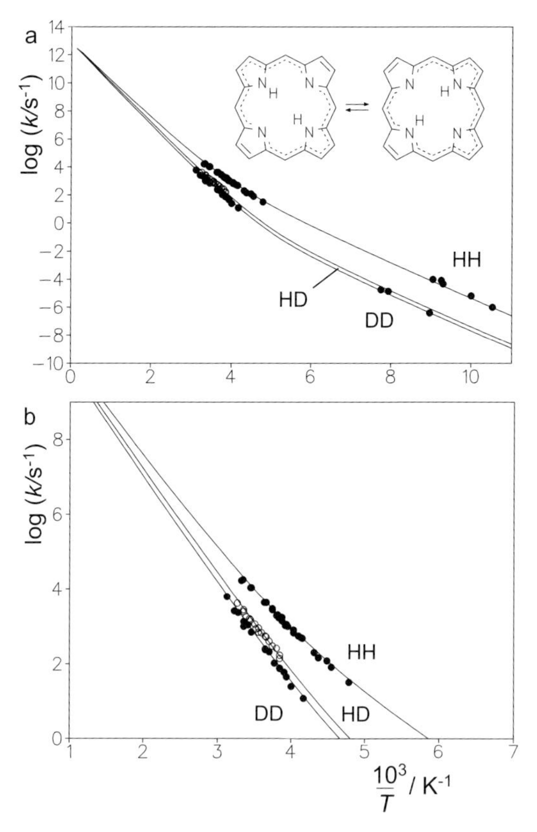

Figure 6.23 Mixed liquid and solid state Arrhenius diagrams of the HH-transfer of porphyrin, adapted from Ref. [18d].

for L = T. The resulting single H/D/T Arrhenius diagram of the porphyrin trans-cis reaction is depicted in Fig. 6.24. This representation allows comparison with the Arrhenius diagram of the tautomerism of the deprotonated unsubstituted porphyrin anion depicted in Fig. 6.25. The tautomerism of the latter was discovered by Braun et al. [23a], and the rate constants k^H were measured for the liquid and the solid state, as well as k^D and k^T for the liquid state [23b]. Whereas the reaction profile for the anion is symmetric, it is asymmetric for the parent compound, as illustrated schematically in Fig. 6.26.

For the parent compound porphyrin, an Arrhenius curve pattern of the type discussed in Fig. 6.9 is observed. Noteworthy is the same low-temperature slope E_m of the Arrhenius curves of the HH and DD reaction in Fig. 6.23, i.e. of the H- and D- reaction in Fig. 6.24. E_m will be mainly caused by the asymmetry of the reaction profile because at least the energy of the cis-intermediate is required for tunneling to occur, but also the reorganization energy of the ring skeleton will contribute. Note also that the low-temperature kinetic H/D isotope effect is smaller

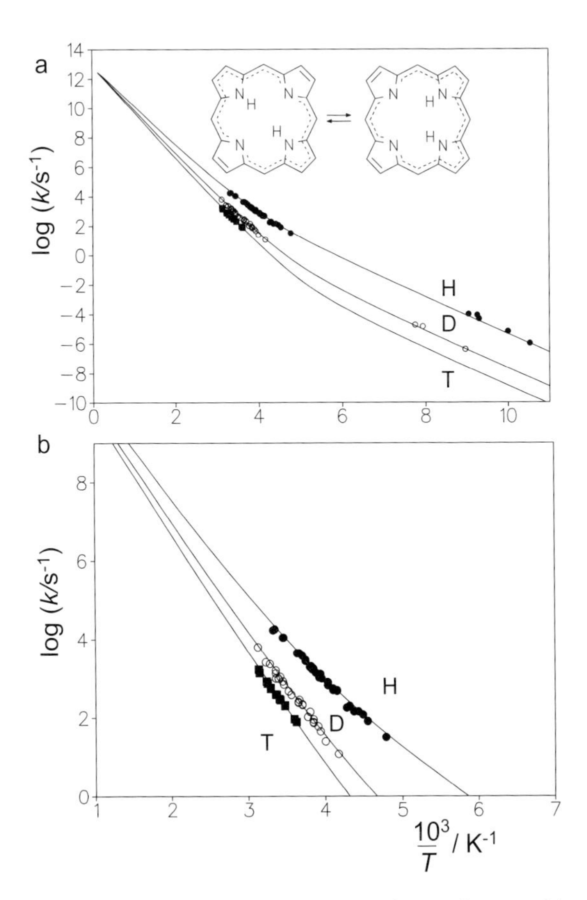

Figure 6.24 Mixed liquid and solid state Arrhenius diagram of the uphill trans–cis H/D/T-transfer of porphyrin. Data from Ref. [18e].

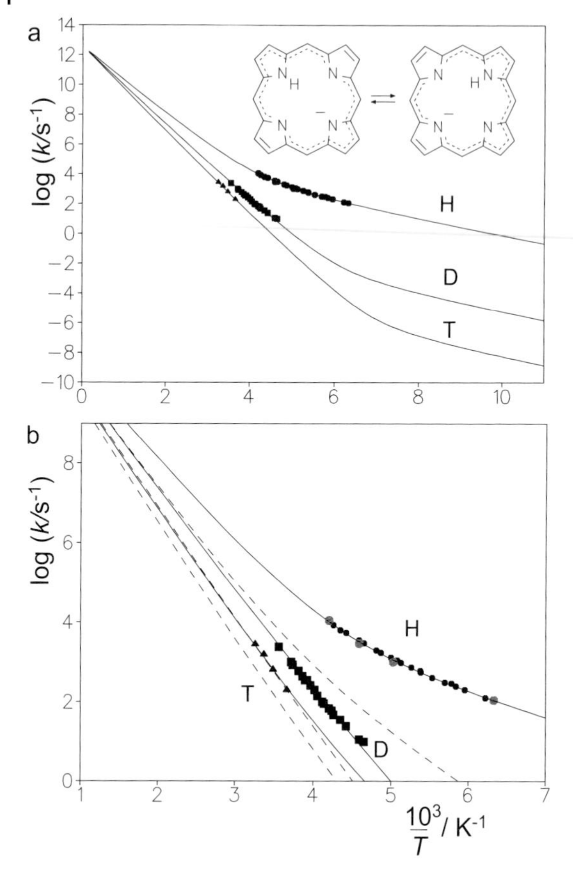

Figure 6.25 Mixed liquid and solid state Arrhenius diagram of the tautomerism of the porphyrin anion. Adapted from Ref. [23b].

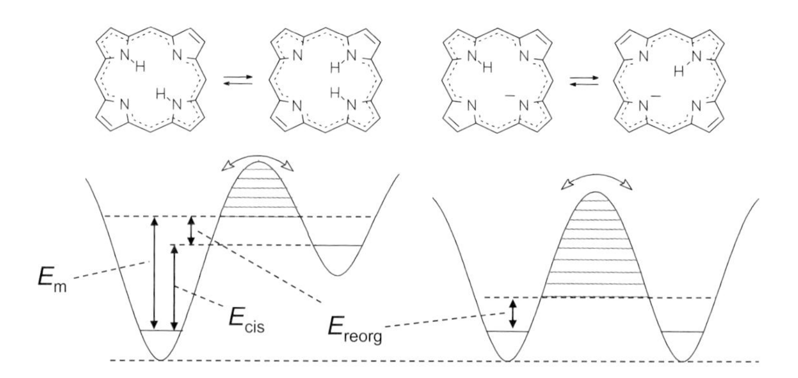

Figure 6.26 Potential curves (shown schematically) of the tautomerism of porphyrin and its mono-deprotonated anion. Adapted from Ref. [18e].

than predicted from the relatively large barrier difference for H and D evaluated at high temperatures. In order to match this effect a relatively high value of Δm for the heavy atom tunneling contribution had to be used in order to reduce the low-temperature isotope effect.

In contrast, this was not necessary in the case of the porphyrin anion where the transfer is degenerate and where the low-temperature kinetic isotope effects are substantially larger than in the parent compound. Therefore, the much smaller value of E_m in the anion is assigned to the reorganization of the porphyrin skeleton preceeding the transfer. As compared to the parent compound, both larger values for the tunneling distance as well as for the differences of the barrier heights of the isotopic reactions are obtained. These findings can be associated with the lack of the reaction asymmetry in the anion, as discussed in the previous section.

6.3.1.2 Unsymmetrically Substituted Porphyrins

In subsequent studies the question arose as to how the kinetics of the tautomerism of porphyrins and porphyrin analogs are affected by a reduction in the molecular symmetry arising from the introduction of single substituents. From a theoretical viewpoint, this question was especially interesting as formal kinetics of the stepwise transfers in Fig. 6.21 predict an evolution of the Arrhenius curves as discussed in Fig. 6.14(d)–(f). When the symmetry of the reaction is perturbed, one of the two barriers of the stepwise transfer is increased and the other decreased, as was illustrated in Fig. 6.13. In the HD reaction the D transfer exhibits the larger barrier and becomes rate-limiting, whereas in the DH reaction H is transferred in the rate-limiting step. Therefore, the HD reaction becomes slower and the DH reaction faster until they coincide with the DD and with the HH rates, as illustrated in Fig. 6.14(f).

This effect was observed by Schlabach et al. [20] for an unsymmetrically substituted acetylporphyrin (ACP, Fig. 6.27, X=CH_3CO) dissolved in CD_2Cl_2. The thermodynamics and the kinetics of the HH, HD, DH and DD reactions could be studied by NMR. It was observed that the acetyl substituted pyrrole ring exhibits a smaller proton affinity as compared to the other pyrrole rings which are substituted with aliphatic substituents (Fig. 6.28(a)). The equilibrium constant was given by

$$K_{AC\rightarrow BD} = 1.14 \times \exp(-5.82 \text{ kJ mol}^{-1}/RT) \tag{6.49}$$

Therefore, reaction pathways involving transition states and intermediates with hydrogen isotopes located on the non-substituted pyrrole rings are favored.

The Arrhenius curve pattern of Fig. 6.28(a) corresponds to the intermediate case between those of Fig. 6.14(d) and (e). It was calculated as follows. It was assumed that only the pathway AC→AD→BD contributes to the reaction rate constants but not AC→BC→BD (Fig. 6.21). Thus, only the first terms in Eq. (6.25) needed to be retained. Neglecting secondary isotope effects, the second hydron in the superscripts of the rate constants could be omitted. Using the substitution A→AC, B→AC and D→BD it was shown that

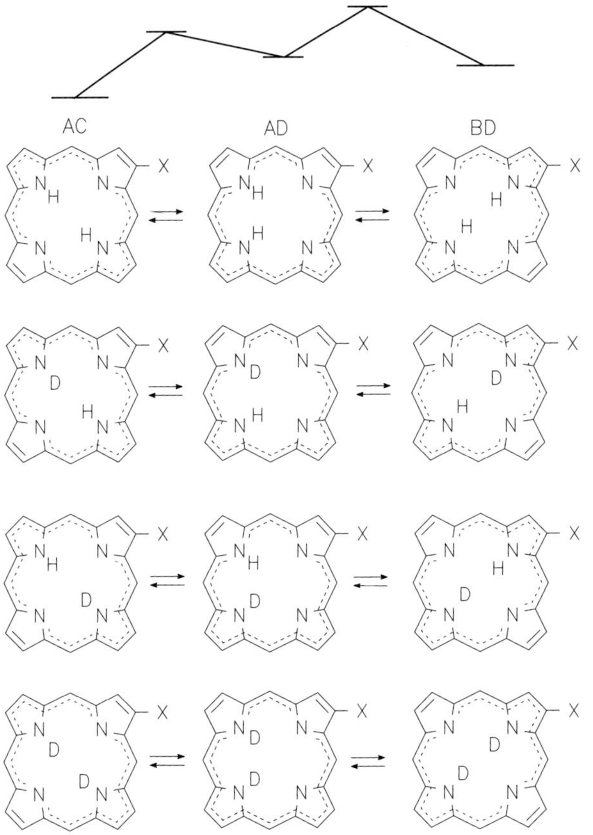

Figure 6.27 Stepwise HH, HD, DH and DD transfer in monosubstituted porphyrins.

$$k^{HH}_{AC\to BD} = \frac{k^{H}_{AC\to AD} K_{AC\to BD} k^{H}_{BD\to AC}}{k^{H}_{AC\to AD} + K_{AC\to BD} k^{H}_{BD\to AC}}, \quad k^{DD}_{AC\to BD} = \frac{k^{D}_{AC\to AD} K_{AC\to BD} k^{D}_{BD\to AC}}{k^{D}_{AC\to AD} + K_{AC\to BD} k^{D}_{BD\to AC}}$$

$$k^{HD}_{AC\to BD} = \frac{k^{H}_{AC\to AD} K_{AC\to BD} k^{D}_{BD\to AC}}{k^{H}_{AC\to AD} + K_{AC\to BD} k^{D}_{BD\to AC}}, \quad k^{DH}_{AC\to BD} = \frac{k^{D}_{AC\to AD} K_{AC\to BD} k^{H}_{BD\to AC}}{k^{D}_{AC\to AD} + K_{AC\to BD} k^{H}_{BD\to AC}} \tag{6.50}$$

Equation (6.50) expresses the experimental rate constants as a function of the single H-transfer forward and backward rate constants $k^{H}_{AC\to AD}$ and $k^{H}_{BD\to AC}$ of the steps defined in Fig. 6.27. Both reaction steps are now characterized by different tunnel parameters listed in Table 6.4, used to calculate the Arrhenius curves of Fig. 6.28(a). The step AC→AD involves a slightly larger barrier energy E_d and a slightly larger minimum energy E_m for tunneling to occur as compared to step BD→AD because of the asymmetry of the reaction. From a quantitative standpoint, the tunnel parameters may be subject to changes if data could be observed over a wider temperature range.

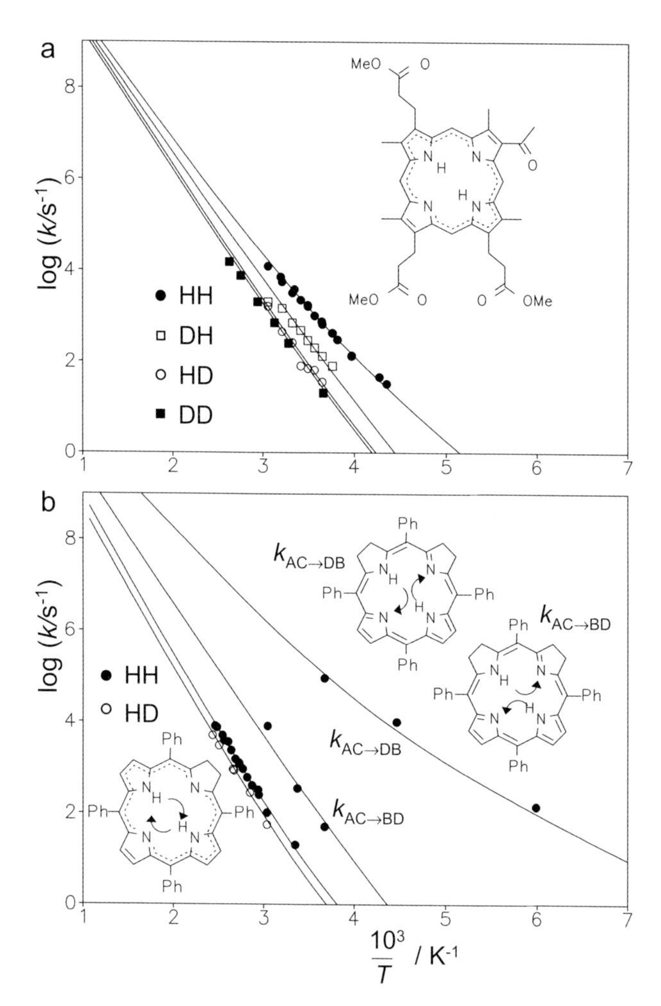

Figure 6.28 Arrhenius diagrams of the HH-transfer (a) in a substituted acetylporphyrin according to Ref. [20] and (b) in *meso*- tetraphenylisobacteriochlorin (upper curves) [19a] and in *meso*-tetraphenylchlorin (lower curves) dissolved in organic solvents [19d].

In conclusion, the theory of formal kinetics developed in the theoretical section for the description of stepwise multiple hydrogen transfers is supported by these experiments, at least for cases with weak hydrogen bonds. Thus, in these systems the assumption that each reaction step can be described in terms of a rate process characterized by rate constants is valid. The hydrogen bonds involved are not strong enough and the barriers not small enough that coherent tunneling states with delocalized protons or hydrogen atoms play an important role in this class of compounds.

6.3.1.3 Hydroporphyrins

Hydroporphyrins consist of porphyrins where one or more pyrrole rings are hydrogenated. The inner hydrogen atoms of porphyrins and substituted porphyrins resonate around –2 ppm which is typical for a Hückel aromatic $4n+2$ electronic π-system. The same was found for the AC and CA tautomers of *meso*-tetraphenylchlorin (TPC, Fig. 6.28(b)) where a single pyrrole ring is hydrogenated, and for *meso*-tetraphenylbacteriochlorin (TPBC, Fig. 6.28(b)) exhibiting two hydrogenated rings in trans-arrangement [19a]. Therefore, is was concluded that the two peripheral double bonds of porphyrin and the peripheral double bond of chlorin are not essential for the aromatic electron delocalization pathway, as illustrated in Fig. 6.21. In other words, porphyrin, chlorin and bacteriochlorin represent Hückel systems with 18 π-electrons in the aromatic pathways depicted in Fig. 6.21. In contrast, the chemical shifts of the inner hydrogen atoms of *meso*-tetraphenylisobacteriochlorin (TPiBC, Fig. 6.21(d)) were shifted substantially to low field, leading to the conclusion that all trans-forms of TPiBC do not represent aromatic 18 π electron systems.

The analysis is different for the intermediate states. TPiBC is predicted to form aromatic cis-intermediates CD and DC which are then lowered in energy as compared to the zwitterionic intermediates AB and BA. Thus, Fig. 6.21c predicts the reactions of iso-bacteriochlorin to be faster than those of porphyrin. On the other hand, the aromatic character of bacteriochlorin is lost in the intermediate states, moreover the trans-tautomers BD and DB of bacteriochlorin exhibit a zwitterionic structure. Thus, one should expect a substantial increase in the barrier height of the exchange between AC and CA in bacteriochlorin as compared to porphyrin.

These predictions were indeed confirmed experimentally. Schlabach et al. [19b] showed that in the case of TPC the trans-tautomers BD and DB cannot be observed directly by NMR; however, rate constants of the HH and the HD reaction could be obtained for the interconversion between AC and CA. According to an analysis similar to that leading to Eq. (6.25) it was shown that in the case of the HH reaction the observed rate constants are given by

$$k^{HH}_{AC\rightarrow CA} = k^{HH}_{AC\rightarrow BD} = k^{H}_{AC\rightarrow AD} >> k^{DD}_{AC\rightarrow CA} = k^{DD}_{AC\rightarrow BD} = k^{D}_{AC\rightarrow AD} \tag{6.51}$$

It has been shown [19b] that for the HD reaction two pathways are possible exhibiting reaction profiles similar to those of Fig. 6.13(c) and (d). The rate constants of the HD reaction are given by

$$k^{HD}_{AC\rightarrow CA} = k^{HD}_{AC\rightarrow BD} = \tfrac{1}{2} k^{H}_{AC\rightarrow AD} + \tfrac{1}{2} k^{D}_{AC\rightarrow AD} \approx \tfrac{1}{2} k^{H}_{AC\rightarrow AD} \tag{6.52}$$

This result can also be directly obtained by inspection of Fig. 6.21(b), setting H_a = H and H_b = D: the pathways of the HD reaction are dominated by the steps where D is transferred from ring C to D to A, so that H is transferred from A to B to C. The latter pathway is rate-limiting and exhibits the same rate constant as the corresponding HH reaction. The factor of $1/2$ in Eq. (6.52) arises from the fact that the alternative pathway where D is transferred via ring B is much slower.

The corresponding Arrhenius curves obtained are included in the lower part of Fig. 6.28(b). Within the margin of error, the kinetic HH/HD isotope effect of about 2 predicted by Eq. (6.52) was confirmed experimentally. Table 6.4 indicates that both E_m and E_d are increased as compared to porphyrin, as expected. The total increase as compared to TPP is about 11 kJ mol^{-1}, an effect which can be attributed to the loss of the aromaticity of TPP in TPC when the tautomerism occurs.

For TPBC no intramolecular tautomerism could be observed up to 140 °C, indicating that the sum of E_m+E_d is larger than about 86 kJ mol^{-1}, assuming a rate constant smaller than 100 s^{-1} at 140 °C as estimated from the linewidths of the ^{1}H signal of the inner protons.

Finally, for TPiBC the rate constants of the processes AD ↔ DB and AD ↔ BD (Fig. 6.21(c)) could be measured [19b]. The results are included in the Arrhenius curves of Fig. 6.28(b). The AD ↔ BD reaction is slower than in TPP which is not surprising, as the molecule is not aromatic. By contrast, the AD ↔ DB reaction is substantially faster than in TPP, an effect which has been associated with the formation of the aromatic cis-intermediate. The reaction rates are similar to those of the porphyrin anion. Although only a few rate constants were measured, one can anticipate with the accepted pre-exponential factor of $10^{12.6}$ s^{-1} a substantial concave curvature of the Arrhenius curves, i.e. a tunneling process occurring at much lower energies as compared to TPP. This is again the consequence of a more symmetric reaction profile as compared to TPP because the energy gap between the non-aromatic initial state AC and the aromatic cis-intermediate DC is substantially reduced.

6.3.1.4 Intramolecular Single and Stepwise Double Hydrogen Transfer in H-bonds of Medium Strength

When the hydrogen bonds become stronger the hydrogen transfer rates increase as the barriers are lowered. This is the case in a series of compounds discussed in this section.

The first system is indigodiimine which exhibits an intramolecular double proton transfer [68] as illustrated in Fig. 6.29(a), together with the corresponding Arrhenius curve (lower line). This process renders the two halves of the molecule equivalent. An even faster NH_2 rotation renders all NH protons equivalent. The rate constants were obtained by performing measurements at low temperatures, using a deuterated liquefied freon mixture $CDCl_3$/$CDFCl_2$/ CDF_2Cl as NMR solvent [68]. The parameters of the calculated Arrhenius curves are included in Table 6.4 but are not further discussed as kinetic isotope effects were not obtained.

The reaction rates are similar in the polycrystalline porphyrin analog dimethyldibenzo-tetraaza[14]annulene (DTAA) [69] representing a 6-membered H-chelate. They are even faster in the porphyrin isomer porphycene (Fig. 6.29(b)) [70] representing a 7-membered H-chelate with an even stronger intramolecular hydrogen bond. Both molecules form two trans-tautomers which are degenerate in the isolated molecules. However, solid state interactions lift this degeneracy. The rate constants of the forward uphill reactions could be measured in the case of DTAA

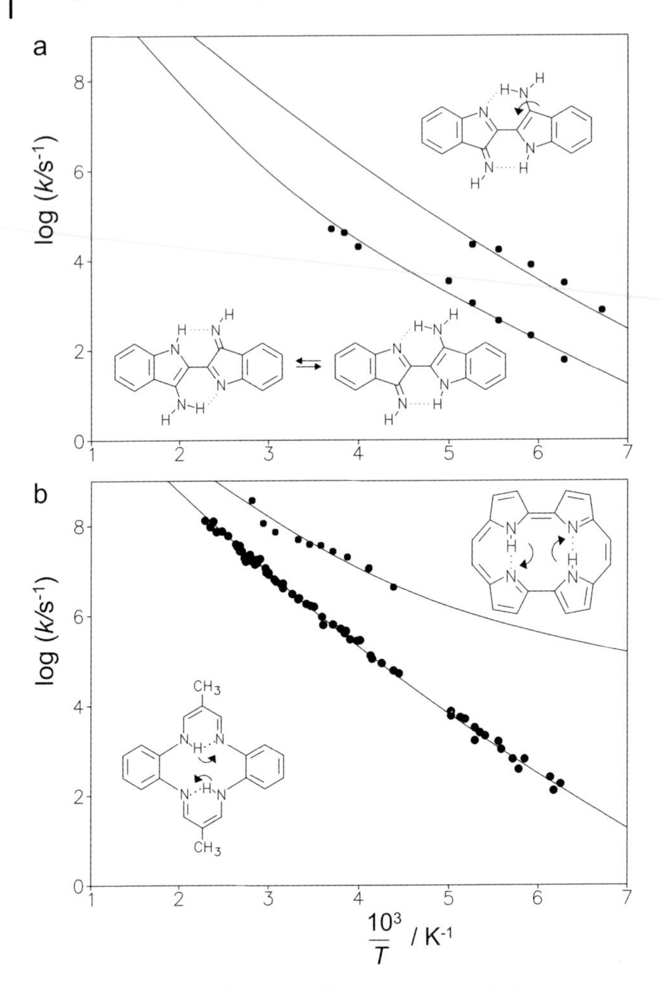

Figure 6.29 Arrhenius diagrams of the tautomerism of (a) indigodiimine [68] dissolved in a $CDCl_3/CDFCl_2/CDF_2Cl$ mixture and of (b) polycrystalline dimethyldibenzotetraaza[14]annulene (DTAA) [69] and of polycrystalline porphycene [70].

using high resolution ^{15}N solid state NMR by a combination of line shape analysis and ^{15}N-T_1 longitudinal relaxation time measurements. In the case of porphycene only the latter method could be used, which allows one to obtain rate constants on the micro- to nanosecond timescale. As kinetic isotope effects could not yet be obtained a detailed analysis of the reaction mechanisms was not yet possible. However, the present data seem to be compatible with stepwise HH tunneling processes where the energies of the cis-intermediates govern the Arrhenius curves at low temperatures.

In polycrystalline tetramethyldibenzotetraaza[14]annulene (TTAA, Fig. 6.30) a related tautomerism was observed [49]. By a combination of solid state ^{15}N and ^{2}H

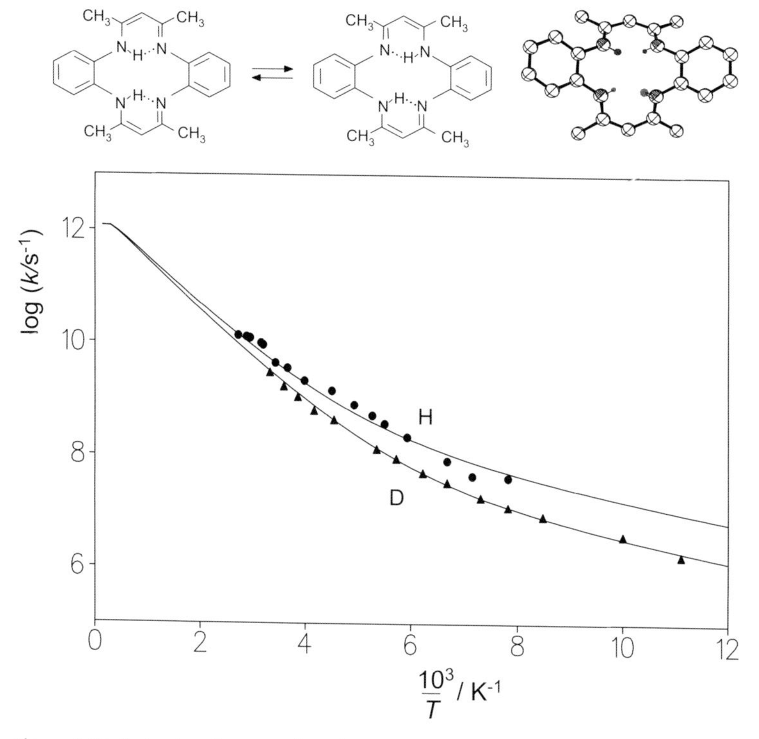

Figure 6.30 Arrhenius diagram of single H-transfer in polycrystalline tetramethyldibenzotetraaza[14]annulene (TTAA) according to Ref. [49].

NMR relaxometry rate constants and kinetic H/D isotope effects of the single H-transfer indicated in Fig. 6.30 could be measured. Evidence was found that the transfer is near-degenerate. Concave Arrhenius curves for the H- and the D-reactions were observed over a large temperature range, exhibiting surprisingly small kinetic H/D isotope effects, which were explained in terms of a relatively large heavy atom contribution to tunneling and a small barrier width. The latter arises from the substantially stronger H-bond in TTAA as compared to porphyrin.

6.3.1.5 **Dependence on the Environment**

The question of how intermolecular interactions perturb the symmetry of a degenerate H-transfer was studied as a function of temperature by Wehrle et al. [71] using TTAA dissolved in glassy polystyrene. In all cases, the transfer was found to be faster than the dynamic range of solid state ^{15}N NMR. The latter gave information about the distribution of the equilibrium constants of H-transfer. The results were rationalized in terms of the scenario of Fig. 6.31.

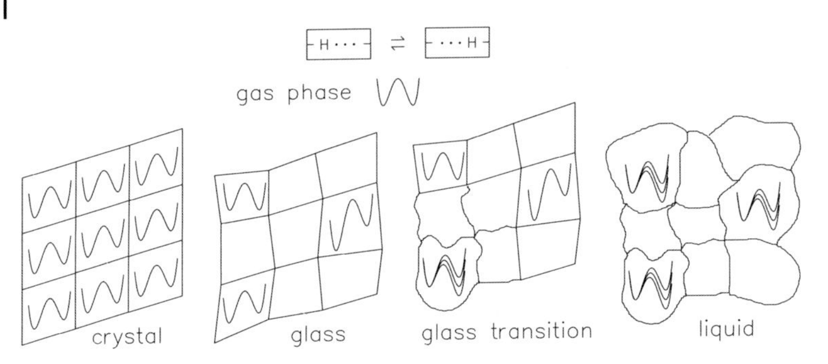

Figure 6.31 Model for the dependence of the proton transfer potential on the environment arising from experimental observations. Adapted from Wehrle et al. [71].

When a molecule exhibiting a symmetric double well for the proton motion in the gas phase is placed in a molecular crystalline environment, the crystal field will induce an energy difference ΔE between the tautomers. Whereas ΔE will be the same for all molecules in a crystal, ΔE will depend, in a disordered system such as a glass, on the local environment, leading to a distribution of ΔE-values. At the glass point, some environments may become mobile leading to an average value of $\Delta E_{av} = 0$, whereas other environments still experience non-zero values. Only well above the glass transition is a situation typical for the liquid reached where all molecules exhibit an average value $\Delta E_{av} = 0$.

6.3.1.6 Intermolecular Multiple Hydrogen Transfer in H-bonds of Medium Strength

The double proton transfer in cyclic dimers of crystalline benzoic acid has been studied by various authors using NMR relaxation techniques [72, 73]. For a recent account of this work the reader is referred to the study of Horsewill et al. [74] who published the correlation times of the HH, HD and DD reactions given by

$$\frac{1}{\tau_c} = k_{12} + k_{21}, \text{where} \quad K = \frac{x_2}{x_1} = \frac{k_{12}}{k_{21}} = \exp(-\Delta E/RT),$$
$$\Delta E = 85\ \text{K} = 0.36\ \text{kJ mol}^{-1} \tag{6.53}$$

Here, ΔE represents the energy difference between the two tautomers whose gas-phase degeneracy is lifted by solid-state interactions. x_1 and x_2 represent the mole-fractions.

The resulting Arrhenius diagram where the forward rate constants k_{12} of Horsewill et al. [74] are plotted as a function of the inverse temperature is depicted in Fig. 6.32. The Arrhenius curves exhibit large concave curvatures as expected for tunneling. Both τ_c as well as the backward reaction k_{21} (not plotted) are almost independent of temperature in this regime. The tunnel parameters used to calculate the Arrhenius curves are included in Table 6.4. The values of the minimum energy for tunneling to occur, E_m, are slightly larger than the energy difference ΔE between the two tautomers.

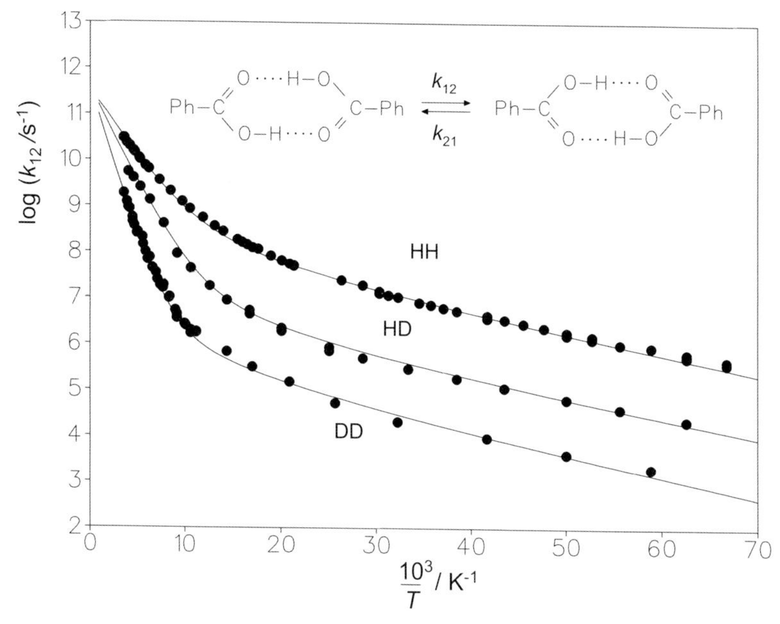

Figure 6.32 Arrhenius curves of the solid state tautomerism of benzoic acid dimers. Data for the HH and HD reactions taken from Ref. [74], and for the DD reaction from Ref. [72c]. The solid lines were calculated using the Bell–Limbach tunneling model with the parameters listed in Table 6.4.

In contrast to the intramolecular HH-transfers discussed above, replacement of each H by D leads to a significant isotope effect. At low temperatures, the HH/HD and the HD/DD isotope effects are 24 and 21, i.e. quite similar. In contrast, the extrapolated values at room temperature are about 3 and 6. As compared to other systems discussed below, usually, either similar values are obtained as was illustrated in Fig. 6.14(a) or the kinetic HH/HD isotope effects are larger than the HD/DD isotope effects, an effect arising from tunneling. Whereas the HH and the HD curves in Fig. 6.32 calculated for this review could be simulated assuming only the usual slight changes in the tunneling mass, barrier height and of the barrier width, a substantially larger barrier height and also a larger value of the minimum energy for tunneling to occur had to be assumed for the DD reaction. At present, it is tempting to associate this finding with the fact that something special has happened with the deuterated crystals. For this discussion remember that deuteration generally leads to a different position of D with respect to the hydrogen bond center as compared to H, and to an increase in the heavy atom distance. Such differences have been observed recently for acetic acid dimer [75] and other hydrogen bonded systems [40]. This would lead to an additional term which increases the barrier height of the DD transfer.

The tautomerism of crystalline pyrazoles which is discussed in the following, is particularly interesting because of the variety of hydrogen bonded complexes formed by this type of compounds in the solid state. Depending on the substituents, one finds non-reactive chains or reactive cyclic dimers, trimers or even tetramers in which degenerate HH, HHH, or HHHH-transfers can take place [76, 77] as depicted in Fig. 6.33 to 6.37. In this series, the influence of crystal fields which can lift the gas phase degeneracy of the transfer processes was not observed within the margin of experimental error.

The Arrhenius diagram of the degenerate HH/HD/DD transfer in the cyclic dimer of crystalline 3,5-diphenyl-4-bromopyrazole (DPBrP) [27] is depicted in Fig. 6.33. The kinetic HH/HD and HD/DD isotope effects are about 5 at room temperature and are similar, i.e. follow the rule of the geometric mean (RGM) as predicted by Fig. 6.14(a). The total HH/DD isotope effect is about 25. Concave Arrhenius curves indicate tunneling at low temperatures. This finding has been interpreted in terms of a single barrier reaction where all H loose zero-point energy in the transition state.

The RGM is also fulfilled in the case of the degenerate HHH-transfer in the cyclic trimer of crystalline state 3,5-dimethylpyrazole (DMP) (Fig. 6.34) [25a]. The individual isotope effects are about 4 at room temperature, and the total isotope effect is around 47, which is typical again for a single-barrier reaction. The barrier height can be varied substantially by removing the bulky methyl groups and by introducing various substituents in the 4-positions as indicated in Fig. 6.35 [25b]. In the next section, the discussion of this effect will be pursued.

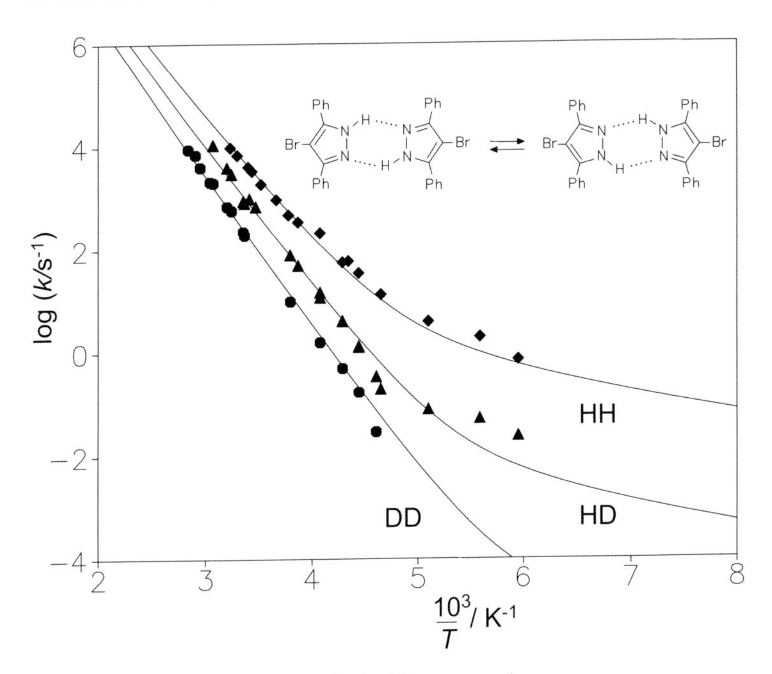

Figure 6.33 Arrhenius diagram for the double proton and deuteron transfer in solid DPBrP. Adapted from Ref. [27].

The kinetics of the HHHH-transfer in the cyclic tetramer of 3,5-diphenyl-4-pyrazole (DPP) has been evaluated recently [27]. The overall kinetic HHHH/DDDD isotope effects were found to be only around 12. This value indicated absence of a single barrier HHHH process where one would expect a larger overall effect. Instead, the Arrhenius pattern depicted in Fig. 6.36 could be explained in terms of a stepwise HH+HH process according to the profile of Fig. 6.17, where two hydrons are transferred in each step, leading to the expected isotope effects depicted in Fig. 6.19(b) and (c). This means that the rate constants of the HHHD and the HDHD reaction are very similar, and also those of the DDHH, DDHD, DDDD reactions. This leads to a very special dependence of the rate constants observed on the deuterium fraction x_D in the mobile proton sites. The mole fractions of all isotopologs according to a statistical distribution are depicted in Fig. 6.37(a), and the sums of mole fractions of the relevant species exhibiting similar rate constants in Fig. 6.37(b). It is clear, that practically only three different species and rate constants are observed in this case.

Ab initio calculations performed on pyrazole clusters reproduced these findings [78] and indicated a switch from concerted double and triple proton transfers in the cyclic dimer and trimer of pyrazole to a stepwise HH+HH mechanism for the

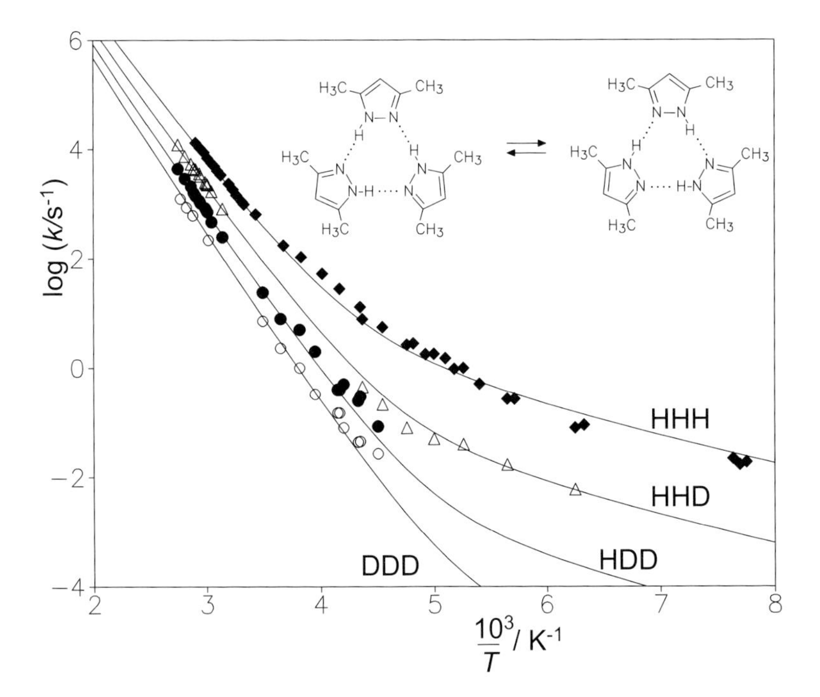

Figure 6.34 Arrhenius diagram for the double proton and deuteron transfer in the cyclic trimers of solid DMP. Adapted from Ref. [25a]. The solid curves were calculated using the Bell–Limbach tunneling model as described in the text.

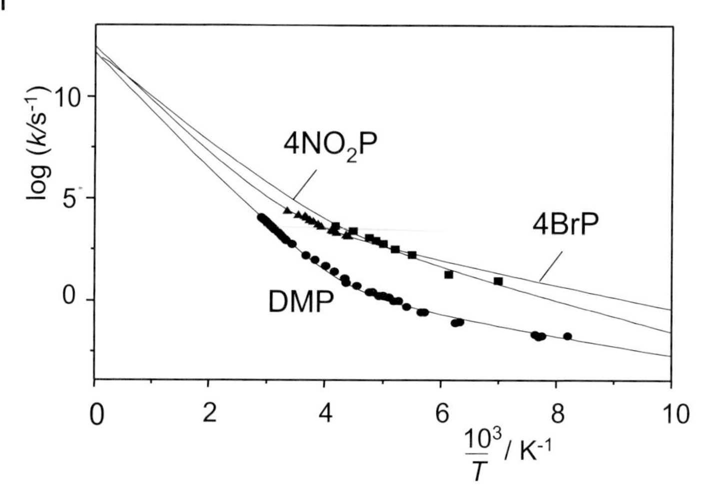

Figure 6.35 Arrhenius diagram for the triple proton transfers in cyclic trimers solid DMP, 4-nitropyrazole and 4-Br-pyrazole. Adapted from Ref. [25b].

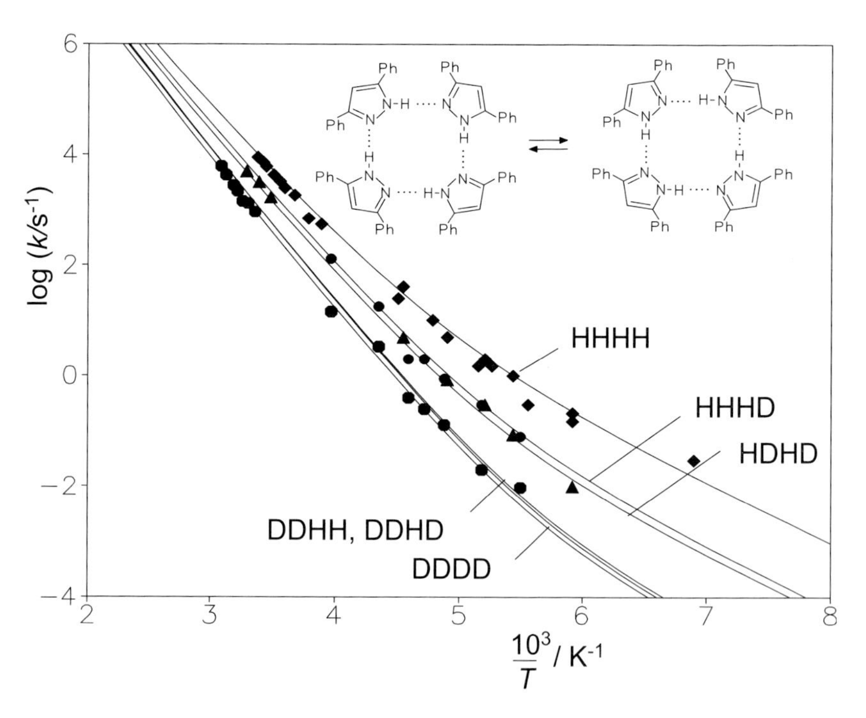

Figure 6.36 Arrhenius diagram for the quadruple proton and deuteron transfer in solid DPP. Adapted from Ref. [27]. The solid curves were calculated using the Bell–Limbach tunneling model.

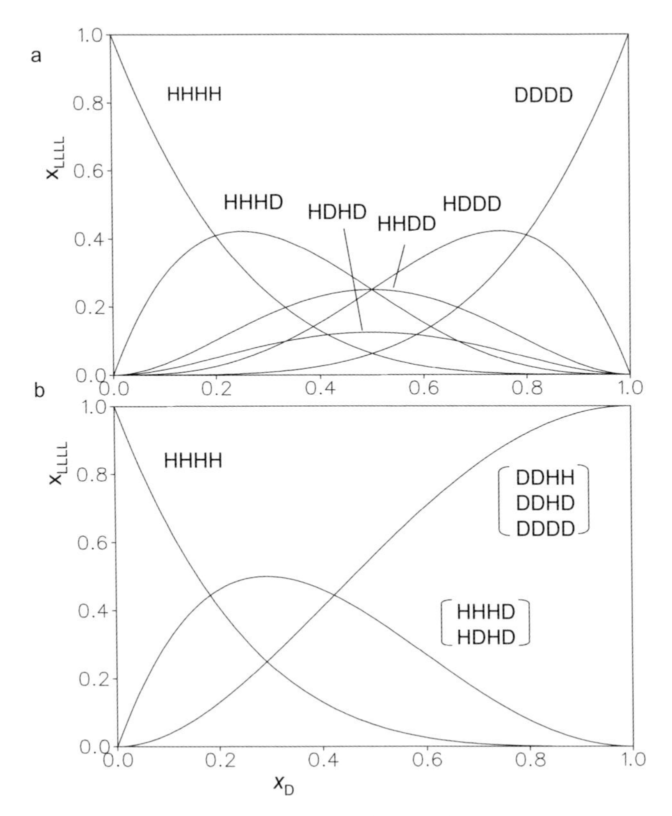

Figure 6.37 Statistical mole fractions of isotopologs (a) and ensembles of isotopologs (b) of the tetrameric DPP. Adapted from Ref. [27].

tetramer, consisting of two consecutive concerted double proton transfers. The concerted mechanism for the dimer was recently confirmed by Rauhut et al. [79].

Finally, note that Horsewill et al. [80] have reported an intramolecular quasi-degenerate quadruple HHHH tunneling process between the OH-groups of solid calix[4]arene, exhibiting temperature-independent rate constants. In a later section the discussion of this process will be pursued.

6.3.1.7 Dependence of the Barrier on Molecular Structure

In Fig. 6.38(a) are depicted the correlated NHN-hydrogen bond coordinates (Table 6.5) of porphyrin, of TTAA, porphycene, of the pyrazoles discussed in the previous section, as well as the calculated values of the transition states of porphyrin [65] and of its mono-deprotonated anion [81]. Also the data point of the double proton transfer in the cyclic *N,N′*-di-(*p*-F-phenyl)amidine dimer was added, which is discussed in the next section as it involves a hydrogen bond pre-equilibrium and a coupling to the reorientation of the aryl groups.

Note that all geometries are located on the NHN-hydrogen bond correlation curve of Fig. 6.3, especially the coordinates of the transition states of porphyrin and of its anion, exhibiting values of 2.60 and 2.66 Å. This means that hydrogen bond compression is the most important heavy atom motion which enables H-transfer; the transition state structures correspond to those expected for the strongest possible NHN-hydrogen bonds, whereas the initial states do not show any sign of hydrogen bonding.

The question arises how the intrinsic barrier of the symmetric H-transfer depends on the hydrogen bond geometries. In Fig. 6.38(b), therefore, the experimental values of the total barrier $E_d + E_m$ are plotted as a function of q_2 (Table 6.5). As a reference, the values of zero for the transition states calculated for porphyrin [65] and for the anion [81] are included. The dotted lines were calculated using the expression

$$E_d + E_m = C(q_2 - q_{2min}) \tag{6.54}$$

with $q_{2min} = 2.60$ Å, $C = 60$, 240, and 155 kJ mol^{-1} Å^{-1}. The calculated curve with the smallest slope reproduces well the experimental data of the H-transfers. The

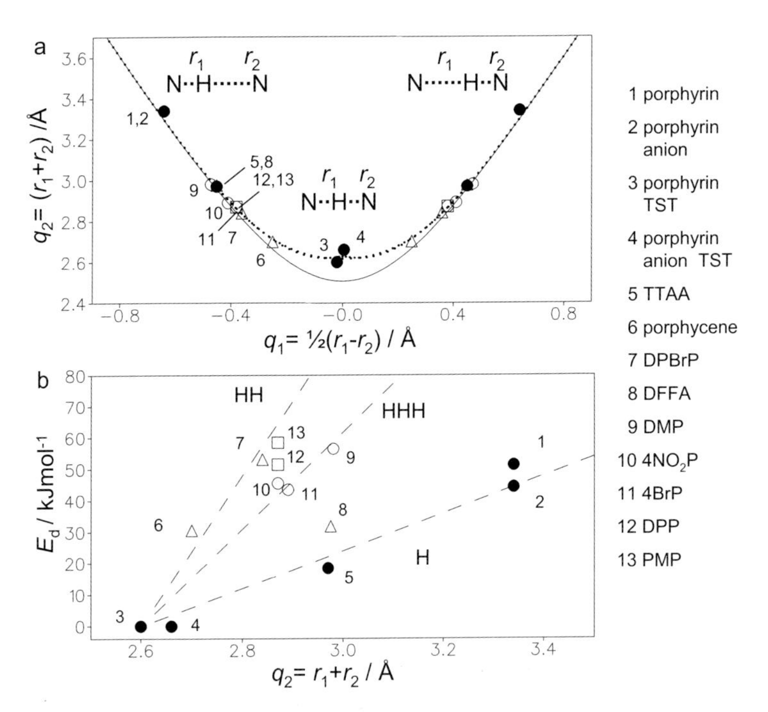

Figure 6.38 (a) Hydrogen bond geometries of various molecular systems containing NHN-hydrogen bonds. (b) Barrier heights of the H-transfers calculated from the Arrhenius curves of the species in (a). The barrier heights of the transition states are set to zero. Values taken from Table 6.5.

Table 6.5 Selected distances in systems with intra- and intermolecular H-transfers.

System		Ref	r_{NN}/Å	r_1/Å	r_2/Å	q_1/Å	q_2/Å	E_d/kJ mol^{-1}	E_m /kJ mol^{-1}
porphyrin	1	81, 18d, 18e	2.89	1.03	2.31	0.64	3.34	28.7	22.7
porphyrin TST[a]	2	56		1.28	1.32	0.02	2.60	0	0
porphyrin anion	3	81		1.03	2.31	0.64	3.34	34.3	10.0
porphyrin anion TST[a]	4	81		1.33	1.33	0.005	2.66	0	0
tetramethyltetraaza-[14]annulene TTAA	5	49	2.97	1.03	1.94	0.45	2.97	15.1	3.4
porphycene	6	70	2.63	1.10	1.60	0.25	2.70	24.7	5.9
diphenyl-p-bromopyrazole DPBrP	7	27	2.84	1.06	1.78	0.45	2.84	47.5	5.6
N,N′-di-(p-F-phenyl)amidine DFFA	12	24c, 82b	2.975	1.03	1.94	0.45	2.975	26.4	5.4
dimethylpyrazole DMP	8	25a, 25b, 27	2.98	1.02	1,96	0.48	2.98	48.1	8.4
4-nitropyrazole $4NO_2P$	9	25b	2.87	1.03	1,96	0.42	2.89	36.0	7.52
4-bromopyrazole 4BrP	10	25b	2.89	1.05	1.82	0.41	2.87	38.0	7.53
diphenylpyrazole DPP	11	27	2.874	1.05	1.82	0.42	2.87	32.5	19.0

a calculated using ab initio methods.

calculated curve assigned to the HH-transfer is tentative, as there is only a single point (7, DPBrP) which is well established. Its slope is substantially larger than the slope of the H-transfer curve. This may arise from the fact that two bonds instead of one have to be broken and reformed. Point 6 (porphycene) is located on this curve; it is tempting to conclude that the tautomerism of this molecule represents a more or less concerted HH-transfer process. However, point 8 refers to the HH-transfer in cyclic dimers of a diarylamidine (Fig. 6.44–6.46) discussed later. This point is located very far from the HH curve. This might indicate that something unusual happens here, for example a mechanism somewhere in between a concerted and a stepwise HH-transfer. It is interesting to note that for a given hydrogen bond geometry, the total barrier increases substantially from the H to the HH reaction, but that the barrier decreases again for a HHH reaction. Points 12 and 13 representing the HHHH reaction in pyrazole tetramers are located close to the HH curve; this is in agreement with the interpretation of consecutive HH+HH reaction.

Let us at this point draw some conclusions based on the systems which have been discussed so far. All systems were "simple" in the sense that heavy atom motions were restricted to changes in bond lengths and angles of the molecular skeletons in which the hydrogen bonds are embedded. Major conformational changes or coupling to solvent molecules were not present. Figure 6.39 summarizes the findings schematically.

Figure 6.39(a) illustrates hydrogen bond compression during a single H-transfer process according to the hydrogen bond correlation of Fig. 6.3. In the initial and final states, the geometric H/D isotope effects imply a shortening of the covalent bond distance and a lengthening of the hydrogen bond [40]. In the transition state the deuterated system can be somewhat more compressed as compared to the protonated system, because the wavefunction of D is sharper than the wavefunction of H, i.e. D is closer to the H-bond center than H. The barrier height is larger for D than for H.

The mechanism of HH-transfer depends on whether the two hydrogen bonds involved are cooperative (Fig. 6.39(b)) or anti-cooperative (Fig. 6.39(c)). In the case of two cooperative H-bonds compression of one bond leads also to a compression of the second bond. Compression of one of two anti-cooperative bonds leads, however, to a lengthening of the other bond. In the case of non-cooperative H-bonds compression of the first bond has no effect on the second bond. When H in one bond is shifted to the H-bond center, assisted by compression of this hydrogen bridge, this compression will also lead to a compression of the second hydrogen bond, which in turn shifts also the hydrogen in this bond to the H-bond center. In other words, cooperative hydrogen bonds seem to favor a concerted or single-barrier HH-transfer. This may be the case in benzoic acid dimer, porphycene and pyrazole dimers and trimers. By contrast, in the case of anti- or non-cooperative H-bonds, only one H-bond can be suppressed but not the other, and only a single H is transferred, leading to a stepwise motion involving a metastable intermediate. This is the case in porphyrins, phthalocyanins, indigodiimine, tetraaza[14]annulenes. In the next section, other examples will follow.

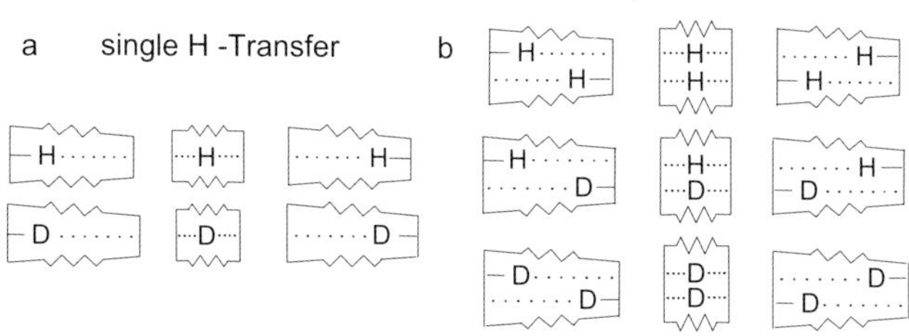

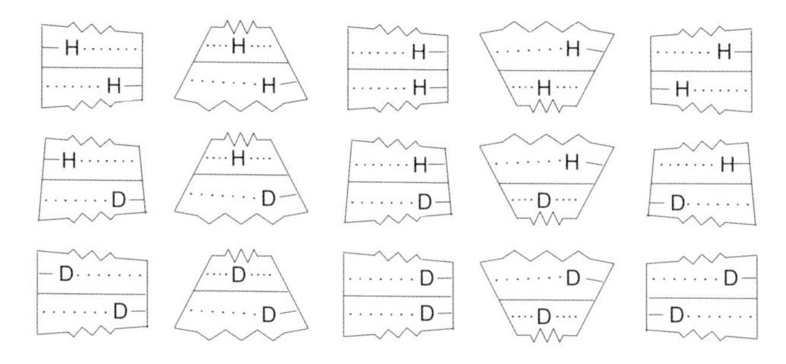

Figure 6.39 Simplified models of H- and D-substituted hydrogen transfer systems. The boxes containing springs symbolize the symmetries and the compressibilities of the hydrogen bonds. (a) Geometric H/D isotope effects during compression assisted H-transfer in a single hydrogen bond. (b) Geometric H/D isotope effects during compression assisted concerted HH-transfer in two cooperatively coupled hydrogen bonds. (c) Geometric H/D isotope effects during compression assisted stepwise HH-transfer in two anticooperatively coupled hydrogen bonds.

6.3.2 H-transfers Coupled to Major Heavy Atom Motions

In many H-transfer reactions in solution the reaction centers have first to form a reactive complex from non-reactive configurations or conformations i.e. they require a major molecular mobility. Therefore, it is understandable that complex H-transfers cannot take place in the solid state. In this section, various experimental published cases will be discussed.

6.3.2.1 H-transfers Coupled to Conformational Changes

Let us first discuss the intramolecular degenerate double proton transfers in azophenine [21] and in oxalamidines [22a]. By liquid state NMR of the ^{15}N labeled compounds the intramolecular pathways of the transfer processes were established and the rate constants k^{HH}, $k^{HD} = k^{DH}$ and when possible k^{DD} were measured. The Arrhenius diagrams are depicted in Fig. 6.40. In all cases, the reactions in solution were suppressed in the solid state, indicating major heavy atom motions in addition to H-bond compression.

The kinetic HH/HD/DD isotope effects are given by Eq. (6.48) and are typical for stepwise degenerate reaction mechanisms involving metastable cis-intermediates reached by single H-transfers as illustrated by Fig. 6.11 and 6.12. In a similar case as described above for porphyrin, the observed rate constants k^{LL} could be converted into the rate constants k^{L} of the uphill single H-transfers. k^{H} and k^{D} were then calculated in terms of the Bell–Limbach tunneling model using the parameters included in Table 6.4 and converted back to k^{LL} using Eq.(6.48).

The reaction rates of tetraphenyloxalamidine (TPOA) dissolved in CD_2Cl_2 are only slightly larger than those of azophenine (AP) dissolved in $C_2D_2Cl_4$. The kinetic isotope effects are larger in the latter; moreover, they depend on temperature, whereas those of TPOA exhibit little temperature dependence. The tautomerism of the bicyclic oxalamidine OA7 is, on the other hand, substantially slower than that of TPOA. In the corresponding 6-membered bicyclic oxalamidine OA6

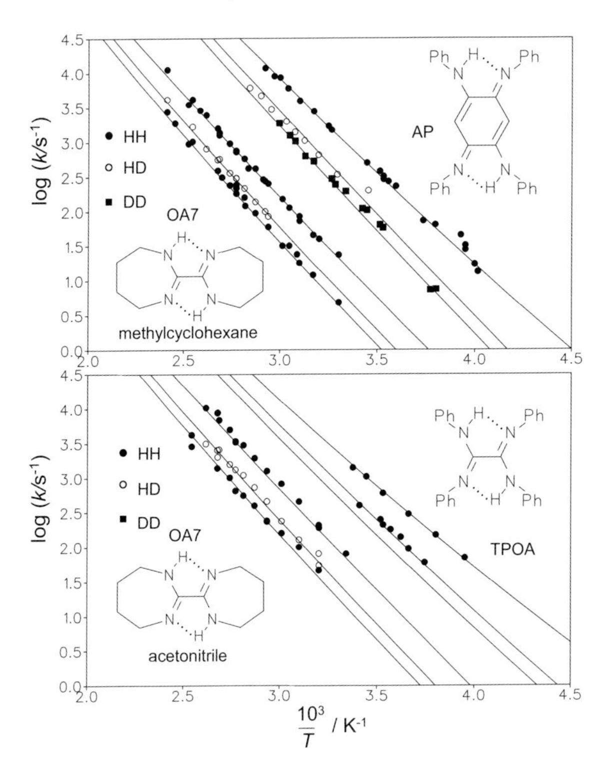

Figure 6.40 (a) Arrhenius diagrams of the tautomerism of azophenine (AP) dissolved in CD_2Cl_2 (top, [22a]) and of the seven-membered bicyclic oxalamidine OA7 (bottom) dissolved in methylcyclohexane [22b, 22c]. (b) Arrhenius diagrams of the tautomerism of tetraphenyloxalamidine (TPOA) dissolved in $C_2D_2Cl_4$ (top, [18b]) and of OA7 dissolved in acetonitrile (bottom, [22b, 22c]). The solid lines were calculated using the parameters listed in Table 6.4 as described in the text.

no double proton transfer was detectable [22c]. On the other hand, the tautomerism of OA7 was substantially faster in acetonitrile (dielectric constant 37.5) as compared to methylcyclohexane (dielectric constant 2.02), as illustrated in Fig. 6.40. These findings supported the formation of a zwitterionic intermediate according to the stepwise mechanism of Fig. 6.11. The small dependence of the kinetic isotope effects on temperature is confirmed for OA7 as the Arrhenius curves of the isotopic reactions are almost parallel. Note that a quantitative discussion of these parameters is difficult as the temperature range of the experimental data was limited. Therefore, the parameter sets obtained are not unique.

However, qualitatively, the above findings and the tunnel parameters obtained can be explained in terms of Fig. 6.41. In all cases, the total barrier heights E_d + E_m for each single reaction step are the same; in addition, it is assumed that the classical kinetic hydrogen/deuterium isotope effects for the over-barrier reactions are the same. Therefore, in the high temperature regime, the associated Arrhenius curves coincide. However, drastic differences are expected at lower temperatures, when tunneling becomes important. In this region, temperature independent kinetic isotope effects are expected, leading to parallel Arrhenius curves. Tunneling can occur only at energies indicated by the hatched areas. In Fig. 6.41(a) and (d) E_m is given by the energy of the intermediate E_i, whereas in Fig. 6.41(b) and (e) an additional reorganization energy E_r is required, mainly to compress the

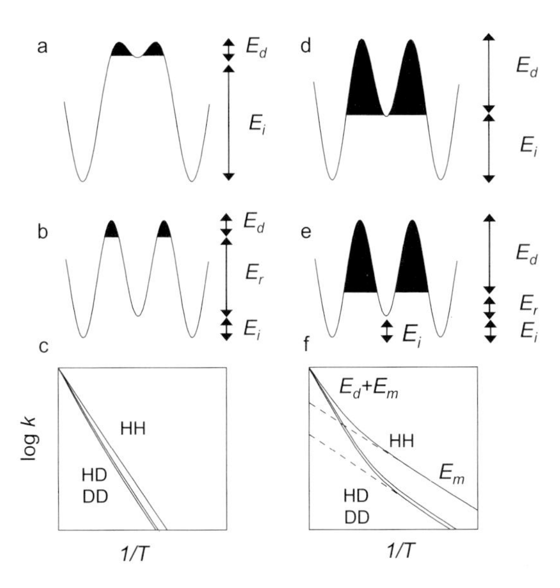

Figure 6.41 Visualization of a modified Bell tunneling model for degenerate, stepwise double proton transfers involving an intermediate. A minimum energy E_m is required for proton tunneling, which can take place only in the hatched regions. E_d barrier energy. (a) and (d) E_m is given by the energy of the intermediate, E_i . (b) and (e) E_m is given by E_i + E_r where E_r is associated with heavy atom reorganization preceding the proton transfer. (c) and (d) Corresponding Arrhenius curves (shown schematically) calculated in terms of the Bell–Limbach tunneling model. Adapted from Ref. [22c].

hydrogen bond, as discussed in the theoretical section. This hydrogen bond compression may involve additional molecular conformational changes. The corresponding Arrhenius curves are depicted in Fig. 6.41(c) and (d). The large changes in the experimental activation energies in the tautomerism of the oxalamidines and of azophenine, and at the same time the small changes in the kinetic isotope effects indicate then that the main differences arise from different values of $E_m = E_r + E_i$. It is plausible that the changes within the oxalamidines are then mainly due to different reorganization energies E_r.

This hypothesis was confirmed by semi-empirical calculations of various oxalamidines [22d]. The results are visualized in Fig. 6.42. In all cases, a substantial heavy atom reorganization precedes the H-transfer, which is strongly dependent on the chemical structure. This reorganization mainly involves a decrease in the nitrogen–nitrogen distances of the hydrogen bond in which the proton transfer takes place, thus lowering the barrier for the tautomerism. In contrast, in all other cases, hydrogen bond compression is associated with major conformational changes, requiring an additional reorganization energy. In TPOA and azophenine (not shown), H-bond compression is associated with a phenyl group reorientation.

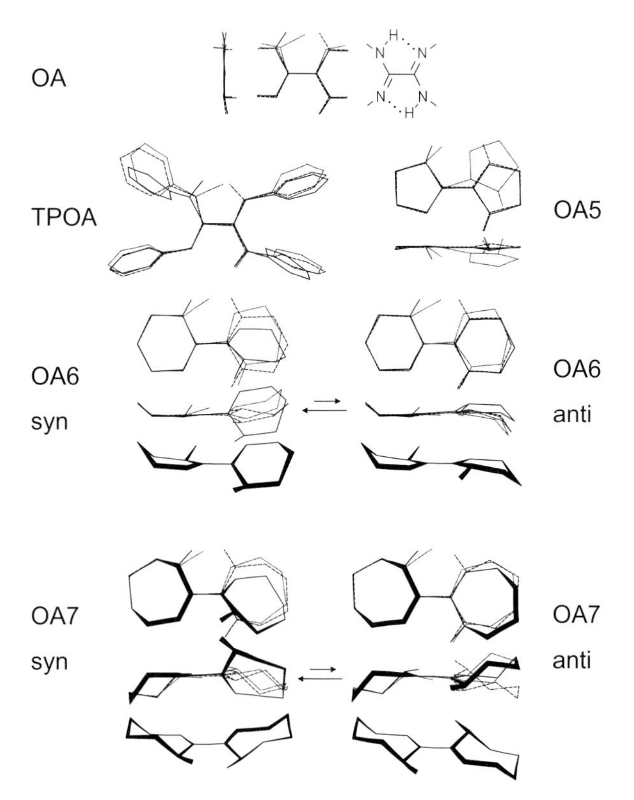

Figure 6.42 Heavy atom reorganization during the HH-transfer in oxalamidines calculated using the semiempirical PM3-MNDO method. Adapted from Ref. [22d].

This reorganization is not possible in the solid state, where only single tautomers are formed [21b, 22c]. The bicyclic oxalamidines also require a ring reorganization for H-bond compression to occur, which is smaller for OA7 than for OA5 and OA6, in accordance with experimental findings. For OA6 and OA7 syn- and anti-conformations were found, which both exhibited similar energies for the transition states. However, note that the ring reorganization did not involve a barrier leading to a pre-equilibrium for the tunneling step, as indicated in Fig. 6.20.

In all cases, the molecular structures do not allow for a simultaneous compression of both hydrogen bonds which would require a very high energy. Therefore, the transfers are stepwise, as indicated by Fig. 6.39(c).

The effects of small changes in the molecular structure can be observed in the case of the related diarylamidines which are the nitrogen analogs of formic acid and which represent models for nucleic acids. In tetrahydrofuran, for *N,N′*-di-(*p*-F-phenyl)amidine (DFFA) three forms were observed by NMR, a solvated s-cis-form and a solvated s-trans-form which is in fast equilibrium with a cyclic dimer in which a HH-transfer takes place [24] as illustrated in Fig. 6.43. Fortunately, at low temperatures, the s-cis- and the s-trans-forms were in slow exchange. The rate constants of the HH, HD and DD reactions were determined by dynamic ^{1}H and ^{19}F NMR as a function of concentration, deuterium fraction in the mobile proton sites and of temperature. The dependence of the observed rate constants of the s-trans-form on concentration is depicted in Fig. 6.44. The solid lines were calculated using Eq. (6.39) from which the rate constants in the dimer as well as the equilibrium constants of the dimer formation could be obtained. The Arrhenius

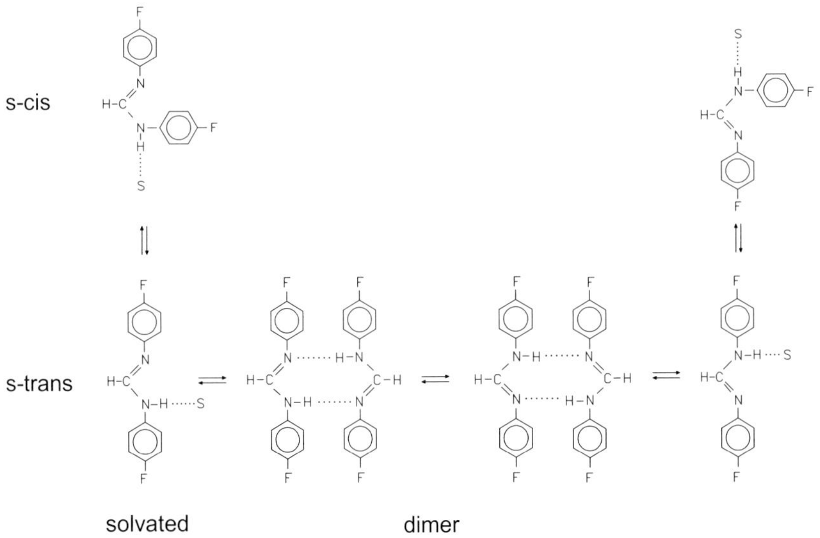

Figure 6.43 Conformational isomerism, hydrogen bond exchange and HH-transfer in *N,N′*-di-(*p*-F-phenyl)amidine (DFFA) dissolved in tetrahydrofuran (S) according to Ref. [24c].

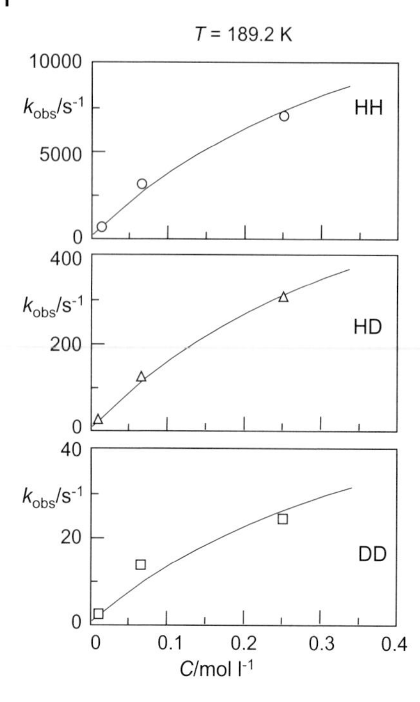

Figure 6.44 Pseudo-first order rate constants of the intermolecular HH, HD and DD transfers in cyclic dimers of *s-trans-N,N'*-di-(*p*-F-phenyl)amidine (DFFA) dissolved in tetrahydrofuran as a function of the concentration. The solid lines were calculated using an equilibrium constant of 1.12 L mol^{-1} of the monomer–dimer equilibrium. Adapted from Ref. [24c].

diagrams obtained are depicted in Fig. 6.45. Two large isotope effects are observed indicating a single barrier reaction according to Fig. 6.14(a). As the rate constants are intrinsic to the dimer, the contribution from the hydrogen bond equilibrium was eliminated from the minimum energy for tunneling to occur, but not contributions from heavy atom rearrangements, in particular from the expected aryl group conformations.

For that reason, symmetric diarylamidines with varying substituents in the *p*-position of the phenyl rings were studied by X-ray crystallography and dynamic solid state ^{15}N NMR [82]. The tendency to form cyclic dimers in the solid state was supported. In most cases, the angles α_N and α_{NH} between the phenyl groups and the molecular skeleton at the imino and the amino nitrogen atoms were different for a given molecule; the aryl ring at the imino nitrogen atom was often found to be coplanar with the molecular skeleton, but a substantial angle was observed at the amino nitrogen. This circumstance can be attributed to steric interactions of aromatic *o*-CH groups and the CH group of the amidine unit. It leads to a large preference for one of the two potentially degenerate tautomers, and suppresses the HH reaction in the solid. A degenerate HH-transfer was observed only in the OCH_3 substituted compound, where the two angles were similar but not coplanar with the molecular skeleton.

In solution the aryl groups of a cyclic dimer will not, therefore, be the same as lead to an asymmetry of the double well for the HH-transfer, as illustrated in Fig. 6.46. Reorientation of the phenyl groups to angles around 50° will symmetrize the

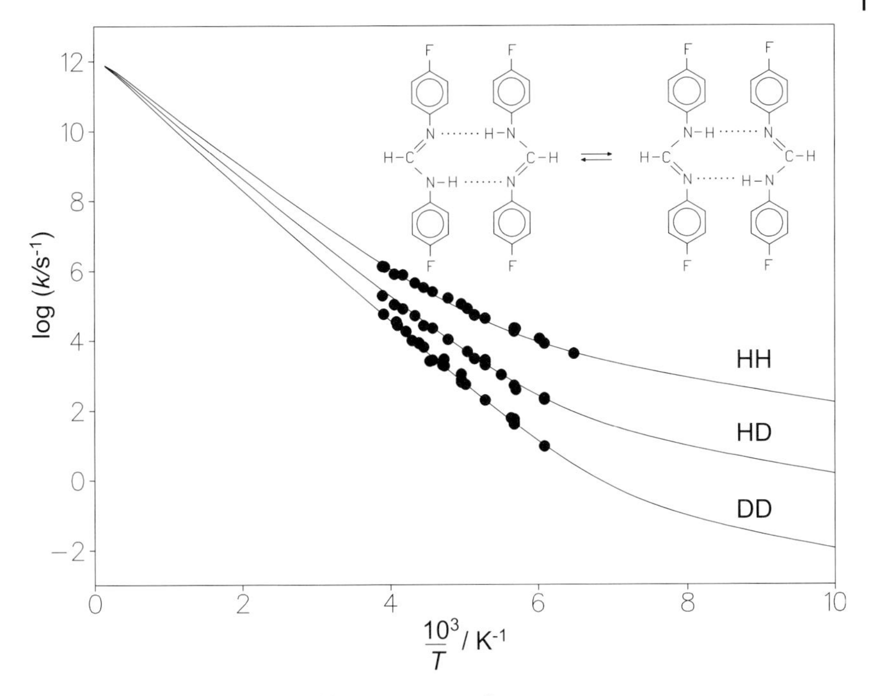

Figure 6.45 Arrhenius diagrams of the tautomerism of DFFA dissolved in THF. Adapted from Ref. [24c].

potential and minimize the barrier height of the HH-transfer. The latter is expected to take place in this configuration. Finally, the process is completed by a reorientation of the aryl groups. This means that the total barrier of the HH reaction in solution within the cyclic dimer will be slightly higher than in the symmetric configuration in the solid state. This is what was indeed observed for the rate constants of the OCH_3 substituted diarylamidine in the solid state.

6.3.2.2 H-transfers Coupled to Conformational Changes and Hydrogen Bond Pre-equilibria

Replacing the CH unit of diarylamidines with imino nitrogens leads to diaryltriazenes. As illustrated in Fig. 6.46, the aryl groups are found to be coplanar with the triazene unit. A consequence is that an intermolecular steric interaction between aromatic CH arises, which prevents the formation of cyclic dimers. Thus, diaryltriazenes are not able to exchange protons without the help of a catalyst, as has been shown recently [33]. In order to obtain more information about catalytic proton exchange, the base catalyzed transfer of 1,3-bis(4-fluorophenyl)[1,3-$^{15}N_2$]triazene was studied in more detail using ^{1}H and ^{19}F NMR. As catalysts dimethylamine, trimethylamine and water were studied, using tetrahydrofuran-d_8 and methylethylether-d_8 as solvents. The latter is liquid down to 130 K.

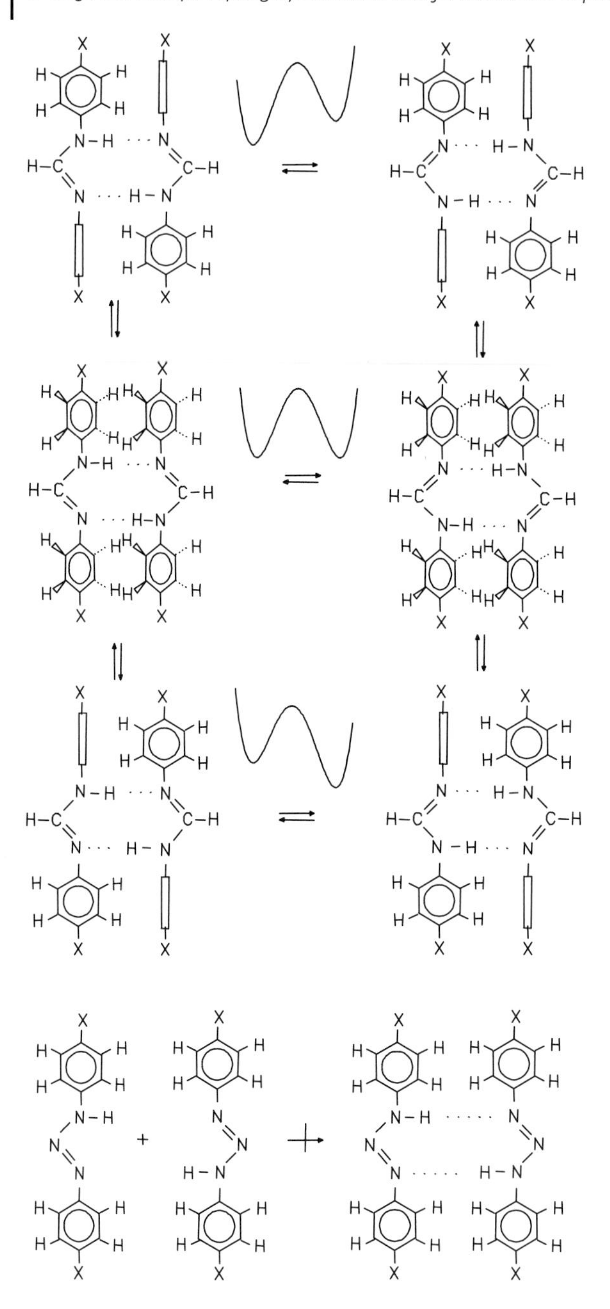

Figure 6.46 Coupling of the HH-transfer in cyclic dimers of diarylformamidines to the rearrangement of the aryl groups. Adapted from Ref. [82].

Surprisingly, both dimethylamine and trimethylamine were able to pick up the mobile proton of the triazene at one nitrogen atom and carry it to the other nitrogen atom, resulting in an intramolecular transfer process catalyzed each time by a different base molecule. Even more surprising is that the intramolecular transfer (Fig. 6.47(a)) catalyzed by dimethylamine is faster than the superimposed intermolecular double proton transfer (Fig. 6.47(b)).

The kinetic H/D isotope effects are small, especially in the catalysis by trimethylamine, indicating a major heavy atom rearrangement and absence of tunneling. This is because of the high asymmetry of the H-transfer from the triazene to the base. Semi-empirical PM3 and *ab initio* DFT calculations indicate a reaction pathway via a hydrogen bond switch of the protonated amine representing the transition state, where the imaginary frequency required by the saddle point corresponds to a heavy atom motion, as was illustrated schematically in Fig. 6.7. Tunneling is absent because of the very high tunneling masses involved, corresponding to the mass of the base.

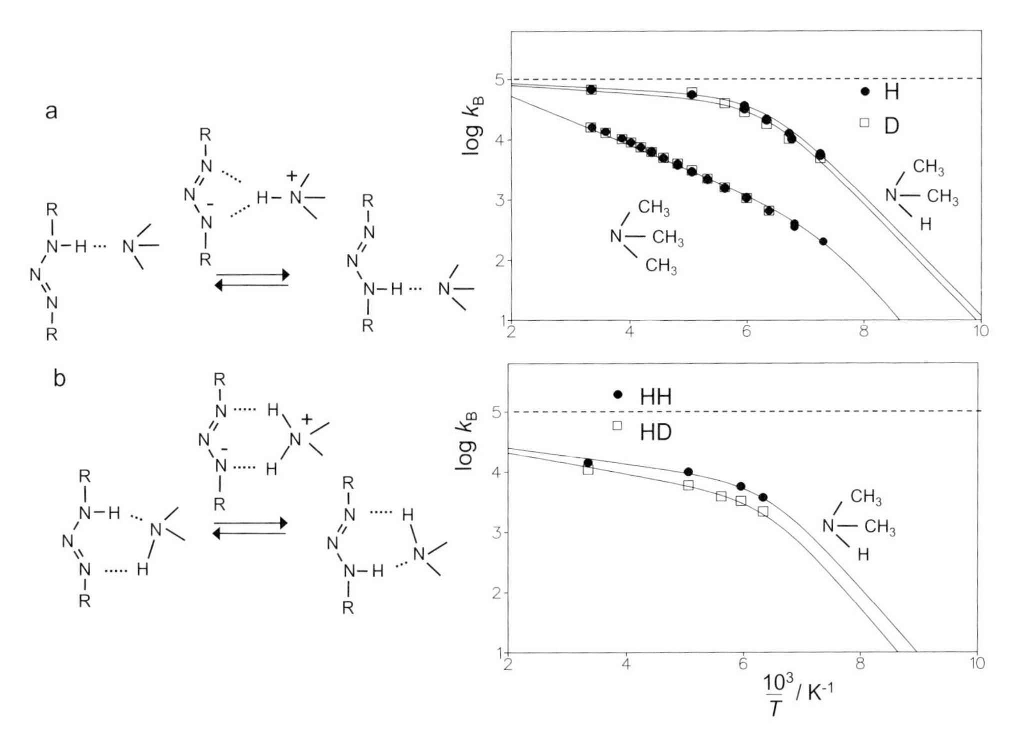

Figure 6.47 (a) Arrhenius diagrams of the intramolecular proton and deuteron transfer in 1,3-bis-(4-fluorophenyl)-[1,3-$^{15}N_2$]triazene dissolved at a concentration of 0.1 mol l^{-1} in methyl ethyl ether-d_8, catalyzed by the bases dimethylamine (0.0028 mol l^{-1} at $x_D = 0$ and 0.0041 mol l^{-1} at $x_D = 0.95$) and trimethylamine (0.02 mol l^{-1} at $x_D = 0$ and $x_D = 0.96$). k_B represents the average inverse life times of the base B between two exchange events. (b) Arrhenius diagrams of the intermolecular proton and deuteron transfer of 1,3-bis-(4-fluorophenyl)-[1,3-$^{15}N_2$]triazene catalyzed by dimethylamine. Adapted from Ref [33b].

The Arrhenius curves of all processes exhibit strong convex curvatures. This phenomenon is explained in terms of the hydrogen bond association of the triazene with the added bases, preceding the proton transfer. At low temperatures, all basic molecules form a hydrogen bonded reactive complex with the triazene, and the rate constants observed equal those of the reacting complex. However, at high temperatures, dissociation of the complex occurs, and the temperature dependence of the observed rate constants is affected also by the enthalpy of the hydrogen bond association according to Eqs. (6.37) and (6.46). As tunneling is not involved, the Arrhenius curves are not further discussed. For that the reader is referred to the original literature [53].

Al-Soufi et al. [83] have followed the kinetics of the intramolecular H- and D-transfer between the keto and the enol form of 2-(2′-hydroxy-4′-methylphenyl) benzoxazole (MeBO) dissolved in alkanes using optical methods. No dependence of the rate constants on the solvent viscosity could be found. The Arrhenius diagram obtained over a very wide temperature range is depicted in Fig. 6.48. At low temperatures, the very rare regime of temperature-independent rate constants is obtained, exhibiting a very large temperature independent kinetic H/D isotope effect of about 1400. At room temperature, still a quite large effect of about 14.5 is obtained.

Al-Soufi et al. [83] mentioned that the experimental pre-exponential factors obtained at high temperature were only about 10^9 s^{-1} instead of about 10^{13} s^{-1} as expected. Therefore, the Arrhenius curves of this reaction were recalculated here using Eq. (6.43), assuming an equilibrium between a reactive form and a non-reactive form. The parameters are listed in Table 6.2. Because of the large body of data all parameters could be determined.

At low temperatures, the intrinsic Arrhenius curves of the H- and the D-transfer, symbolized by the dashed lines, coincide with the observed ones, represented by the solid lines, as $k_{obs} = k$. However, at high temperature it was assumed that a non-reactive form of the molecule dominates because of its more positive entropy, leading to $k_{obs} = kK$. Thus, both the observed rate constants and the observed pre-exponential factors are smaller than expected.

Note that, at low temperatures, a very small minimum energy E_m for tunneling to occur is found, which refers to the reactive complex. This value could, therefore, be determined in addition to the values of ΔH and ΔS of the pre-equilibrium. In other cases, as discussed later, only the sum of $\Delta H + E_m$ can be determined. The barrier for the transfer is similar to that found for TTAA. The difference between the barriers for H and D is substantially large, of the order of that found for porphyrin. In addition, a contribution for heavy atom tunneling is observed.

At present, one can only speculate about the structure of the postulated non-reactive form. It could be that at low temperatures, the keto form may exhibit a zwitterionic aromatic character, and at high temperature a less polar but quinoid structure. Both structures are normally limiting structures. However, the zwitterionic structure is highly solvated and will exhibit, therefore, a much more negative entropy as compared to the zwitterionic structure. The entropy decrease is expected to be especially large in the case of apolar but polarizable solvents as has been shown by Caldin et al. [84]. Another possibility could be the formation of an

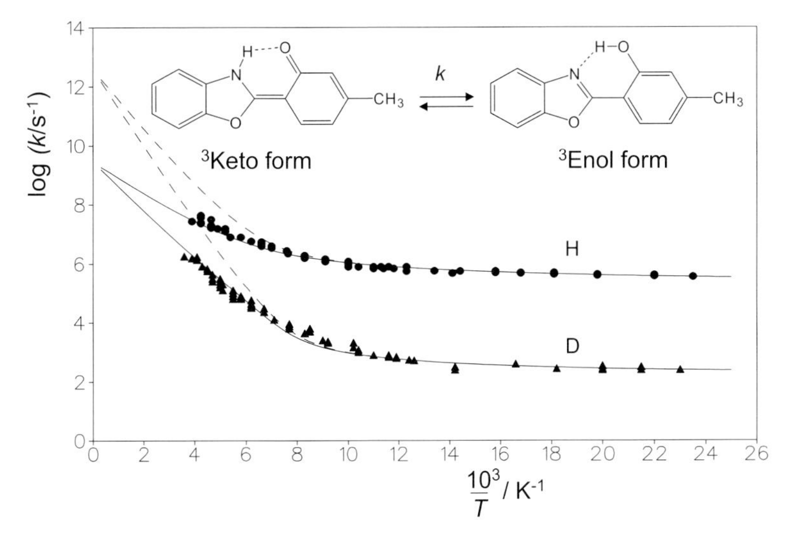

Figure 6.48 Arrhenius plot of the triplet state tautomerism of 2-(2′-hydroxy-4′-methylphenyl) benzoxazole (Me-BO, upper curve) and its deuterated analog (lower curve) dissolved in alkanes. The kinetic data were taken from Al-Soufi et al. [83]. The solid lines were calculated using the parameters listed in Table 6.4.

enolic conformer exhibiting an intramolecular OHO- instead of an OHN-hydrogen bond. However, further spectroscopic and kinetic measurements are necessary to clarify this problem.

Let us now discuss the well-studied case of the isomerization of the 2,4,6-tri-*tert*-butylphenyl radical to 3,5-di-*tert*-butylneophyl in apolar organic solvents, depicted in Fig. 6.49 which has been studied by Brunton et al. [8]. Various barrier types were used by these authors for Bell-type semiclassical tunneling calculations. It was shown that an inverted parabola could not give a satisfactory fit. The pre-exponential factors found for other barrier types were of the order of 8 to 12. As depicted in Fig. 6.49, a solution to the problem can be obtained in terms of an equilibrium where again a reactive form dominates at low and a non-reactive form at high temperature, as in the preceding case of Me-BO. In the case of the 2,4,6-tri-*tert*-butylphenyl radical one may interpret the reactive form with a configuration where the C–H bonds of the methyl groups are pointing in the direction of the aromatic acceptor carbon atom. Such a configuration could have a more negative entropy as compared to the non-reactive forms with unfavorable transfer geometries, dominating at high temperatures. The tunnel parameters used to calculate the Arrhenius curves are included in Table 6.4. No anomaly can be detected; the high barrier can be explained by the little capability for the formation of CHC-hydrogen bonds.

Let us discuss now some examples where non-reactive states are present at low temperatures, and reactive states at high temperatures.

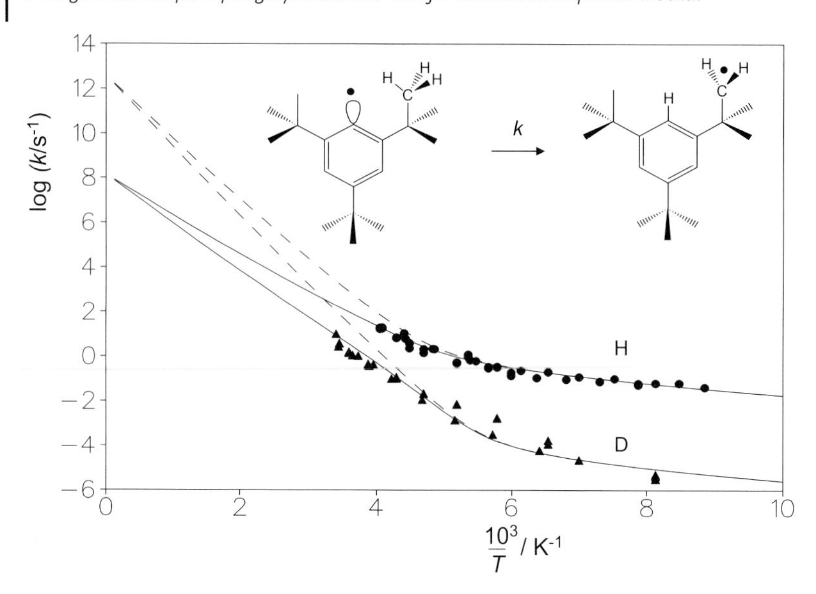

Figure 6.49 Arrhenius curves of the isomerization of the 2,4,6-tri-*tert*-butylphenyl radical to 3,5-di-*tert*-butylneophyl in apolar organic solvents. The solid and dashed lines were calculated as described in the text using the parameters listed in Table 6.4. Data from Brunton et al. [8].

The first two examples are the H-transfers in 2-hydroxyphenoxyl radicals which have been studied using dynamic EPR spectroscopy. When 3,6-di-*tert*-butyl-2-hydroxyphenoxyl and its deuterated analog are dissolved in heptane the Arrhenius diagram of Fig. 6.50 was obtained by Bubnov et al. [85]. The kinetic isotope effect is about 10 at room temperature. Setting the pre-exponential factor to $10^{12.6}$ (Table 6.4), leads to the concave Arrhenius curve depicted as solid lines. In contrast, Fig. 6.51 depicts the kinetic data of the parent compound 2-hydroxyphenoxyl in CCl_4/CCl_3F to which 0.11 mol l^{-1} dioxane had been added to increase the solubility [86]. Now, a kinetic isotope effect of about 56 is obtained at room temperature. This large difference between both molecules had been noted already some time ago by Limbach et al. [87]. In particular, it was noted that the two Arrhenius curves of the H and the D reaction are almost parallel. Application of the Bell–Limbach tunneling leads to unusually large pre-exponential factors of 10^{18} s^{-1}. As shown in Fig. 6.20(b) and the parameters of Table 6.4, the use of Eq. (6.43) improves the analysis, although the interpretation is similar to that obtained before.

The dashed lines in Fig. 6.51 indicate the intrinsic Arrhenius curve of the transfer, whereas the solid line indicates the one including the pre-equilibrium. The reduction of the rate constants as compared to the di-*tert*-butyl radical is explained by the formation of a non-reactive species at low temperatures, which is hydrogen bonded to the added dioxane. Thus, for the reaction to occur, the intramolecular H-bonded species has first to be formed, which exhibits a higher energy but also a

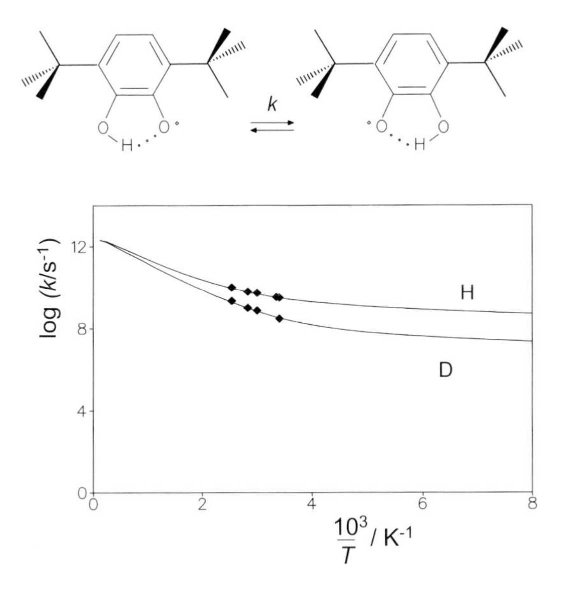

Figure 6.50 Arrhenius curves of the tautomerism of 3,6-di-*tert*-butyl-2-hydroxyphenoxyl dissolved in heptane according to Bubnov et al. [85]. The solid lines were calculated using the parameters listed in Table 6.4.

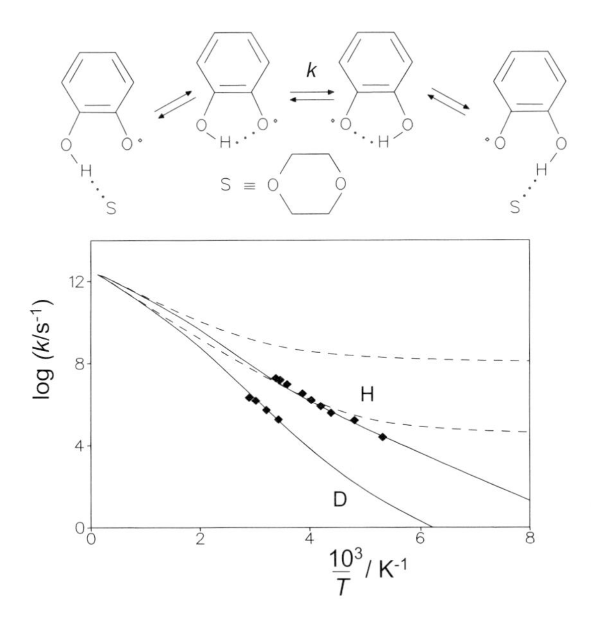

Figure 6.51 Arrhenius curves of the tautomerism of 2-hydroxyphenoxyl dissolved in CCl_4/CCl_3F/dioxane according to Loth et al. [86]. The solid lines were calculated using the parameters listed in Table 6.4.

more positive entropy. A comparison of the Arrhenius curves in Fig. 6.20(b) indicates that the desolvated intramolecular H-bonded species is never dominant over the whole temperature range, as the interaction with dioxane is stronger because of the linear intermolecular H-bond, in comparison with the weaker intramolecular H-bond.

The larger kinetic H/D isotope effects in the parent radical can be explained in terms of the higher symmetry of the parent radical as compared to the di-*tert*-butyl radical. In the latter, the methyl groups on both sides of the ring are not ordered, leading to effective asymmetric double well potentials of the H-transfer. These examples show how subtle structural effects can lead to very different H-transfer properties.

A related solvent effect was found for the proton exchange between acetic acid and methanol in tetrahydrofuran by Bureiko et al [88] and by Gerritzen et al. [9]. Hydrogen bonding to the solvent prevents the formation of the cyclic complexes in which the proton exchange takes place. Unfortunately, these complexes could not be seen directly. The rate constants were measured as a function of concentra-

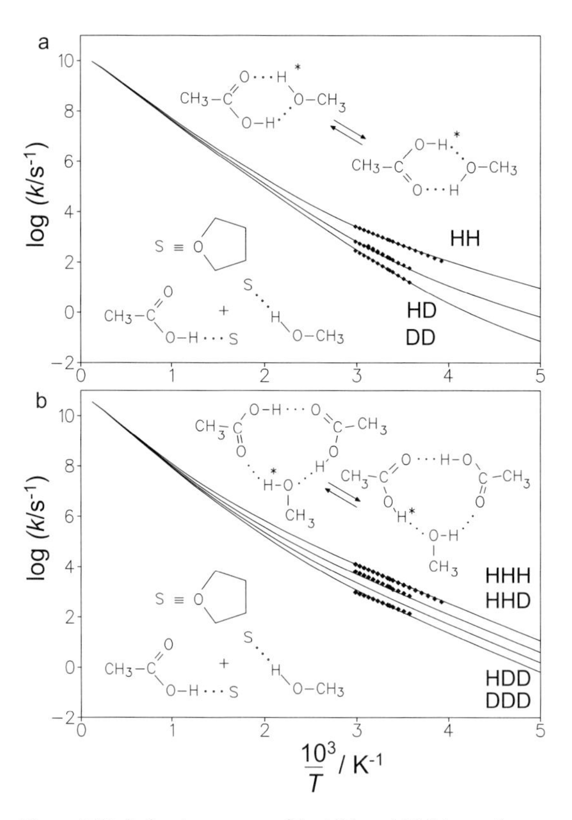

Figure 6.52 Arrhenius curves of the HH and HHH-transfer between acetic acid and methanol dissolved in tetrahydrofuran. Adapted from Ref. [9].

tion. At low concentration a second-order rate law was obtained indicating a HH-transfer in a cyclic 1:1 hydrogen bonded complex between acetic acid and methanol. At higher concentrations, the rate law changed indicating the participation of two acetic acid molecules, i.e. a HHH process. The multiple kinetic isotope effects were measured as a function of the inverse temperature as illustrated in Fig. 6.52.

For the double proton transfer two large kinetic HH/HD and HD/DD isotope effects of about 5 and 3 were observed, consistent with the pattern of Fig. 6.14(a) expected for a single barrier process. In the latter, tunneling was not yet included, which leads to the observed deviation from the Rule of the Geometric Mean. Recently, this reaction has been modeled using the instanton approach by Fernández-Ramos et al. [89]. The Arrhenius curves could be reproduced. The calculated geometries of the initial and the transition state are depicted in Fig. 6.53. In the latter, a proton is shifted towards the oxygen atom of methanol, but it is not completely transferred, rather, a strong hydrogen bond is formed.

The Arrhenius diagram of the HHH-transfer in the 2:1 complex is depicted in Fig. 6.52(b). The kinetic isotope effects are similar to those expected for a single barrier process according to Fig. 6.16(a). They exhibit little dependence on temperature, indicating a rather narrow barrier. Unfortunately, the reacting complex could not be observed directly and its structure studied in more detail. Note, however, that this complex represents a model for the catalytic sites of proteases as proposed by Northrop [90].

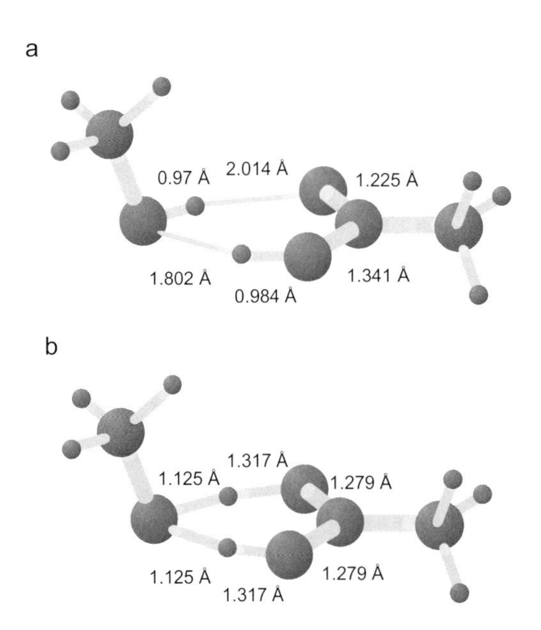

Figure 6.53 Structures of the initial state (a) and of the transition state (b) of the 1:1 complex between acetic acid and methanol according to Fernández-Ramos et al. [89]. The geometries were fully optimized at QCISD (quadratic configuration interaction including single and double substitutions) level of theory.

6.3.2.3 H-transfers in Complex Systems

Although the model systems of the previous section already involved major molecular motions, the latter can be even much more complex in living systems. Here, only two extreme examples are considered, i.e. hydride transfer in an enzyme and proton transfer in pure methanol.

6.3.2.3.1 The Case of H-transfer in Thermophilic Alcohol Dehydrogenase (ADH)

Firstly, let us discuss the example of a thermophilic alcohol dehydrogenase from *Bacillus stearothermophilus* (*bs*ADH) studied by Kohen et al. [91, 92]. This enzyme catalyzes the abstraction of a hydride to the nicotinamide cofactor NAD^+ as depicted in Fig. 6.54. The Arrhenius diagram is depicted in Fig. 6.54(a); a sudden decrease in the apparent slope and the apparent intercept of the Arrhenius curves is observed around room temperature (Fig. 6.54(b)). The puzzling observation is that the kinetic isotope effects are independent of temperature in the high-temperature regime but dependent on temperature in the low-temperature regime.

The solid lines in Fig. 6.54 were calculated recently [54] assuming the simple reaction network of Fig. 6.55. It is assumed that the enzyme adopts two different

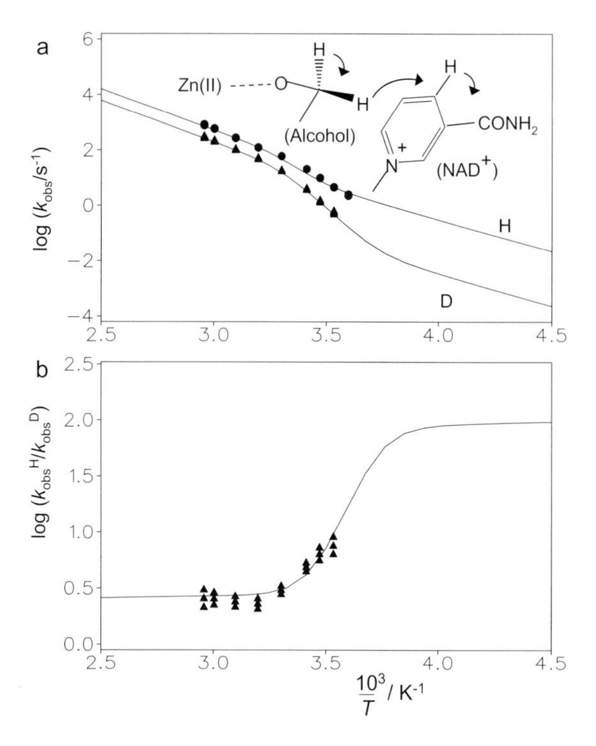

Figure 6.54 Arrhenius curves (a) and kinetic H/D isotope effects (b) of the intrinsic H-transfer in a thermophilic alcohol dehydrogenase (ADH) according to Kohen et al. [91]. The solid lines were calculated using the parameters listed in Table 6.4.

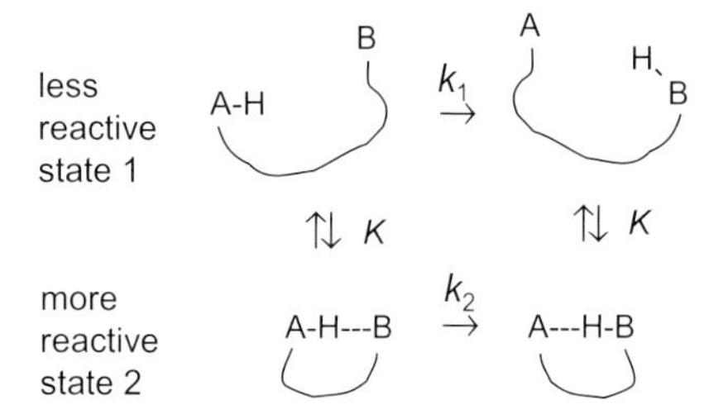

Figure 6.55 Conformational dependent H-transfer in a biomolecule.

states 1 and 2 at equilibrium (K), where 1 is less reactive than 2. In the less reactive state 1, dominating at lower temperatures, the rate constant of H-transfer is given by k_1, but in the more reactive state, dominating at higher temperatures, it is given by k_2. Assuming again that the H-transfer is slower than the conversions between the states the following expression is obtained by modification of Eq. (6.43), i.e.

$$k = x_1 k_1 + x_2 k_2 = k_1 \frac{1}{1+K} + k_2 \frac{K}{1+K} \tag{6.55}$$

x_1 and x_2 correspond to the mole fractions of states 1 and 2 and K is again the equilibrium constant of the formation of state 2 from state 1. According to Table 6.4, state 2 dominates at higher temperatures, in spite of its higher energy, because of its very large positive entropy. This state could be one where the protein has become ideally flexible for proper activity, in contrast to the low-temperature regime. This conclusion is in accordance with the fact that this *bs*ADH was evolved to function at ~65 C and with qualitative suggestions proposed in the past to rationalize the curved Arrhenius plot [91].

The tunnel parameters included in Table 6.4 indicate a ground state tunneling situation at high and at low temperatures, with temperature-independent kinetic isotope effects. The apparent temperature dependence observed at low temperatures is then the result of the transition between the two regimes, but does not arise from intrinsic temperature-dependent kinetic isotope effects. Note that in both states the pre-exponential factor of $10^{12.6}$ s^{-1} employed throughout this study was consistent with the data. The minimum energy for tunneling to occur is larger in the high-temperature state 2, but the barrier height and the barrier width are smaller than in the low temperature state 1. Thus, there seems to be a substantial change in the barrier parameters upon flexibilization of the enzyme at higher temperatures.

6.3.2.3.2 The Case of H-transfer in Pure Methanol and Calix[4]arene

As a second case of high complexity, let us discuss how protons are exchanged in pure protic liquids. This problem has been studied by Gerritzen et al. [93] who studied the inverse proton lifetimes τ_{AH}^{-1} of $CH_3OH \equiv AH$ in the pure liquid and of CH_3OH present as isotopic impurity in chemically pure $CH_3OD \equiv AD$ as a function of temperature. A mechanism involving the transfer of a small number n of protons in relatively stable cyclic hydrogen bonded intermediates $(AH)_n$ according to Fig. 6.10 to 6.18 was discarded. One reason was the finding that the addition of an organic solvent to methanol immediately quenches the proton exchange observed for the pure liquid. This would not be the case if the transfer takes place in cyclic hydrogen bonded intermediates.

Therefore, an autoprotolysis mechanism was proposed consisting of dissociation:

$$AH + AH \underset{k_n}{\overset{k_d}{\rightleftarrows}} A^- + AH_2^+ \tag{6.56}$$

$$\text{cation propagation } AH_2^+ + AH \overset{k_1}{\rightleftarrows} AH + AH_2^+ \tag{6.57}$$

$$\text{anion propagation } AH^- + AH \overset{k_2}{\rightleftarrows} AH^- + AH \tag{6.58}$$

$$\text{neutralization } A^- + AH_2^+ \underset{k_d}{\overset{k_n}{\rightleftarrows}} AH + AH \tag{6.59}$$

The concentration of methoxonium and methoxide ions in the pure liquid is given by the autoprotolysis constant

$$K^H = \frac{k_d}{k_n} = c_{AH_2^+} c_{AH^-} \tag{6.60}$$

where K^H represents the autoprotolysis constant, which is 2.76×10^{-17} mol^2 l^{-2} at 298 K, i.e. $c_{AH_2^+} = c_{AH^-} = 5.2 \times 10^{-9}$ mol l^{-1} [94]. The autoprotolysis mechanism is immediately suppressed by adding an organic solvent as it reduces the dielectric constant and hence the autoprotolysis constant.

The following expression for the inverse proton life times follows from the autoprotolysis mechanism in CH_3OH a straightforward way [93]

$$\tau_{AH}^{-1} = (k_1^H c_{AH_2^+} + k_2^H c_{AH^-}) = (k_1^H + k_2^H)\sqrt{K^H} \tag{6.61}$$

For 1% CH_3OH in CH_3OD it was shown that

$$\tau_{AH}^{-1}(CH_3OD) = (k_1^H(CH_3OD) + k_2^H(CH_3OD))\sqrt{K^D} \tag{6.62}$$

where K^D represents the autoprotolysis constant of CH_3OD. Let E_{a1}^H and E_{a2}^H be the energies of activation describing the temperature dependence of k_1^H and k_2^H. By assuming that $E_{a1}^H \cong E_{a2}^H$ it follows then from Eq. (6.61) that the effective energy of activation of proton exchange is given by

$$E_a^H = E_{a1}^H + \frac{\Delta H^H}{2} \tag{6.63}$$

where ΔH^H represents the enthalpy of autoprotolysis of CH_3OH. In other words, if the cation and the anion are not created by autoprotolysis but stem from acid and basic impurities, their concentration will be temperature independent and E_a^H is equal to E_{a1}^H. When the impurities are, however, removed, E_a^H is substantially increased. This is indeed observed, as illustrated in the Arrhenius plot of Fig. 6.56(a). The solid lines are given by

$$\begin{aligned} \tau_{AH}^{-1}(CH_3OH) &= 10^{6.1}\exp(-28.5\,\mathrm{kJmol}^{-1}/RT) \\ \tau_{AH}^{-1}(CH_3OD) &= 10^{5.7}\exp(-29.3\,\mathrm{kJmol}^{-1}/RT) \end{aligned} \quad (6.64)$$

The kinetic isotope effects observed are given by [93]

$$\frac{\tau_{AH}^{-1}(CH_3OH)}{\tau_{AH}^{-1}(CH_3OD)} = \frac{(k_1^H(CH_3OH)+k_2^H(CH_3OH))}{(k_1^H(CH_3OD)+k_2^H(CH_3OD))} \times \sqrt{\frac{K^H}{K^D}} = 3.2 \text{ at } 298\text{ K} \quad (6.65)$$

This overall kinetic isotope effect represents the product of an average kinetic isotope effect of the two propagation steps times the equilibrium isotope effect of the

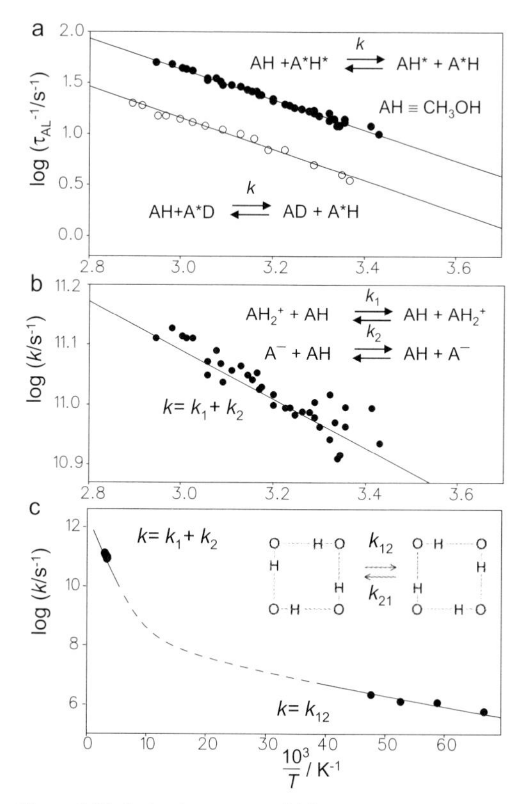

Figure 6.56 Arrhenius curves of (a) proton exchange in pure CH_3OH and CH_3OD. Adapted from Gerritzen et al. [93]. (b) Arrhenius curves of the elementary proton transfers in methanol calculated from the data of Fig. 6.56(a) and the known ionization constant. (c) Kinetic data of the HHHH-transfer in solid *p-tert*-butyl calix[4]arene reported by Horsewill et al. [80] combined with those of Fig. 6.56(a).

autoprotolysis. The value for water solutions was shown to be $K^H/K^D = 5$ at 298 K [95], whereas for methanol solutions a value of 6.5 was predicted [96]. Thus, the average kinetic isotope effect of the propagation is between 1.3 and 1.4. This effect corresponds to the usual isotope ratio expected for the reorientation of CH_3OH and CH_3OD.

Using Eq. (6.61), the inverse life times $\tau_{AH}^{-1}(CH_3OH)$ were converted into the sum $k_1^H + k_2^H = k_1^H(CH_3OH) + k_2^H(CH_3OH)$ plotted in Fig. 6.56(b) as a function of the inverse temperature. The solid line was given by [93]

$$k_1^H + k_2^H = 10^{12.3}\exp(-7.7\,\mathrm{kJmol^{-1}}/RT),\ k_1^H + k_2^H = 9.3\times10^{10}\mathrm{s^{-1}}\ \text{at 298 K} \qquad (6.66)$$

It follows that $\Delta H^H \cong 40$ kJ mol^{-1}. Note that Grunwald et al. [97] measured values of $k_1 = 8.8 \times 10^{10}$ s^{-1} and $k_2 = 1.85 \times 10^{10}$ s^{-1} for buffered solutions; their sum is in very good agreement with the value obtained for pure methanol.

The values of $k_1^H + k_2^H$ for water are very close to those of methanol. Using the known dependence of K^H as a function of temperature, the proton lifetimes in pure water were estimated [93]

$$\tau_{AH}^{-1}(H_2O) = 10^{9.9}\exp(-39\,\mathrm{kJmol^{-1}}/RT) = 15000\ \mathrm{s^{-1}}\ \text{at 298 K} \qquad (6.67)$$

Thus, proton exchange is faster than in pure methanol because of the larger autoprotolysis constant.

For comparison, let us compare the methanol data with those obtained by Horsewill et al. [80] who have reported an intramolecular quadruple proton transfer in the solid state between the four OH groups of solid calix[4]arene. Almost temperature independent rate constants were observed, which are again indicative of tunneling. An Arrhenius curve can be calculated using reasonable parameters (Table 6.4) which can reproduce both the pure methanol and the calix[4]arene data. It would be interesting to know more about the kinetic isotope effects in both systems.

6.4 Conclusions

In this chapter, the Bell–Limbach tunneling model has been applied to describe the Arrhenius curves of a number of single and multiple hydrogen transfer reactions. This model contains a number of parameters which can be obtained by simulation of the Arrhenius curves when enough experimental data are available. It is proposed to describe concerted multiple H-transfers in terms of a single barrier process. Multiple kinetic isotope effects of stepwise transfers can be treated in terms of formal kinetic reaction theory, where in each step one or more protons can be transferred, again in a concerted way. This approach is justified if each step can be described in terms of rate constants. This may not be the case for very strong hydrogen bonds, where H can be delocalized. Each reaction step can be treated as a first approach in terms of the Bell–Limbach or any other tunneling model, or in terms of a more sophisticated quantum-mechanical theory.

A main result of the examples discussed above is that pre-exponential factors of H-transfers coupled to only minor heavy atom motions are of the order of $kT/h \cong 10^{12.6}$ s^{-1}, the value predicted by Eyring's transition state theory for the high-pressure limit [5]. Deviations are then a first and important diagnostic tool for detecting so far unrecognized heavy atom motions and pre-equilibria such as conformational isomerism and hydrogen bond equilibria. Two types of heavy atom motions are considered, i.e. those which precede the H tunnel process and those which take place during the tunnel process. The latter give rise to an increased tunneling mass which reduces kinetic H/D isotope effects arising from tunneling. Thus, this model helps experimentalists to interpret their kinetic data, but does not preclude further quantum-mechanical studies of the hydrogen transfer steps.

Acknowledgements

This research has been supported by the Deutsche Forschungsgemeinschaft, Bonn, and the Fonds der Chemischen Industrie (Frankfurt). I am indebted to Professors Maurice Kreevoy, Minneapolis, Minnesota, USA; R. L. Schowen, Lawrence, Kansas, USA and G. S. Denisov, St. Petersburg, Russian Federation for stimulating discussions over three decades. I also thank Professors G. S. Denisov and R. L. Schowen for carefully reading the manuscript and for their helpful comments.

References

1 E. Caldin, V. Gold (Eds.), *Proton Transfer Reactions*, Chapman and Hall: London, 1975.
2 R. A. Marcus, *J. Chem. Phys.* **1966**, *45*, 4493–4499.
3 P. Schuster, G. Zundel, C. Sandorfy (Eds.), *The Hydrogen Bond*, North Holland, Amsterdam, 1976, Vol 1–3.
4 (a) H. H. Limbach, Dynamic NMR Spectroscopy in the Presence of Kinetic Hydrogen/ Deuterium Isotope Effects, in *NMR Basic Principles and Progress, Deuterium and Shift Calculation*, Springer-Verlag, Heidelberg 1991, Vol. 23, pp. 66–167; (b) H. H. Limbach, Dynamics of Hydrogen Transfer in Liquids and Solids, in *Encyclopedia of Nuclear Magnetic Resonance*, D. M. Grant, R. K. Harris (Eds.), *Advances in NMR*, John Wiley & Sons, Chichester, 2002, Suppl. Vol. 9, pp. 520–531.
5 S. Glasstone, K. J. Laidler, H. Eyring, *The Theory of Rate Processes*. McGraw-Hill, New York 1941.
6 (a) J. Bigeleisen, *J. Chem. Phys.* **1949**, *17*, 675–678; (b) J. Bigeleisen, *J. Chem. Phys.* **1955**, *23*, 2264–2267; (c) L. Melander, W. H. Saunders, *Reaction Rates of Isotopic Molecules*, Krieger, Malabar, FL, 1987.
7 (a) R. P. Bell, *The Proton in Chemistry*, 2nd edn. Chapman and Hall, London, **1973**; (b) R. P. Bell, *The Tunnel Effect*, Chapman and Hall, London, **1980**.
8 G. Brunton, J. A. Gray, D. Griller, L. R. C. Barclay, K. U. Ingold, *J. Am. Chem. Soc.* **1978**, *100*, 4197–4200.

9 (a) D. Gerritzen, H. H. Limbach, *J. Am. Chem. Soc.* **1984**, *106*, 869–879; (b) H. H. Limbach, W. Seiffert, *J. Am. Chem. Soc.* **1980**, *102*, 538–542.

10 J. Basran, S. Patel, M. J. Sutcliffe, N. S. Scrutton, *J. Biol. Chem.* **2001**, *276*, 6234–6242.

11 (a) N. Bruniche-Olsen, J. Ulstrup, *J. Chem. Soc., Faraday Trans. 1* **1979**, *75*, 205–226; (b) E. D. German, A. M. Kuznetsov, R. R. Dogonadze, *J. Chem. Soc., Faraday Trans. 2* **1980**, *76*, 1128–1146.

12 (a) A. M. Kuznetsov, J. Ulstrup, *Can. J. Chem.* **1999**, *77*, 1085–1096; (b) A. M. Kuznetsov, J. Ulstrup, *Proton Transfer and Proton Conductivity in Condensed Matter Environment.* In: *Isotope Effects in the Biological and Chemical Sciences*, A. Kohen, H. H. Limbach (Eds.) Taylor & Francis, Boca Raton FL, 2005, Ch. 26, pp. 691–724.

13 M. J. Knapp, K. Rickert, J. P. Klinman, *J. Am. Chem. Soc.* **2002**, *124*, 3865–3874.

14 (a) W. Siebrand, T. A. Wildman, M. Z. Zgierski, *J. Am. Chem. Soc.* **1984**, *106*, 4083–4089; (b) W. Siebrand, T. A. Wildman, M. Z. Zgierski, *J. Am. Chem. Soc.* **1984**, *106*, 4089–4096.

15 D. Truhlar, Variational Transition-State Theory and Multidimensional Tunneling for Simple and Complex Reactions in the Gas Phase, Solids, Liquids, and Enzymes, in Ref. [12b], Ch. 22, pp. 579–620.

16 Z. Smedarchina, W. Siebrand, A. Fernández-Ramos, Kinetic Isotope Effects in Multiple Proton Transfer, in Ref. [12b], Ch. 22, pp. 521–548.

17 O. Brackhagen, Ch. Scheurer, R. Meyer, H. H. Limbach, *Ber. Bunsenges. Phys. Chem.* **1998**, *102*, 303–316.

18 (a) H. H. Limbach, J. Hennig, D. Gerritzen, H. Rumpel, *Faraday Discuss., Chem. Soc.* **1982**, *74*, 229–243; (b) M. Schlabach, B. Wehrle, H. Rumpel, J. Braun, G. Scherer, H. H. Limbach, *Ber. Bunsenges. Phys. Chem.* **1992**, *96*, 821–833; (c) B. Wehrle, H. H. Limbach, M. Köcher, O. Ermer, E. Vogel, *Angew. Chem.* **1987**, *99*, 914–917; *Angew. Chem. Int. Ed. Engl.* **1987**, *26*, 934–936; (d) J. Braun, M. Schlabach, B. Wehrle, M. Köcher, E. Vogel, H. H. Limbach, *J. Am. Chem. Soc.* **1994**, *116*, 6593–6604; (e) J. Braun, H. H. Limbach, P. G. Williams, H. Morimoto, D. Wemmer, *J. Am. Chem. Soc.* **1996**, *118*, 7231–7232.

19 (a) M. Schlabach, H. Rumpel, H. H. Limbach, *Angew. Chem.* **1989**, *101*, 84–87; *Angew. Chem. Int. Ed. Engl.* **1989**, *28*, 76–79; (b) M. Schlabach, G. Scherer, H. H. Limbach, *J. Am. Chem. Soc.* **1991**, *113*, 3550–3558.

20 (a) M. Schlabach, B. Wehrle, H. H. Limbach, E. Bunnenberg, A. Knierzinger, A. Shu, B. R. Tolf, C. Djerassi, *J. Am. Chem. Soc.* **1986**, *108*, 3856–3858; (b) M. Schlabach, H. H. Limbach, E. Bunnenberg, A. Shu, B. R. Tolf, C. Djerassi, *J. Am. Chem. Soc.* **1993**, *115*, 4554–4565.

21 (a) H. Rumpel, H. H. Limbach, *J. Am. Chem. Soc.* **1989**, *111*, 5429–5441; (b) H. Rumpel, H. H. Limbach, G. Zachmann, *J. Phys. Chem.* **1989**, *93*, 1812–1818.

22 (a) G. Otting, H. Rumpel, L. Meschede, G. Scherer. H. H. Limbach, *Ber. Bunsenges. Phys. Chem.* **1986**, *90*, 1122–1129; (b) G. Scherer, H. H. Limbach, *J. Am. Chem. Soc.* **1989**, *111*, 5946–5947; (c) G. Scherer, H. H. Limbach, *J. Am. Chem. Soc.* **1994**, *116*, 1230–1239; (d) G. Scherer, H. H. Limbach. *Croat. Chem. Acta*, **1994**, *67*, 431–440.

23 (a) J. Braun, C. Hasenfratz, R. Schwesinger, H. H. Limbach, *Angew. Chem.* **1994**, *106*, 2302–2304; *Angew. Chem. Int. Ed. Engl.* **1994**, *33*, 2215–2217; (b) J. Braun, R. Schwesinger, P. G. Williams, H. Morimoto, D. E. Wemmer, H. H. Limbach, *J. Am. Chem. Soc.* **1996**, *118*, 11101–11110.

24 (a) L. Meschede, D. Gerritzen, H. H. Limbach, *Ber. Bunsenges. Phys. Chem.* **1988**, *92*, 469–485; (b) H. H. Limbach, L. Meschede, G. Scherer, *Z. Naturforsch.a* **1989**, *44*, 459–471; (c) L. Meschede, H. H. Limbach, *J. Phys. Chem.* **1991**, *95*, 10267–10280.

25 (a) F. Aguilar-Parrilla, O. Klein, J. Elguero, H. H. Limbach, *Ber. Bunsenges. Phys. Chem.* **1997**, *101*, 889–901; (b) O. Klein, M. M. Bonvehi, F. Aguilar-Parrilla, J. Elguero, H. H. Limbach, *Isr. J. Chem.* **1999**, *34*, 291–299.

26 H. H. Limbach, O. Klein, J. M. Lopez Del Amo, J. Elguero, *Z. Phys. Chem.* **2004**, *217*, 17–49.
27 O. Klein, F. Aguilar-Parrilla, J. M. Lopez, N. Jagerovic, J. Elguero, H. H. Limbach, *J. Am. Chem. Soc.* **2004**, *126*, 11718–11732.
28 J. Brickmann, H. Zimmermann, *Ber. Bunsenges. Phys. Chem.* **1966**, *70*, 157–165; **1966**, *70*, 521–524; **1967**, *71*, 160–1964; J. Brickmann, H. Zimmermann, *J.Chem.Phys.* **1969**, *50*, 1608–1618.
29 P. M. Morse, E. C. G. Stückelberg, *Helv. Phys. Acta* **1931**, *4*, 335–354; R. L. Somorjai, D. F. Hornig, *J.Phys.Chem.* **1962**, *36*, 1980–1987; J. Laane, *Appl. Spectrosc.* **1970**, *24*, 73–80.
30 W. F. Rowe, R. W. Duerst, E. B. Wilson, *J. Am. Chem. Soc.* **1976**, *98*, 4021–4023; S. L. Baughcum, R. W. Duerst, W. F.Rowe, Z. Smith, E. B. Wilson, *J. Am. Chem. Soc.* **1981**, *103*, 6296–6393; S. L. Baughcum, Z. Smith, E. B. Wilson, R. W. Duerst, *J. Am. Chem. Soc.* **1984**, *106*, 2260–2265.
31 R. L. Redington, R. L. Sams, *J. Phys. Chem. A* **2002**, *106*, 7494–7511.
32 F. Madeja, M. Havenith, *J. Chem. Phys.* **2002**, *117*, 7162–7168.
33 D. W. Firth, P. F. Barbara, H. P. Trommsdorff, *Chem.Phys.* **1989**, *136*, 349–360.
34 J. D. McDonald, *Annu. Rev. Phys. Chem.* **1979**, *30*, 29–50.
35 D. Borgis, J. T. Hynes, *J. Chem. Phys.* **1991**, *94*, 3619–3628; D. Borgis, J. T. Hynes, *Chem. Phys.* **1993**, *170*, 315–346.
36 H. H. Limbach, G. Buntkowsky, J. Matthes, S. Gründemann, T. Pery, B. Walaszek, B. Chaudret, *Chem. Phys. Chem.* **2006**, *7*, 551–554; G. Buntkowsky, B. Walaszek, A. Adamczyk, Y. Xu, H. H. Limbach, B. Chaudret, *Phys. Chem. Chem. Phys.* **2006**, *8*, 1929–1935.
37 R. P. Bell, *J. Chem. Soc., Faraday Discuss.* **1965**, *39*, 16–24.
38 (a) L. Pauling, *J. Am. Chem. Soc.* **1947**, *69*, 542–553; (b) I. D. Brown, *Acta Crystallogr., Sect. B* **1992**, *48*, 553–572.
39 (a) T. Steiner, *J. Chem. Soc., Chem. Commun.* **1995**, 1331–1332; (b) T. Steiner, *J. Phys. Chem. A* **1998**, *102*, 7041–7052; (c) T. Steiner, *Angew. Chem. Int. Ed. Engl.* **2002**, *41*, 48-76.
40 H. H. Limbach, G. S. Denisov, N. S. Golubev, Hydrogen Bond Isotope Effects Studied by NMR, in Ref [12b], Ch. 7, pp. 193–252.
41 (a) H. S. Johnston, *Gas Phase Reaction Rate Theory. Modern Concepts in Chemistry*, The Ronald Press Company, New York 1966; (b) D. G. Truhlar, *J. Am. Chem. Soc.* **1972**, *94*, 7584–7586; (c) N. Agmon, *Chem. Phys. Lett.* **1977**, *45*, 343–345.
42 H. H. Limbach, M. Pietrzak, H. Benedict, P. M. Tolstoy, N. S. Golubev, G. S. Denisov, *J. Mol. Struct.* **2004**, *706*, 115–119.
43 H. H. Limbach, M. Pietrzak, S. Sharif, P. M. Tolstoy, I. G. Shenderovich, S. N. Smirnov, N. S. Golubev, G. S. Denisov, *Chem. Eur. J.* **2004**, *10*, 5195–5204.
44 K. F. Wong, T. Selzer, S. J. Benkovic, S. Hammes-Schiffer, *Proc. Nat. Acad. Sci. USA* **2005**, *102*, 6807–6812.
45 F. H. Westheimer, *Chem. Rev.* **1961**, *61*, 265–273.
46 R. A. More O'Ferrall, *Substrate Isotope Effects.* in Ref. [1], Ch. 8, pp. 201–262.
47 (a) G. Wentzel, *Z. Phys.* **1926**, *138*, 518–529; (b) H. A. Kramers, *Z. Phys.* **1926**, *39*, 828–840; (c) L. Brillouin, *C. R. Acad. Sci*, **1926**, *153*, 24–26.
48 Ref. [7a], p. 275.
49 U. Langer, L. Latanowicz, Ch. Hoelger, G. Buntkowsky, H. M. Vieth, H. H. Limbach, *Phys. Chem. Chem. Phys.* **2001**, *3*, 1446–1458.
50 S. N. Smirnov, H. Benedict, N. S. Golubev, G. S. Denisov, M. M. Kreevoy, R. L. Schowen, H. H. Limbach, *Can. J. Chem.* **1999**, *77*, 943–949.
51 S. Schweiger, G. Rauhut, *J. Phys. Chem. A* **2003**, *107*, 9668–9678.
52 M. Eigen, *Angew. Chem. Int. Ed. Engl.* **1964**, *3*, 1–23.
53 (a) F. Männle, H. H. Limbach, *Angew. Chem. Int. Ed. Engl.* **1996**, *35*, 441–442; (b) H. H. Limbach, F. Männle, C. Detering, G. S. Denisov, *Chem. Phys.* **2005**, *319*, 69–92.

54 H. H. Limbach, J. M. Lopez del Amo, A. Kohen, *Philos. Trans. B.* **2006**, *361*, 1399–1415.

55 R. M. Claramunt, J. Elguero, A. R. Katritzky, *Adv. Heterocycl. Chem.* **2000**, *77*, 1–50.

56 D. K. Maity, T. N. Truong, *J. Porph. Phthal.* **2001**, *5*, 289–299.

57 C. B. Storm, Y. Teklu, *J. Am. Chem. Soc.* **1974**, *94*, 1745–1747; C. B. Storm, Y. Teklu, *Ann. N. Y. Acad. Sci.* **1973**, *206*, 631–640.

58 (a) J. Hennig, H. H. Limbach, *J. Chem. Soc., Faraday Trans.2* **1979**, *75*, 752–766; (b) H. H. Limbach, J. Hennig, R. Kendrick, C. S. Yannoni, *J. Am. Chem. Soc.* **1984**, *106*, 4059–4060.

59 P. Stilbs, M. E. Moseley, *J. Chem. Soc., Faraday Trans. 2* **1980**, *76*, 729–731.

60 J. Hennig, H. H. Limbach, *J. Magn. Reson.* **1982**, *49*, 322–328.

61 A. A. Bothner-By, R. L. Stephens, J. Lee, C. D. Warren, R. W. Jeanloz, *J. Am. Chem. Soc.* **1984**, *106, 811–813.*

62 A. Bax, D. G. Davis, *J. Magn. Reson.* **1985**, *63*, 207–213.

63 H. H. Limbach, J. Hennig, *J. Chem. Phys.* **1979**, *71*, 3120–3124; H. H. Limbach, J. Hennig, J. Stulz, *J. Chem. Phys.* **1983**, *78*, 5432–5436 ; H. H. Limbach, *J. Chem. Phys.* **1984**, *80*, 5343–5345.

64 A. Sarai, *Chem. Phys. Lett.* **1981**, *83*, 50–54 ; A. Sarai, *J. Chem. Phys.* **1982**, *76*, 5554–5563; A. Sarai, *J. Chem. Phys.* **1984**, *80*, 5341–5343.

65 D. K. Maity, R. L. Bell, T. N. Truong, *J. Am. Chem. Soc.* **2000**, *122*, 897–906.

66 (a) R. D. Kendrick, S. Friedrich, B. Wehrle, H. H. Limbach, C. S. Yannoni, *J. Magn. Reson.* **1985**, *65*, 159–161; (b) B. Wehrle, H. H. Limbach, *Chem. Phys.* **1989**, *136*, 223–247.

67 T. J. Butenhoff, C. B. Moore, *J. Am. Chem. Soc.* **1988**, *110*, 8336–8341.

68 J. D. Thoburn, W. Lüttke, C. Iber, H. H. Limbach, *J. Am. Chem. Soc.* **1996**, *118*, 12459–12460.

69 C. G. Hoelger, B. Wehrle, H. Benedict, H. H. Limbach, *J. Phys. Chem.* **1994**, *98*, 843–851.

70 U. Langer, C. Hoelger, B. Wehrle, L. Latanowicz, E. Vogel, H. H. Limbach, *J. Phys. Org. Chem.* **2000**, *13*, 23–34.

71 B. Wehrle, H. Zimmermann, H. H. Limbach, *J. Am. Chem. Soc.* **1988**, *110*, 7014–7024.

72 (a) B. H. Meier, F. Graf, R. R. Ernst, *J. Chem. Phys.* **1982**, *76*, 767–774; (b) A. Stöckli, B. H. Meier, R. Kreis, R. Meyer, R. R. Ernst, *J. Chem. Phys.* **1990**, *93*, 1502–1520; (c) A. Heuer, U. Haeberlen, *J. Chem. Phys.* **1991**, *95*, 4201–4124.

73 (a) J. L. Skinner, H. P. Trommsdorff, *J. Phys. Chem. A* **1988**, *89*, 897–907; (b) R. Meyer, R. R. Ernst, *J. Chem. Phys.* **1990**, *93*, 5518–5532; (c) Y. Kim, *J. Am. Chem. Soc.* **1996**, *118*, 1522–1528; (d) Y. Kim, *J. Phys. Chem. A* **1998**, *102*, 3025–3036.

74 Q. A. Xue, A. J. Horsewill, M. R. Johnson, H. P. Trommsdorff, *J. Chem. Phys.* **2004**, *120*, 11107–11119.

75 P. M. Tolstoy, P. Schah-Mohammedi, S. N. Smirnov, N. S. Golubev, G. S. Denisov, H. H. Limbach, *J. Am. Chem. Soc.* **2004**, *126*, 5621–5634.

76 F. Aguilar-Parrilla, G. Scherer, H. H. Limbach, M. C. Foces-Foces, F. H. Cano, J. A. S. Smith, C. Toiron, J. Elguero, *J. Am. Chem. Soc.* **1992**, *114*, 9657–9659.

77 (a) A. Baldy, J. Elguero, R. Faure, M. Pierrot, E. J. Vicent, *J. Am. Chem. Soc.* **1985**, *107*, 5290–5291; (b) J. A. S. Smith, B. Wehrle, F. Aguilar-Parrilla, H. H. Limbach, M. C. Foces-Foces, F. H. Cano, J. Elguero, A. Baldy, M. Pierrot, M. M. T. Khurshid, J. B. Larcombe-McDouall, *J. Am. Chem. Soc.* **1989**, *111*, 7304–7312; (c) F. Aguilar-Parrilla, C. Cativiela, M. D. Diaz de Villegas, J. Elguero, M. C. Foces-Foces, J. I. G. Laureiro, F. H. Cano, H. H. Limbach, J. A. S. Smith, C. Toiron, *J. Chem. Soc., Perkin 2* **1992**, 1737–1742; (d) J. Elguero, G. I. Yranzo, J. Laynez, P. Jiménez, M. Menédez, J. Catalán, J. L. G. De Paz, F. Anvia, R. W. Taft, *J. Org. Chem.* **1991**, *56*, 3942; (e) J. Elguero, F. H. Cano, M. C. Foces-Foces, A. Llamas-Saiz, H. H. Limbach, F. Aguilar-Parrilla, R. M. Claramunt, C. Lopez, *J. Heterocycl. Chem.* **1994**, *31*, 695–700; (f) F. Aguilar-Parrilla, F. Männle, H. H. Limbach, J. Elguero, N. Jagerovic, *Magn. Reson. Chem.* **1994**,

32, 699–702; (g) A. Llamaz-Saiz, M. C. Foces-Foces, F. H. Cano, P. Jimenez, J. Laynez, W. Meutermans, J. Elguero, H. H. Limbach, F. Aguilar-Parrilla, *Acta Crystallogr., Sect. B* **1994**, *50,* 746–762; (h) F. Aguilar-Parrilla, H. H. Limbach, M. C. Foces-Foces, F. H. Cano, N. Jagerovic, J. Elguero, *J. Org. Chem.* **1995**, *60,* 1965–1970; (i) J. Elguero, N. Jagerovic, M. C. Foces-Foces, F. H. Cano, M. V. Roux, F. Aguilar-Parrilla, H. H. Limbach, *J. Heterocycl. Chem.* **1995**, *32,* 451–456; (j) C. Lopez, R. M. Claramunt, A. Llamas-Saiz, M. C. Foces-Foces, J. Elguero, I. Sobrados, F. Aguilar-Parrilla, H. H. Limbach, *New J. Chem.* **1996**, 20, 523–536; (k) C. Hoelger, H. H. Limbach, F. Aguilar-Parrilla, J. Elguéro, O. Weintraub, S. Vega, *J. Magn. Res.* **1996**, *A120,* 46–55; (l) J. Catalan, J. l. M. Abboud, J. Elguero, *Adv. Heterocycl. Chem.* **1987**, *41,* 187; (m) F. Toda, K. Tanaka, M. C. Foces-Foces, A. Llamas-Saiz, H. H. Limbach, F. Aguilar-Parrilla, R. M. Claramunt, C. Lopez, J. Elguero, *J. Chem. Soc., Chem. Commun.* **1993**, 1139–1142; (n) F. Aguilar-Parrilla, R. M. Claramunt, C. Lopez, D. Sanz, H. H. Limbach, J. Elguero, *J. Phys. Chem.* **1994**, *98,* 8752–6760; (o) C. Foces-Foces, A. Echevarría, N. Jagerovic, I. Alkorta, J. Elguero, U. Langer, O. Klein, M. Minguet-Bonvehí, H. H. Limbach, *J. Am. Chem. Soc.* **2001**, *123,* 7898–7906.

78 J. L. G. de Paz, J. Elguero, M. C. Foces-Foces, A. Llamas-Saiz, F. Aguilar-Parrilla, O. Klein, H. H. Limbach, *J. Chem. Soc. Perkin Trans. 2* **1997**, 101–109.

79 (a) S. Schweiger, B. Hartke, G. Rauhut, *J. Phys. Chem. A* **2003**, *107,* 9668–9878; (b) S. Schweiger, B. Hartke, G. Rauhut, *Phys. Chem. Chem. Phys.* **2005**, 7, 493–500.

80 D. F. Brougham, R. Caciuffo, A. J. Horsewill, *Nature* **1999**, *397,* 241–243.

81 T. Vangberg, A. Ghosh, *J. Phys. Chem. B* **1997**, *101,* 1496–1497.

82 (a) F. Männle, I. Wawer, H. H. Limbach, *Chem. Phys. Lett.* **1996**, *256,* 657–662; (b) R. Anulewicz, I. Wawer, T. M. Krygowski, F. Männle, H. H. Limbach, *J. Am. Chem. Soc.* **1997**, *119,* 12223–12230.

83 W. Al-Soufi, K. H. Grellmann, B. Nickel, *J. Phys. Chem.* **1991**, *95,* 10503–10509.

84 E. F. Caldin, S. Mateo, *J. Chem. Soc., Faraday Trans.1* **1975**, *71,* 1876–1904.

85 (a) A. I. Prokofiev, N. N. Bubnov, S. P. Solodnikov, M. I. Kabachnik, *Tetrahedron Lett.* **1973**, 2479–2480; (b) N. N. Bubnov, S. P. Solodnikov, A. I. Prokofiev, M. I. Kabachnik, *Russ. Chem. Rev.* **1978**, *47,* 549–571.

86 K. Loth, F. Graf, H. Günthardt, *Chem. Phys.* **1976**, *13,* 95–113.

87 H. H. Limbach, D. Gerritzen, *Faraday Discuss. Chem. Soc.* **1982**, *74,* 279–296.

88 S. F. Bureiko, G. S. Denisov, N. S. Golubev, I. Y. Lange, *React. Kinet. Catal. Lett.* **1979**, *11,* 35–38.

89 A. Fernández-Ramos, Z. Smedarchina, J. Rodríguez-Otero, *J. Chem. Phys.* **2001**, *114,* 1567–1574.

90 D. B. Northrop, *Acc. Chem. Res.* **2001**, *34,* 790–797.

91 A. Kohen, R. Cannio, S. Bartolucci, J. P. Klinman, *Nature* **1999**, *399,* 496–499.

92 A. Kohen, *Kinetic Isotope Effects as Probes for Hydrogen Tunneling in Enzyme Catalysis.* In *Isotope Effects in the Biological and Chemical Sciences,* A. Kohen, H. H. Limbach (Eds.) Taylor & Francis, Boca Raton FL 2005, Ch. 28, pp. 743–765.

93 D. Gerritzen, H. H. Limbach, *Ber. Bunsenges. Phys. Chem.* **1981**, *85,* 527–535.

94 C. S. Leung and E. Grunwald, *J. Phys. Chem.* **1970**, *74,* 696–701.

95 E. J. King, in *Physical Chemistry of Organic Solvent Systems,* A. Covington, T. Dickinson (Eds.), Plenum Press, New York 1973, Ch. 3, pp. 331–403.

96 V. Gold, S. Grist, *J. Chem. Soc. B,* **1971**, 1665–1670.

97 E. Grunwald, C. F. Jumper, S. Meiboom, *J. Am. Chem. Soc.* **1963**, *84,* 4664–4671.

7
Intra- and Intermolecular Proton Transfer and Related Processes in Confined Cyclodextrin Nanostructures

Abderrazzak Douhal

7.1
Introduction and Concept of Femtochemistry in Nanocavities

In this chapter, we discuss the recent progress made in studying intramolecular and intermolecular reactions of proton (or hydrogen atom) transfer using cyclodextrins as host for selected systems undergoing these kinds of photoinduced reaction. The experiments establish the ultrafast nature of proton motion to convert a trapped reactant to a trapped photoproduct. Upon caging a molecule in a molecular pocket (nanochamber), it produces a confined structure with interesting physical and chemical properties. It reduces the degrees of freedom available to the molecule to move along the reaction coordinates, and confines the wavepacket in a small area of propagation [1, 2]. Therefore, by reducing the space for molecular relaxation, it makes the system robust and immune to transferring "damage" or heat over long distances.

Following an ultrafast electronic excitation of the embedded guest in a nanochamber, the nascent wavepacket is trapped in a small area of the potential-energy surface (caged wavepacket), and its evolution along this surface and the subsequent relaxation dynamics to other states (or through chemical reactions) will funnel and will be controlled by the restricted confined geometry. The system has open only a few channels to move along the reduced potential-energy surface, and the neighboring water molecules may direct its evolution. A similar situation occurs in semiconductors where the conduction electrons are not only particles but also waves. So, trapped in a confined area, electrons can only have energies dictated by the present wave patterns that will fit in this small region. Therefore, in a similar but easy way, a free electron can then possibly be trapped by a molecule-chamber entity such as cyclodextrins, calixarenes, Cram boxes, zeolites, as has been realized in solution and in finite clusters. Cooling (or vibrational relaxation) due to a fast (picosecond regime) exchange of heat with the environment (caging medium) might be also controlled by changing the nature or the size of the cage. It is well known that the nature of the solvent plays a key role in the issue of a chemical reaction, in bulk solvent and in a cavity. Understanding the ultrafast dynamics for different cages may help one to understand better the cata-

Hydrogen-Transfer Reactions. Edited by J. T. Hynes, J. P. Klinman, H. H. Limbach, and R. L. Schowen

ISBN: 978-3-527-30777-7

lytic mechanism in these cavities, like those involved in enzymes and zeolites. In addition, an interesting phenomenon in studying nanocavity solvation is the slow solvation of water confined in a nanospace such as those offered by cyclodextrins. This special water is reminiscent of that located at the surface of biological molecules [3–23].

Here, we will focus on fast (picosecond regime) and ultrafast (femtosecond regime) dynamics of proton (or H-atom) transfer and related events that may occur before or/and after the atomic rearrangement in selected systems trapped in cyclodextrin cavities. The information is relevant for a better understanding of many systems where confinement is important for reactivity and function. To this end, we first give a short overview of the photochemistry and photophysics of CD complexes.

7.2
Overview of the Photochemistry and Photophysics of Cyclodextrin Complexes

The ability of cyclodextrins (CDs) (Fig. 7.1), oligosaccharides of six, seven, or eight D-glucopyranose ($C_6H_{10}O_5$) units (the well known ones are α-, β-, and γ-CD with a diameter of ~ 5.7, 8 and 9.5 Å, respectively), to encapsulate organic and inorganic molecules has led to intensive studies of their inclusion complexes [24–32]. The relatively hydrophobic interior and hydrophilic exterior of their molecular pockets make them suitable hosts for supramolecular chemistry and for studying the spectroscopy and dynamics of several molecular systems [33–75]. Therefore, the hydrophobic nanocavity of such a host offers an opportunity for studying size-controlled nanoenvironment effects like reduced degrees of freedom of the guest and modified coupling to the heat reservoir.

Several studies on CD complexes with aromatic molecules using steady-state and nanosecond spectroscopy have been reported. These studies aimed to understand the photophysical and photochemical behavior of organic guests such as fluorescence and phosphorescence enhancement, excimer/exciplex formation, photocleavage, charge and proton transfer, energy hopping, and *cis-trans* photo-

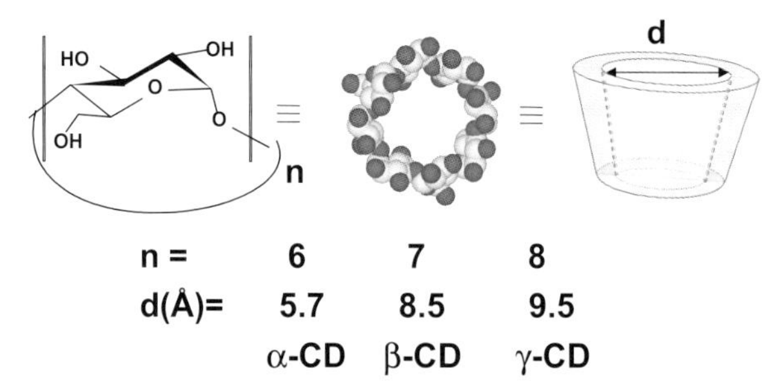

Figure 7.1 Structures of cyclodextrins (CDs) and approximate values of the largest diameter of their nanocages.

isomerization. Most of these reports describe the effect of molecular restriction due to the cavity size of the host and protection of the guest (from quenchers such as oxygen molecules and H-bonding interaction with the medium – water) provided by the CD cavity and its low polarity relative to that of water, on the photophysical and photochemical properties of the encapsulated guest. The presence of H-bonding, electron accepting and donating groups, and twisting groups influences the electronic properties of the encapsulated guest. The value of the inclusion equilibrium constant depends on several parameters where H-bonding, polarity, relative size of the guest to that of the cavity, play an important role for the stability of the relative population of the confined system. Both enthalpic and entropic terms determine the energetic balance between the free and encapsulated guest. Some of the studies have reported the formation of higher stochiometries (1:2, 2:1 and 2:2) or even the formation of nanotubes where a large number of CD capsules are involved [3–33, 60–68]. A general rule to predict the effect of CD on the absorption and emission properties of dyes upon encapsulation cannot be found as the static and dynamic interactions are specific to the studied systems and may change markedly in the excited state [44–75].

7.3 Picosecond Studies of Proton Transfer in Cyclodextrin Complexes

7.3.1 1′-Hydroxy,2′-acetonaphthone

One of the molecules which shows, upon photoexcitation, an internal proton transfer and a twisting motion in a confined nanostructure of CD is 1′-hydroxy-2′-acetonaphthone (HAN) (Fig. 7.2A) [2]. HAN has been studied in gas [76–79], liquid [80–82], polymers [80], and in CD nanocavities [82–84]. The excited dye undergoes an ultrafast (less than 30 fs) excited-state intramolecular proton-transfer (ESIPT) reaction followed by a subsequent twisting motion along the C–C bond when the medium allows it (Fig. 7.2A). The formed excited phototautomer, K*, a keto-type structure where the carbonyl group is on the aromatic ring, may experience an internal rotation producing a twisted keto rotamer (KR*). The C–C bond linking the aromatic part to the protonated acetyl group shows double bond character due to the transfer of the proton and electronic charge rearrangement, as suggested by the ground- and excited-state *ab initio* calculations [85, 86]. In the presence of CD, the stochiometry of the inclusion complex formed depends on the nature and size of the cage. For β- and γ-CD, the complex has a 1:1 stochiometry, while for α-CD, the stochiometry is 1:2, HAN:$(\alpha\text{-CD})_2$ (Fig. 7.2B).

The spectroscopy and dynamics of the complexes have also been found to depend on the nature of the cage (Fig. 7.3) [83, 84]. Compared to the observation using water or tetrahydrofuran (considered as a solvent with a polarity comparable to that of CD) the following picture has been provided. Due to the small size of the α- and β-CD cages, the rotation of the protonated acetyl group of the guest is

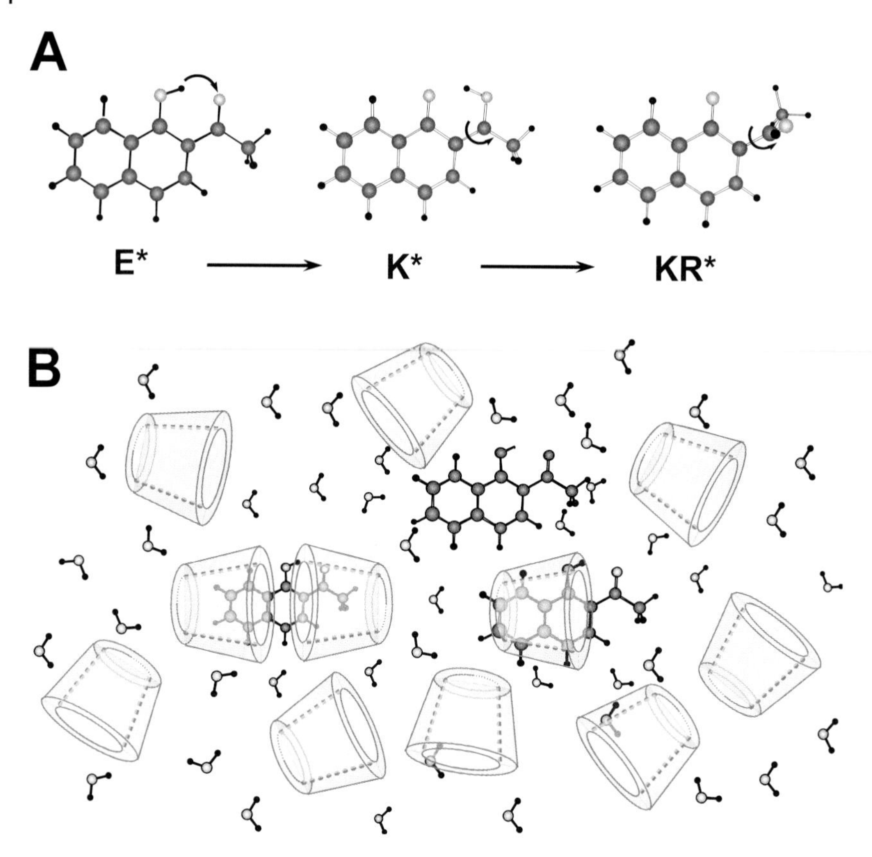

Figure 7.2 (A) Schematic representation of photoinduced proton-transfer reaction in the enol form (E*) and twisting motion in the keto-type (K*) structure to generate the keto-rotamer (KR*) of excited 1′-hydroxy,2′-acenaphthone (HAN). (B) Illustration of HAN molecule in water and complexed with one or two CD nanocavities.

restricted, giving rise to a stronger K* emission (460 nm). For γ-CD, with a larger cage, the conversion of K* to KR* is allowed and the resulting emission is at 500 nm and the phototautomers are more fluorescent due to the protection (from quenching by formation of an H-bond with water and O_2) provided by the cage. The lifetime of caged KR* is in the ns regime, in contrast to that of free K* (about 90 ps). That of caged K* is on the sub-ns or ns time scale. Depending on the size of the CD cavity, the protonated acetyl group can be found inside or outside the cage. So, its twisting to produce KR* and thus connecting with nonradiative channels depends on the degree of confinement and stoichiometry of the complex (Fig. 7.3). The reorientation times of the guest and of the guest:host complexes have been examined using emission anisotropy experiments. For 1:1 complexes, the reorientation times are about 70 and 50 ps for β-CD and γ-CD, respectively.

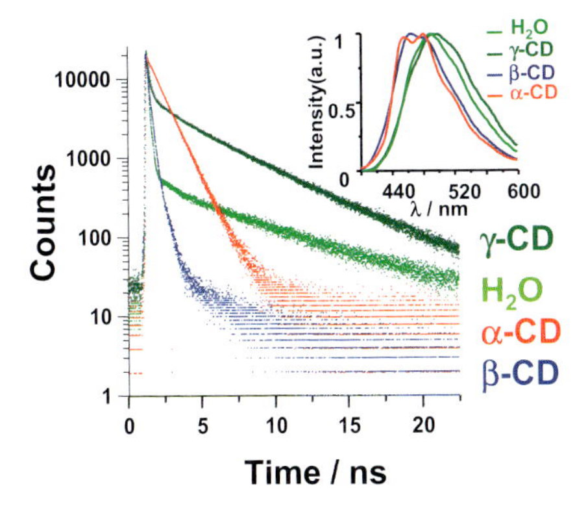

Figure 7.3 Magic-angle emission decays of HAN in water and in the presence of mM of CDs. The inset shows the emission spectra of HAN in water and complexed to CDs [82–84].

Therefore, within a cage of small size (β-CD) the internal molecular rotation in the trapped guest is restricted and its reorientation time is longer. For a larger cage, internal rotation can occur, producing a caged KR* rotamer, in full agreement with the emission spectral position, and fluorescence lifetime (Fig. 7.3). As expected, the overall rotational time of the complex (HAN:CD) increases with the size of the caging entity: 745 ps and 1.1 ns for β-, and γ-CD complexes, respectively [83]. For, α-CD complexes (the smallest cavity amongst the CDs) the situation is different as the complexes involve one or two H-bonded linked CD cavities (Fig. 7.2) [84]. Caged and photoproduced K* in 1:1 complexes has an emission lifetime of ~ 90 ps, similar to that of free K* in water, while that found in the 1:2 complex is about 1 ns. For 1:1 complexes, the protonated acetyl group of K* is found outside the CD cavity and it should not experience any restriction for twisting to produce caged KR* (which was observed). For 1:2 complexes, the restriction dictated by the cavity of 2 CD does not allow the formation of KR*, but enhances the ns-emission of caged K*, as it is now protected from quenchers and twisting nonradiative channels. The emission behavior of caged HAN in α-CD led to a 40 nm blue shift (shortest wavelengths) in the time-resolved emission spectra when the gating time increased (Fig. 7.4). The $COHCH_3$ rotation of K* to produce KR* involves an energy gain of about 16 kJ mol^{-1}, close to the energy gap (~ 24 kJ mol^{-1}) between these structures in the gas phase obtained using theoretical calculations [86]. Rotational times of the complexes have also been measured (Fig. 7.4). In pure solvents, like water and tetrahydrofuran, the rotational times are 70 and 35 ps, respectively. These values show the role of H-bonding interactions with the solvent (water) and the involvement of the solvation shell (water) in the friction dynamics of the dye. The rotational time of the caged phototautomers in α-CD cages depends on the emission wavelength and varies from 50 to 180 ps. The global rotational time of the 1:2 complex is almost constant, ~ 950 ps. This value accords with that estimated for two linked CD using the hydrodynamic theory under stick conditions [87]. The variation of the shortest time with the wave-

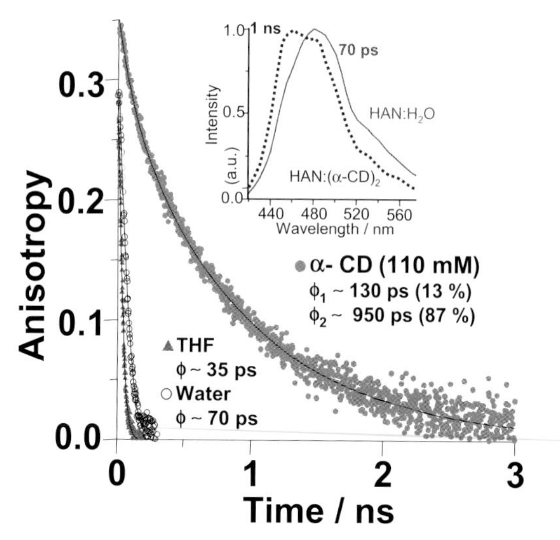

Figure 7.4 Anisotropy decays of HAN in water, tetrahydrofuran (THF) and in the presence of α-CD. The solid lines are the best fits using a single or a bi-exponential function giving the indicated rotational times. The inset also displays the emission spectra of the inclusion complexes gated at 70 ps and 1 ns [84].

length of emission indicates the existence of several rotamers (or conformers), and thus agrees with the involvement of 1:1 and 1:2 complexes. The result indicates that the size (space domain) of the nanocavity of the host (1 or 2 CD) determines the photodynamics (time domain) from ps to ns regime, and emission spectroscopy (shift by about 40 nm) of the nanostructure.

7.3.2
1-Naphthol and 1-Aminopyrene

While HAN shows an ESIPT reaction between two groups both located on the molecular frame of the dye and does not need solvation for the occurrence of internal proton motion, 1- and 2-naphthol, which are among the most studied aromatic systems, show an excited-state intermolecular proton transfer to the medium (water, alcohols) [88–96]. The produced anionic structure emits at the blue side of the normal form. The effects of CD on the intermolecular proton-transfer reaction from 1-naphthol (1-NP) [97, 98] to water and from water to 1-aminopyrene (1-AP) [98] have been studied by emission spectroscopy. For 1-NP in pure water, the decay of the 360 nm emission band (that of the neutral reactive species leading to the anionic one emitting at the blue side, 460 nm) was fitted with a 36 ps exponential component [98]. In the presence of β-CD, the decay at 370 nm needed two exponential functions with time constants of 700 ps (75%) and 1600 ps (25%) [98]. The average time constant for deprotonation of 1-NP in the presence of β-CD (1:1

stochiometry) is 930 ps. The difference in the deprotonation times for free (36 ps) and caged (930 ps) guest was interpreted as a signature of different mechanisms for proton transfer reactions from 1-NP to water [98]. The time for deprotonation of the caged 1-NP in other media such as micelles (600 ps, 1900 ps) [78], and polymer-surfactant aggregate (1600 ps, 5300 ps) [99, 100] was found to be much slower (or almost blocked) compared to that of uncomplexed dye [99].

For 1-AP, the rate of the proton-transfer reaction from water to the dye was found to increase upon formation of 1:1 complexes involving β-CD [98]. To explain this observation, the authors suggest using geometrical factors for the confined structures (Fig. 7.5) which influence the dynamics of the proton-transfer reaction. For geometries where the amino group is found near the hydroxy rims of β-CD, the rate of proton transfer is affected by the microenvironment due to the cage and is enhanced by a factor of 2 [98]. This rate is similar to that observed in a mixture of water : simple alcohols with comparable molar ratio to water. For geometries where the amino group is further away from the hydroxy rims, the amino group is then surrounded by water molecules. The rate constant of the proton-transfer reaction is not affected by complexation and is similar to that of uncomplexed 1-AP. Under the experimental conditions of the study (dye:CD ratio, pH and ps-time resolution) the authors suggest that the former confinement is favored by a ratio of 1.5:1 [128]. By comparing to the behavior when using ethanol–water mixtures, the authors proposed to model the CD effect on the proton-transfer rate constant of 1-NP and of 1-AP by that of suitable homogeneous mixtures of water and organic solvents. However, such a molecular model cannot be realistic as it does not take into account several microscopic differences like those of hydrophobic interactions, solvation of the transition state, the proton jump mechanism once the atom was ejected, diffusion and geminate recombination of the anion and cation, different H-bond networks with different cooperativities, and the different friction factors of both media, to cite a few. It has been found that the lifetime of the H-bond of a water molecule with that of an organized media (polar headgroup of a micelle) is about 13 times longer than that of the water–water H-bond, and 3.5 kcal mol^{-1} is needed to "liberate" the bound water molecule from the polar head of the micelle [101]. Furthermore, careful analysis of the emission decay of 1-NP in water leads tono exponential behavior [95], and change in dielectric constant [102], solvation time and diffusion should be taken into account, even in organized media [94]. Thus, to understand the origin of the slow component of deprotonation of 1-NP in CD or in organized media, and most

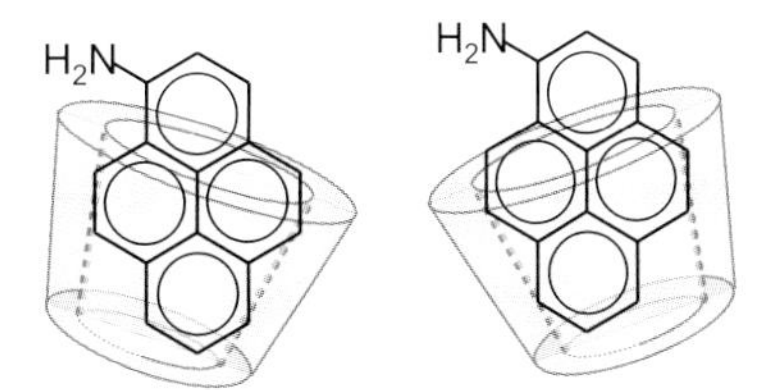

Figure 7.5 Proposed confined geometries of 1-aminopyrene:β-CD complexes [98].

probably of other dyes showing slow solvation dynamics (time constant longer than 300 ps or so), one has to take into account several factors influencing the microscopic and heterogeneous solvation, the self motions of the guest and host for an efficient encounter (sub-nanosecond to microsecond regime) and related H-bond networks dynamics [23]. Therefore, the slowest time constant (more than 100 ps) assigned to solvation dynamics might contain strong contributions from the above factors which would not be included in solvation processes in a classical restricted terminology.

7.4 Femtosecond Studies of Proton Transfer in Cyclodextrin Complexes

7.4.1 Coumarins 460 and 480

Coumarins have been the subject of many steady-state and ultrafast studies [103–113]. They have been also studied in CD cavities for different purposes [26, 113–122]. For coumarin 480 (C480) and coumarin 460 (C460) in aqueous solutions, the blue shift in the emission spectrum and the increase in the fluorescence lifetimes in the presence of γ-CD have been assigned to the formation of an inclusion complex with CD [120–122]. Molecular mechanics calculations for the C460/γ-CD complex suggest that the most stable inclusion complex has the carbonyl end of the guest near the narrower, primary-OH end of the host, and the amino end group of C460 near the secondary-OH side of the cavity [122]. The calculations also suggest the formation of an H-bond between the carbonyl group of the guest and two hydroxy groups of the host. The time-dependent Stokes shift of the emission of γ-CD:coumarin complexes has been studied [122]. The solvation time of the excited complex ranges from < 50 fs to 1.2 ns, and most of the solvation takes place within 1 ps. The fast component of the solvation decay for both complexes (C480:γ-CD and C460:γ-CD) is similar to that in bulk water suggesting that the first solvation shell does not dominate the solvent response, and it is mainly due to a collective response of water molecules found near the guest and outside the cavity. However, at longer times the solvation dynamics in γ-CD is at least 3 orders of magnitude slower when compared to that observed in water [122]. For C480 : γ-CD three components of the slow relaxation were observed: 13, 109 and 1200 ps. For C480 : γ-CD and C460 : γ-CD entities, molecular dynamics simulations suggest the presence of 13 and 16 water molecules, respectively, inside the cavity [122]. These water molecules are symmetrically distributed around the guest, with one molecule of water near the oxygen of the carbonyl group of the guest. The authors suggested that the slower relaxation components may be due to motion of the guest in and out of the restrictive host, fluctuations of the CD ring, or orientation of highly constrained water molecules [122]. A comparable number of water molecules in the first solvation shell of atomic solutes in pure water have been calculated [123].

7.4.2
Bound and Free Water Molecules

To get a better insight into the solvation dynamics of coumarin within CD a multi-shell continuum model and molecular hydrodynamic theory have been used [124]. The theory can explain solvation dynamics having time constants of less than ~ 100 ps or so where the contribution of self-motion of the (large) probe (and of a large host) is still not important in the solvation dynamics of the probe. The theoretical results indicated that the contribution of the translational component is small, while the orientational component governs the polarization relaxation. For a short time scale (1 ps), inclusion of the intermolecular vibrational mode in the dielectric model leads to good agreement between the calculated and experimental solvation time correlation functions [124]. For a longer time scale (more than 100 ps or so), although the theory does not reproduce well the experimental observation, it suggests that the slower component is controlled by the rotational motion which is affected by the freezing of translational modes of water inside the restricting nanocavity. The models used for the simulations cannot describe properly the effects of the CD rings on the solvent dynamical modes [124]. Note also that at least two kinds of water molecules can be found within CD or at the gate and proximity of cyclodextrins: water molecules bound to the hall of the cage where two types of intermolecular H-bond may act due to the H-bond donating and accepting nature of water and CD, and structurally different networks of water molecules in the vicinity of the host. To some extent, this prevents them playing a more important role in the ultrafast solvation dynamics. Comparable situations have been suggested for water at the interface of micelles [2, 14–16, 21–23, 124]. Two geometrical relaxations are key elements for the H-bonds' cooperativity and related solvation: the H-bond coordination of water molecules (the number of water molecules involved in the process) and the O···O distance shrinkage [125]. The number of water molecules trapped inside CD (up to seven molecules for crystalline hydrated β-CD) [126] or bound to the gates is limited and the energy to cooperate is increased when compared to the situation in bulk water. Neutron scattering experiments on β-CD at room temperature [127] showed that the disorder of water molecules inside the cavity is dynamic in nature, and involves jump time constants of about 10–100 ps. The conformational flexibility of the glycopyranoside units and the dynamics of disorder of water can influence the solvation dynamics of CD. Molecular dynamic simulations of water diffusion in CD with different degrees of hydration, and in water, have been performed [128]. The calculations show that water molecules found outside the cavity have access to the main diffusion pathway. The results also suggest that the diffusion constant for transport of water molecules along the main diffusion pathway (0.007 $Å^2$ ps^{-1}), parallel to the crystallographic axis b is about 1/30 of the value in bulk water at room temperature, and 1/53 at 320 K [128]. The water molecules outside the main diffusion pathway have an easy and fast (ps time scale) access to this channel. Interestingly, no significant change in the diffusion constants and pathways with the relative humidity of crystalline β-CD was found. For the coumarin-480 trapped in CD, the water molecule H-bonded to the carbonyl group of the coumarin, as suggested by calculations [122], can be involved in an H-bond network formed by water molecules at the gate of the host and may cause intramolecular charge-transfer dynamics leading to slower solvation dynamics. Furthermore, the slow motion of water molecules at both gates of cyclodextrins has been observed for guests able to show excited-state proton-transfer reactions [129]. For 2-naphthol and 7-hydroxyquinoline [129], the change in the proton-transfer rate constant upon encapsulation by CD is clear evidence that the H-bond network at the gates of CD is special and plays a crucial role in reaction and solvation dynamics of the trapped guest. Furthermore, molecular dynamics simulations have shown a nearly 50% decrease in the dielectric constant of water confined in a nanocavity [130]. Such a decrease will influence the electrostatic field around the guest and may therefore cause large changes in its dynamics, especially when the probe suffers significant change in its dipole moment upon electronic excitation as in the case of the coumarins. Indeed, several studies of water involving micelles, biological systems and channels have shown significant decrease in polarity [131–136] and slowing down of the rate constant for the relaxation of water molecules inside these media [137–152]. Water molecules confined in such environments exhibit a larger degree of spatial and orientational order than in the bulk phase with the formation of well defined molecular layers in the vicinity of the cavity or channel surface [137–152]. The water at this special layer (comparable to biological water) plays a crucial role in the activity of many biological molecules such as enzymes, proteins and DNA. The current view of protein–water interactions is associated with a variety of functional roles, some of which are specific to a given system, whereas others are general to all proteins. The water molecules associated with the surface of proteins are in constant exchange with the bulk solvent, and the related nonlinear dynamics plays a key factor in the function and stability: water is a rate-limiting partner in biological and biochemical processes. For recent reviews on this topic, Ref. [7] reviews protein–water interaction in a slow dynamic world; and Ref. [21] reviews femtosecond dynamics of macromolecular hydration where biological water is critical to the stability of the structure and function of the biological system.

Using fs resolution, two residence times of water at the surface of two proteins have been reported (Fig. 7.6) [21]. The natural probe tryptophan amino acid was used to follow the dynamics of water at the protein surface. For comparison, the behavior in bulk water was also studied. The experimental result together with the theoretical simulation-dynamical equilibrium in the hydration shell, show the direct relationship between the residence time of water molecules at the surface of proteins and the observed slow component in solvation dynamics. For the two biological systems studied, a "bimodal decay" for the hydration correlation function, with two primary relaxation times was observed: an ultrafast time, typically 1 ps or less, and a longer one typically 15–40 ps (Fig. 7.7) [21]. Both times are related to the residence period of water at the protein surface, and their values depend on the binding energy. Measurement of the OH librational band corresponding to intermolecular motion in nanoscopic pools of water and methanol confined in reverse micelles has been reported [153]. The result shows that the librational band, which has its maximum at 670 cm^{-1} in the bulk liquids, shifts to lower frequencies and its shape changes considerably as the size of the reverse micelle decreases. A two-state model based on bound and free water fractions of water (or methanol) was used to fit the shape of the librational band [153]. Using large-scale atomistic molecular dynamics simulation, it has been proposed, for an aqueous micelle solution of cesium perfluorooctane, that water molecules at the

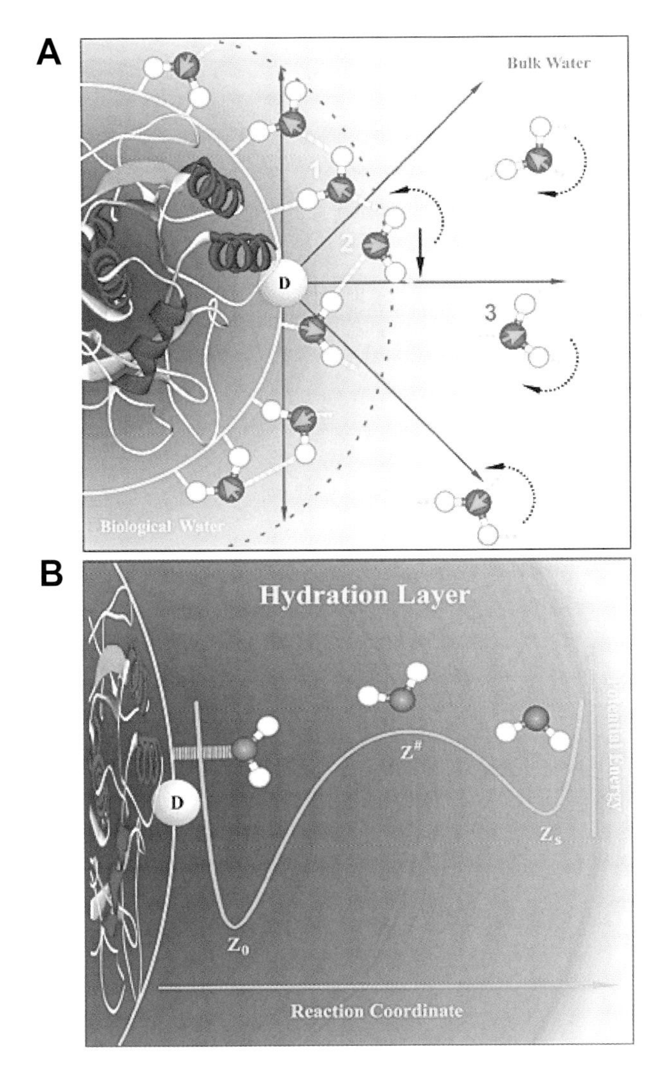

Figure 7.6 (A) An illustration of the dynamic equilibrium of water molecules at the hydration layer of a protein, with bound (1), *quasi-free* (2) and free water molecules (3). (B) The potential energy for the exchange [21].

interface fall into two categories: bound and free, with a ratio of 9:1 [22]. The water molecules bound to the hall can be further classified on the basis of the number of H-bonds linking them to the polar headgroups of the micelle. The entropy contribution is found to be critical in determining the relative populations of the free and bound water molecules. It has been shown also that the H-bond dynamics of

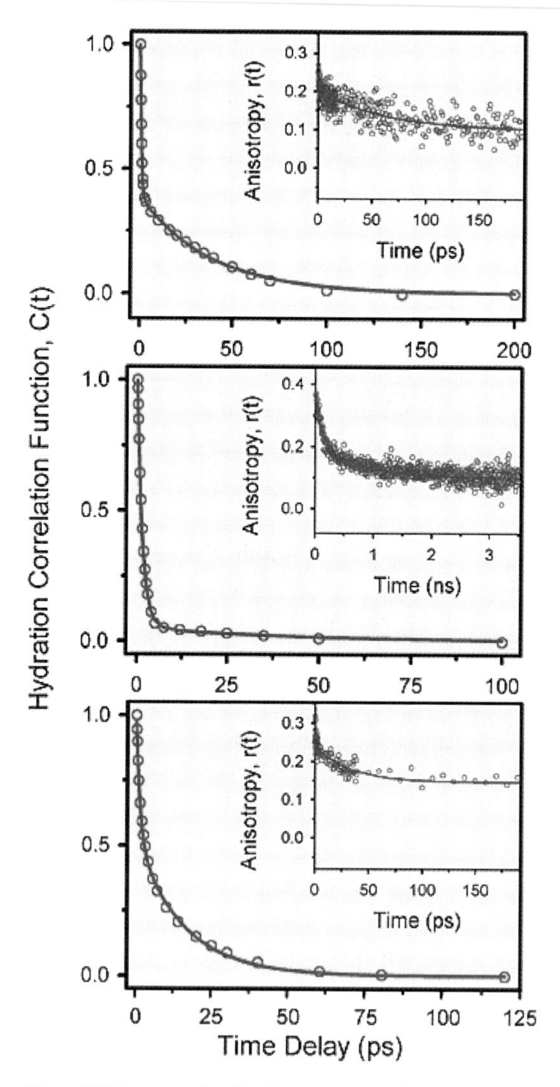

Figure 7.7 Time-resolved hydration process for the proteins Sublitisin *Carlsberg* (SC) and Monellin (Mn). The time evolution of the constructed correlation function is shown for the protein SC (top), the Dansyl dye bonded SC (middle), and for the protein Mn (bottom). The inset of each part shows the corresponding time-resolved anisotropy $r(t)$ decay [21].

two tagged water molecules bound to the polar heads of a micellar surface is almost 13 times slower than that found in bulk solution. Water molecules can remain bound to the micellar surface for more than 100 ps [22]. In general, interfacial water is energetically more stable than bulk water, and this will affect its ability to translate, and possibly to reorient. The finding may help to explain the origin of the universal slow relaxation at complex aqueous interfaces of several systems where water resides at their surface or interfaces.

Using ultrafast optical Kerr effects spectroscopy, the orientational dynamics of liquids (including water) in nanoporous sol–gel glasses has been studied [16]. In the pore, a ~ 3 ps time constant was observed and assigned to orientational relaxation of water in a hydrophobic pore, in agreement with a previous report [17]. At the pore surface, the orientational relaxation time of water was found to be much slower (15–35 ps) and is dependent on the size of the pore (Fig. 7.8). However, for all pore sizes, the water relaxation at the surfaces is faster for hydrophobic sites than for hydrophilic ones. Thus, the nature of the surface (involving hydrophilic or hydrophobic sites) is a key factor in the dynamics of the confined water molecules as it will influence the number of H-bonds involved in the networks. As noted [16], the rate of relaxation at the surfaces depends on the pore diameter, suggesting that water relaxation at these sites is highly cooperative, and may extend out to zones significantly larger than that defined by a single water molecule. An average solvation time of 220 ps for C-480 in a sol–gel glass of pore size 10–20 Å has been reported [154]. Taking into account the size of the probe and that of the pores, the rotational mobility of the guest should be restricted. Furthermore, the presence of the polar guest inducing an electric field within the pores, and enhancing the local polarization field, should slow down the motion of the trapped water molecules and thus cause a slowing in solvation dynamics of C-480 in the sol–gel matrix. For ethanol in sol–gel glass the average nonexponential behavior of the solvation time of Nile Blue A is about 19 ps and 36 ps within 75 and 50 Å average pores, respectively [155]. Within a polyacrylamide hydrogel having larger

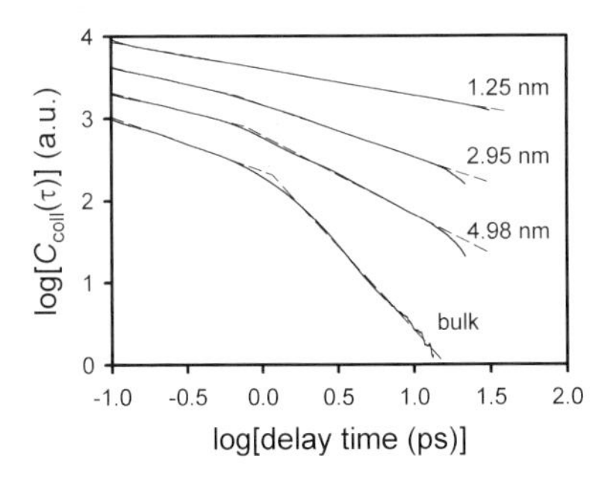

Figure 7.8 log–log plot (solid lines) of the collective orientational correlation functions C_{coll} for water confined in pores of different diameters (inserted number) and power-law fits (dashed lines). Lower and upper traces are for hydrophilic and hydrophobic pores, respectively [16].

pores, a solvation time shorter than 50 ps has been reported [156]. However, because of the limited time resolution of the ps-technique used, the fast component of the solvation dynamics of the above system is missing, and comparison of the result with those based on ultrafast techniques should be taken with caution. For CD, the water molecules of interest will H-bond with the hydrophilic gate of CD, leading to a decrease in the number of water–water H-bonds, and then to a slower dynamics. Inside the CD capsule, the surface is hydrophobic, but the small number (seven molecules for crystalline *β*-CDs) of water molecules which might be trapped there does not help to accelerate the caged dynamics, and then slow solvation is observed. In addition, a (polar) guest with strong local electric field may induce important local polarization of the trapped water molecules which decreases their orientational and translational motions, resulting in longer solvation times for confined structures when compared to bulk solutions.

7.5.3 2-(2′-Hydroxyphenyl)-4-methyloxazole

Noncovalent interactions which govern the ligand-binding process during guest:-host complexation have been studied using 2-(2′-hydroxyphenyl)-4-methyloxazole (HPMO) as a probe (guest) and CD, micelles and human serum albumin (HSA) proteins as hosts (Fig. 7.9) [56–59]. The interest in the HSA protein caging effect

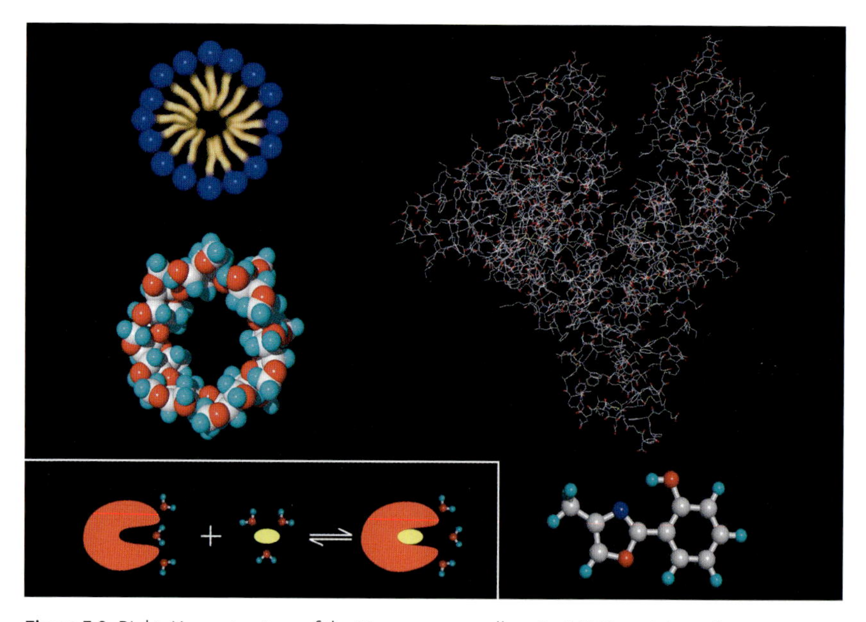

Figure 7.9 Right: X-ray structure of the Human serum albumin (HSA) protein and molecular structure of the used ligand HPMO. Left: Schematic representation of a normal micelle structure (left top); X-ray structure of *β*-cyclodextrin (left middle), and illustration of a protein–ligand recognition process (left bottom) [58].

lies in the important biological function of this protein in body carriers and in the discovery of new drugs and the development of phototherapy [157–159].

The first report on a fs-dynamics study of HMPO in aqueous solutions of CD showed the confinement affect on the ultrafast dynamics of HPOM caged into a β-CD nanocavity (Fig. 7.10) [58]. After an ultrafast intramolecular proton-transfer reaction in the guest, a subsequent twisting motion can take place while the guest is restricted in the nanocavity. The produced keto-type phototautomer emits a large Stokes shifted band (~ 10 000 cm^{-1}), and the wavelength of the maximum emission depends on the caging medium [57]. The blue transient behavior of HPMO in β-CD (430 nm) is different from that observed in 3-methyl pentane (3MP) (420 nm) [56]. Therefore, after the fs-excitation, the excited enol structure suffers a loss of aromaticity in the six-member ring and the electronic charge rapidly redistributes. This constitutes the driving force for the fs-proton transfer from the N···H–O to the N–H···O configuration [57, 160]. The charge redistribution barely changes the direction of the transition moment, as suggested by the ob-

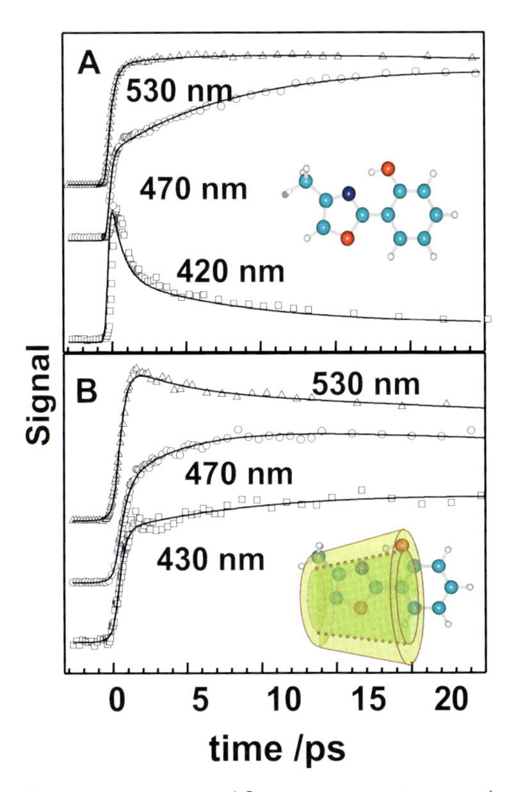

Figure 7.10 Femtosecond-fluorescence transients gated at different wavelengths for HPMO in (A) 3-methyl pentane and (B) in water solution containing β-CD. The structures of HPMO and the 1:1 complex with β-CD are indicated. The observation wavelengths are inserted [56].

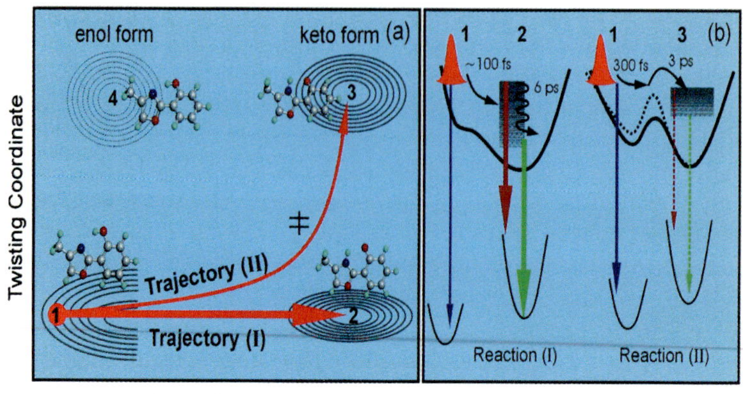

Figure 7.11 (a) Schematic illustration for direct proton-transfer reaction (trajectory I) and the one involving twisting motion (trajectory II). (b) Potential-energy curves along the reaction coordinate for trajectories I and II with the observed emission as indicated schematically. The dashed line in (b) at S_1 represents the potential energy for the protein environment. Note that rotamers of the non-planar enol ground state can undergo proton transfer on the upper surface following an initial twist towards planarity [58].

served initial anisotropy (0.34) which is close to the ideal one (0.4) involving a parallel transition. A more detailed study of HPMO in dioxane and in different cavities has been reported [58]. Although the observed patterns of the fs-transients are similar, they have been divided into two groups of time-resolved fluorescence emission transients overlapping at 430 nm. For wavelengths shorter than 430 nm, all the transients show fast decays, while at wavelengths longer than 430 nm, the transient shows rise and decay. To explain the observed behavior, two trajectories of fs-dynamics of the guest have been proposed (Fig. 7.11) [58]. A direct one (I) in which the wavepacket moves quickly along the proton coordinate without any barrier, and a second one (II) where the system evolves along the repulsive potential toward the keto-type structure involving two types of motions: one is the proton motion at earlier times and the other is the twisting motion of the heterocyclic moieties at a later time. This last motion involves an energy barrier, as also predicted by calculations [160], and this will increase upon confinement by CD or by HSA protein, making the time for barrier crossing longer in these cavities. The times for both motions along trajectory II become significantly longer in the cavity of HSA protein. While the barrier crossing dynamics in dioxane occurs in 3 ps, in HSA it depends on the interrogated wavelength: 8, 20, and 37 ps at 415, 420 and 430 nm, respectively [58, 59].

To get more insight into the effect of confinement on the binding between HPMO and the host, time-resolved anisotropy measurements have been carried out [58]. The result (Fig. 7.12) shows a remarkable difference in the anisotropy decays, especially for the HSA protein case. While in dioxane, the rotational time constant (45 ps) is close to the expected one using hydrodynamic theory [58], this time increases with the rigidity of the host (97 ps for a micelle, 154 ps for β-CD

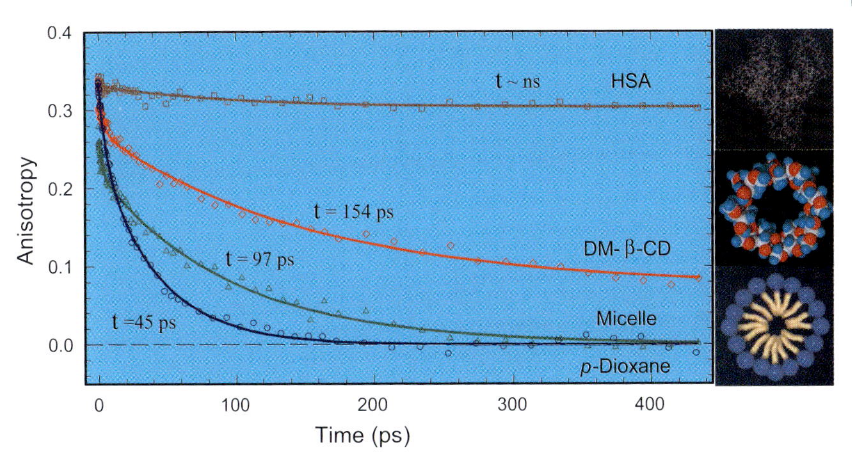

Figure 7.12 Femtosecond-resolved fluorescence anisotropy decays of HPMO in Dioxane, normal micelle, DM-β-CD and HAS protein [58]. Note the effect of confinement on the rotational motion of the probe.

and ~ ns for HSA) indicating the increase of confinement. While the orientation relaxation of the guest is almost complete in the micelle after 500 ps, it persists for longer times in CD and HSA protein indicating the slowing down (CD) or absence (HSA) of diffusive orientational motion in these cavities. Proton NMR studies of HPOM in CD solutions suggested that the oxazole ring of the guest is fully inserted into the cavity of the host [58]. Using initial and final anisotropy values, the calculated average change in direction of the transition moment of the guest inside CD is about 30°. For HSA, the high value of the initial anisotropy and the lack of any initial ultrafast decay indicate a strong hydrophobic interaction between the guest and the probe, the hindrance of molecular structure relaxation, and the rigidity of the local nanoenvironment where the guest is trapped. The observed result is consistent with X-ray structural studies showing a strong hydrophobic interaction between the ligand's aromatic ring and the residues located at site I of the HSA protein [61].

7.5.4 Orange II

A study of the femtosecond dynamics of Orange II (OII) encapsulated by CD has been reported [162]. In the presence of CD, OII only shows the formation of the 1:1 inclusion complex in which the confined structure (mode of penetration of the dye inside the cavity) depends on the nature of the host (Fig. 7.13). For OII:β-CD, the benzenesulfonate moiety is not included in the cavity, while for OII:γ-CD, this part is protected by the molecular cage [163, 164]. The transient made from the ultrafast transient lens (UTL) measurements carried out in the absence and presence of β-, and γ-CD showed a remarkable slowing down of the long component

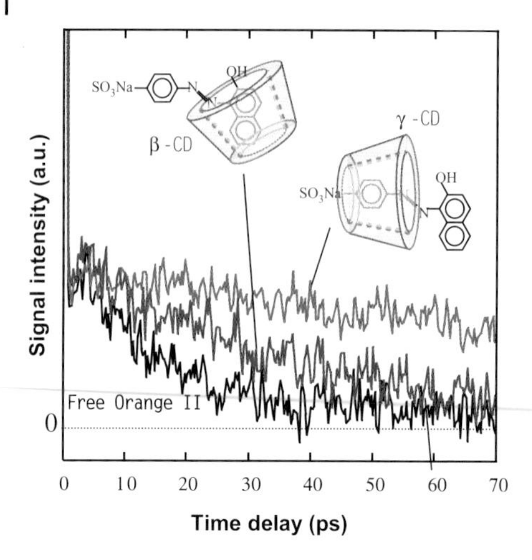

Figure 7.13 Decays of ultrafast transient lens (UTL) signal of Orange II in neutral water and in the presence of β- and γ-CD [162].

in comparison with the dynamics of water solution (Fig. 7.13) [162]. This component changes from 13 ps in water to 28 and 140 ps in β- and γ-CD, respectively. The difference was interpreted as a result of the formation of an H-bond between the OH groups of the guest and those of the cyclodextrin glucopyranose unit, hindering the free motion of the guest inside the nanocavity, lengthening the time for the intermediate state to isomerise, and slowing down the cooling process.

7.6 Concluding Remarks

In this chapter, we have examined the effect of the CD nanocavity on the fast and ultrafast events of a molecule trapped inside such a cavity. The results show that the degree of confinement which is reflected by the structure, the orientation of the guest, the docking and rigidity of the complex, and the polarity of the cage are the main factors that determine the issue of a created wavepacket in the cage, and therefore the spectral and dynamical behavior of the guest. Water molecules located inside and at both gates of CD have special properties reminiscent of biological water. Their restricted dynamics is slower than that found in bulk water. This abnormal behavior plays a key role in many chemical and biological processes, and one can consider taking advantage of this relatively slow response to explore new directions for research and potential applications in nano- and biotechnology. Finally, besides the studies of molecules showing proton-transfer reactions in CDs reviewed here, several proton-transfer studies have been carried out using other nanospaces provided by membranes, normal or reverse micelles, polymers, lipid vesicles, liquid crystals, sol–gels, dendrimers, proteins, DNA, zeolites and nanotubes.

Acknowledgment

This work was supported by the MEC and the JCCM (Spain) through projects CTQ2005-00114/BQU, MAT-2002-00301and PBI-05-046, respectively.

References

1 Zewail, A. H. *J. Phys. Chem. A* **2000**, *104*, 5660.
2 Douhal, A. *Chem. Rev.* **2004**, *104*, 1955.
3 Douhal, A. *Acc. Chem. Res.* **2004**, *37*, 3496.
4 Nandi, N., Bhattacharyya, K., Bagchi, B. *Chem. Rev.* **2000**, *100*, 2013.
5 Ringe, D. *Curr. Opin. Struct. Biol.* **1995**, *5*, 825.
6 Teeter, M. M., Yamano, A., Stec, B., Mohanty, U. *Proc. Natl. Acad. Sci. U.S.A.* **2001**, *98*, 11242.
7 Mattos, C. *Trends Biochem. Sci.* **2002**, *27*, 203.
8 Pratt, L. R., Pohorille, A. *Chem. Rev.* **2002**, *102*, 2671.
9 Marchi, M., Sterpone, F., Ceccarelli, M. *J. Am. Chem. Soc.* **2002**, *124*, 6787.
10 Ruffle, V., Michalarias, I., Li, J., Ford, R. C. *J. Am. Chem. Soc.* **2002**, *124*, 565.
11 Sarkar, N., Dutta, A., Das, S., Bhattacharyya, K. *J. Phys. Chem.* **1996**, *100*, 15483.
12 Riter, R. E., Willard, D. M., Levinger, N. E. *J. Phys. Chem. B* **1998**, *102*, 2705.
13 Mandal, D., Sen, S., Sukul, D., Bhattacharyya, K. *J. Phys. Chem. B* **2002**, *106*, 10741.
14 Faeder, J., Ladanyi, B. M. *J. Phys. Chem. B* **2001**, *105*, 11148.
15 Levinger, N. E. *Science* **2002**, *298*, 1722.
16 Farrer, R. A., Fourkas, J. T. *Acc. Chem. Res.* **2003**, *36*, 605.
17 Winkler, K., Lindler, J., Bursing, H., Vohringer, P. *J. Chem. Phys.* **2000**, *113*, 4674.
18 Otting, G. in *Biological Magnetic Resonance*, Ramakrishna, N., Berliner, L. J., (Eds.), Kluwer Academic/Plenum, New York **1999**, Vol. 17, p. 485.
19 Nandi, N., Bagchi, B. *J. Phys. Chem. B* **1997**, *101*, 10954.
20 Bizzarri, A. R., Cannistraro, S. *J. Phys. Chem. B* **2002**, *106*, 6617.
21 Pal, S. K., Peon, J., Bagchi, B., Zewail, A. H. *J. Phys. Chem. B* **2002**, *106*, 12376.
22 Pal, S., Balasubramanian, S., Bagchi, B. *Phys. Rev. E* 67, **2003**, 61502.
23 Bhattacharyya, K. *Acc. Chem. Res.* **2003**, *36*, 95.
24 Thoma, J. A., Stewart, L. in *Starch: Chemistry and Technology*; Whistler, R. L., Paschall, E. F. (Eds.), Academic Press, New York **1965**, p. 209.
25 Bender, M. L., Komiyama, M. (Eds.), *Cyclodextrin Chemistry*, Springer-Verlag, New York **1978**.
26 Szejtli, J. L. (Ed.), *Cyclodextrins and Their Inclusion Complexes*, Akademiai Kiado, Budapest **1982**.
27 Sjetli, J. L. (Ed.), *Cyclodextrin Technology*; Kluwer Academic Publishers, Dordrecht, The Netherlands **1988**.
28 Wenz, G. *Angew. Chem., Int. Ed. Engl.* **1994**, *33*, 803.
29 Szejtli, J. L. in *Comprehensive Supramolecular Chemistry*, Atwood, J. L., Davies, J. E. D., MacNicol, D. D., Vogtle, (Eds.), Pergamon, New York **1996**, Vol. 3.
30 Lehn, J. M. (Ed.) *Supramolecular Chemistry: Concepts and Perspectives*, VCH Publishers, New York **1995**.
31 Balzani, V., Scandolla, F. (Eds.), *Supramolecular Chemistry*, Ellis Horwood, London **1991**.
32 See the special issue of *Chem. Rev.* **1998**, *98*, 1743.
33 Bortolus, P., Monti, S. in *Advances in Photochemistry*, Neckers, D. C., Volman, D. H., von Bunau, G. (Eds.), John Wiley & Sons, New York **1996**, Vol. 21, p.1.
34 Nag, A., Dutta, R., Chattopadhyay, N., Bhattacharyya, K. *Chem. Phys. Lett.* **1989**, *157*, 83.
35 Flamigni, L. *J. Phys. Chem.* **1993**, *97*, 9566.

36 Monti, S., Kohler, G., Grabner, G. *J. Phys. Chem.* **1993**, *97*, 13011.
37 Nakamura, A., Sato, S., Hamasaki, K., Ueno, A., Toda, F. *J. Phys. Chem.* **1995**, *99*, 10952.
38 Turro, N. J., Okubo, T., Weed, G. C. *Photochem. Photobiol.* **1982**, *35*, 325.
39 Hamai, S. *J. Phys. Chem.* **1989**, *93*, 6527.
40 Catena, G. C., Bright, F. V. *Anal. Chem.* **1989**, *61*, 905.
41 Yorozu, T., Hoshino, M., Imamura, M. *J. Phys. Chem.* **1982**, *86*, 4426.
42 Street, K. W. Jr., Acree, W. E. Jr. *Appl. Spectrosc.* **1988**, *43* (7), 1315.
43 Agbaria, R. A., Butterfield, M. T., Warner, I. M., *J. Phys. Chem.* **1996**, *10, 17133.*
44 Al-Hassan, K. A. *Chem. Phys. Lett.* **1994**, *227*, 527.
45 Jiang, Y.-B. *J. Photochem. Photobiol. A, Chem.* **1995**, *88*, 109.
46 Hamai, S. *J. Phys. Chem. B* **1999**, *103*, 293.
47 Grabner, G., Rechthaler, K., Köhler, G., Rotkiewicz, K. *J. Phys. Chem. A* **2000**, 104, 13656.
48 Barros, T. C., Stefaniak, K., Holzwarth, J. F., Bohne, C. *J. Phys. Chem. A* **1998**, *102*, 5639.
49 Park, H.-R., Mayer, B., Wolschann, P., Kohler, G. *J. Phys. Chem.* **1994**, *98*, 6158.
50 Grabner, G., Monti, S., Marconi, G., Mayer, B., Klein, Ch. Th., Kohler, G. *J. Phys. Chem.* **1996**, *100*, 20069.
51 Dondon, R., Frey-Forgues, S. *J. Phys. Chem. B* **2001**, *105*, 10715.
52 Pastor, I., Di Martino A., Mendicutti, F. *J. Phys. Chem. B* **2002**, *106*, 1995.
53 Matsushita, Y., Suzuki, T., Ichimura, T., Hikida, T. *Chem. Phys.* **2003**, *286*, 399.
54 Douhal, A., Amat-Guerri, F., Acuña, A. U. *Angew. Chem. Int. Ed. Engl.* **1997**, 36, 1514.
55 Douhal, A. *Ber. Bunsenges. Phys. Chem.* **1998**, *102*, 448.
56 Douhal, A., Fiebig, T., Chachisvilis, M., Zewail, A. H. *J. Phys. Chem. A* **1998**, *102*, 1657.
57 García-Ochoa, I., Diez López, M. A., Viñas, M. H., Santos, L., Martínez-Ataz, E., Amat-Guerri, F., Douhal, A. *Chem. Eur. J.* **1999**, *5*, 897.
58 Zhong, D. P., Douhal, A., Zewail, A. H. *Proc. Natl. Acad. Sci. U.S.A.* **2000**, *97*, 14052.
59 Douhal, A. in *Femtochemistry*, De Schryver, F. C., De Feyter, S., Schweitzer, G. (Eds.), Wiley-VCH, Weinheim **2001**, Ch. 15, p.267.
60 Li, G., McGown, L. B. *Science* **1994**, *264*, 249.
61 Pistolis, G., Malliaris, A. *J. Phys. Chem.* **1996**, *100*, 155623.
62 Gibson, H. W., Bheda, M. C., Engen, P. T. *Progr. Polym. Sci.* **1994**, *19*, 843.
63 Harada, A., Li, J., Kamachi, M. *Nature* **1992**, *356*, 325.
64 Harada, A., Li, J., Kamachi, M. *Nature* **1993**, *364*, 516.
65 Born, M., Ritter, H. *Angew. Chem., Int. Ed. Engl.* **1995**, *34*, 309.
66 Harada, A. *Supramol. Sci.* **1996**, *3*, 19.
67 Pistolis, G., Malliaris, A. *J. Phys. Chem. B*, **1998**, *102*, 1095.
68 Raymo, F. M., Stoddart, F. J. *Chem. Rev.* **1999**, *99*, 1643.
69 Eskin, B. A., Grotkowski, C. E., Connolly, C. P., Ghent, W. R. *Biol. Trace Element Res.* **1995**, *49*, 9.
70 Fujimoto, T., Nakamura, A., Inoue, Y., Sakata, Y., Kneda, T. *Tetrahedron Lett.* **2001**, *42*, 7987.
71 Julien, L., Canceill, J., Valeur, B., Bardez, E. , Lefèvre, J.-P. , Lehn, J.-M., Mrachi-Artner, V., Pansu, R. *J. Am. Chem. Soc.* **1996**, *118*, 5432.
72 Berberan-Santos, M. N., Choppinet, P., Fedorov, A., Jullien, L., Valeur, B. *J. Am. Chem. Soc.* **2000**, *122*, 11876.
73 Cacialli, F., Wilson, J. S., Michels, J. J., Daniel, C., Silva, C., Friend, R. H., Severin, N., Samori, P., Rabe, J. P., O'connell, J. M., Taylor, P. N., Anderson, H. L. *Nature Mater.* **2002**, *1*, 160.
74 Kabashin, A. V., Meunier, M., Kingston, C., Luong, J. H. T. *J. Phys. Chem. B*, **2003**, *107*, 4527.
75 Balabai, N., Linton, B., Napper, A., Priyadarsky, S., Sukharevsky, A. P., Waldeck, D. H. *J. Phys. Chem. B* **1998**, *102*, 9617.
76 Douhal, A., Lahmani, F.; Zehnacker-Rentien, A. *Chem. Phys.* **1993**, *178*, 493.
77 Douhal, A.; Lahmani, F., Zewail, A. H. *Chem. Phys.* **1996**, *207*, 477.
78 Lu, C., Hsieh, R.-M- R., Lee, I-R., Cheng, P-Y. *Chem. Phys. Lett.* **1999**, *310*, 103.

79 Lochbrunner, S., Schultz, T., Shaffer, J. P., Zgierski, M. Z., Stolow, A. *J. Chem. Phys.* **2001**, *114*, 2519.

80 Tobita, S., Yamamoto, M., Kurahayashi, N., Tsukagoshi, R., Nakamura, Y., Shizuka, H. *J. Phys. Chem. A.* **1998**, *102*, 5206.

81 Lochbrunner, S., Stock, K., De Waele, V., Riedle, E. in *Femtochemistry and Femtobiology: Ultrafast Dynamics in Molecular Science*, Douhal, A., Santamaria, J. (Eds.), World Scientific, Singapore **2002**, p. 202.

82 Organero, J. A., Santos, L., Douhal, A. in *Femtochemistry and Femtobiology: Ultrafast Dynamics in Molecular Science*, Douhal, A., Santamaria, J. (Eds.), World Scientific, Singapore, **2002**, p. 225.

83 Organero, J. A., Tormo, L., Douhal, A. *Chem. Phys. Lett.* **2002**, *363*, 409.

84 Organero, J. A., Douhal, A. *Chem. Phys. Lett.* **2003**, *373*, 426.

85 Organero, J. A., García-Ochoa, I., Moreno, M., Lluch, J. M., Santos L., Douhal, A. *Chem. Phys. Lett.* **2000**, *328*, 83.

86 Organero, J. A., Moreno, M., Santos, L., Lluch J. M., Douhal, A. *J. Phys. Chem. A*, **2000**, *104*, 8424.

87 Hu, C., Zwanzing, R, *J. Chem. Phys.* **1974**, *60*, 4354.

88 Smith, K. K., Kaufmann, *J. Phys. Chem.* **1981**, *85*, 2895.

89 Syage, J. A. *Faraday Discuss. Chem. Soc.* **1994**, *97*, 401.

90 Douhal, A., Lahmani, F., Zewail, A. H. *Chem. Phys.* **1996**, *207*, 477.

91 Limbach H., Manz J. (Eds.), *Hydrogen Transfer: Experiment and Theory, Ber. Bunsengens. Phys. Chem.* **1998**, *102, Special Issue.*

92 Genosar, L., Cohen, B., Huppert, D. *J. Phys. Chem. A* **2000**, *104, 6689.*

93 Cohen, B., Huppert, D. *J. Phys. Chem. A* **2002**, *106, 1946.*

94 Cohen, B., Huppert, D., Solnstev, K. M., Tsfadia, Y., Nachliel, E., Gutman, M. *J. Am. Chem. Soc.* **2002**, *1124*, 7539.

95 Tolbert, L. M., Solnstev, K. M. *Acc. Chem. Res.* **2002**, *35*, 19.

96 Mandal, D., Pal, S. K., Bhattacharyya, K. *J. Phys. Chem. B.* **1998**, *102*, 9710.

97 Abgaria, R. A., Uzan, B., Gill, D. *J. Phys. Chem.* **1989**, *93*, 3855.

98 Hansen, J. E., Pines, E., Fleming, G. R. *J. Phys. Chem.* **1992**, *96*, 6904.

99 Sukul, D., Pal, S. K., Mandal, D., Sen, S., Bhattacharyya, K. *J. Phys. Chem. B.* **2000**, *104*, 6128.

100 Dutta, P., Halder, A.; Mukherjee, P., Sen, S., Bhattacharyya, K. *Langmuir* **2000**, *18*, 7867.

101 Balasubramanian, S., Pal, S., Bagchi, B. *Phys. Rev. Lett.* **2002**, *89*, 115505.

102 Senapati, S., Chandra, A. *J. Phys. Chem. B* **2001**, *105*, 5106.

103 Simon, J. D. *Acc. Chem. Res.* **1988**, *21*, 21.

104 Bagchi, B. Chandra, A. *Adv. Chem. Phys.* **1991**, *1*, 80.

105 Ladanyi, B., Skaf, M. S. *Annu. Rev. Phys. Chem.* **1993**, *335*, 44.

106 Suppan, P. *J. Chem. Soc., Faraday Trans. 1* **1987**, *83*, 495.

107 Ferreira, J. A. B., Coutinho, P. J. G., Costa, S. M. B., Martinho, J. M. G. *J. Chem. Phys.* **2000**, *262*, 453.

108 Cichos, F., Brown, R., Rempel, U., Von Borczyskowski, C. *J. Phys. Chem. A* **1999**, *103*, 2506.

109 Shirota, H., Castner, E. W. *J. Chem. Phys.* **2000**, *112*, 2367.

110 Frolicki, R., Jarzeba, W., Mostafavi, M., Lampre, I. *J. Phys. Chem. A* **2002**, *106*, 1708.

111 Molotsky, T., Huppert, D. *Phys. Chem. A*, **2002**, *106*, 8525.

112 Christie, R. M. *Rev. Prog. Color.* **1993**, *23*, 1.

113 Hoult, J. R. S., Paya, M. *Gen. Pharmac.* **1996**, *27*, 713.

114 Pitchumani, K., Velusamy, P., Srinivasan, C. *Tetrahedron* **1994**, *45*, 12979.

115 Yamaguchi, H., Higashi, M. *J. Inclusion Phenom. Mol. Recognit. Chem.* **1990**, *9*, 51.

116 Bergmark, W. R., Davis, A., York, C., Macintosh, A., Jones, G. *J. Phys. Chem.* **1990**, *94*, 5020.

117 Asimov, M. M., Rubinov, A. N. *J. Appl. Spectrosc.* **1995**, *62*, 353.

118 Ishiwata, S., Kamiya, M. *Chemosphere* **1998**, *37*, 479.

119 Karnik, N. A., Prankerd, R. J., Perrin, J. H. *Chirality* **1991**, *3*, 124.

120 Nag, A., Chakrabarty, T., Bahattacharyya, K. *J. Phys. Chem.* **1990**, *94*, 4203.

121 Bergmark, W. R., Davies, A., York, C., Jones, G., II. *J. Phys. Chem.* **1990**, *94*, 5020.

122 Vajda, S., Jimenez, R., Rosenthal, S. J., Fidler, V., Fleming, G. R., Castner, E. W.,

Jr. *J. Chem. Soc., Faraday Trans.* **1995**, *91*, 867.

123 Maroncelli, M., Fleming, G. R. *J. Chem. Phys.* **1988**, *88*, 5044.

124 Richmond, G. L. *Chem. Rev.* **2002**, *102*, 2693.

125 Ohmine, I., Saito, S. *Acc. Chem. Res.* **1999**, *32*, 741.

126 Zabel, V., Saenger, W., Manson, S. A. *J. Am. Chem. Soc.* **1986**, *108*, 3664.

127 Steiner, T., Saenger, W., Lechner, R. E. *Mol. Phys.* **1991**, *72*, 1211.

128 Braesicke, K., Steiner, T., Saenger, W., Knapp, W. W. *J. Mol. Graphics Mod.* **2000**, *118*, 143.

129 García-Ochoa, I., Diez Lopez, M.-A., Viñas, M. H., Santos, L., Martínez Ataz, E.; Sánchez, F., Douhal, A. *Chem. Phys. Lett.* **1998**, *296*, 335.

130 Senapti, S., Chandra, A. *J. Phys. Chem. B* **2002**, *105*, 5106.

131 Luisi, P. L., Straub, B. E., (Eds.), *Reverse Micelles*, Plenum Press, New York **1984**.

132 Backer, C. A., Whitten, D. G. *J. Phys. Chem.* **1987**, *91*, 865.

133 Keh, E., Valeur, B. *J. Colloid Interface Sci.* **1981**, *79*, 465.

134 Belletete, M., Lachapelle, M., Durocher, G. *J. Phys. Chem.* **1990**, *94*, 5337.

135 Guha Ray, J., Sengupta, P. K. *Chem. Phys. Lett.* **1994**, *230*, 75.

136 Zhu, D.-M., Wu, X., Schelly, Z. A. *J. Phys. Chem.* **1992**, *96*, 7121.

137 Mashimo, S., Kuwabara, S., Yagihara, S., Higasi, K. *J. Phys. Chem.* **1987**, *91*, 6337.

138 Quist, P. O., Halle, B. *J. Chem. Soc. Faraday Trans. 1* **1988**, *84*, 1033.

139 Brown, D., Clarke, J. H. R. *J. Phys. Chem.* **1988**, *92*, 2881.

140 Fukazaki, M., Miura, N., Shinyashiki, N., Kurita, D., Shioya, S., Haida, M., Mashimo, S. *J. Phys. Chem.* **1995**, *99*, 431.

141 Cho, C. H., Chung, M., Lee, J., Nguyen, T., Singh, S., Vedamuthu, M., Yao, S., Zhu, J.-B., Robinson, G. W. *J. Phys. Chem.* **1995**, *99*, 7806.

142 Riter, R. E., Willard, D. M., Levinger, N. E. *J. Phys. Chem. B* **1998**, *102*, 2705.

143 Das, S., Datta, A., Bhattacharyya, K. *J. Phys. Chem. A* **1997**, *101*, 3299.

144 Belch, A. C., Berkowitz, M. *Chem. Phys. Lett.* **1985**, *113*, 278.

145 Sanson, M. S. P., Kerr, I. D., Breed, J., Sankararamakrishnan, R. *Biophys. J.* **1996**, *70*, 693.

146 Zhang, L., Davis, H. T., Kroll, D. M., White, H. S. *J. Phys. Chem.* **1995**, *99*, 2878.

147 Lynden-Bell, R. M., Rasaiah, J. C. *J. Chem. Phys.* **1996**, *105*, 9266.

148 Linse, P. *J. Chem. Phys.* **1989**, *90*, 4992.

149 Faeder, J., Ladanyi, B. M. *J. Chem. Phys.* **2000**, *104*, 1033.

150 Chiu, S. W., Jakobsson, E., Subramanian, S., McCammon, J. A. *Biophys. J.* **1991**, *60*, 273.

151 Thompson, W. H. *J. Chem. Phys.* **2002**, *117*, 6618.

152 Senapati, S., Berkowitz, M. L. *J. Chem. Phys.* **2003**, *118*, 1937.

153 Venables, D. S., Huang, K., Schmuttenmaer, C. A. *J. Phys. Chem. B* **2001**, *105*, 9132.

154 Pal, S.K., Sukul, D., Mandal, D., Sen, S., Bhattacharyya, K. *J. Phys. Chem. B.* **2002**, *104*, 2613.

155 Baumann, R., Ferante, C., Deeg, F. W., Brauchle, C. *J. Chem. Phys.* **2001**, *114*, 5781.

156 Datta, A., Das, S., Mandal, D., Pal, S., Bhattacharyya, K. *Langmuir*, **1997**, *13*, 6922.

157 Gellman, S. H. (Ed.), *Chem. Rev.* **1997**, *97*, 1231.

158 Berde, C. B., Hudson, B. S., Simon, R. D., Sklar, L. A. *J. Biol. Chem.* **1979**, 254, 391.

159 Rosenoer, V. M., Oratz, M., Rothschild, M. A. *Albumin Structure, Function, and Uses*, Pergamon Press, Oxford **1977**.

160 Guallar, V., Moreno, M., Lluch, J. M., Amat-Guerri, F., Douhal, A. *J. Phys. Chem.* **1996**, *100*, 19789.

161 He, X. M., Carter, D. C. *Nature (London)* **1992**, *358*, 209.

162 Yui, H., Takei, M., Hirose, Y., Sawada, T. *Rev. Sci. Instrum.* **2003**, *74*, 907.

163 Suzuki, M., Yasaki, Y. *Chem. Pharm. Bull.* **1979**, *27*, 1343.

164 Suzuki, M., Yasaki, Y. *Chem. Pharm. Bull.* **1984**, *32*, 832.

8
Tautomerization in Porphycenes

Jacek Waluk

8.1
Introduction

Porphyrins and metalloporphyrins are objects of intense studies in both basic and applied sciences. Owing to their crucial role in biological processes, such as photosynthesis, oxygen transport and activation, porphyrins have been labeled "Pigments of Life" [1]. The accuracy of this characterization is additionally confirmed by the use of porphyrins in medicine as phototherapeutic agents [2–6]. The potential of porphyrins as building blocks is enormous and includes, for example, artificial light-harvesting and photosynthetic systems [7–9], molecular memories [10, 11], photovoltaic devices [12], and many other advanced materials [13–18]. The prospect of applications stimulated a rapid development of synthetic procedures that resulted in obtaining new classes of compounds, such as expanded, contracted, or inverted porphyrins [19, 20]. Of particular interest in this area are constitutional isomers of porphyrin (also dubbed "reshuffled" porphyrins) [20], tetrapyrrole macrocycles which differ from the parent molecule in the way of linking the pyrrole rings by methine groups (Fig. 8.1). The research in this field started with the synthesis of porphycene in 1986 [21]. Since then, three more "nitrogen-in" isomers have been obtained: corrphycene [22], hemiporphycene [23, 24], and isoporphycene [25, 26]. Closely related to this class is an "inverted" ("N-confused") porphyrin [27, 28].

All these molecules have been shown to be planar and aromatic. They also exhibit electronic spectra characteristic for porphyrins, with the lower intensity Q transitions in the red part of the visible region followed by stronger Soret bands in the near-UV (Fig. 8.2). However, the relative intensities of the Q and Soret transitions differ significantly among the series. The lowest intensity of Q bands is observed in porphyrin and corrphycene. Hemiporphycene reveals stronger Q bands, and porphycene even stronger ones. This behavior can be rationalized upon inspecting the energies of the frontier π orbitals. The analysis of HOMO and LUMO splitting based on the perimeter model [29] has led to correct predictions not only for relative absorption intensities, but also for signs and patterns in magnetic circular dichroism. These predictions were experimentally confirmed for porphycene [30], corrphycene [31], and hemiporphycene [32].

Hydrogen-Transfer Reactions. Edited by J. T. Hynes, J. P. Klinman, H. H. Limbach, and R. L. Schowen

ISBN: 978-3-527-30777-7

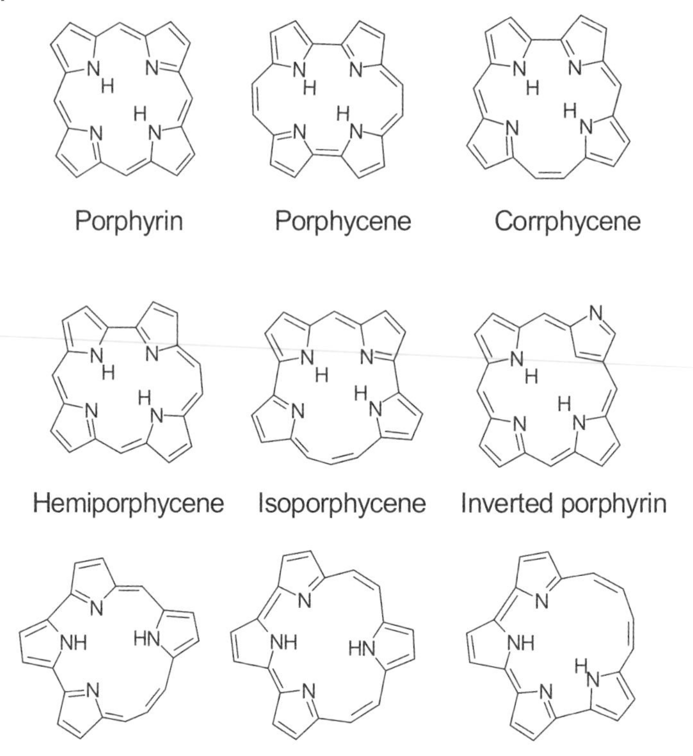

Figure 8.1 Constitutional isomers of porphyrin. The names are given for the compounds which have been synthesized to date.

Of all the porphyrin isomers, porphycene and its derivatives have been studied most thoroughly. DFT calculations predict [33] that porphycene **1** as a free base is the most stable of all isomers, including porphyrin **2**. The reason for this exceptional stability of **1** is a very strong intramolecular double NH···N hydrogen bond (HB), due to, first, the rectangular shape of the inner cavity that leads to a nearly linear arrangement of the three atoms, and, second, a small N–N distance (2.63 Å in **1** [21] as compared to 2.90 Å in **2** [34]). In metal complexes, the relative stabilities of porphycene and porphyrin are reversed, since the larger, square-shaped cavity in the latter is much better suited to the accommodation of a metal ion.

As will be demonstrated below, these differences in the cavity shape and distances have a huge impact on both the thermodynamics and kinetics of tautomerization. Moreover, it has been shown that the porphycene cavity dimensions are very sensitive to peripheral substituents, even such "mild" ones as alkyl groups. For variously alkylated porphycenes, the N–N distances can be varied from an extremely small value of 2.53 Å, resembling that in proton sponges [35] to 2.80 Å, close to the value for porphyrin. This provides a unique opportunity to study distance dependences in a class of very similar molecules. The material presented in this chapter will often be based on such comparisons and on the relationship with

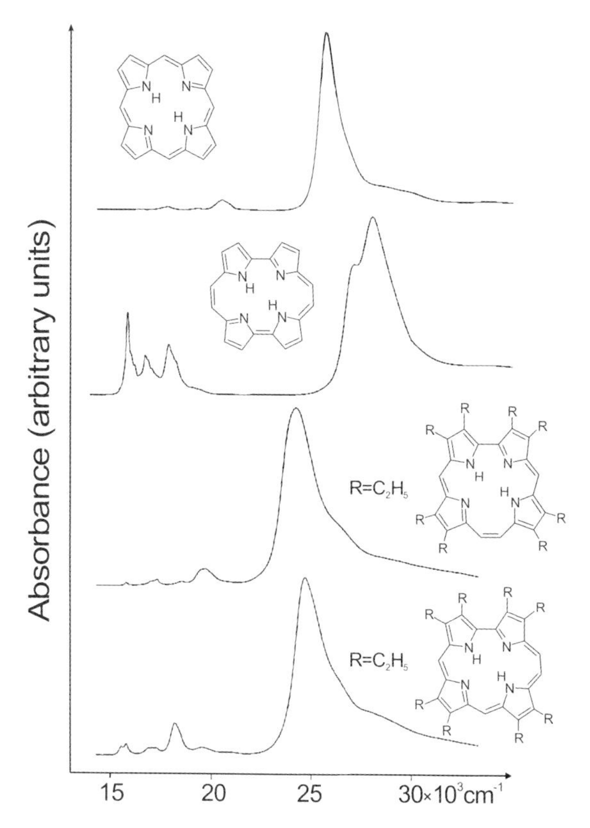

Figure 8.2 Room temperature electronic absorption spectra. From top to bottom: porphyrin, porphycene, 2,3,6,7,11,12,17,18-octaethylcorrphycene, and 2,3,6,7,11,12,16,17-octaethylhemiporphycene.

porphyrin. The tautomerization characteristics will be discussed separately for the ground state and for the lowest excited singlet and triplet states.

8.2 Tautomerization in the Ground Electronic State

8.2.1 Structural Data

Different tautomeric forms can be envisaged for both **1** and **2**. In the trans configuration, the protons are located on the opposite nitrogen atoms, whereas in the cis structure they are positioned on the adjacent ones (Fig. 8.3). Of the two cis forms of **1**, the vicinal arrangement of two protons on the same bipyrrole unit (cis-1) is energetically much more favorable than that of the other possible species

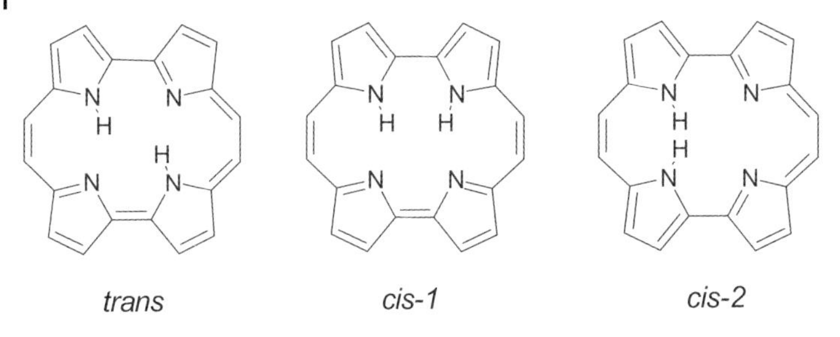

Figure 8.3 Possible tautomeric forms of **1**.

(cis-2). In principle, the structure with both protons shared equally by two nitrogen atoms is also possible, but the calculations place it at higher energies [33].

The crystal structure of free base porphyrin [34, 36, 37] clearly reveals the trans configuration. Interestingly, two inner protons maintain the same distance from both unprotonated nitrogen atoms, which results in D_{2h} symmetry. This type of symmetry has been independently confirmed by linear dichroism experiments on **2** photooriented in low-temperature rare gas matrices [38, 39]. Thus, each proton participates in two intramolecular hydrogen bonds. Due to unfavorable geometry (N–N separation of 2.89 Å and the NHN angle of only 116° [40]) these bonds are rather weak. The IR active NH stretching frequency is observed in rare gas matrices at around 3320 cm^{-1} [38].

In porphycene, the situation is completely different. In the X-ray structure, the protons are equally delocalized over all four nitrogen atoms. The distance between hydrogen-bonded nitrogen atoms is 2.63 Å [21]. Interestingly, this value lies in between those calculated for the trans and cis forms of **1** (Table 8.1), suggesting the presence of both species in the crystal. The two intramolecular hydrogen bonds are strong not only because of the small N–N separation, but also due to a nearly linear arrangement, with an NHN angle of only 152° [40]. Comparison between the cavity geometries in **1** and **2** is presented in Fig. 8.4.

Strong hydrogen bonds in **1** should result in the NH vibrations being shifted to low frequencies. Actually, the NH stretching vibration has not yet been identified in the IR spectrum of **1**, even though the calculations predict that it should correspond to the most intense band [41, 42]. The IR and Raman experiments could

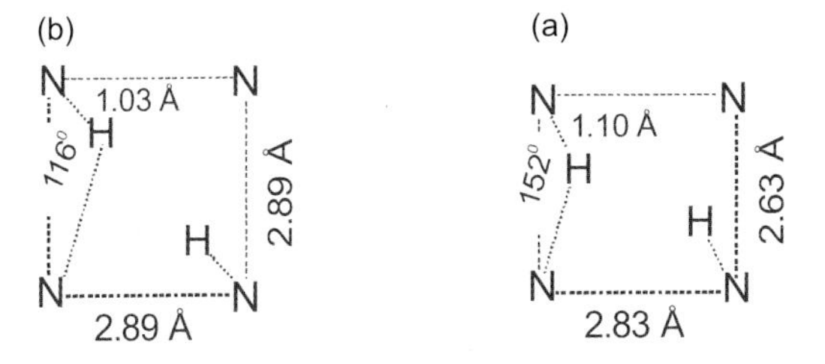

Figure 8.4 The inner cavity and hydrogen bond parameters in **1** (a) and **2** (b) determined from solid state ^{15}N NMR studies [40].

Tab. 8.1 Calculated relative energies (kcal mol^{-1}) of ground state trans and cis forms, trans–cis transition states (TS), trans–trans second-order saddle points (SS), and the inner cavity dimensions [Å]. The results for the higher energy cis tautomer of **1** (cis-2) are not shown.

Compound	Method of calculation	ΔE(trans–cis) [a]	E(TS) [a]	E(SS) [a]	d_{NN}trans [b]	d_{NN}cis [c]
1	B3LYP/TZ2P [d]	2.4 (1.9)	4.9 (1.6)	7.6 (1.6)	2.67/2.82	2.63/2.82/2.88
1	MP2/6-31G(d)// B3LYP/6-31G(d)[d]	5.1 (4.6)	5.7 (2.4)	7.6 (1.6)	2.68/2.83	2.64/2.83/2.89
1	BLYP/6-31G(d,p)[e]	2.0			2.66/2.87	2.62/2.86/2.92
1	B3LYP/6-31G(d,p)	2.2 (1.6)	4.1 (1.0)	6.1 (0.6)	2.66/2.84	2.61/2.84/2.89
1a	B3LYP/6-31G(d,p)	1.4 (0.7)	2.4 (–0.5)	3.6 (-1.7)	2.58/2.91	2.55/2.91/2.95
2	B3LYP/TZ2P[f]	8.3 (8.1)	16.2 (13.1)	24.4 (18.3)	2.93	
2	MP2/6-31G(d)// B3LYP/6-31G(d)[f]	8.9 (8.7)	17.0 (13.9)	24.8 (18.7)		

a Energies including zpve correction in parentheses;
b first entry, the distance between hydrogen-bonded, second entry, between non-bonded nitrogen atoms (vertical and horizontal N–N distances in Fig. 8.3, respectively);
c third entry, the distance between the nitrogen atoms on which the protons are located (upper horizontal N–N distance in Fig. 8.3);
d Ref. [41];
e Ref. [33];
f Ref. [92].

not definitely establish the structure of the ground state tautomer. On the one hand, the mutual exclusion principle seems to be obeyed, which would indicate the trans form as the dominant species. However, calculations for the cis structure show that the intensities of the vibrational transitions that become IR-active due to lack of an inversion center should be very weak [42].

The in-plane rigidity of the porphycene skeleton is probably not very high, since the cavity dimensions are significantly influenced by alkyl substituents. The substitution at the ethylene bridge carbon atoms shortens the N–N distance to an extremely small value of 2.53 Å in both 9,10,19,20-tetramethyl porphycene **1a** and the tetra-n-propyl derivative **1b**. Substitution by four alkyl groups in positions 2, 7, 12, 17 does not strongly influence the cavity dimensions, as illustrated by the n-propyl (**1c**) and tetra-*tert*-butyl (**1d**) derivatives. On the other hand, 2,3,6,7-12,13,16,17-octaethylporphycene **1e** reveals the largest separation of the hydrogen-bonded nitrogen atoms and the reversal of the long and short rectangle sides as compared to all other porphycenes. The cavity dimensions for several porphycenes are presented in Table 8.2.

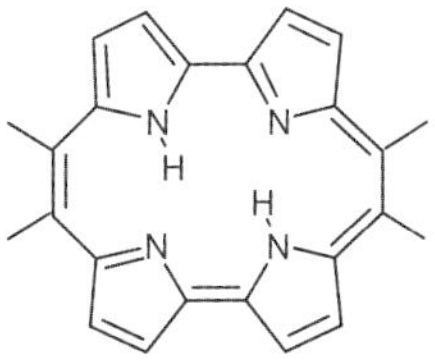

1a

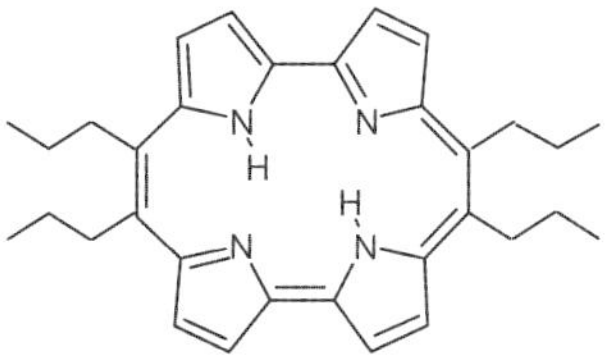

1b

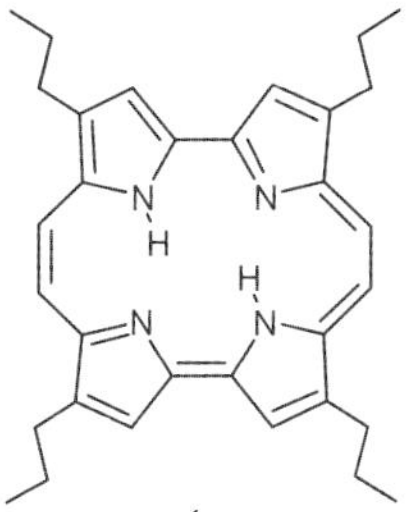

1c

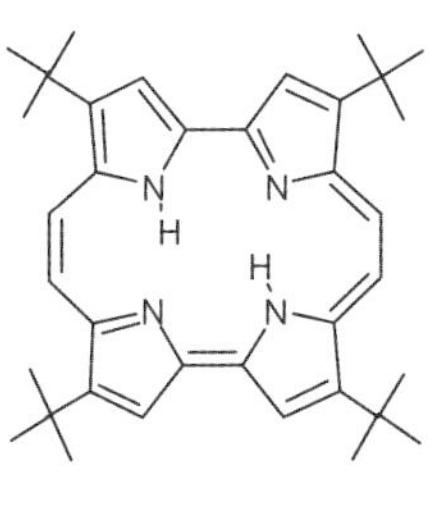

1d

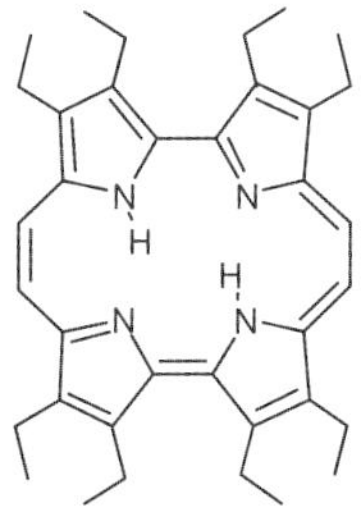

1e

The differences in HB strength in **1** and **2** are nicely reflected in the values of the NMR chemical shifts of the inner protons (Table 8.2). The values for porphycene are shifted downfield from those of porphyrin, and correlate clearly with the N–N separation. The chemical shifts of the peripheral protons are very similar in both isomers.

Tab. 8.2 ^{1}H NMR chemical shifts Δ and the distances between hydrogen-bonded nitrogen atoms (*d*) in several porphycenes and in **2**.

	1	**1a**	**1b**	**1c**	**1e**	**2**
Δ [ppm]	3.15	6.67	6.82	3.04	0.65	–3.76
d [Å]	2.63	2.53	2.53	2.62	2.80	2.90

8.2.2
NMR Studies of Tautomerism

Numerous NMR experiments have been reported for variously substituted porphyrins, both in solution [43–66] and in the solid phase [58, 61, 63, 67, 68]. At elevated temperature, the trans–trans interconversion is fast on the NMR time scale. For example, for ^{15}N-enriched porphyrin in toluene-d_8 the rate constant at 298 K is more than 2×10^4 s^{-1}; at 254 K, k_{HH} = 1300 s^{-1} [63]. Further lowering of the temperature results in a complete localization of protons. Below about 230 K, the two trans tautomers do not interconvert in the dark. They can still be transformed into each other, however, upon photoirradiation, even at cryogenic temperatures [38, 39].

Initial NMR studies of porphyrin suggested a concerted double hydrogen transfer pathway. However, careful and elegant analysis of hydrogen/deuterium/tritium isotope effects on tautomerization rates measured at different temperatures [61, 63, 65] has led to the conclusion that the ground state reaction in porphyrin involves a stepwise mechanism. The experimental data were interpreted using a slightly modified one-dimensional tunneling model of Bell [69]. The transfer of the first hydrogen atom, creating the cis form, occurs via tunneling from a level that has to be thermally activated. The required minimum energy needed for the tunneling to occur, E_m = 24.0 kJ mol^{-1} [63] consists of two contributions: (i) the difference between the cis and trans forms; (ii) the reorganization energy of the molecular skeleton. The value of E_m is much smaller than that of the classical barrier, estimated as about 57 kJ mol^{-1} [70]. From the cis structure, the system may either go back to the substrate or, by the transfer of the second hydrogen atom, achieve the other trans tautomeric form. One should note that the cis structure of porphyrin has never been experimentally detected. This is most probably due to its short lifetime, for which the estimates range from less than 10 ps [70] to 10^{-8} s [63].

In porphycene, the rates, barriers and the overall tautomerization mechanism seem to be completely different. Comparison of the reaction for **1** and **2** in the crystalline state was performed using ^{15}N CPMAS NMR (Fig. 8.5) [40, 68]. At elevated temperatures both molecules reveal only one peak, characteristic of a rapid inner hydrogen exchange. For porphyrin, lowering of the temperature from 356 to 192 K leads to line broadening and, finally, to the separation of =N– and NH

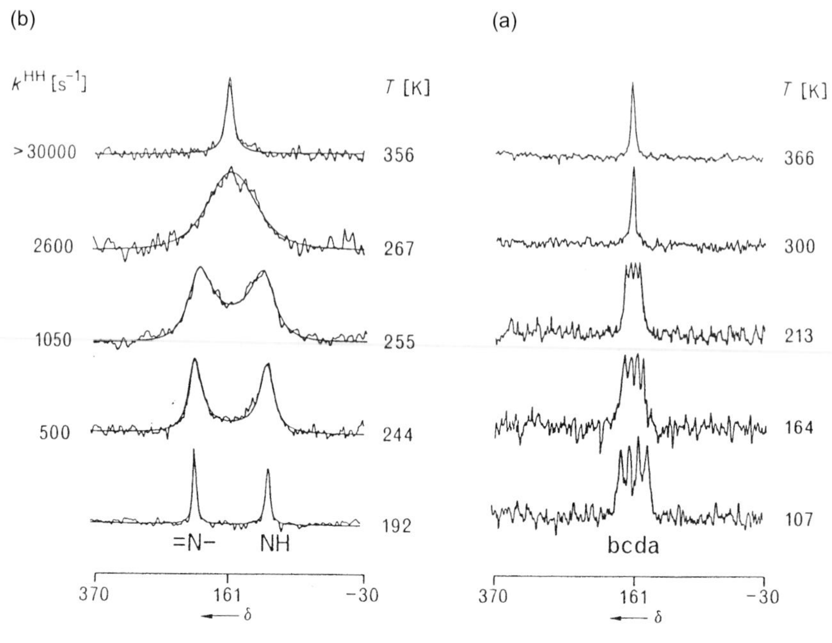

Figure 8.5 Solid state ^{15}N NMR spectra of **1** (a) and **2** (b). Reprinted with permission from Ref. [68].

peaks, showing that the process becomes frozen. On the contrary, porphycene does not reveal any broadening. Below 213 K, four narrow peaks are observed, assigned to two nonequivalent asymmetric proton transfer systems. The tautomerization in each of them is extremely rapid, even at temperatures as low as 107 K. Two possibilities were considered: (i) two nonequivalent porphycene molecules in the crystal, each containing two proton transfer systems; (ii) each molecule in the crystal exhibiting four different tautomeric forms. The former case would involve only trans–trans equilibria, whereas for the latter, both trans and cis structures should be present. The degeneracy in both trans and cis forms is removed by intermolecular interactions in the crystal. Below 50 K, only the trans forms are observed [40]. Case (ii) seems more probable, since the room temperature X-ray structure of **1** shows only one type of molecule in the crystal, and also because the experimentally determined N–N distance lies in between the values calculated for the trans and cis forms.

Three alkyl derivatives of porphycene, **1b**, **1c**, and **1e** have also been studied using ^{15}N CPMAS NMR [71]. As in the case of parent **1**, the tautomerization was found to be so rapid that the rate constants could not be determined from the line shape analysis. It was thus not possible to establish a correlation between the cavity parameters and tautomerization dynamics. **1b** and **1c** revealed narrow doublets, whereas **1e** showed one line that did not broaden, even at 173 K. **1e** has the largest N– N distance among porphycenes and should therefore exhibit the weakest hydrogen bonding. Even for this molecule, ground state tautomerization is very rapid on the NMR time scale.

The ^{15}N CPMAS NMR work on crystalline porphycene [40], coupled with the analysis of the ^{15}N T_1 relaxation times, resulted in the determination of the rate constants of the hydrogen transfer process. They varied from 3.66×10^8 s^{-1} at 355 K to 4.24×10^6 s^{-1} at 228 K. Strong coupling between the two hydrogen bonds was found and interpreted as indication that the tautomerization in porphycene occurs as a correlated process. The temperature dependence of the rate indicated the major role of tunneling, with an effective barrier of 31.8 kJ mol^{-1} and the value of $E_m = 5.9$ kJ mol^{-1}, much smaller than that in porphyrin. The evidence for tunneling agrees with the results obtained for **1** using a completely different experimental technique – optical excitation of a molecule isolated in a supersonic jet [72].

8.2.3
Supersonic Jet Studies

The fluorescence excitation spectrum of parent porphycene isolated in an ultracold supersonic jet consists of doublets, separated by 4.4 cm^{-1} (Fig. 8.6). This behavior contrasts with that of porphyrin, which exhibits "normal" behavior, with single peaks corresponding to particular vibronic transitions [73]. Upon replacing one or two inner protons with deuterons, the splitting disappears (which, given the experimental resolution, means that it becomes less than 0.1 cm^{-1}). Adding

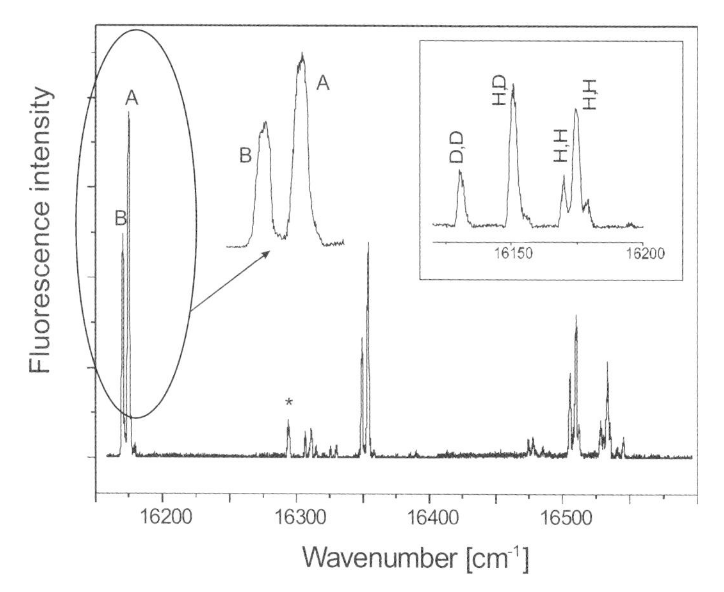

Figure 8.6 Fluorescence excitation spectra of **1** isolated in a supersonic jet. Inset, the 0–0 transitions observed for undeuterated, singly, and doubly deuterated **1**. The peak marked with an asterisk corresponds to the complex of **1** with water.

water or alcohol to the sample results in the detection of complexes that do not reveal the doublet structure. Therefore, the splitting has been attributed to the ground state tunneling of two inner hydrogen atoms. The fluorescence excitation spectrum in singly and doubly deuterated **1** is shifted to the red, by approximately the same amount per each substituted proton. This indicates that the stabilization of the deuterated species is larger in S_1 than in S_0, which implies that the hydrogen bond is weaker in the excited state. Most probably, the molecule expands upon excitation, and the cavity becomes larger (or, at least, the shorter side of the rectangle increases). This finding, along with the observation that all the vibronic peaks reveal the same separation of the doublet components, suggests that the observed value of 4.4 cm^{-1} can be assigned to ground state tunneling splitting (Fig. 8.7). One can estimate the barrier to double hydrogen tunneling using a simple one-dimensional model [69, 74] that leads to the formula relating the observed splitting, ΔE, to the effective mass, m, barrier width, d, and barrier height, V:

$$\Delta E = \frac{h\nu}{\pi} \exp\left[-\frac{\sigma d}{\hbar} \sqrt{2m(V - E_0)} \right] \tag{8.1}$$

σ is the shape parameter, close to unity and equal to $\pi/4$ for parabolic barriers. The values of m and d can be estimated within reasonable limits. The vibrational frequency, ν, is more difficult to assess, because of the lack of experimental assignment for the NH stretching vibration. One can use the formula obtained for tunneling in symmetrical double-minimum potentials via a harmonic oscillator approximation [75]:

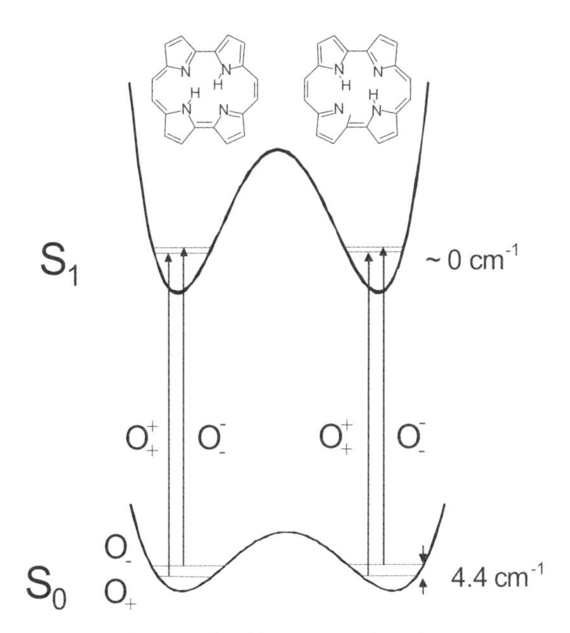

Figure 8.7 The ground and lowest excited singlet state potential energy profiles along the tautomerization coordinate in **1**.

$$\Delta E = \frac{ha^{3/2}d}{2m^{5/2}} \exp(-ad^2) \tag{8.2}$$

where $a = 2\ mv/\hbar$. Varying m between 1 and 2 and d between 0.55 and 0.65 Å, yields values of v ranging between 185 and 673 cm^{-1}. These values, in turn, result in the estimated barrier heights of 217–1360 cm^{-1} (2.6–16.0 kJ mol^{-1}). The barrier is evidently much lower than in porphyrin. The evaluation of the barrier for **1** using ^{15}N NMR yielded the value of 25.9 kJ mol^{-1} [40]. It has to be remembered, however, that these measurements were done in the crystalline phase and are explicitly based on a slightly unsymmetrical double-minimum potential.

For alkyl-substituted porphycenes **1a** and **1b**, with smaller N–N separation and thus stronger hydrogen bonds, one would expect a larger tunneling splitting. The experiments in the supersonic jets reveal a quite complicated pattern, different from that of parent porphycene [76]. Two peaks, separated by 12.5 cm^{-1} are observed in the 0–0 region of the fluorescence excitation spectrum (Fig. 8.8). This doublet structure does not disappear for singly and doubly deuterated compounds. Each of the peaks consists of doublets in undeuterated species and is associated with a different vibronic structure. Two low frequency modes (32 and

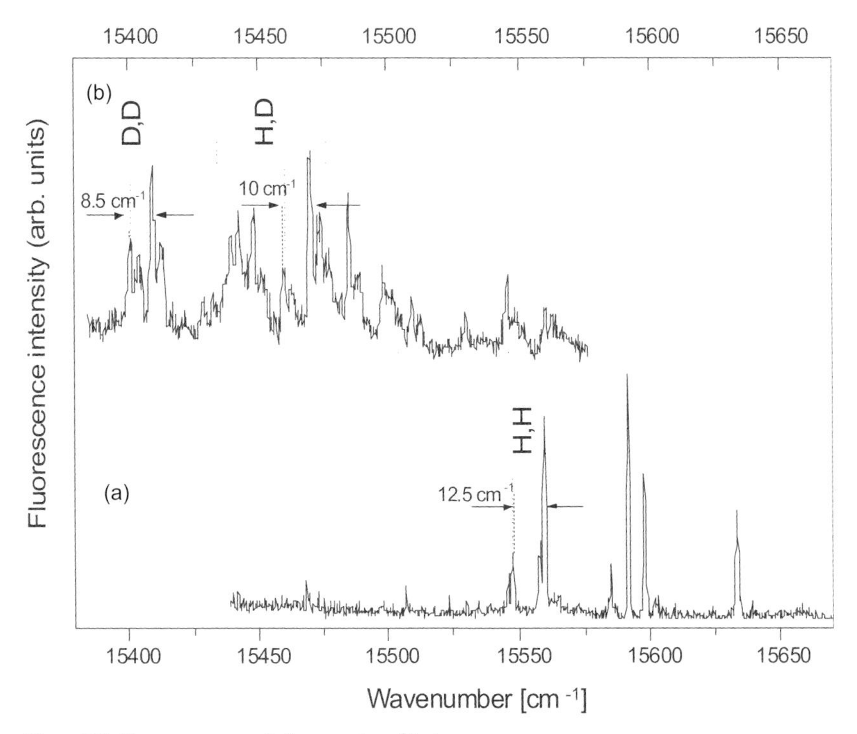

Figure 8.8 Fluorescence excitation spectra of **1a** in a supersonic jet. (a) Undeuterated species; (b) features assigned to the compound with one (H,D) or two (D,D) inner protons replaced by deuterons.

38 cm^{-1}) are associated with the upper peak, whereas only one vibration (38 cm^{-1}) related to the lower component is detected. Hole-burning experiments in the jets were performed to secure these assignments. These results can be interpreted as evidence for two different ground state structures, trans and cis tautomers. The calculations suggest that the two forms should have very similar energies (Table 8.1). The presence of two species in **1a** and **1b** has also been detected, via biexponential fluorescence decay, in solutions, glasses, polymers and rare gas matrices [77].

The values of tunneling splitting were obtained for ground and excited state cis and trans forms of undeuterated **1a**. The splittings, similar for cis and trans species, indicate a larger barrier to tautomerization in S_1 than in S_0.

8.2.4
The Nonsymmetric Case: 2,7,12,17-Tetra-n-propyl-9-acetoxyporphycene

In asymmetrically-substituted porphycenes, the two trans (or cis) forms should no longer be degenerate. Indeed, the studies of 2,7,12,17-tetra-n-propyl-9-acetoxyporphycene **1f** [78] reveal a splitting of the peak that corresponds to the S_0–S_1 transition in molecule **1c**, which lacks the acetoxy substituent (Fig. 8.9). A doublet appears, with the intensity of each component about half that in **1c**. This strongly suggests that the two peaks correspond to two trans tautomeric forms of **1f**, **1f′** and **1f″**. This assignment was corroborated by fluorescence polarization studies that monitored the anisotropy of **1f″** emission while exciting both forms to their S_1 and S_2 states.

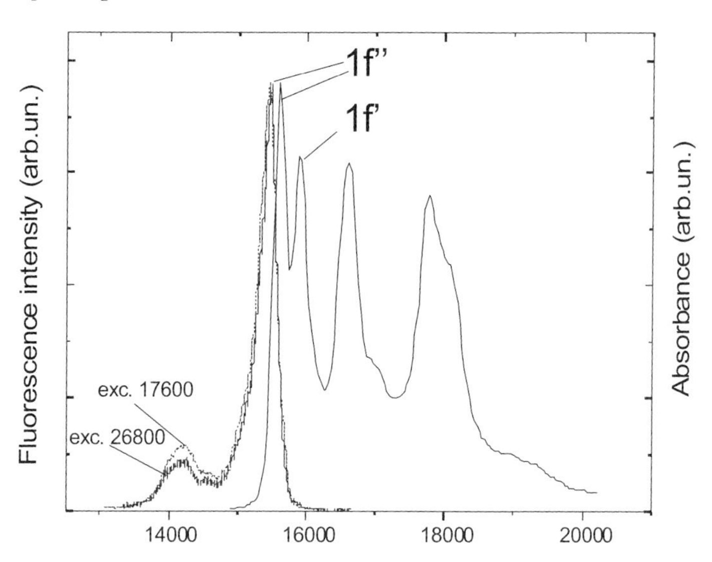

Figure 8.9 Electronic absorption and fluorescence spectra of **1f** measured at 20 K in poly(vinyl butyral) film. The two emission curves correspond to excitation of different fractions of the ground state tautomeric forms.

1f'　　　1f''

Measurements of the temperature dependence of the absorption in solution, in glasses and in polymer films showed that practically no change in the intensity ratio of the peaks corresponding to the different tautomeric forms occurs in the low temperature region (20–100 K). Analysis of the absorption and fluorescence reveals that both forms should have very similar energies in the ground state, whereas in S_1 they differ by about 1 kcal mol^{-1}. In contrast to the absorption, fluorescence occurs practically only from one form, assigned to **1f''**. This is true for both high and low temperatures (at 293 K, a weak shoulder assigned to **1f'** can

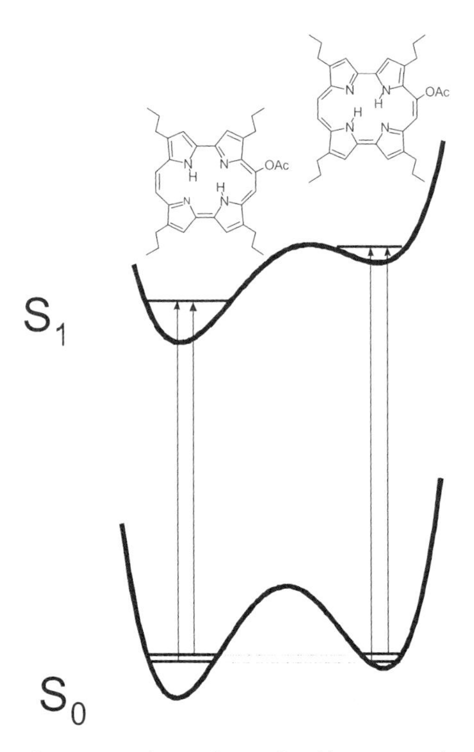

Figure 8.10 Scheme of ground and lowest excited singlet state potential energy profiles along the tautomerization coordinate in **1f**.

also be detected). The picture that emerges is that of a nearly symmetric double-minimum ground state potential, and an asymmetric shape in S_1 (Fig. 8.10). Most importantly, the constant ratio of the two forms in the low temperature range suggests tunneling as the mechanism of interconversion between the two nonequivalent trans tautomers.

8.2.5
Calculations

Table 8.1 shows the results of energy calculations of ground state trans and cis forms, as well as the energies of the transition states for single and double hydrogen transfer. For comparison, some results for porphyrin **2** have also been included. For both **1** and **2**, inclusion of electron correlation is necessary to obtain a proper symmetry.

The computational results confirm the large differences between **1** and **2** observed experimentally. Although the trans forms are always predicted as the most stable species, the cis tautomers are much closer in energy in porphycenes. Actually, for **1a** both forms are predicted to be practically isoenergetic, which agrees nicely with the experimentally observed presence of two forms in this molecule.

The differences in barrier heights are even more dramatic. This is best exemplified by calculations on porphycenes **1** and **1a**: the inclusion of zero point energy results in the transition state energy being lower than that of the cis form in **1** and of both forms in **1a**. These results show that the harmonic approximation is not appropriate for the vibrations involved in the hydrogen transfer path in porphycenes (NH stretch, in particular) and that the barriers to tautomerization must be very low.

In summary, both experiment and calculations demonstrate low barriers for tautomerization in porphycene as compared to porphyrin. This may explain the tunneling effects readily observed for **1** and its derivatives, as well as the change in the reaction mechanism, from stepwise in **2** towards synchronous in **1**. Moreover, the cis structures, postulated, but never observed for porphyrin, may be present in porphycenes, their population increasing with the strength of the intramolecular hydrogen bond, i.e., with the decrease in the distance between hydrogen-bonded nitrogen atoms.

8.3
Tautomerization in the Lowest Excited Singlet State

Fluorescence of porphycene embedded in rigid media was found to be depolarized, both at room temperature in the poly(vinyl butyral) matrix [79] and at low temperatures in glasses [30, 80] (Fig. 8.11). In a rigid environment, where the reorientation of an excited chromophore is not possible, the direction of the S_0–S_1 transition moment should be the same in absorption and emission, leading, for excitation into S_1, to the anisotropy value of 0.4. Instead, the observed values

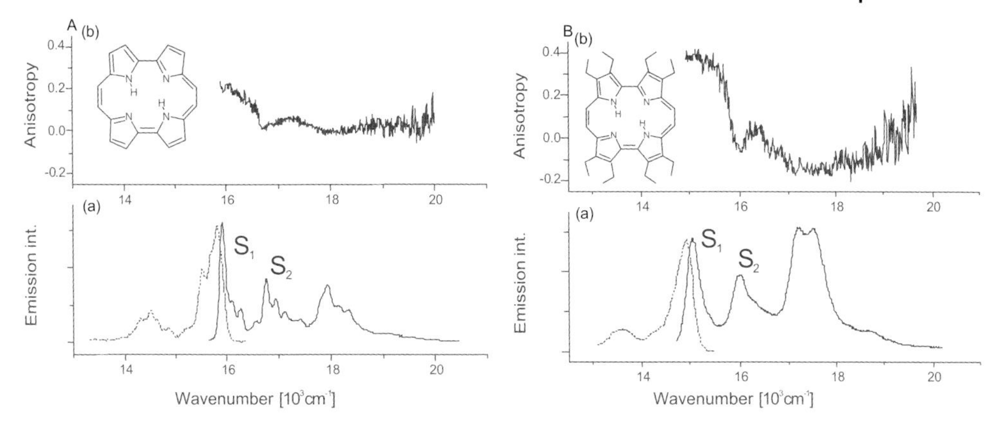

Figure 8.11 (a) Fluorescence (dotted line) and fluorescence excitation spectra; (b) anisotropy of fluorescence excitation of **1** (A) and **1e** (B). Both samples were measured at 113 K in 1-propanol glass. In the excitation spectra, the emission was monitored at 15900 cm^{-1} (**1**) and at 14900 cm^{-1} (**1e**).

barely exceed 0.1. The depolarization is also observed upon excitation to S_2. According to calculations, for this state the transition moment should be nearly orthogonal to that of the emitting S_1 state, and the anisotropy should be close to –0.2. The observed value, however, is close to zero.

The depolarization of fluorescence has been observed at temperatures around 100 K not only for **1**, but also for the two derivatives, **1b** and **1c**. in contrast, the measurements performed under the same conditions for **1e** revealed no sign of depolarization. The “textbook” values of the anisotropy were obtained, i.e. 0.4 and about –0.2 for excitation into S_1 and S_2, respectively (Fig. 8.11). The octaethyl derivative **1e** is the porphycene with the largest separation (2.80 Å) between the hydrogen-bonded nitrogen atoms and should therefore exhibit the slowest tautomerization kinetics. It was thus concluded that the reduced anisotropy values observed in three different porphycenes are caused by excited state tautomerization [30, 80]. As shown in Fig. 8.12, the interconversion between the two trans tautomers changes the direction of the transition moment. Therefore, only a part of the excited state population emits fluorescence polarized parallel to that of the transition moment in S_0–S_1 absorption (this fraction should approach 0.5 if the tautomerization is fast compared to the excited state lifetime). For the remaining frac-

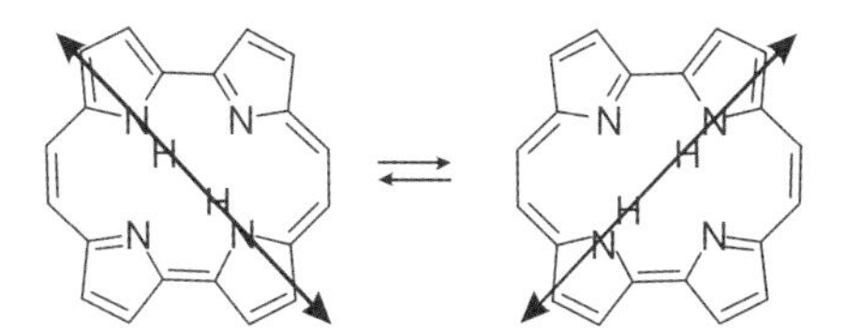

Figure 8.12 S_0–S_1 transition moment direction changes as a result of trans–trans conversion in **1**.

tion of excited molecules, the angle between absorbing and emitting transition moments is large, which leads to decreased anisotropy values. In **1e**, the excited state reaction is too slow to occur during the S_1 lifetime, about 10 ns.

The analysis of fluorescence anisotropy turned out to be a very powerful tool. First, depolarization provides a direct proof of excited state tautomerization that occurs on the timescale of the S_1 lifetime (or faster), which is, at 293 K, of the order of 10 ns for **1** and **1c** and ps in the case of **1a** and **1b**. It should be noted that the cis–cis tautomerization would not change the transition moment direction and thus cannot lead to depolarization. Cis–trans conversion, in turn, would result in much smaller changes in the anisotropy. The observed anisotropy values thus prove that the rapid excited state tautomerization in porphycene involves mostly trans–trans interconversion. A small fraction of cis structures may also be present, but it cannot be dominant. The same is true for the ground state, as revealed by simulations of expected anisotropy values assuming various fractions of cis tautomers in the ground and lowest excited singlet states [30, 82]. This is a very important result, given that previous structural assignments were mostly based on calculations, since they were impossible to obtain from X-ray data, or, in the case of NMR [40, 68], IR and Raman measurements [42], not unequivocal.

Second, careful analysis of anisotropy data is useful not only with respect to investigation of structural and kinetic aspects of tautomerization, but also as a means to obtain detailed spectroscopic information about transition moment directions. Both procedures will be described below in more detail.

8.3.1
Tautomerization as a Tool to Determine Transition Moment Directions in Low Symmetry Molecules

The directions of transition moments in every chromophore are dictated by molecular symmetry. For the cis tautomers of porphycene (C_{2v} point group), only three mutually orthogonal transition moment directions are allowed. On the other hand, the trans form is of C_{2h} symmetry and, therefore, any direction in the molecular plane is possible, as well as the direction perpendicular to the plane. The determination of transition moment directions in such low symmetry molecules is not an easy task. However, in the case of "narcissistic" type of reactions exemplified by trans–trans conversion in **1**, one can take advantage of an additional symmetry element introduced by the tautomerization process. Double hydrogen transfer converts the molecule into its image, with the horizontal and vertical mirror symmetry planes perpendicular to the molecular plane (Fig. 8.13). Thus, tautomerization results in the rotation of each in-plane transition moment direction. The angle of rotation is twice the value of the angle formed by a particular transition moment with the horizontal (or vertical) in-plane axis. It can be shown [80] that, for a fast excited state process, which results in equal population of both trans tautomers, the measured fluorescence anisotropy r will be expressed by the formula:

$$r(a,\beta) = \{\ 3[\cos^2(\beta - a/2) + \cos^2(\beta + a/2)] - 2\ \}/10 \qquad (8.3)$$

where β and $a/2$ are the angles between the molecular horizontal axis and the transition moment in absorption and emission, respectively.

Measuring anisotropy upon excitation into S_1 leads to the determination of $\beta_1 = a_1/2$, from the expression:

$$r(a, a/2) = [3\cos^2(a) + 1]/10 \qquad (8.4)$$

Knowledge of a enables one to obtain the directions of the moments of transitions to higher excited states. In order to determine the absolute values of the angles, the sign of at least one of them has to be known or assumed. Here, even approximate calculations are usually sufficient.

This procedure can be extended to a general model that takes into account the possible presence of both trans and cis forms, each of them differently populated in S_0 and S_1. The resulting formulas are:

$$r(a,\beta,\gamma,\delta) = (1-e)r(a,\beta) + gr(\delta-\gamma) + (e-g)r(\beta-\gamma),\quad g < e \qquad (8.5)$$

$$r(a,\beta,\gamma,\delta) = (1-g)r(a,\beta) + er(\delta-\gamma) + (g-e)r(a,\delta),\quad g > e \qquad (8.6)$$

where g is the fraction of ground state cis tautomers being excited, while e indicates the fraction of the cis form in the excited state; δ and γ denote the angles between the horizontal axis and the transition moments in absorption and emission, respectively, in the cis form. Due to symmetry, these angles can only assume values of 0° or 90°.

Application of this model to **1** resulted in the determination of the transition moments for the four lowest $\pi\pi^*$ transitions, responsible for the Q and Soret bands. It was found that for the lowest excited singlet state the transition moment is approximately parallel to the line connecting the protonated nitrogen atoms, whereas the S_0–S_2 transition moment is nearly orthogonal to this direction. This is similar to the case of free base porphyrin, where the molecular symmetry (D_{2h})

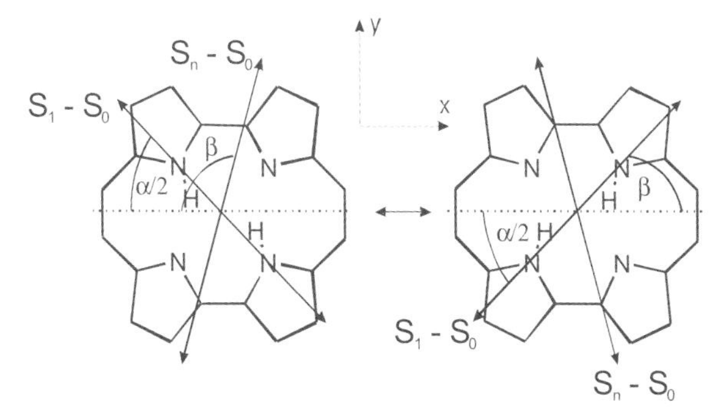

Figure 8.13 General scheme of tautomerism in **1** and the angles relevant for the anisotropy values.

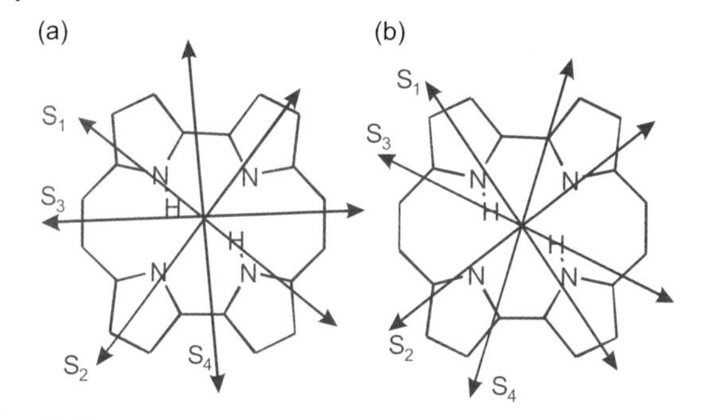

Figure 8.14 Transition moment directions for the lowest singlet excited states of **1**. (a) Experimental values, (b) the results of TD-DFT(B3LYP/6-31G(d,p)) calculations. S_1 and S_2 correspond to Q bands, S_3 and S_4 to Soret transitions (cf. also Fig. 8.2).

dictates that only these two in-plane transition moment directions are allowed. Interestingly, the anisotropy data for another low-symmetry (C_s) porphyrin isomer, 2,3,6,7,11,12,17,18-octaethylcorrphycene reproduce the same pattern [81]. The moments of the transitions corresponding to the two Soret bands in **1** lie approximately along the lines bisecting the two angles formed by the moments of transitions to S_1 and S_2 (Fig. 8.14).

An independent verification of this procedure was possible for **1b** [82]. The four n-propyl substituents ensure that this molecule can be oriented to a high degree in stretched polymer sheets. Linear dichroism (LD) measurements on the aligned samples resulted in the determination of transition moment directions. A pattern similar to that of parent **1** was obtained. Both LD and emission anisotropy procedures yielded similar values, which, additionally, confirmed that the assignment of fluorescence depolarization to excited state tautomerization was correct.

8.3.2
Determination of Tautomerization Rates from Anisotropy Measurements

Fluorescence anisotropy measurements can also be used to obtain the rates of the excited state tautomerization. Two variants can be applied. The first is based on the analysis of time-resolved anisotropy curves. These are constructed from measurements of the fluorescence decay recorded with different positions of the polarizers in the excitation and emission channels. The anisotropy decay reflects the movement of the transition moment and thus, the hydrogen exchange. For molecules with a long-lived S_1 state, the anisotropy decay can also be caused by rotational diffusion. In order to avoid depolarization effects due to molecular rotation, the experiments should be carried out in rigid media, such as polymers or glasses. When the S_1 lifetime is short compared to that of rotational diffusion, tautomerization rates can be determined even in solution. This is the case for **1b**, for which time-resolved anisotropy measurements have been performed at 293 K, using a fluorescence setup with single picosecond resolution [77]. Fluorescence was found to be depolarized from the very onset of the signal, which puts the lower limit of the tautomerization rate at values larger than 10^{11} s^{-1}. This extremely high rate agrees with expectations based on the very strong hydrogen bond in this molecule. For both **1a** and **1b**, biexponential fluorescence decays provide evidence for the presence of two forms, even at low temperatures. At the same time, fluorescence remains depolarized. It implies that the rate of excited trans–trans conversion may be faster than that of the cis–trans reaction. In other words, simultaneous transfer of two hydrogen atoms is favored over a single hydrogen transfer process.

The above method requires quite long measurement times and a proper matching of decay curve amplitudes. Moreover, the noise increases rapidly at longer delay times. Another procedure, experimentally less demanding, based on the steady state anisotropy values also allows determination of the trans–trans interconversion rates. Figure 8.15 shows the anisotropy of fluorescence excitation curves obtained for **1** in polyvinyl butyral films in the temperature range 20–293 K. Three temperature regions can be distinguished: (i) a high-temperature, fast-reaction regime, when the tautomerization proceeds on a time scale much shorter than the S_1 decay; (ii) an intermediate region, where the rates of the reaction and of S_1 deactivation are comparable; (iii) a low-temperature range, when the tautomerization is frozen, at least on the time scale of the S_1 lifetime. The ob-

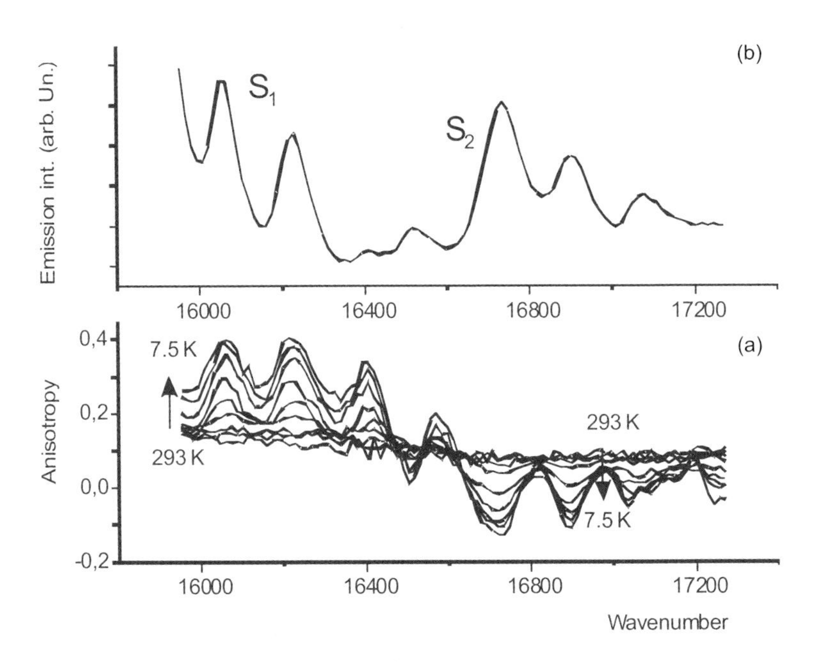

Figure 8.15 (a) Steady state anisotropy of fluorescence excitation of **1** recorded in poly(vinyl butyral) film as a function of temperature. The spectra were taken at 293, 250, 215, 165, 125, 85, 65, 45, and 7.5 K. (b) The excitation spectrum at 85 K monitored at 15900 cm^{-1}.

served fluorescence anisotropy r and the reaction rate k_{PT} are related by the formula:

$$k_{PT} = \frac{1}{\tau}\left[\frac{r(0) - r}{2r - r(a) - r(0)}\right] \quad (8.7)$$

where $r(0)$ and $r(a)$ are the anisotropies of the initially excited and tautomeric species, respectively, which can be obtained from measurements performed in the low and high-temperature regimes; τ is the value of the fluorescence decay time. The same rates are assumed for forward and backward reaction. Thus, the only quantity required, other than the anisotropy values, is the fluorescence lifetime, which can usually be measured with high accuracy. Figure 8.16 illustrates the application of this procedure in the study of the temperature dependence of tautomerization in porphycene dissolved in poly(vinyl butyral) film in the temperature range 7.5–293 K. Arrhenius type behavior is observed. The apparent activation energy for tautomerization is about 0.55±0.05 kcal mol^{-1} (192±20 cm^{-1}), a very small value compared to the barrier of several kcal mol^{-1}, estimated on the basis of tunneling splitting [72] and molecular geometry. A tentative interpretation is to postulate that the hydrogen exchange occurs as a thermally activated synchronous tunneling process, and the activation corresponds to exciting a vibration that lowers the barrier for the process. Indeed, the calculations for **1** predict an a_g vibration around 180 cm^{-1} [41, 42] of which the experimental counterpart is observed both in the ground and S_1 states [42, 83]. The form of this mode corresponds exactly to what can be expected to facilitate the concerted transfer of two hydrogen atoms: the distance between the two pairs of hydrogen-bonded nitrogen atoms is simultaneously decreased, with both NH···N bonds becoming more linear (Fig. 8.17).

An alternative, but less probable explanation would be to assume a mechanism analogous to that of porphyrin – thermally activated tunneling of a single hydrogen atom. However, to account for the fact that no significant fraction of cis tauto-

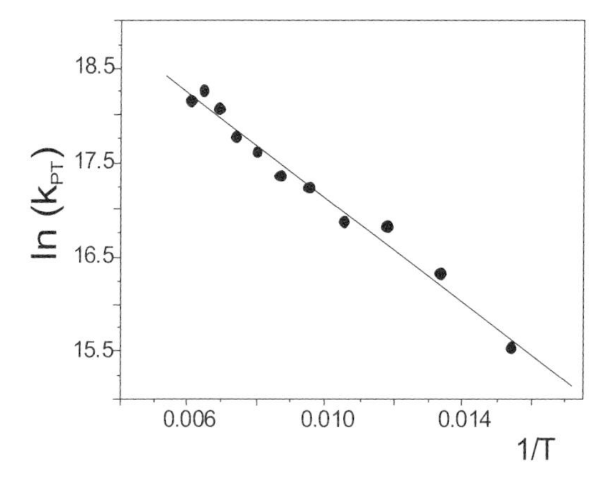

Figure 8.16 Arrhenius plot for tautomerization rates obtained for **1** in polyvinyl butyral (see also Fig. 8.15).

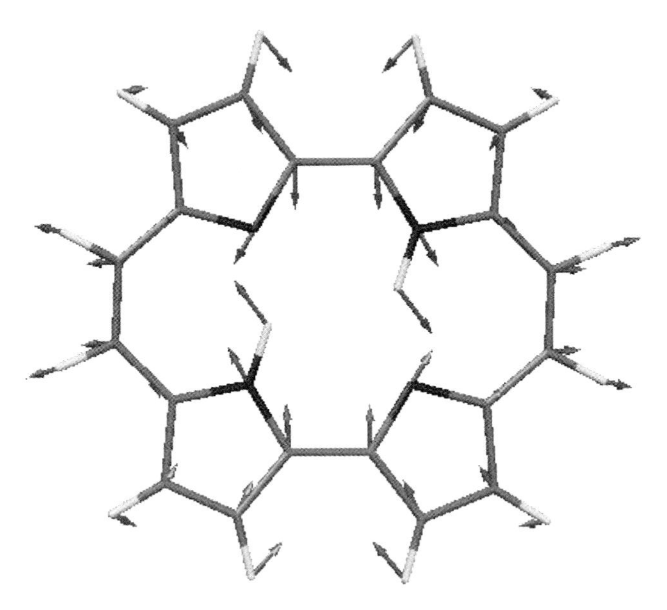

Figure 8.17 The form of the normal mode responsible for lowering of tautomerization barrier for simultaneous double hydrogen transfer (calculated using B3LYP/6-31G(d,p)).

mers could be observed in **1**, it is necessary that the tunneling of the first hydrogen atom is followed by a rapid transfer of the second hydrogen. In the limit of the latter process being much faster than the transfer of the first hydrogen atom, the synchronous mechanism is recovered.

8.4 Tautomerization in the Lowest Excited Triplet State

The tautomerism of **1** in the triplet state has been studied by time-resolved electron paramagnetic resonance (TR EPR) and electron spin echo spectroscopy in glassy matrices in the temperature range 4–100 K [84, 85]. Remarkable temperature variations have been observed. Two different triplet state species were detected at higher temperatures, but only one in the low temperature region. This was interpreted as evidence for two tautomeric species, assigned to the trans and cis forms. Since both are observed at 100 K, it was concluded that the energy difference is about 0.8 kJ mol^{-1}. This value is slightly higher than the value obtained by NMR for the ground state of crystalline **1**. In the latter case, only trans forms were detected below 50 K [40].

The exchange of hydrogens between the two tautomers was found to be slow on the EPR time scale at 100 K. The upper limit of the reaction rate was estimated at this temperature as 7×10^8 s^{-1}.

It may be instructive to compare the excited state behavior of **1** with that of **2**. It is well known that photoexcitation of free base porphyrins leads to trans–trans conversion [86]. This process occurs even at liquid helium temperatures, with the efficiency estimated as about 1% [87]. The mechanism is still under debate. Most probably, photoinduced tautomerization does not occur in S_1; it has been suggested that the reactions can proceed in T_1 [88] or even via higher excited triplet states [87]. It was also postulated that the cis structure in T_1 can be involved [89]. However, no such structure has yet been detected.

The differences in tautomerization in the lowest excited singlet and triplet states of **1** and **2** reflect the pattern observed for the ground state. In contrast to porphyrin, both trans and cis structures are detected in porphycenes, separated by a very small energy gap. The tautomerization barriers are also much lower in **1**. Finally, at least in the case of the lowest excited singlet state, synchronous transfer of two hydrogen atoms seems to be preferred over a stepwise mechanism, even when both forms are present (**1a** and **1b**).

8.5
Tautomerization in Single Molecules of Porphycene

Numerous variants of single molecule spectroscopy [90] are based on fluorescence detection. In particular, analysis of the spatial patterns of the emission from a single molecule makes it possible to determine its orientation in three dimensions. In this procedure, molecules are treated as dipoles emitting electromagnetic radiation. During the experiment, which lasts for seconds to minutes, photons are collected from a single chromophore immobilized in a rigid environment. For porphycene, it is obvious that, due to tautomerization, the molecule cannot be considered as a single dipole. In the simplest case of trans–trans interconversion, the fluorescence can be envisaged as occurring from two nearly orthogonal dipoles (Fig. 8.12). If a fraction of cis forms is also present, a third dipole should be added, its direction bisecting the angle formed by the moments of the transitions in the two trans species. It is to be expected that the spatial pattern of the emission should in such cases be different from that due to a single dipole. This was confirmed in the experiment that resulted in the detection of fluorescence from well above 60 single porphycene molecules [91]. The so-called azimuthal polarization mode of the exciting light was used. Various spatial patterns of the emission were observed (Fig. 8.18). In many cases they consisted of nearly perfect rings. It is not possible to obtain such a pattern in the case of a single dipole. The simulations of intensity patterns led to the picture of two dipoles forming an angle of about 70°, in perfect agreement with the results obtained from fluorescence anisotropy studies of bulk porphycene in glassy matrices [80]. Some molecules revealed double lobe emission intensity patterns. This could be interpreted in two ways. The first assumes a single dipole and thus, no tautomerization. The other, more probable, explanation is that these patterns are due to molecules that are oriented perpendicularly to the sample surface: for such orien-

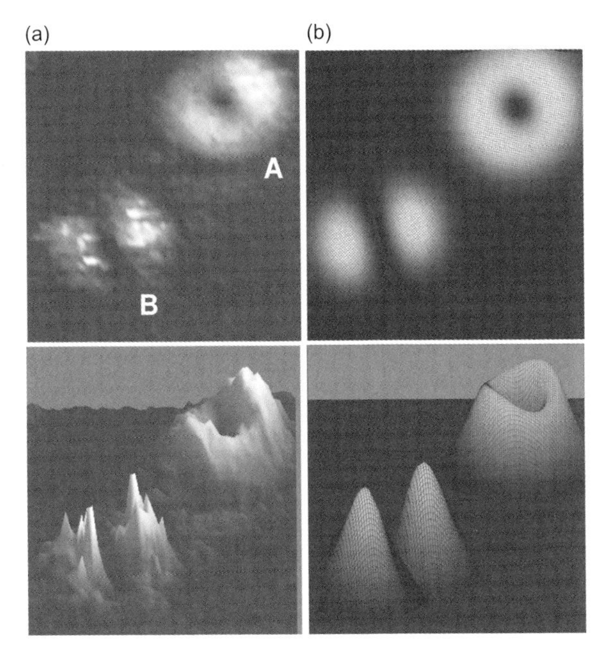

Figure 8.18 Spatial patterns of fluorescence of single porphycene molecules A and B immobilized in poly(methyl methacrylate) at 293 K. (a) Experiment, (b) simulation [91].

tation, it is not possible to distinguish between the single and the multiple in-plane transition dipole case.

Thus, the molecular symmetry of free base porphycene that dictates a rotation of transition moment directions upon tautomerization has enabled the observation of a basic chemical reaction on a single molecule level. The procedure can be extended to studies of more complicated porphyrinoids, such as sapphyrin, a pentapyrrolic macrocycle in which three inner hydrogen atoms can migrate within the cavity formed by five nitrogen atoms.

8.6 Summary

Even though the studies of tautomerism in porphycenes have not yet reached the stage of definitive answers and final conclusions, the accumulated material is rich, diverse and fascinating. Many of the observed effects deserve to be characterized in more detail. These include tunneling splitting due to double hydrogen transfer, coexistence of trans and cis forms, the exact mechanism of trans–trans, cis–trans, and, possibly, also cis–cis conversions, energy differences and barriers between trans and cis structures, location of NH vibrations, etc. The available data demonstrate that it should be possible to investigate all these problems for three different electronic states, S_0, S_1, and T_1. Proper characterization of structure and dynamics, including tunneling, for the excited states in particular, is also a challenge for computational methods.

An important property of porphycenes is their ability to undergo significant changes in the dimensions of the inner cavity under the influence of such "delicate" substituents as alkyl groups. This provides a rare opportunity for systematic studies of the rate vs. distance (in general, rate–geometry) relationship. The observations gathered so far for the lowest excited singlet state behavior of several porphycenes show that a change of 0.25 Å in the N–N separation may lead to orders of magnitude variations in the tautomerization rates, which clearly points to the crucial role of tunneling. Such studies may also contribute to the understanding of the correlation between the inner protons, or even provide a model for their quantum entanglement.

The data presented in this work demonstrate large differences between the tautomerization characteristics in porphycene and its parent isomer, free base porphyrin. The origin for these dissimilarities may be due to different electronic structure or to distinct geometries. It seems at present that the latter are mainly responsible for the different patterns of tautomerism in **1** and **2**. In particular, strong, nearly linear hydrogen bonds in porphycenes lead to low barriers and rapid tautomerization rates.

The electronic structure seems to play a less important role, as may be speculated on the basis of similar values of NMR chemical shifts of peripheral hydrogen atoms, or by considering analogies in the pattern of the electronic spectra. In order to elucidate in more detail the role of electronic structure in tautomerism, we are now studying porphyrins in which peripheral substitution leads to inner cavity parameters similar to those of porphycenes.

Acknowledgments

The author expresses his gratitude to Emanuel Vogel and Josef Michl, who exposed him to the beauty of porphycenes. Special thanks go to many coworkers that have contributed to the results presented in this work: Michał Gil, Alexander Starukhin, Alexander Kyrychenko, Natalia Urbańska, Oksana Pietraszkiewicz, Marek Pietraszkiewicz, Yuriy Stepanenko, Jerzy Sepioł, Hubert Piwoński, Andrzej Mordziński, Alexander Vdovin, Alfred Meixner, Clemens Stupperich, Achim Hartschuh, Kristine Birklund-Andersen, Alexander Gorski, Jacek Dobkowski, Paweł Borowicz and Grażyna Orzanowska. Financial support from the Polish Committee for Scientific Research (grant 3T09A 113 26) is appreciated.

References

1 A. R. Battersby, C. J. R. Fookes, G. W. J. Matcham, E. McDonald, *Nature* **1980**, *285*, 17–21.

2 R. Bonnett, *Chem. Soc. Rev.* **1995**, *24*, 19–33.

3 T. J. Dougherty, C. J. Gomer, B. W. Henderson, G. Jori, D. Kessel, M. Korbelik, J. Moan, Q. Peng, *J. Natl. Cancer Inst.* **1998**, *90*, 889–905.

4 S. B. Brown, E. A. Brown, I. Walker, *Lancet Oncol.* **2004**, *5*, 497–508.

5 M. R. Detty, S. L. Gibson, S. J. Wagner, *J. Med. Chem.* **2004**, *47*, 3897–3915.

6 T. Maisch, R. M. Szeimies, G. Jori, C. Abels, *Photochem. Photobiol. Sci.* **2004**, *3*, 907–917.

7 S. L. Gould, G. Kodis, R. E. Palacios, L. de la Garza, A. Brune, D. Gust, T. A. Moore, A. L. Moore, *J. Phys. Chem. B* **2004**, *108*, 10566–10580.

8 D. Gust, T. A. Moore, A. L. Moore, *Acc. Chem. Res.* **2001**, *34*, 40–48.

9 W. M. Campbell, A. K. Burrell, D. L. Officer, K. W. Jolley, *Coord. Chem. Rev.* **2004**, *248*, 1363–1379.

10 A. Renn, U. P. Wild, A. Rebane, *J. Phys. Chem. A* **2002**, *106*, 3045–3060.

11 C. M. Carcel, J. K. Laha, R. S. Loewe, P. Thamyongkit, K. H. Schweikart, V. Misra, D. F. Bocian, J. S. Lindsey, *J. Org. Chem.* **2004**, *69*, 6739–6750.

12 T. Hasobe, P. V. Kamat, M. A. Absalom, Y. Kashiwagi, J. Sly, M. J. Crossley, K. Hosomizu, H. Imahori, S. Fukuzumi, *J. Phys. Chem. B* **2004**, *108*, 12865–12872.

13 Y. Kobuke, K. Ogawa, *Bull. Chem. Soc. Jpn.* **2003**, *76*, 689–708.

14 C. M. Drain, J. D. Batteas, G. W. Flynn, T. Milic, N. Chi, D. G. Yablon, H. Sommers, *Proc. Natl. Acad. Sci. U. S. A.* **2002**, *99*, 6498–6502.

15 I. Washington, C. Brooks, N. J. Turro, K. Nakanishi, *J. Am. Chem. Soc.* **2004**, *126*, 9892–9893.

16 J. L. Sessler, S. Camiolo, P. A. Gale, *Coord. Chem. Rev.* **2003**, *240*, 17–55.

17 I. Okura, *J. Porphyr. Phthalocy.* **2002**, *6*, 268–270.

18 J. P. Collman, J. T. McDevitt, G. T. Yee, C. R. Leidner, L. G. McCullough, W. A. Little, J. B. Torrance, *Proc. Natl. Acad. Sci. U. S. A.* **1986**, *83*, 4581–4585.

19 J. L. Sessler, S. J. Weghorn, *Expanded, Contracted and Isomeric Porphyrins*, Elsevier Science, Oxford, **1997**.

20 *The Porphyrin Handbook*, K. M. Kadish, K. M. Smith, R. Guilard, (Eds.), Vol. 2, Academic Press, New York, **2000**.

21 E. Vogel, M. Köcher, H. Schmickler, J. Lex, *Angew. Chem. Int. Ed. Engl.* **1986**, *25*, 257–259.

22 J. L. Sessler, E. A. Brucker, S. J. Weghorn, M. Kisters, M. Schafer, J. Lex, E. Vogel, *Angew. Chem. Int. Ed. Engl.* **1994**, *33*, 2308–2312.

23 H. J. Callot, A. Rohrer, T. Tschamber, B. Metz, *New J. Chem.* **1995**, *19*, 155–159.

24 E. Vogel, M. Broring, S. J. Weghorn, P. Scholz, R. Deponte, J. Lex, H. Schmickler, K. Schaffner, S. E. Braslavsky, M. Muller, S. Porting, C. J. Fowler, J. L. Sessler, *Angew. Chem. Int. Ed. Engl.* **1997**, *36*, 1651–1654.

25 E. Vogel, M. Broring, C. Erben, R. Demuth, J. Lex, M. Nendel, K. N. Houk, *Angew. Chem. Int. Ed. Engl.* **1997**, *36*, 353–357.

26 E. Vogel, P. Scholz, R. Demuth, C. Erben, M. Broring, H. Schmickler, J. Lex, G. Hohlneicher, D. Bremm, Y. D. Wu, *Angew. Chem. Int. Ed. Engl.* **1999**, *38*, 2919–2923.

27 P. Chmielewski, L. Latos-Grazyński, K. Rachlewicz, T. Głowiak, *Angew. Chem. Int. Ed. Engl.* **1994**, *33*, 779–781.

28 H. Furuta, T. Asano, T. Ogawa, *J. Am. Chem. Soc.* **1994**, *116*, 767–768.

29 J. Michl, *Tetrahedron* **1984**, *40*, 3845–3934.

30 J. Waluk, M. Müller, P. Swiderek, M. Köcher, E. Vogel, G. Hohlneicher, J. Michl, *J. Am. Chem. Soc.* **1991**, *113*, 5511–5527.

31 A. Gorski, E. Vogel, J. L. Sessler, J. Waluk, *J. Phys. Chem. A* **2002**, *106*, 8139–8145.

32 A. Gorski, E. Vogel, J. L. Sessler, J. Waluk, *Chem. Phys.* **2002**, *282*, 37–49.

33 Y. D. Wu, K. W. K. Chan, C. P. Yip, E. Vogel, D. A. Plattner, K. N. Houk, *J. Org. Chem.* **1997**, *62*, 9240–9250.
34 L. E. Webb, E. B. Fleischer, *J. Chem. Phys.* **1965**, *43*, 3100–3111.
35 E. Bartoszak, M. Jaskólski, T. Gustaffson, I. Olovsson, E. Grech, *Acta Crystallogr., Sect. B* **1994**, *50*, 358–353.
36 B. M. L. Chen, A. Tulinsky, *J. Am. Chem. Soc.* **1972**, *94*, 4144–4157.
37 A. Tulinsky, *Ann. N.Y. Acad. Sci.* **1973**, *206*, 47–69.
38 J. G. Radziszewski, J. Waluk, J. Michl, *Chem. Phys.* **1989**, *136*, 165.
39 J. Radziszewski, J. Waluk, J. Michl, *J. Mol. Spectrosc.* **1990**, *140*, 373.
40 U. Langer, C. Hoelger, B. Wehrle, L. Latanowicz, E. Vogel, H. H. Limbach, *J. Phys. Org. Chem.* **2000**, *13*, 23–34.
41 P. M. Kozlowski, M. Z. Zgierski, J. Baker, *J. Chem. Phys.* **1998**, *109*, 5905–5913.
42 K. Malsch, G. Hohlneicher, *J. Phys. Chem. A* **1997**, *101*, 8409.
43 E. D. Becker, R. B. Bradley, C. J. Watson, *J. Am. Chem. Soc.* **1961**, *83*, 3743–3748.
44 C. B. Storm, Y. Teklu, **1972**, *94*, 1745–1747.
45 C. B. Storm, Y. Teklu, E. A. Sokoloski, *Ann. N.Y. Acad. Sci.* **1973**, *206*, 631–640.
46 R. J. Abraham, G. E. Hawkes, K. C. Smith, *Tetrahedron Lett.* **1974**, *16*, 1483–1486.
47 S. S. Eaton, G. R. Eaton, *J. Am. Chem. Soc.* **1977**, *99*, 1601–1604.
48 C. S. Irving, A. Lapidot, *J. Chem .Soc., Chem. Comm.* **1977**, 184–186.
49 H. J. C. Yeh, *J. Magn. Reson.* **1977**, *28*, 365.
50 J. Hennig, H. H. Limbach, *J. Chem.Soc., Faraday Trans. 2* **1979**, 752–766.
51 P. Stilbs, M. E. Moseley, *J. Chem.Soc., Faraday Trans. 2* **1980**, *76*, 729–731.
52 J. Hennig, H. H. Limbach, *J. Magn. Reson.* **1982**, *49*, 322–328.
53 H. H. Limbach, J. Hennig, D. Gerritzen, H. Rumpel, *Faraday Discuss., Chem. Soc* **1982**, *74*, 229–243.
54 J. Hennig, H. H. Limbach, *J. Am. Chem. Soc.* **1984**, *106*, 292–298.
55 P. Stilbs, *J. Magn. Reson.* **1984**, *58*, 152.
56 M. Schlabach, B. Wehrle, H. H. Limbach, E. Bunnenberg, A. Knierzinger, A. Y. L. Shu, B.-R. Tolf, C. Djerassi, *J. Am. Chem. Soc.* **1986**, *108*, 3856–3858.
57 M. J. Crossley, L. D. Field, M. M. Harding, S. Sternhell, *J. Am. Chem. Soc.* **1987**, *109*.
58 L. Frydman, A. Olivieri, L. E. Diaz, B. Frydman, F. G. Morin, C. Mayne, D. M. Grant, A. D. Adler, *J. Am. Chem. Soc.* **1988**, *110*, 336–342.
59 M. Schlabach, H. Rumpel, H. H. Limbach, *Angew. Chem. Int. Ed. Engl.* **1989**, *28*, 76–79.
60 M. Schlabach, G. Scherer, H. H. Limbach, *J. Am. Chem. Soc.* **1991**, *113*, 3550–3558.
61 M. Schlabach, B. Wehrle, H. Rumpel, J. Braun, G. Scherer, H. H. Limbach, *Ber. Bunsenges. Phys. Chem.* **1992**, *96*, 821–833.
62 M. Schlabach, H. H. Limbach, E. Bunnenberg, A. Y. L. Shu, B.-R. Tolf, C. Djerassi, *J. Am. Chem. Soc.* **1993**, *115*, 4554–4565.
63 J. Braun, M. Schlabach, B. Wehrle, M. Köcher, E. Vogel, H. H. Limbach, *J. Am. Chem. Soc.* **1994**, *116*, 6593–6604.
64 J. Braun, C. Hasenfratz, R. Schwesinger, H. H. Limbach, *Angew. Chem. Int. Ed. Engl.* **1994**, *33*, 2215–2217.
65 J. Braun, H. H. Limbach, P. G. Williams, H. Morimoto, D. E. Wemmer, *J. Am. Chem. Soc.* **1996**, *118*, 7231–7232.
66 J. Braun, R. Schwesinger, P. G. Williams, H. Morimoto, D. E. Wemmer, H. H. Limbach, *J. Am. Chem. Soc.* **1996**, *118*, 11101–11110.
67 H. H. Limbach, J. Hennig, R. Kendrick, C. S. Yannoni, *J. Am. Chem. Soc.* **1984**, *106*, 4059–4060.
68 B. Wehrle, H. H. Limbach, M. Kocher, O. Ermer, E. Vogel, *Angew. Chem. Int. Ed. Engl.* **1987**, *26*, 934–936.
69 R. P. Bell, *The Tunnel Effect in Chemistry*, Chapman and Hall, London, **1980**.
70 Z. Smedarchina, M. Z. Zgierski, W. Siebrand, P. M. Kozlowski, *J. Chem. Phys.* **1998**, *109*, 1014–1024.
71 B. Frydman, C. O. Fernandez, E. Vogel, *J. Org. Chem.* **1998**, *63*, 9385–9391.
72 J. Sepioł, Y. Stepanenko, A. Vdovin, A. Mordziński, E. Vogel, J. Waluk, *Chem. Phys. Lett.* **1998**, *296*, 549–556.
73 U. Even, J. Jortner, *J. Chem. Phys.* **1982**, *77*, 4391–4399.

74 H. P. Trommsdorff, *Adv. Photochem.* **1998**, *24*, 147–204.
75 M. D. Harmony, *Chem. Phys. Lett.* **1971**, *10*, 337–340.
76 A. Vdovin, J. Sepioł, N. Urbańska, M. Pietraszkiewicz, A. Mordziński, J. Waluk, *J. Am. Chem. Soc.* **2006**, *128*, 2577–2586.
77 M. Gil, P. Borowicz, D. Marks, H. Zhang, M. Glasbeek, J. Waluk, in preparation.
78 M. Gil, J. Jasny, E. Vogel, J. Waluk, *Chem. Phys. Lett.* **2000**, *323*, 534–541.
79 A. Kyrychenko, J. Herbich, M. Izydorzak, M. Gil, J. Dobkowski, F. Y. Wu, R. P. Thummel, J. Waluk, *Isr. J. Chem.* **1999**, *39*, 309–318.
80 J. Waluk, E. Vogel, *J. Phys. Chem.* **1994**, *98*, 4530–4535.
81 A. Gorski, E. Vogel, J. Waluk, in preparation.
82 K. Birklund-Andersen, E. Vogel, J. Waluk, *Chem. Phys. Lett.* **1997**, *271*, 341.
83 A. Starukhin, E. Vogel, J. Waluk, *J. Phys. Chem. A* **1998**, *102*, 9999–10006.
84 C. W. M. Kay, K. Möbius, *Mol. Phys.* **1998**, *95*, 1013–1019.
85 C. W. M. Kay, G. Elger, K. Möbius, *Phys. Chem. Chem. Phys.* **1999**, *1*, 3999–4002.
86 O. N. Korotaev, R. I. Personov, *Opt. Spectrosc. (Engl. Transl.)* **1972**, *32*, 479–480.
87 S. Völker, J. H. van der Waals, *Mol. Phys.* **1976**, *32*, 1703–1718.
88 V. A. Kuz'mitskii, *Chem. Phys. Rep.* **1996**, *15*, 1783–1797.
89 S. S. Dvornikov, V. A. Kuz'mitskii, V. N. Knyukshto, K. N. Solov'ev, *Dokl. Akad. Nauk SSSR* **1985**, *282*, 362–366.
90 R. Rigler, M. Orrit, T. Basche, *Single Molecule Spectroscopy*, Springer, Berlin, **2002**.
91 H. Piwoński, C. Stupperich, A. Hartschuh, J. Sepioł, A. Meixner, J. Waluk, *J. Am. Chem. Soc.* **2005**, *127*, 5302–5303.
92 J. Baker, P. M. Kozlowski, A. A. Jarzecki, P. Pulay, *Theor. Chem. Acc.* **1997**, *97*, 59–66.

9
Proton Dynamics in Hydrogen-bonded Crystals

Mikhail V. Vener

9.1
Introduction

In this chapter we will consider molecular crystals with normal hydrogen bonds in which the donor A:H interacts with an acceptor :B. The so-called "bifurcated" and "trifurcated" H-bonds [1] as well as the new multiform unconventional H-bonds [2] are beyond the scope of the present chapter. We will focus on the proton dynamics in molecular crystals with strong and moderate H-bonds [3] in the ground electronic state. Attention will be focused on the interpretation of the structural and spectroscopic manifestations of the dynamics of the bridging proton as established in X-ray, neutron diffraction, infrared, and inelastic neutron scattering (INS) studies of H-bonded crystals.

Various theoretical approaches have been developed for the description of the structure, spectral properties, and proton tunneling in H-bonded systems [4–7]. Computations for particular H-bonded species in the gas phase have been performed [8]. Due to strong environmental effects the applicability of gas-phase calculations to the proton dynamics in H-bonded crystals is questionable. Many theoretical approaches are based on oversimplified models (harmonic potentials and one-dimensional treatment of proton tunneling) and they usually contain parameters obtained from the experiment to be interpreted. This is why a consistent view on hydrogen bonding phenomenon in molecular crystals is still far from being achieved.

The aims of this article are:

1. To show that a uniform and noncontradictory description of the specific properties of molecular crystals with quasi-linear H-bonds can be obtained in terms of a two-dimensional (2D) treatment assuming strong coupling between the proton-transfer coordinate and a low-frequency vibration.
2. To interpret experimental structural and spectroscopic regularities of crystals with a quasi-symmetric A···H···A fragment using a model 2D potential energy surface (PES).

Hydrogen-Transfer Reactions. Edited by J. T. Hynes, J. P. Klinman, H. H. Limbach, and R. L. Schowen

ISBN: 978-3-527-30777-7

3. To describe quantitatively the effects of a crystalline environment on the structure, PES, and vibrational spectra of strong H-bonds in terms of density functional theory (DFT) calculations with periodic boundary conditions.

9.2
Tentative Study of Proton Dynamics in Crystals with Quasi-linear H-bonds

A qualitative description of the structural and spectroscopic properties of H-bonded crystals requires the use of a simple two-mode linear model (9.1).

$$
\begin{array}{l}
\mathrm{B}\cdots\cdots\bullet\cdots\mathrm{H{-}A} \\
\rightarrow\!|\ s\ |\!\leftarrow \\
|\!\leftarrow R \rightarrow\!|
\end{array}
\tag{9.1}
$$

Here A and B are "heavy" molecular fragments and their internal structure is not specified in detail now. The centers of gravity of the B, A and H particles lie on the same straight line. The proton coordinate (s) is measured from the center of gravity (•) of the whole system (9.1). The distance between the terminal atoms (rather than between the centers of gravity of fragments B and A) is usually taken as the coordinate R. Typical values of the reduced masses (m and M) and characteristic frequencies (ν_{as} and ν_s) corresponding to the s and R coordinates are: $m \sim 1$ and $M > 10$ a.u.; $\nu_{as} > 1000$ and $\nu_s < 200$ cm^{-1}.

Figure 9.1 presents a 2D PES $U(s,R)$ for a symmetrical system B = A. A specific feature of this PES is a strong coupling between the two coordinates. Due to the

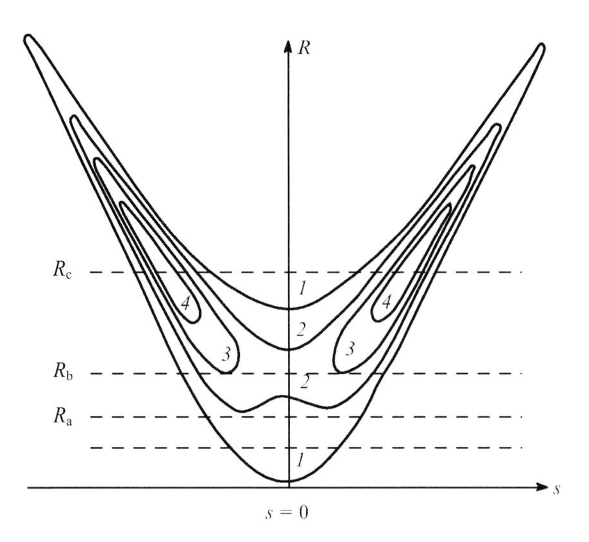

Figure 9.1 Model 2D PES $U(s,R)$ of the symmetrical A···H···A fragment. Closed loops (1–4) are isoenergy contour lines in arbitrary units.

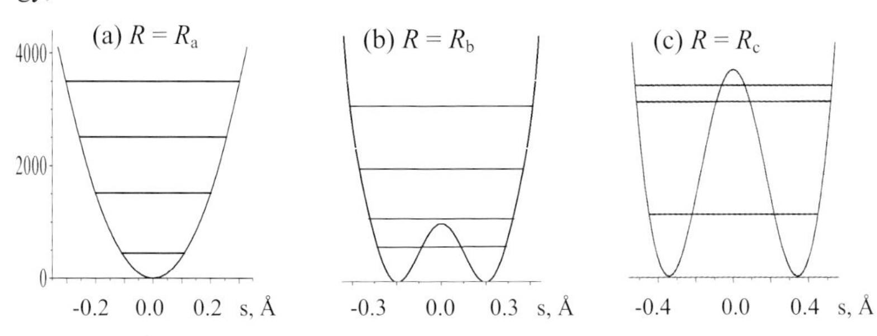

Figure 9.2 Slices through the model 2D PES $U(s,R)$ for three fixed R values: (a) $R = R_a$, (b) $R = R_b$, (c) $R = R_c$. For the sake of comparison the $W(R)$ function, see Eq. (9.6), is set equal to zero. The four lowest proton adiabatic levels are shown but the lower pair is not resolved in (c).

coupling the *shape* of the profile along the proton coordinate parametrically depends on R. At $R = R_a$ there is a one-well profile (Fig. 9.2(a)). At $R = R_b$ there appears a small barrier (Fig. 9.2(b)), and at $R = R_c$ a double-well profile with a relatively high barrier exists (Fig. 9.2(c)). The point is that the difference between R_c and R_a for the considered H-bonded crystals is around 0.2 Å, that is of the order of the amplitude of zero-point vibration along the coordinate R.

The R coordinate is a typical example of a so-called promoting mode [7]. It does not change (B = A) or very slightly changes (B ≠ A) when the bridging proton shifts from A to B. However, it appreciably changes in the transition state and thus modifies the height of the potential barrier.

9.2.1
A Model 2D Hamiltonian

The 2D model of the linear B···H–A fragment, assuming a strong coupling between the proton (AH stretch) and low-frequency (B···A stretch) coordinates, was introduced by Stepanov [9, 10]. It seems to be the simplest model enabling one to interpret the different specific features of H-bonded systems [11–16]. In terms of this model, the vibrational wave function of the H-bonded system is written as

$$\Phi(s,R,q_1,\dots,q_{N-8}) \approx \Psi(s,R)\phi(q_1)\dots\phi(q_{N-8}) \tag{9.2}$$

Here N is the number of degrees of freedom; $\phi(q_i)$ is a harmonic wave function, and $\Psi(s,R)$ is the eigenfunction of the following 2D Hamiltonian:

$$H = -\frac{\hbar^2}{2m}\frac{\partial^2}{\partial s^2} - \frac{\hbar^2}{2M}\frac{\partial^2}{\partial R^2} + U(s,R) \tag{9.3}$$

Kinematic coupling is supposed to be nil and reduced masses are written as:

$$m = \frac{M_H(M_A + M_B)}{M_H + M_A + M_B} \approx M_H; \qquad M = \frac{M_A M_B}{M_A + M_B} \gg m \tag{9.4}$$

Here M_H, M_A and M_B stand for the mass of the corresponding particle. In the case of intramolecular H-bonds A and B may be treated as the terminal atoms, in the case of intermolecular H-bonds A and B may be considered as the "heavy" fragments.

The 2D PES $U(s,R)$ of the B⋯H–A fragment in a crystal can be obtained using DFT calculations with periodic boundary conditions (Section 9.3.2) or extracted from experimental data (Section 9.2.2). In the case of the model 2D PESs the equilibrium value of the B⋯A distance is usually defined by structural data. The barrier height along the proton coordinate V_0 is the main control parameter whose variation allows one to reproduce the spectroscopic data.

There are two possible approaches to solution of the Schrödinger equation with the model Hamiltonian (9.3). The first is based on direct solution using grid [17] and basis [18] methods or the so-called multiconfiguration time-dependent Hartree method [19]. (The latter method is a combination of grid and basis set methods in the sense that the time-dependent basis functions are represented on a suitable grid). The second approach uses a Born–Oppenheimer type separation between the motions of the light and heavy nuclei (the method of adiabatic separation between vibrational variables) [20]. In the second approach the wave function $\Psi(s,R)$ is written as:

$$\Psi(s,R) \approx \Psi_{vn}(s,R) = \varphi_v(s,R)\chi_{vn}(R) \tag{9.5}$$

Here $\varphi_v(s,R)$ is the wave function of the "fast" subsystem (proton or deuteron, with quantum number v) depending on R as a parameter. $\chi_{vn}(R)$ is the wave function of the "slow" subsystem (with quantum number n). The 2D PES can be written as:

$$U(s,R) = V(s,R) + W(R) \tag{9.6}$$

Here $V(s,R)$ is the proton (deuteron) potential energy, and $W(R)$ is the B⋯A interaction energy. The wave equation for the motion of the "fast" subsystem is given by

$$\left[-\frac{\hbar^2}{2m}\frac{\partial^2}{\partial s^2} + V(s,R)\right]\varphi_v(s,R) = e_v(R)\varphi_v(s,R) \tag{9.7}$$

where the adiabatic proton (deuteron) terms $e_v(R)$ depend on R as a parameter. The *shape* of $e_v(R)$ depends on v and is different for H and D. The wave equation for the B⋯A stretch is

$$\left[-\frac{\hbar^2}{2M}\frac{\partial^2}{\partial R^2} + e_v(R) + W(R)\right]\chi_{vn}(R) = E_{vn}\chi_{vn}(R) \tag{9.8}$$

The "effective" potential for the B··A stretch is different for H and D and depends on the value of ν. Various methods of numerical solution of the 1D vibrational Schrödinger equation have been suggested in the literature, for example see Refs. [21, 22].

Attractive features of the adiabatic separation of vibrational variables are its relative simplicity and its provision for a clear picture in terms of modes (see the next section).

As a result of solving the Schrödinger equation for the model Hamiltonian (9.3) one gets anharmonic frequencies and relative IR intensities (if the dipole moment function is available) in the *particular* region of the IR spectrum [23–25]. It should be noted, however, that in many theoretical studies only the energy spectrum of the model 2D Hamiltonian was considered [26, 27].

9.2.2
Specific Features of H-bonded Crystals with a Quasi-symmetric O···H···O Fragment

In many crystals with strong or moderate H-bonds and a quasi-symmetric A···H···A fragment the proton is known to have two equilibrium positions separated by a potential barrier. The most convincing evidence for this can be provided by neutron diffraction methods. For crystals containing O···H···O fragments with an O···O equilibrium distance (R_e) in the range 2.40–2.70 Å, neutron diffraction studies reveal the existence of two protonic density maxima separated by 0.1–0.8 Å, see Table 1 in Ref. [28]. (Following results presented in Refs. [29, 30] it was assumed that the lower limit of the length of the O···O hydrogen bond is 2.40 Å).

Molecular crystals with strong or moderate H-bonds demonstrate several structural and spectroscopic regularities. (i) The O···O equilibrium distance changes upon deuteration to give the so-called Ubbelohde effect. $\Delta R = R_e(D) - R_e(H)$ is negative for $R_e \sim 2.40$ Å. With increasing R_e the ΔR value becomes positive to reach a maximum at $R_e \sim 2.55$ Å, after which it decreases [31]. (ii) The proton stretching vibrational frequency (asymmetric vibration of the O···H···O fragment, or the OH stretch) ν^H is remarkably decreased when R_e decreases from ~ 2.70 to 2.40 Å. The decrease is ~ 2700 cm^{-1} that is from ~ 3200 to ~ 500 cm^{-1} [32, 33]. (For the sake of simplicity, the subscript "as" is omitted in this Section). (iii) The dependence of the isotopic frequency ratio, $\gamma = \nu^H/\nu^D$, on ν^H is non-monotonic. Upon frequency lowering (that is shortening of the O···O distance) γ decreases to unity (and below) and then sharply increases. Novak [32] obtained this result for crystals with the O···H···O fragment, and Grech et al. [34] observed it for compounds with the N···H···N fragment.

To interpret these regularities the model 2D PES $U(s,R)$ was suggested [28, 33, 35] for crystals with a quasi-symmetric O···H···O fragment. It was assumed that strong coupling between the s and R coordinates is manifested in a special dependence of the barrier height V_0 on R. With V_0 taken as zero at $R = 2.40$ Å, it rises exponentially in the region $2.40 < R < 2.60$ Å. At larger R the increase is much slower. Representative profiles along the proton coordinate at three "crucial" values of R are given in Fig. 9.2, with details provided by specific examples in the remainder of this section.

R ~ 2.40 Å (Fig. 9.2(a), $H_5O_2^+$ in crystals, see Section 9.3.2) The PES profile has the shape of a single symmetric well with the bridging proton (deuteron) localized at the center symmetry. The proton (deuteron) stretching frequencies are described as $|0\rangle \rightarrow |1\rangle$ transitions with ν^H ~ 1100 cm^{-1} and ν^D ~ 800 cm^{-1} leading to the usual value of γ (~ 1.35). When *R* is larger than 2.40 Å but the barrier remains small (with the $|0\rangle$ level still above the barrier) the proton vibrational amplitude increases and the proton kinetic energy accordingly drops in every quantum state. This results in lowering of the proton $|\nu\rangle$ levels, more for higher ν values. Figures 9.2(a) and (b) (and extrapolation of Fig. 9.2(c) to higher levels) show the strong decrease for levels above the barrier maximum.

R ~ 2.47 Å (Fig. 9.2(b), sodium hydrogen bis(4-nitrophenoxide)dihydrate [36]) The lowest proton level is slightly below the central barrier. The $|0\rangle$ and $|1\rangle$ levels draw together and the $|0\rangle \rightarrow |1\rangle$ transition is around 600 cm^{-1}, in accord with the experimental data [36]. The separation of the corresponding levels for the deuteron decreases faster with *R* growth than does the proton levels. This is why the γ value is very high (~ 2).

R ~ 2.56 Å (Fig. 9.2(c), chromous acid [33]) The barrier grows very fast in this region and the $|0\rangle \rightarrow |1\rangle$ transition frequency is very low for the H and D species. At the temperature of the experiment the $|1\rangle$ level is sufficiently populated that the $|1\rangle \rightarrow |2\rangle$ transitions appear in the IR spectra. There may be strong scattering of the experimental stretching frequencies because of the two different frequencies arising from the asymmetric stretch of the O···H···O fragment. In principle, there exists one more IR active transition in the considered *R* region. This is $|0\rangle \rightarrow |3\rangle$, which has a frequency that may differ strongly from the frequency of the $|1\rangle \rightarrow |2\rangle$ transition. The $|0\rangle \rightarrow |3\rangle$ transition seems to be realized only for systems with the O···D···O fragment [33]. As a result, the γ values are ~ 1.3 for the case of $\nu^H(|1\rangle \rightarrow |2\rangle)/\nu^D(|1\rangle \rightarrow |2\rangle)$ and ~ 1.0 for the case of $\nu^H(|1\rangle \rightarrow |2\rangle)/\nu^D(|0\rangle \rightarrow |3\rangle)$.

R > 2.60 Å (Benzoic acid dimer, see Section 9.2.3.1) The barrier along the proton coordinate at the equilibrium O···O distance is very high (> 7000 cm^{-1} [37]) and the energy gap between levels $|2\rangle$ and $|3\rangle$ is very small (several wavenumbers). The wells become isolated and the hydrogen bond becomes identical with the asymmetric OH···O H-bonds of moderate strength, ν^H > 2700 cm^{-1} and γ ~ 1.34 [11].

To interpret the Ubbelohde effect, the *W*(*R*) is assumed to be harmonic. According to Eqs. (9.6) and (9.7), the total energy of the O···H···O system at v = 0, $E_0(R)$, is given by

$$E_0(R) = e_0(R) + (1/2)\, k_{\sigma\sigma}(R - R_0)^2 \tag{9.9}$$

For simplicity, $k_{\sigma\sigma}$ is a constant and R_0 is some parameter. Using the equilibrium condition $dE_0/dR = 0$ and Eq. (9.9), we obtain for the isotopic change in the O···O distance

$$\Delta R\,(\, R_e(D) - R_e(H) \approx (1/k_{\sigma\sigma})\,[(de_0{}^H(R)/dR)_0 - (de_0{}^D(R)/dR)_0] \tag{9.10}$$

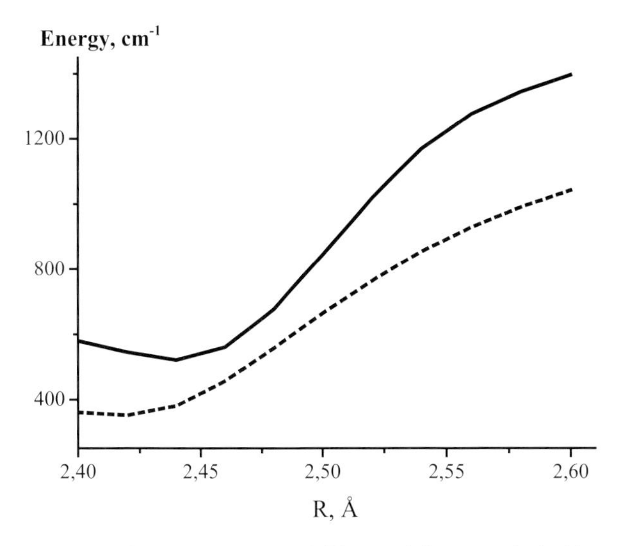

Figure 9.3 The ground proton (solid line) and deuteron (dashed line) energy level $e_0(R)$, see Eq. (9.7), as function of R.

here the subscript "0" after the parenthesis refers to R_0. Hence, knowing $e_0{}^{H}(R)$ and $e_0{}^{D}(R)$ and computing derivatives at various points R_0, ΔR can be found for any R_e.

According to Fig. 9.3, the proton and deuteron adiabatic levels $e_0{}^{H}$ and $e_0{}^{D}$ show a different dependence on R. As a result, ΔR is negative for $R_e \sim 2.40$ Å, with increasing R_e it becomes positive, reaches a maximum at some R_e, and then decreases.

9.2.3 Proton Transfer Assisted by a Low-frequency Mode Excitation

Due to the relatively small mass of the proton it is assumed that tunneling effects play a crucial role in many proton and hydrogen atom reactions, and that they determine many specific properties of H-bonded systems in the gas and condensed phases [38]. It should be noted, however, that proton tunneling has been proved experimentally for only a relatively small number of H-bonded systems in the gas phase and in crystals.

In systems with equivalent initial and final states proton tunneling may be manifested as observable splitting of some vibrational states. Tunneling splittings of the ground and several excited vibrational states have been measured for systems with intramolecular [39–42] and intermolecular [43, 44] H-bonds in the gas phase. For some substances (9-hydroxyphenalenone derivatives) the tunneling splitting of the ground vibrational state was measured in the molecular crystals [45, 46]. For other systems (malonaldehyde) tunneling does not occur in the molecular crystals due to the crystal field effect, which produces asymmetry of the PES along the proton-transfer coordinate [47].

In many molecular crystals that are formally symmetric according to their crystal structure, the double-minimum PESs are asymmetric [48–50]. As a rule, the asymmetry is of the order of 50–100 cm^{-1} and its lifetime is much longer than the duration of X-ray diffraction or spectroscopic experiments. This fact is of fundamental importance for the proton transfer phenomenon because it implies that the transfer of a charged (proton or deuteron) particle must be associated with a considerable energy of medium repolarization. The asymmetry makes it impossible to use the model by which proton tunneling occurs at the occasional moments when vibrational levels in two asymmetric wells become nearly degenerate under the action of environmental fluctuations (for example see Refs. [51, 52]). An alternative model for the proton transfer mechanism was used in Refs. [14, 35, 53, 54]. According to this model the proton transfer between two wells, being asymmetric, occurs by quantum jumps between vibrational levels under the action of random forces from the environment. It was applied to crystalline benzoic acid to describe the temperature dependence of the nuclear spin–lattice relaxation time, which was measured in the neutral dimer [48, 55] and in its deuterated analog.

In solids there are many low-frequency modes which can influence the proton movement and thus affect its magnetic relaxation. Only the stretching and rocking intracomplex vibration modes seem to be sufficiently strongly coupled with the proton coordinate to account for relaxation. The O···O stretching vibration was used in Refs. [35, 56] as the low-frequency (promoting) mode.

9.2.3.1 Crystals with Moderate H-bonds

Crystals with the quasi-symmetric O···H···O fragment and moderate H-bonds are characterized by the following parameter values: $2.60 < R_e < 2.75$ Å and $2500 < \nu^H < 3400$ cm^{-1} [1, 11, 32]. A typical example is the molecular crystal of benzoic acid dimer. At the equilibrium O···O distance ($R_e = 2.64$ Å) the barrier along the s coordinate is very high (the second doublet of the OH stretch lies under the peak). Because of the above-mentioned asymmetry in the crystal the energy difference between the two lowest energy levels ΔE can be written as [57]:

$$\Delta E = \sqrt{J^2 + A^2} \tag{9.11}$$

Here A is the energy asymmetry of the potential energy minima and J is the tunneling splitting of the ground vibrational state (in the considered double-well potential when A equals zero). In benzoic acid crystals $A \sim \Delta E$, because $J \sim 0.2$ cm^{-1} [58] and $\Delta E \sim 65$ [48].

Asymmetry was introduced to the model 2D PES, see Eq. (9.6), by adding a new term $A(s)$, describing the energy difference between the depths of the two wells. The solution of the 2D Schrödinger equation was found by the method of adiabatic separation, see Section 9.2.1. In so doing, only the symmetric coordinate s of a two-proton system was taken into account

$$s = (1/2)(s_1 - s_2) \tag{9.12a}$$

Here s_i is the coordinate of the ith ($i = 1, 2$) proton, measured from the midpoint of the hydrogen bond bridge. The antisymmetric coordinate a was set to be zero:

$$a = (1/2)(s_1 + s_2) \qquad (9.12b)$$

Hence, the double-proton transfer was assumed to proceed via a concerted mechanism, with the one-dimensional kinetic-energy matrix appropriate to one particle of twice the proton mass. A similar model is widely used to analyze the vibrations of cyclic hydrogen-bonded dimers (for example, see Ref. [59]). The O···O coordinate in this case symbolizes the intermolecular stretching of a doubly H-bonded system. The problem of the double-proton transfer for the benzoic acid dimer is thereby reduced to a problem conceptually similar to that for a single O···H···O entity except for the denoted light atom mass and consequences of the asymmetric 2D PES.

Calculations show that only the two lowest proton states, with $\nu = 0$ and $\nu = 1$ and the energy separation $\Delta E \sim 65\ \text{cm}^{-1}$, have to be taken into account. The vibrational level of the second excited proton state $\nu = 2$ is too high (> 2700 cm^{-1}) to be populated. Due to the PES asymmetry the proton in the ground state ($\nu = 0$) is located in the left well (an initial state), and in the first excited state ($\nu = 1$) the proton is located in the right well (a final state). In other words, the effective potential $e_0(R) + W(R)$, see Eq. (9.8), describes O···O levels in the left well (quantum number m), while the $e_1(R) + W(R)$ potential describes O···O levels in the right well (quantum number n). ($W(R)$ was assumed to be an anharmonic function in Ref. [35]). Quantum transitions between different vibrational levels are induced by random forces acting from the surroundings. One can distinguish between two kinds of transition:

1. The O···O thermal excitation and de-excitation (intrawell transitions)

$$|\nu,m\rangle \leftrightarrow |\nu,m'\rangle, \ (\nu = 0, \text{ left well}, \ m \neq m') \qquad (9.13a)$$

$$|\nu,n\rangle \leftrightarrow |\nu,n'\rangle, \ (\nu = 1, \text{ right well}, \ n \neq n') \qquad (9.13b)$$

2. The tunneling (interwell) transitions:

$$|\nu,m\rangle \leftrightarrow |\nu',n\rangle, \ (\nu \neq \nu'; \ \nu,\nu' = 0,1) \qquad (9.14)$$

Calculations show that the proton transfer assisted by low-frequency vibration excitation predominates over the "pure tunneling" transition $|0,0\rangle \leftrightarrow |1,0\rangle$. The probability of this particular transition is nearly 10^3 times smaller than the probability for the transition driven by the O···O vibrations (e.g., $|0,2\rangle \leftrightarrow |1,3\rangle$).

To interpret this result one has to consider values of the $\langle 0m|R|0m\rangle$ matrix elements for different m.

$$\langle 0m|R|0m\rangle = \iint \varphi_0(s,R)\chi_{0m}(R)R\varphi_0(s,R)\chi_{0m}(R)dsdR = \int \chi_{0m}(R)R\chi_{0m}(R)dR \qquad (9.15)$$

In terms of the model used, see Eq. (9.13a), it gives the effective value of R at different O···O levels when the tunneling particle is located in the left well (the initial state). One can see (Table 9.1) that excitation of the O···O vibrations on two quanta leads to the decrease in the effective value of R from 2.64 Å (the equilibrium O···O distance) to 2.61 Å.

Tab. 9.1 Values of the matrix element <0*m*|*R*|0*m*> computed as a function of *m* for two R_e distances.

Matrix element	Moderate H-bonded crystal $R_e = 2.64$ Å, $V_0(R_e) > 7000$ cm^{-1}	Strong H-bonded crystal $R_e = 2.52$ Å, $V_0(R_e) \sim 2500$ cm^{-1}
<00|*R*|00>	2.64	2.520
<01|*R*|01>	2.618	2.580
<02|*R*|02>	2.610	2.589
<03|*R*|03>	2.638	2.602

It has already been mentioned (see Section 9.2.2) that the main characteristic feature of the model 2D PES [28, 33, 35, 56] is the special dependence of the barrier height V_0 on R. V_0 rises exponentially in the region $2.40 < R < 2.60$ Å and at larger R values it increases much more slowly. This is why a relatively small decrease in the effective value of R leads to a strong decrease in the effective value of the barrier. As a result, the tunneling probability from the second excited O···O level |02> is much larger than that from the |00> level.

The temperature dependence of the proton and deuteron spin relaxation times were calculated in Ref. [35]. It was assumed that (i) intrawell transitions, see Eqs. (9.13a) and (9.13b), proceed much more rapidly than those between wells, see Eq. (9.14); and (ii) that a factor characterizing the coupling of the proton to modes of the surrounding medium (excluding the O···O mode) is a function of the temperature only. Theoretical curves, see Figs. 5 and 6 in Ref. [35], are in reasonable agreement with experiment. The apparent activation energy of proton transfer is ~ 800 cm^{-1}, while the potential barrier height in the proton coordinate is much larger (> 7000 cm^{-1}). The discrepancy between these values is caused by the fact that the probability of the interwell transition depends non-monotonically on m (Figs. 2 and 3 in Ref. [35]). The probability reaches its maximum at the second excited O···O stretch level and then decreases for $m > 3$. This is why the temperature dependence of the total probability for interwell transitions increases at low temperatures, but levels off near room temperature.

The described model resembles that used in Refs. [14, 55] to evaluate the temperature dependence of the nuclear spin–lattice relaxation time in the benzoic acid dimer and its deuterated analog. The rocking mode was used as the low-frequency mode in Refs. [14, 55] and the potential barrier in the proton coordinate was set to be ~ 1800 cm^{-1}. As a result, the energy gap between the two lowest proton levels (pair of levels) calculated with this PES (~ 1100 cm^{-1}) is much lower than the experimental OH stretching frequency observed for carboxylic acid dimers (> 2700 cm^{-1}) [32].

The quantum dynamics of proton transfer in the H-bonds of carboxylic acid dimers was also investigated using magnetic field-cycling NMR and optical spectroscopy in the low temperature region, and quasi-elastic neutron scattering at higher temperatures [60, 61]. The interpretation of these results has been done in terms of theoretical methods based on a perturbative instanton approach [61].

According to Refs. [35, 56] the mechanism of proton tunneling assisted by the excitation of low-frequency modes in molecular crystals with moderate H-bonds differs strongly from that in the gas phase [62]. In the isolated formic acid dimer the tunneling splitting increases monotonically with the O···O excitation, while in crystals the dependence of the probability of proton tunneling on the excitation low-frequency modes is non-monotonic.

9.2.3.2 **Crystals with Strong H-bonds**

Crystals with a quasi-symmetric O···H···O fragment and strong H-bonds are characterized by the following parameter values: $2.40 < R_e < 2.60$ Å and $\nu^H <$ 2500 cm^{-1} [1, 32, 33]. The barrier along the proton coordinate varies from 0 to ~ 5000 cm^{-1}. We will consider a model system with $R_e = 2.52$ Å. In terms of the model 2D PES [35, 56] the second pair of the adiabatic proton levels $|2\rangle$ and $|3\rangle$ locate in the vicinity of the potential barrier maximum at this R_e distance. At $R_e =$ 2.52 Å $V_0 \sim 2500$ cm^{-1} the tunneling splitting is $J \approx 85$ cm^{-1}, i.e. of the same order of magnitude as the energy asymmetry of potential, $A \sim 30$ cm^{-1} [49, 50, 63]. At this R_e value only 60% of the proton density is localized in the left well in the ground state $|00\rangle$. Hence, the transition $|0,0\rangle \leftrightarrow |1,0\rangle$, see Eq. (9.14), is not a complete proton tunneling but only a redistribution of the proton density over two wells. This effect can be observed in principle in some physical experiments (NMR [64], dielectric relaxation [65] etc.).

Proton transfer at $R_e = 2.52$ Å can be treated using the approach developed in the previous section, see Eqs. (9.13) and (9.14). According to Refs. [35, 56] for a model H-bonded system with $R_e \sim 2.52$ Å no vibrational assistance of the proton tunneling occurs. This effect can be explained using the $\langle 0m|R|0m\rangle$ matrix element for different m, see Eq. (9.15). The excitation of the low-frequency vibrations leads to an increase in the effective O···O distance (Table 9.1), i.e. to an increase in the effective barrier along the proton coordinate.

To summarize, in molecular crystals with strong H-bonds the proton transfer is not assisted by low-frequency mode excitation. In other words, proton dynamics is apparently not coupled to the low-frequency vibrations, in agreement with the results of INS studies of different crystals with strong H-bonds [49, 50, 63].

According to these papers "... the proton transfer is totally decoupled from the heavy atom dynamics ...".

9.2.3.3 Limitations of the Model 2D Treatment

The 2D model [28, 33, 35, 56] enables one to get a qualitative agreement with the available structural and spectroscopic regularities established for molecular crystals with a quasi-symmetric O···H···O fragment. However, a strong dispersion of the experimental points implies that the considered properties do not depend on one variable (R_e), but on several other characteristics; there are first bending of the O···H···O fragment, the chemical composition of the compound, and the crystal structure. To get a *semi-quantitative*, and in some cases *quantitative*, agreement with experiment, one has to increase the dimensionality of the PES, for example see [66–68], and to take into account effects of the crystalline environment, see Section 9.3.

In some H-bonded systems, for example the hydrogen maleate ion [69], vibrations of the O···H···O fragment are mixed strongly with the other vibrations. This implies that Eq. (9.2) is not applicable to such systems and the Ψ function depends on other vibrational coordinates, in particular, the CO stretch mode.

In the case of synchronous proton transfer in cyclic H-bonded clusters consisting of a single type of monomer the s and R coordinates can be treated as "totally symmetric" OH and O···H···O stretches [70]. In order to compute anharmonic frequencies of the IR active asymmetric OH stretch the asymmetric coordinate a, see Eq. (9.12b), is to be taken into consideration, for example see Refs. [62, 71].

It was mentioned in Section 9.2.3.1 that two different model 2D PESs were used to evaluate the temperature dependence of the nuclear spin–lattice relaxation time in benzoic acid dimer. The problem arising in this connection is how to choose values for the fitting parameters entering the model PESs. In most investigations they are defined from the experiment to be interpreted. Unfortunately the potentials determined this way are as a rule unable to properly describe other experimental regularities and data (spectroscopic, structural) known for the same compounds. Moreover the agreement between theory and experiment disappears if the fitting parameter values are taken, for example, from spectroscopic data [72].

In Sections 9.2.3.2 and 9.2.3.3 it was assumed that the dynamic asymmetry caused by the environment has a long lifetime and that this asymmetry can be considered as static. In some cases a reorganization of the surroundings has to be included explicitly in the model Hamiltonian via the introduction of a so-called "reorganization" mode [7]. The energy gap between two wells, corresponding to the initial and final states, is a typical example of a reorganization coordinate [73]. Calculation procedures including the use of this variable for the description of proton transfer in condensed phase (polar solvents) are well developed (compare numerical modeling by molecular dynamics simulations [74] with the use of continuum approximation [75]).

9.2.4 Vibrational Spectra of H-bonded Crystals: IR versus INS

Vibrational spectroscopy is widely used in experimental studies of molecular crystals with H-bonds. In some cases low-temperature INS experiments are not in complete agreement with IR absorption spectra and potential functions determined with IR (or Raman) spectroscopy are quite different from those derived from INS measurements [76]. The inconsistencies may be arranged into two groups. (i) Low-temperature INS experiments on CrOOH are not in complete agreement with IR absorption spectra in the 1500–2500 cm^{-1} region [77]. (ii) In the IR spectra of sodium bifluoride (NaFHF) the ν_{as}(FHF) band starts at 1300 and extends up to 2000 cm^{-1} with some structure in the region due to combination bands of the type $\nu_{as}(FHF) + n\nu_s(F\cdots F)$, $n = 0,1,\ldots$ [78]. In contrast, the INS spectrum of NaFHF shows a much narrower band due to ν_{as}(FHF) (full width half-height ~ 159 cm^{-1}) and no combinations [79]. Careful considerations of problems (i) and (ii) enable one to account for these apparently conflicting phenomena.

(i) Spectroscopic studies of molecular crystals with strong and moderate H-bonds have been carried out under widely varying temperature conditions. The INS spectra are usually obtained at low temperatures (< 30 K) in the 0–2000 cm^{-1} region, while the IR or Raman absorption spectra of the same crystals at low temperatures are often not available and, moreover, the IR absorption is very weak in the low frequency region (< 300 cm^{-1}). In the considered crystals the energy difference between the two lowest proton levels ΔE, see Eq. (9.11), is ~ 100 cm^{-1} and the thermal population of the first excited proton level is negligible at $T < 30$ K. This is why in the INS spectra practically all of the bands are due to transitions from the *ground* vibrational state. In the IR spectrum at $T = 100$–300 K the asymmetric stretch of the O···H···O fragment may be due to the transitions from the *first excited* proton level $|1\rangle \rightarrow |2\rangle$ (Section 9.2.2). The other possible transition, namely $|0\rangle \rightarrow |3\rangle$ is located above 2000 cm^{-1} (Fig. 9.2(b) and (c)). INS spectra are characterized by a relatively poor resolution in this high frequency region.

Due to relatively small changes in the dipole moment associated with O···H···O bendings, these fundamentals often have relatively small intensities in the IR spectra. However, these bands correspond to large vibrational displacements of the proton, and they are usually clearly seen in the INS spectrum [30, 49]. Summarizing, the number of bands corresponding to the fundamental vibrations of the O···H···O fragment, and their frequencies in the considered crystals, may be different in the IR and INS spectra.

(ii) Electric anharmonicity is one of the most important factors shaping absorption profiles in the IR spectra of systems with strong and moderate H-bonds, see for example Refs. [23, 66]. As a result, a large number of a relatively intensive nonfundamental transitions may appear in the IR spectra. In particular, the broad IR band associated with the asymmetric stretch of the O···H···O fragment is due to combinations with the O···O stretch. The change in the dipole moment function plays no role in the INS spectra and this is why the ν_{as}(OHO) bands in INS spectra are usually narrow and contain no combinations.

Special attention should be paid to the spectroscopic manifestations of a crystalline environment on proton transfer and on vibrations of the O···H···O fragment. Direct evidence for these effects may come from comparison of the INS spectrum of the considered H-bonded system in the gas phase and in crystals. While this is not possible due to technical problems, the comparisons may be done theoretically using DFT calculations with periodic boundary conditions (Section 9.3). Comparison of the computed INS spectrum of the isolated $H_5O_2^+$ ion with the computed spectrum of the ion in $H_5O_2^+ClO_4^-$ crystals (Fig. 9.7, later) shows that these spectra differ in the 1200–1600 cm^{-1} frequency region. This result implies that environmental effects may strongly affect the INS spectrum in the region of the asymmetric stretching and bending vibrations of the O···H···O fragment.

A possible way to detect strong coupling between the asymmetric and symmetric stretches of the O···H···O fragment is to compare the low-temperature INS spectrum of the crystal with that obtained at relatively high temperature, that is when the first excited state of the O···O stretch is sufficiently populated. To our knowledge, such a comparison has not been done yet.

Effective 1D potentials along the proton coordinate are often used to interpret the structure of spectroscopic regularities of H-bonded crystals. In this case the coupling with other vibrational modes is manifested through differences between the effective H and D potentials [80]. For example, effective 1D potentials were derived from band shape analysis of the OH (OD) IR and Raman stretching modes of $KHCO_3$ ($KDCO_3$) [81] and subsequently used for the interpretation of the INS spectrum of $KHCO_3$ crystals [50]. The difference between the effective H and D potentials (see Fig. 3 in Ref. [50]) is due to coupling with some low-frequency modes.

9.3
DFT Calculations with Periodic Boundary Conditions

At present three different codes are widely used for calculations of the structural and spectroscopic properties of H-bonded crystals, for example see Refs. [82–85]. The Car–Parrinello molecular dynamics (CPMD) program [86] and the Vienna *ab initio* simulation program (VASP) [87, 88] use a plane wave basis set, while an atom centered set is used with periodic boundary conditions in the CRYSTAL package [89]. Different exchange-correlation functionals are implemented in these codes.

In the first step the positions of all atoms in the cell are optimized. Cell parameters are usually borrowed from experiment. In some cases they are optimized [84] and in some cases not [85]. Harmonic frequency calculations verify that the computed structure corresponds to the global PES minimum. In the second step the anharmonic OH stretching [83, 84] frequency is estimated using 1D potential curves calculated as a function of the displacement for the hydrogen atom. In the third step classical molecular dynamics (MD) simulations are performed. The IR [85] or vibrational spectrum [82, 83] of the crystal is computed from the Fourier transform of the corresponding time correlation function (see Section 9.3.1).

To compare the structural and spectroscopic properties of the H-bonded system in the crystal with those in the gas phase, one has to calculate a quasi-isolated system in a simulation box (CPMD and VASP), and to compute the system in the gas phase (CRYSTAL) with the same functional and basis set.

9.3.1 Evaluation of the Vibrational Spectra Using Classical MD Simulations

The IR spectrum is obtained as the Fourier transform of the autocorrelation function of the classical dipole moment **M** [90], calculated at every point of the MD trajectory:

$$I(\omega) = \omega\left(\frac{4}{\varepsilon_0 chn}\right)\tanh\left(\frac{h\omega}{4\,k_B T}\right)\mathrm{Re}\int_0^\infty e^{i\omega t}\langle M(t)M(0)\rangle dt \tag{9.16}$$

where $I(\omega)$ is the relative IR absorption at frequency ω, T is the temperature, k_B is the Boltzmann constant, c is the speed of light in vacuum, ε_0 is the dielectric constant of the vacuum, and n is the refractive index (which was treated as constant).

The INS spectrum is calculated from the Fourier transform velocity autocorrelation functions of the atoms (Eq. (9.16) with velocities instead of the dipole moment), weighted by their inelastic neutron scattering cross-sections [91]. Since the value of the INS cross-section of the H atom is at least one order of magnitude larger than that of the other atoms, only the H atom velocities are usually taken into account. As a result, the $I(\omega)$ values in the computed INS spectrum are related to the mean-square-displacements of the hydrogen atoms. However, the observed INS spectrum has its intensity given by the "scattering law", $S(Q,\omega)$, which depends on the momentum transfer Q, for example, see Eq. (1) in Ref. [30]. Our approach has avoided the impact of the momentum transfer. The difference between the simple simulation using displacements only and one including momentum transfer were considered in Ref. [92]. Comparison of Fig. 6 and 4b in that paper shows that the intensities are damped for larger wavenumbers (this would fit the present case) and the features broaden a lot. However, the peaks are at about the same position. Therefore, we hope that this will also be the case for the present study.

The observed intensities also depend on the refractive index, which in general is frequency dependent [93]. This dependence is unknown in most cases and has not been considered. We note, however, that for liquid water the refractive index is virtually constant between 300 and 3500 cm^{-1} [94]. The dipole autocorrelation function is calculated classically and quantum corrections [95, 96] are introduced through the factor $2/[1+\exp(-\hbar\omega/2\pi k_B T)]$. Eq. (9.16) for the absorption spectrum has previously been applied in calculations of the far- and mid-IR spectra of liquid water [90, 97] and crystals [85]. The quantum correction damps the intensities of the low frequency motions and more sophisticated schemes [98] may lead to more severe damping of the low frequencies – as found for liquid water [99].

9.3.2 Effects of Crystalline Environment on Strong H-bonds: the $H_5O_2^+$ Ion

$H_5O_2^+$ is a prototype for strongly H-bonded systems which plays an important role in condensed phase [7, 100, 101] and enzymatic reactions [102]. The $H_5O_2^+$ structure in crystalline $H_5O_2^+ClO_4^-$ was determined by X-ray diffraction [103]. IR, Raman, and INS spectra are also available for this system [104, 105]. The point is that the different experimental studies were performed at the same temperature ($T \sim 84$ K). The IR spectrum of gas phase $H_5O_2^+$ in the 800–2500 cm^{-1} region has only recently been measured [106, 107]. We use CPMD to simulate the structure, the PES of the O···H···O fragment, and the INS and IR spectra of $H_5O_2^+$ in the $H_5O_2^+ClO_4^-$ crystal for comparison with results for the $H_5O_2^+$ gas-phase species [108, 109]. The discussion addresses the following issues:

1. Environmental effects on the structure, the PES of the O···H···O fragment, and the vibrational spectrum of $H_5O_2^+$.
2. The relation between the IR and INS spectra: specifically the broad IR bands compared to the relatively narrow INS bands for the O···H···O vibrations [110].

The CPMD calculations use the BLYP functional [111] with Trouillier–Martins [112] pseudo-potentials for core electrons. The kinetic energy cutoff for the plane wave basis set is 80 Ry. This value yields structures and harmonic frequencies of the isolated $H_5O_2^+$ and ClO_4^- ions that are comparable to BLYP results obtained with large Gaussian basis sets (aug-cc-pVTZ [113]). The isolated $H_5O_2^+$ ion was computed using the TURBOMOLE package [114–116] in BLYP/aug-cc-pVTZ approximation. The details of the computations may be found in Ref. [108].

9.3.2.1 The Structure and Harmonic Frequencies

Dimensions of the unit cell (which contains four formula units of $HClO_4{\cdot}2H_2O$) and positions of the heavy atoms were taken from Ref. [103]. Optimization of all atomic positions yields a structure with four equivalent $H_5O_2^+ClO_4^-$ units. In accord with the experimental data, the CPMD results suggest a structure in which the proton of the perchloric acid has been transferred to water and is shared by two water molecules, forming the $H_5O_2^+$ ions. These $H_5O_2^+$ ions are hydrogen bonded to the ClO_4^- ions. There are no hydrogen bonds between different $H_5O_2^+$ ions. The theoretical crystal structure is orthorhombic, space group $P2_12_12_1$. This was verified by a vibrational analysis which does not show imaginary frequencies. It should be noted that the X-ray study [103] considered the non-centrosymmetric $Pn2_1a$ structure as a possible choice, but the centrosymmetric space group $Pnma$ was accepted as most suitable. We find that the latter corresponds to a transition structure with one imaginary frequency (~ $78i$ cm^{-1}). The transition

$$P2_12_12_1 \rightarrow Pnma \rightarrow P2_12_12_1 \tag{9.17}$$

corresponds to a simultaneous shift of the four bridging protons along the O···O line. The O···H distances in the O···H···O fragment of $H_5O_2^+$ are equal in the transition structure, but differ in the minimum energy structure. The corresponding Cl–O distances in the two structures differ by less than 0.001 Å, while the O···O distances differ by up to 0.03 Å. The energy difference between the two structures is less than k_BT at the experimental temperature. The large-amplitude vibration of the bridging H, see Eq. (9.17), complicates the determination of the crystal space group.

The computed distances of different H-bonds in the crystal agree fairly well with the experimental values. The O···O distance in $H_5O_2^+$ is 2.426 Å (2.424 Å) and H-bonds between the $H_5O_2^+$ and ClO_4^- ions vary from 2.72 to 2.78 Å (2.784, 2.788 Å). Experimental values [103] are given in parenthesis.

In the crystal and gas phase the $H_5O_2^+$ ion can be considered as a nonrigid molecule because the bridging H undergoes large-amplitude vibrations. The symmetry point group of $H_5O_2^+$ in the crystal is C_{2h}[1]. Approximate C_{2h} symmetry of the $H_5O_2^+$ ion has also been found in X-ray studies of crystalline hydrates of different acids [117]. The computed equilibrium O···O distances are 2.431 Å in the isolated $H_5O_2^+$ ion and 2.426 Å in the crystal, i.e. they are virtually the same. The O···H···O fragment is slightly nonlinear in the gas phase (172°) and becomes linear in the crystal. Due to the strength of the O···H···O bond, the crystalline environment changes the O···O distance only slightly, however, it changes the mutual orientation of the water molecules in $H_5O_2^+$ as compared to the isolated species, see Fig. 9.4. As a result, in the crystal the bending potential along Y increases much faster than along X, and the frequencies of the two bending vibrations of the O···H···O fragment differ strongly in the crystal. In contrast, in the gas phase the bending potentials in the X- and Y-directions are more similar and the two bending modes are virtually degenerate, see Table 9.2.

The harmonic frequencies of the isolated $H_5O_2^+$ ion and the $H_5O_2^+$ ion in the $H_5O_2^+ClO_4^-$ crystal are given for the region above 1000 cm^{-1} in Table 9.2. (The four fundamentals of the ClO_4^- anion and their multiple components are located

1) The symmetry of the $H_5O_2^+$ ion in the $Pnma$ structure depends on the accuracy of the symmetrization procedure, ε. For $\varepsilon = 10^{-5}$ a.u. one gets C_i; for $\varepsilon = 5 \times 10^{-2}$ a.u. – C_{2h}. The latter value is of the same order of magnitude as the bond distances correction used in the X-ray study [103], < 5.2×10^{-2} a.u.

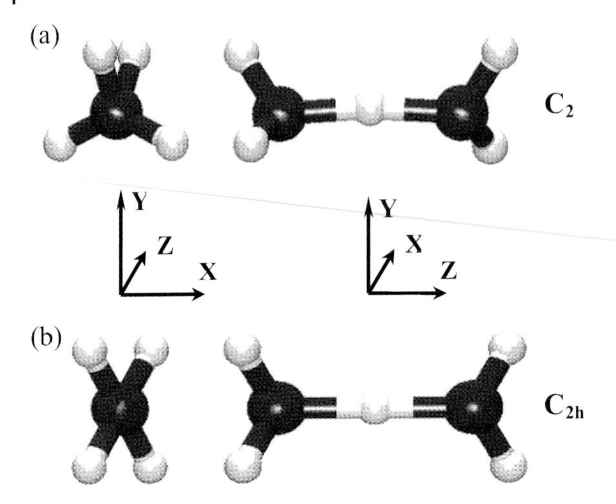

Figure 9.4 Structure of $H_5O_2^+$ in the gas phase (a) and in the crystal (b). The coordinate origin is at the midpoint between the oxygen atoms on the O···O line, *Y* is the C_2 axis and *Z* is the O···O line.

Tab. 9.2 Selected harmonic frequencies of the isolated $H_5O_2^+$ ion computed using TURBOMOLE, and those obtained for the crystal using CPMD. Units are cm^{-1}. (IR relative intensities are given in parentheses).

Assignment	Isolated $H_5O_2^+$		$H_5O_2^+$ in $H_5O_2^+ClO_4^-$ crystal
	C_2	TURBOMOLE[a]	CPMD[b]
O···H$^+$···O asymm. stretch	B	1019 (1.0)	1023, 1130, 1180, 1237
O···H$^+$···O bending	B	1411 (0.1)	1037, 1183, 1190, 1198
O···H$^+$···O bending	A	1500 (0.03)	1710, 1713, 1716, 1727
H_2O symm. bending	A	1652 (0.0)	1622, 1625, 1633, 1643
H_2O asymm. bending	B	1707 (0.33)	1671, 1683, 1634, 1703
O–H stretch	B	3594 (0.03)	3214, 3221, 3229, 3234
O–H stretch	A	3609 (0.04)	3301, 3305, 3311, 3313
O–H stretch	B	3680 (0.08)	3357, 3358, 3358, 3366
O–H stretch	A	3681 (0.10)	3377, 3385, 3389, 3428

a BLYP/aug-cc-pVTZ, force constants calculated numerically from analytic gradients (step size 0.02 a.u.).
b BLYP/plane-wave (80 Ry), force constants calculated numerically, the step length for the finite difference calculations was 0.01 a.u. in both directions.

below 1000 cm^{-1} [104]). For the $H_5O_2^+$ ion in the $H_5O_2^+ClO_4^-$ crystal the stretching and bending vibrations of the lateral water molecules are in the ranges 3214–3428 cm^{-1} and 1671–1703 cm^{-1}, respectively. The corresponding bands in the experimental IR spectrum (Fig. 3a in Ref. [104]) have maxima at 3200/3300 cm^{-1} and at 1700 cm^{-1}, respectively. Harmonic frequencies of the asymmetric stretch of the O···H···O fragment vary from 1023 to 1237 cm^{-1} in the crystal. This may be explained by strong electrostatic interactions between neighbouring O···H···O fragments. (The distance between the nearest bridging protons is around 5.3 Å). Frequencies for the lower of the two bending vibrations are between 1037 and 1198 cm^{-1} in a range overlapping with the range of the asymmetric stretches of the O···H···O fragment. Frequencies for the higher bending vibrations are in the narrow range of 1727–1740 cm^{-1} – very close to the bending vibrations of the terminal water molecules. Our calculations do not support a previous tentative assignment of the bands [104] observed around 1080 cm^{-1}, and in the 1700–1900 cm^{-1} range, to the bending and asymmetric stretching vibrations of the O···H···O fragment, respectively. According to the present calculations the bands between 1000 and 1400 cm^{-1} are due to the asymmetric stretch and bending vibrations of the O···H···O fragment, while the bands between 1600 and 1700 cm^{-1} are due to asymmetric bending vibrations of the terminal water molecules and to bending vibrations of the O···H···O fragment. For other crystals with strong H-bonds, $K_3H(SO_4)_2$ and $Rb_3H(SO_4)_2$, two intense INS bands near 1140 and 1550 cm^{-1} are assigned to the out-of-plane and in-plane bending modes of the O···H···O fragment [49].

9.3.2.2 The PES of the O···H···O Fragment

Profiles of the PES of the O···H···O fragment along the *Z* coordinate (Fig. 9.4) for the isolated ion, and for the ion in the crystal, were evaluated at two O···O distances: the equilibrium R_e value, and R_e + 0.1 Å. In the CPMD calculations the positions of atoms in the considered O···H···O fragment were fixed, while positions of all other atoms in the cell were optimized. Calculations of the profiles in the isolated ion were performed in accord with the prescriptions of Ref. [118]. The potential along the *Z* coordinate is extremely flat in the vicinity of the global minimum, see Fig. 9.5, and for the larger O···O distance it turns into a double minimum function. In the crystal the potential is even flatter than in the isolated ion, and the barrier along the *Z* coordinate grows more rapidly than that in the gas phase ion. A possible explanation is as follows: When the bridging H shifts from the midpoint between the oxygen atoms the dipole moment increases rapidly. Due to electrostatic interactions with the crystalline environment, the nonsymmetric position of the H appears to be more stable than the symmetric location. This effect, evidently, is absent in the gas phase.

The barrier height was calculated as a function of the O···O distance for the isolated ion, and for the ion in the $H_5O_2^+ClO_4^-$ crystal, see Fig. 9.6. In gas-phase H-bonded systems with a symmetric A···H···A fragment, where A = O or N, the potential barrier height is a function of R^2 [119]. According to Fig. 9.6, the barrier

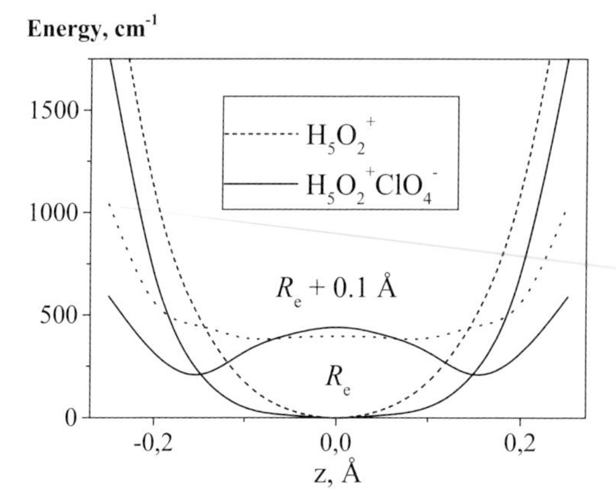

Figure 9.5 Profiles of the PES of the O···H···O fragment along the Z coordinate (Fig. 9.4) in the isolated ion (dashed line) and the ion in the crystal (solid line) computed for two O···O distances: the equilibrium R_e value and $R_e + 0.1$ Å. $R_e =$ 2.426 Å (the ion in the crystal), $R_e = 2.431$ Å (the isolated ion).

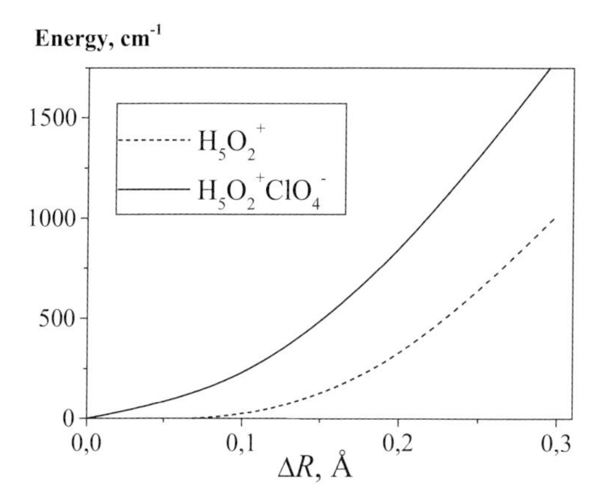

Figure 9.6 The barrier height computed as a function of the O···O distance for the isolated ion (dashed line) and for the ion in the crystal (solid line).

in the crystal increases with increasing R much faster than for the gas phase system. This result supports a key assumption of the model 2D PES (Section 9.2). This is, namely, that in crystals with a quasi-symmetric O···H···O fragment the barrier rises exponentially in the region $2.40 < R_e < 2.60$ Å.

9.3.2.3 Anharmonic INS and IR Spectra

CPMD simulations were performed at constant volume and $T = 84$ K, which is the average temperature of different experimental studies of crystalline $H_5O_2^+ClO_4^-$ [103–105]. In dynamic runs a fictitious electron mass of 600 au and a time step of 0.12 fs were employed. The BLYP equilibrium geometry was used as a starting point. The trajectory length was 4.5 ps.

Born–Oppenheimer molecular dynamics simulations with BLYP forces [120] of the isolated $H_5O_2^+$ ion were made using the TURBOMOLE package. The simulations were performed at constant energy with $T \sim 100$ K, defined as the average of the kinetic energy. The equilibrium structure was used as a starting point. Inertial momenta were distributed randomly according to the desired kinetic temperature. The time step was 0.5 fs and the trajectory length was 3 ps. Theoretical spectra were calculated by averaging over two trajectories, obtained with opposite signs of the starting velocities.

The computed INS spectrum of the $H_5O_2^+ClO_4^-$ crystal agrees well with the experimental one, see panels D and E in Fig. 9.7. In accord with experiment [105], there are two groups of bands below 850 cm^{-1} and two bands in the region between 1000 and 2000 cm^{-1}. Because the scattering cross section of the proton exceeds that of other atoms by one order of magnitude, INS selectively probes modes in which H atoms are involved. Separate calculation of the velocity autocorrelation function of the bridging H only (Fig. 9.7 C) allows the identification of bands that are due to the O···H···O fragment. Comparison of spectra D and C shows that these are the two bands around 1150 and 1600 cm^{-1}. Further information for the assignment comes from the harmonic frequencies and normal modes of $H_5O_2^+$ in the $H_5O_2^+ClO_4^-$ crystals (Table 9.2). The intense band around 1150 cm^{-1} is caused by vibrations of the bridging proton (O···H···O asymmetric stretch and O···H···O bending), while the group of intense bands around 1600 cm^{-1} is caused by both bending vibrations of the O···H···O fragment and bending vibrations of the terminal water molecules. Comparison of spectra D and C (Fig. 9.7) shows that the bending vibrations of the terminal water molecules are at the low wavenumber side of the band around 1600 cm^{-1}. Comparison of the computed INS spectrum of the isolated ion (panel B) with the spectrum of the ion in the crystal (panel D) shows that these spectra differ in the 1200–1600 cm^{-1} frequency region. Separate calculation of the velocity autocorrelation function of the bridging H only (Fig. 9.7A) indicates that the bands in this region correspond to vibrations of the bridging H. It implies that environmental effects change the INS spectrum strongly in the region of the asymmetric stretching and bending vibrations of the O···H···O fragment.

Figure 9.8B shows the IR spectrum of the crystal obtained as the Fourier-transform of the dipole autocorrelation function, Eq. (9.16), in the 800–2000 cm^{-1} frequency region. Following assignment of the INS spectra (Figure 9.7) and the harmonic normal mode calculations (Table 9.2), the bands in the region between 1600 and 1700 cm^{-1} are due to asymmetric bending vibrations of the water molecules and the bending vibrations of the O···H···O fragment. The bands in the region between 1000 and 1300 cm^{-1} are assigned to the asymmetric stretch and

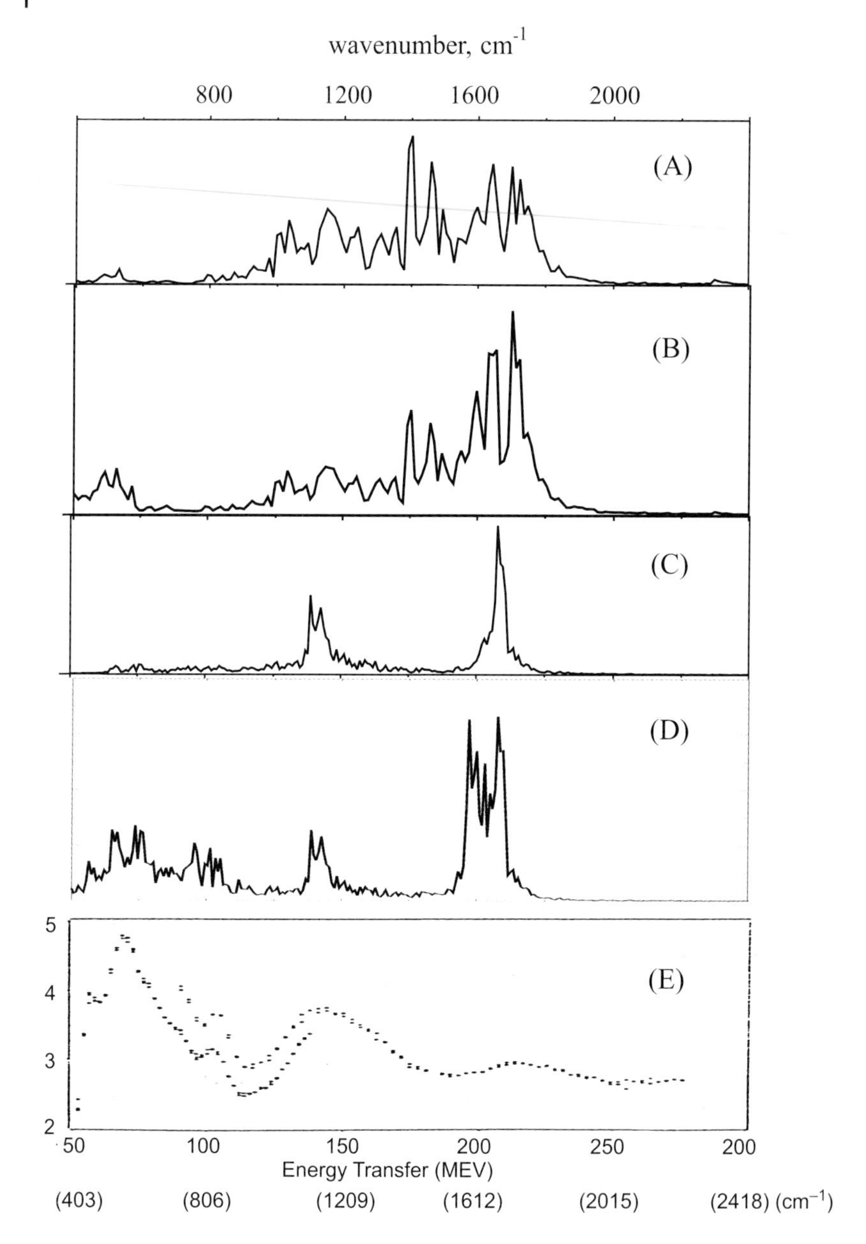

Figure 9.7 Experimental (E, Ref. [105]) and computed (D) INS spectrum of crystalline $H_5O_2^+ClO_4^-$. Panel C shows the INS spectrum of crystalline $H_5O_2^+ClO_4^-$ computed from the velocity autocorrelation function of the bridging H only. Panels B and A show the computed INS spectra of the isolated $H_5O_2^+$ ion and the bridging H, respectively. Intensities are in arbitrary units.

bending vibrations of the O···H···O fragment (Table 9.2). A very broad and flat absorption is observed for the $H_5O_2^+ClO_4^-$ crystal [103] in this region (upper panel in Fig. 9.8A). This is also true for the dihydrates of hydrogen chloride and hydrogen bromide [121] as well as for the dihydrates of halogenometallates of dioxonium [122].

According to experiment [104] shown at the top of Fig. 9.8 the IR intensities of the ClO_4^- fundamentals (< 1000 cm^{-1}) are lower than those of the asymmetric stretches of the O···H···O fragment, while the present calculations using the "Dipole Dynamics" procedure [86] predict the largest IR intensities occur for the low frequency ClO_4^- vibrations. The reasons were discussed in Section 9.3.1. For the sake of comparison, computed and experimental IR spectra of the isolated $H_5O_2^+$ ion are given in Fig. 9.8C and D, respectively. In contrast to the INS spectrum the effects of the crystalline environment are practically not seen in the 1200–1600 cm^{-1} region of the IR spectrum.

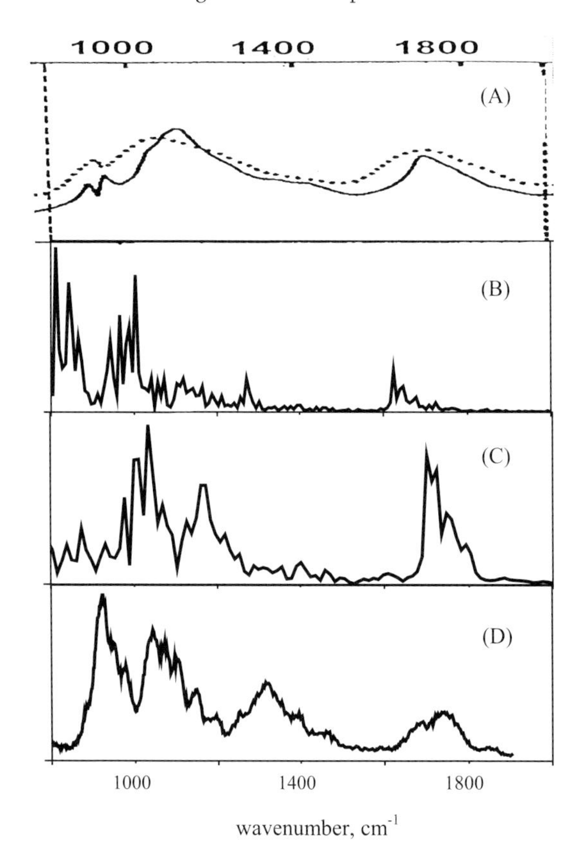

Figure 9.8 Computed IR (B) spectra of crystalline $H_5O_2^+ClO_4^-$ compared to the experimental IR spectrum (A, Ref. [104]). The dotted line corresponds to the IR spectrum in the liquid phase at room temperature [104]. The experimental (D, [106]) and computed (C) gas phase spectra in the region between 800 and 2000 cm^{-1} are also shown. Intensities are in arbitrary units.

Comparison of the theoretical INS and IR spectra of the $H_5O_2^+ClO_4^-$ crystal shows that the O···H···O band around 1150 cm^{-1} is much broader in the IR spectrum than in the INS one. Band broadening in the IR spectrum is caused by electric anharmonicities [23, 66] in addition to mechanical ones. Electric anharmonicities are absent in the INS spectrum and this explains the narrower bands.

Summarizing, the assignment of bands in the INS and IR spectra of the $H_5O_2^+$ ion in the $H_5O_2^+ClO_4^-$ crystal may be done in harmonic approximation.

9.4 Conclusions

A qualitative description of proton dynamics in H-bonded crystals requires a two-dimensional treatment assuming strong coupling between the proton-transfer coordinate and a low-frequency vibration. Due to environmental effects, this coupling is much stronger in crystals than in isolated H-bonded species. DFT calculations with periodic boundary conditions show the barrier along the proton transfer coordinate increases with increase in the O···O distance much faster in crystals than in the gas phase system.

Strong coupling between the two coordinates accounts for major structural and spectroscopic regularities experimentally obtained for molecular crystals with strong H-bonds and quasi-linear A···H···A fragment: (i) Large values for the Ubbelohde effect (the change in the O···O equilibrium distance upon deuteration); (ii) low frequencies (< 1000 cm^{-1}) for the asymmetric stretch of the O···H···O fragment, ν^H; (iii) large variations (from 1 to 2) for the isotopic frequency ratio, $\gamma = \nu^H/\nu^D$.

The mechanism of proton tunneling assisted by low-frequency mode excitation in molecular crystals is clarified. (i) In crystals with moderate H-bonds it differs strongly from the mechanism of proton tunneling in the gas phase entity. In isolated H-bonded species the tunneling splitting increases monotonically with the O···O excitation, while in crystals the dependence of the probability of proton tunneling on a low-frequency mode excitation is non-monotonic. (ii) In molecular crystals with strong H-bonds proton transfer is not assisted by low-frequency mode excitation. In other words, proton dynamics is apparently not coupled to the low-frequency vibrations in strong H-bonded crystals.

Acknowledgment

The author thanks the Russian Federal Agency for Education (Program "Development of the Highest-School Scientific Potential: 2006–2008" Project 2.1.1.5051) and Alexander von Humboldt Foundation for financial support.

References

1 T. Steiner, *Angew. Chem. Int. Ed.* **2002**, *41*, 48.
2 L. M. Epstein, E. S. Shubina, *Coord. Chem. Rev.* **2002**, *231*, 165.
3 G. A. Jeffrey, *An introduction to Hydrogen Bonding*, Oxford University Press, Oxford, 1997.
4 *The Hydrogen Bond. Recent Developments in Theory and Experiment*, Vol. 1, P. Schuster, G. Zundel, C Sandorfy (Eds.), North-Holland, Amsterdam, 1976.
5 *Chem. Phys.* **1995**, *170*(3), Special Issue: Tunneling in Chemical Reactions, V. A. Benderskii, V. I. Goldanskii, J. Jortner (Eds.).
6 *Theoretical Treatments of Hydrogen Bonding*, D. Hadzi (Ed.), Willey, Chichester, 1997.
7 M. V. Basilevsky, M. V. Vener, *Usp. Khim.* **2003**, *72*, 3 (in Russian); M. V. Basilevsky, M. V. Vener, *Russ. Chem. Rev.* **2003**, *72*, 1 (Engl. Transl.).
8 S. Scheiner, *Hydrogen Bonding. A Theoretical Perspective*, Oxford University Press, Oxford, 1997.
9 B. I. Stepanov, *Zh. Fiz. Khim.* **1946**, *20*, 907 (in Russian).
10 B. I. Stepanov, *Nature* **1946**, *157*, 808.
11 N. D. Sokolov, V.A. Savel'ev, *Chem. Phys.* **1977**, *22*, 383.
12 T. Carrington, W. H. Miller, *J. Chem. Phys.* **1986**, *84*, 4364.
13 N. Sato, S. Iwata, *J. Chem. Phys.* **1988**, *89*, 2932.
14 R. Meyer, R. R. Ernst, *J. Chem. Phys.* **1990**, *93*, 5518.
15 H. Sekiya, Y. Nagashima, Y. Nishimura, *J. Chem. Phys.* **1990**, *92*, 5761.
16 J. L. Herek, S. Pedersen, L. Banares, A. H. Zewail, *J. Chem. Phys.* **1992**, *97*, 9046.
17 Z. Bažic, J. C. Light, *Annu. Rev. Phys. Chem.* **1989**, *40*, 469.
18 J. M. Bowman, B. Gazdy, *J. Chem. Phys.* **1991**, *94*, 454.
19 M. H. Beck, A. Jäckle, G. A. Worth, H.-D. Meyer, *Phys. Rep.* **2000**, *324*, 1.
20 M. V. Vener, N. D. Sokolov, *Chem. Phys. Lett.* **1997**, *264*, 429.
21 B. W. Shore, *J. Chem. Phys.* **1973**, *59*, 6450.
22 B. R. Johnson, *J. Chem. Phys.* **1977**, *67*, 4086.
23 Y. Bouteiler, *Chem. Phys. Lett.* **1992**, *198*, 491.
24 L. Ojamae, I. Shavitt, S. J. Singer, *Int. J. Quantum Chem., Quantum Chem. Symp.* **1995**, *29*, 657.
25 J. Stare, J. Mavri, *Comput. Phys. Commun.* **2002**, *143*, 222.
26 M. V. Vener, S. Scheiner, *J. Phys. Chem.* **1995**, *99*, 642.
27 J. E. Del Bene, M. J. T. Jordan, *J. Mol. Struct. Theochem.* **2001**, *573*, 11.
28 N. D. Sokolov, M. V. Vener, and V. A. Savel'ev, *J. Mol. Struct.* **1988**, *177*, 93.
29 I. Olovson, P. G. Jonsson, in *The Hydrogen Bond. Recent Developments in Theory and Experiment*, P. Schuster, G. Zundel, C Sandorfy (Eds.), North-Holland, Amsterdam, 1976, Vol. 2, p. 426.
30 J. Howard, J. Tomkinson, J. Eckert, J. A. Goldstone, A. D. Taylor, *J. Chem. Phys.* **1983**, *78*, 3150.
31 M. Ichikawa, *Acta Crystallogr. Sect. B.* **1978**, *34*, 2074.
32 A. Novak, *Struct. Bond.* **1974**, *18*, 177.
33 N. D. Sokolov, M. V. Vener, V. A. Savel'ev, *J. Mol. Struct.* **1990**, *222*, 365 and references therein.
34 E. Grech, Z. Malarski, L. Sobczyk, *Chem. Phys. Lett.* **1986**, *128*, 255.
35 V. P. Sakun, M. V. Vener, N. D. Sokolov, *J. Chem. Phys.* **1996**, *105*, 379.
36 M. M. Kreevoy, S. Marimanikkuppam, V. G. Young, J. Baran, M. Szafran, A. J. Schultz, F. Trouw, *Ber. Bunsenges. Phys. Chem.* **1998**, *102*, 370.
37 Z. Smedarchina, A. Fernández-Ramos, W. Siebrand, *J. Chem. Phys.* **2005**, *122*, 134309.
38 R. P. Bell, *The Tunnel Effect in Chemistry*, Chapman & Hall, London, 1980.
39 S. L. Baughcum, R. W. Duerst, W. F. Rowe, Z. Smith, E. B. Wilson, *J. Am. Chem. Soc.* **1981**, *103*, 6296.
40 V. E. Bondybey, R. C. Haddon, J. H. English, *J. Chem. Phys.* **1984**, *80*, 5432.

41 K. Tanaka, H. Honjo, T. Tanaka, H. Kohguchi, Y. Ososhima, Y. Endo, *J. Chem. Phys.* **1999**, *110*, 1969.
42 R. L. Redington, R. L. Sams, *J. Phys. Chem. A.* **2002**, *106*, 7494.
43 C. C. Costain, G. P. Srivastava, *J. Chem. Phys.* **1964**, *41*, 1620.
44 F. Madeja, M. Havenith, *J. Chem. Phys.* **2002**, *117*, 7162.
45 T. Matsuo, K. Kohno, M. Ohama, T. Mochida, A. Isuoka, T. Sugewara, *Europhys. Lett.* **1999**, *47*, 36.
46 T. Matsuo, S. Baluja, Y. Koike, M. Ohama, T. Mochide, T. Sugewara, *Chem. Phys. Lett.* **2001**, *342*, 22.
47 D. F. Firth, P. F. Barbara, H. P. Trommsdorff, *Chem. Phys.* **1989**, *136*, 349.
48 T. Agaki, F. Imashiro, T. Terao, N. Hirota, S. Hayashi, *Chem. Phys. Lett.* **1987**, *139*, 331.
49 F. Fillaux, A. Lautié, J. Tomkinson., G. J. Kearley, *Chem. Phys.* **1991**, *154*, 135.
50 F. Fillaux, B. Nicolai, M.H. Baron, A. Lautie, J. Tomkinson, G. J. Kerley, *Ber. Bunsenges. Phys. Chem.* **1998**, *102*, 384.
51 J. Ulstrup, *Charge Transfer Processes in Polar Media*, Springer-Verlag, Berlin, 1979.
52 R. I. Cukier, M. Morillo, *J. Chem. Phys.* **1990**, *93*, 2364.
53 P. E. Parris, R. Silbey, *J. Chem. Phys.* **1985**, *83*, 5619.
54 P. Bonacky, *Chem. Phys.* **1986**, *109*, 307.
55 A. Stockli, B. H. Meier, R. Kreis, R. Meyer, R. R. Ernst, *J. Chem. Phys.* **1990**, *93*, 1502.
56 N. D. Sokolov, M. V. Vener, *Chem. Phys.* **1992**, *168*, 29.
57 J. L. Skinner, H. P. Trommsdorff, *J. Chem. Phys.* **1988**, *89*, 897.
58 A. Oppenländer, C. Rambaut, H. P. Trommsdorff, J.-C. Vial, *Phys. Rev. Lett.* **1989**, *63*, 1432.
59 S. Kushida, K. Nakamoto, *J. Chem. Phys.* **1964**, *41*, 1558.
60 A. J. Horsewill, D. F. Brougham, R. I. Jenkinson, C. J. McGloin, H. P. Trommsdorff, M. R. Johnson, *Ber. Bunsenges. Phys. Chem.* **1998**, *102*, 317.
61 M. A. Neumann, S. Craciun, A. Corval, M. R. Johnson, J. A. Horsewill, V. A. Benderskii, H. P. Trommsdorff, *Ber. Bunsenges. Phys. Chem.* **1998**, *102*, 325.
62 N. Shida, P.F. Barbara, J. Almlöf, *J. Chem. Phys.* **1991**, *94*, 3633.
63 F. Fillaux, J. Tomkinson, *Chem. Phys.* **1991**, *158*, 113.
64 P. M. Tolstoy, S. N. Smirnov, I. G. Shenderovich, N. S. Golubev, G. S. Denisov, H.-H. Limbach, *J. Mol. Struct.* **2004**, *700*, 19.
65 M. Polomska, T. Pawlowski, L. Szcsesniak, B. Hilczer, L. Kirpichnikova, *Ferroelectrics* **2002**, *272*, 81.
66 M. V. Vener, O. Kühn, J. Sauer, *J. Chem. Phys.* **2001**, *114*, 240.
67 C. Emmeluth, M. A. Suhm, D. Luckhaus, *J. Chem. Phys.* **2003**, *118*, 2242.
68 N. Došlic, O. Kühn, *Z. Phys. Chem.* **2003**, *217*, 1507.
69 F. Avbeij, M. Hodošcek, D. Hadži, *Spectrochim. Acta Part A* **1985**, *41*, 89.
70 M. V. Vener, J. Sauer, *J. Chem. Phys.* **2001**, *114*, 2623.
71 M. V. Vener, O. Kühn, J. M. Bowman, *Chem. Phys. Lett.* **2001**, *349*, 562.
72 W. Siebrand, T. A. Widman, M. Z. Zgierski, *Chem. Phys. Lett.* **1983**, *98*, 108.
73 A.Warshel, J. K. Hwang, *J. Chem. Phys.* **1986**, *84*, 4938.
74 A. Staib, D. Borgis, J. T. Hynes, *J. Chem. Phys.* **1995**, *102*, 2487.
75 M. V. Basilevsky, A. V. Soudackov, M. V. Vener, *Chem. Phys.* **1995**, *200*, 87.
76 F. Fillaux, *Int. Rev. Phys. Chem.* **2000**, *19*, 553.
77 M. C. Lawrence, G. N. Robertson, *J. Chem. Phys.* **1987**, *87*, 3375.
78 E. Spinner, *Aust. J. Chem.* **1980**, *33*, 933.
79 T. C. Waddington, J. Howard, K. P. Brierly, J. Tomkinson, *Chem. Phys.* **1982**, *64*, 193.
80 R. G. Delaplane, J. A. Ibers, J. R. Ferraro, J. J. Rush, *J. Chem. Phys.* **1969**, *50*, 1920.
81 F. Fillaux, *Chem. Phys.* **1983**, *74*, 405.
82 C. A. Morrison, M. M. Siddick, P. J. Camp, C. C. Wilson, *J. Am. Chem. Soc.* **2005**, *127*, 4042.
83 F. Fontaine-Vive, M. R. Johnson, G. J. Kearly, J. A. K. Howard, S. F. Parker, *J. Chem. Phys.* **2006**, *124*, 234503.

84 M. Merawa, B. Civalleri, P. Ugliengo, Y. Noel, A. Lichanot, *J. Chem.Phys.* **2003**, *119*, 1045.
85 S. V. Churakov, B. Wunder, *Phys. Chem. Miner.* **2004**, *31*, 131.
86 J. Hutter, M. Parrinello, A. Alavi, D. Marx, M. Tuckerman, W. Andreoni, A. Curioni, E. Fois, U. Röthlisberger, P. Giannozzi, T. Deutsch, D. Sebastiani, A. Laio, J. VandeVondele, A. Seitsonen, S. Billeter, *CPMD, 3.5.2*, IBM Research Laboratory and MPI für Festkörperforschung, Stuttgart, 1995–2001.
87 G. Kresse, J. Hafner, *Phys. Rev. B.* **1993**, *47*, 558; *Phys. Rev. B.* 1994, *49*, 14251.
88 G. Kresse, J. Futhmüller, *Comput. Math. Sci.* **1996**, *6*, 15; *Phys. Rev. B.* 1996, *54*, 11169.
89 V. R. Saunders, R. Dovesi, C. Roetti, M. Causa, N. M. Harrison, R. Orlando, C. M. Zicovich-Wilson, *CRYSTAL 98 User's Manual*, Universita di Torino, Torino, **1998**.
90 W. B. Bosma, L. E. Fried, S. Mukamel, *J. Chem. Phys.* **1993**, *98*, 4413.
91 B. Hudson, A. Warshel, G. R. Gordon, *J. Chem. Phys.* **1974**, *61*, 2929.
92 H. Jobek, A. Tuel, M. Krossner, J. Sauer, *J. Phys. Chem.* **1996**, *100*, 19545.
93 D. A. McQuarrie, *Statistical Mechanics*, Harper Collins, NewYork, 1973, Section 21-1.
94 V. M. Zolotarev, B. A. Mikhailov, L. I. Alperovich, S. I. Popov, *Opt. Spectrosc.* **1969**, *27*, 430 (in Russian).
95 M.-P. Gaigeot, M. Sprik, *J. Phys. Chem. B.* **2003**, *107*, 10344.
96 R. Ramirez, T. Lopez-Ciudad, P. Kumar, D. Marx, *J. Chem. Phys.* **2004**, *121*, 3973.
97 B. Guillot, *J. Chem. Phys.* **1991**, 95, 1543.
98 J. Borysow, M. Moraldi, L. Frommhold, *Mol. Phys.* **1985**, *56*, 913.
99 H. Ahlborn, B. Space, P. B. Moore, *J. Chem. Phys.* **2000**, *112*, 8083.
100 G. V. Yukhnevich, E. G. Tarakanova, V. D. Mayorov, N. B. Librovich, *Usp. Khim.* **1995**, *64*, 963 (in Russian); G. V. Yukhnevich, E. G. Tarakanova, V. D. Mayorov, N. B. Librovich, *Russ. Chem. Rev.* **1995**, *64*, 901 (Engl. Transl.).
101 G. Zundel, *Adv. Chem. Phys.* **2000**, *111*, 1.
102 C. L. Perrin, J. B. Nielson, *Annu. Rev. Phys. Chem.* **1997**, *48*, 511.
103 I. Olovsson, *J. Chem. Phys.* **1968**, *49*, 1063 and references therein.
104 A. C. Pavia, P. A. Giguère, *J. Chem. Phys.* **1970**, *52*, 3551.
105 D. J. Jones, J. Roziere, J. Penfold, J. Tomkinson, *J. Mol. Struct.* **1989**, *195*, 283.
106 K. R. Asmis, N. L. Pivonka, G. Santambrogio, M. Brümmer, C. Kaposta, D. M. Neumark, L. Wöste, *Science.* **2003**, *299*, 1375.
107 T. D. Fridgen, T. B. McMahon, L. MacAleese, J. Lemaire, P. Maitre, *J. Phys. Chem. A* **2004**, *108*, 9008.
108 M. V. Vener, J. Sauer, *Phys. Chem. Chem. Phys.* **2005**, *7*, 258.
109 M. V. Vener, J. Sauer, *Khim. Fiz.* **2005**, *24*, No. 6, 39 (in Russian).
110 G. J. Kearley, F. Fillaux, M. H. Baron, S. Bennington, J. Tomkinson, *Science.* **1994**, *264*, 1285.
111 A. D. Becke, *Phys. Rev. A.* **1988**, *38*, 3098; C. Lee, W. Yang, R. G. Parr, *Phys. Rev. B* **1988**, *37*, 785.
112 N. Troullier, J. L. Martins, *Phys. Rev. B* **1991**, *43*, 1993.
113 R. A. Kendall, T. H. Dunning Jr., R. J. Harrison, *J. Chem. Phys.* **1992**, *96*, 6796.
114 R. Ahlrichs, M. Bär, M. Häser, H. Horn, C. Kölmel, *Chem. Phys. Lett.* **1989**, *162*, 6796.
115 O. Treutler, R. Ahlrichs, *J. Chem. Phys.* **1995**, *102*, 346.
116 M. von Arnim, R. Ahlrichs, *J. Comput. Chem.* **1998**, *19*, 1746.
117 R. Minkwitz, S. Schneider, A. Kornath, *Inorg. Chem.* **1998**, *37*, 4662; and references therein.
118 M. V. Vener, J. Sauer, *Chem. Phys. Lett.* **1999**, *312*, 591.
119 S. Scheiner, *J. Am. Chem. Soc.* **1981**, *103*, 315.
120 S. D. Elliott, R. Ahlrichs, O. Kampe, M. M. Kappes, *Phys. Chem. Chem. Phys.* **2000**, *2*, 3415.
121 A. S. Gilbert, N. Sheppard, *Chem. Commun.* **1971**, *7*, 337.
122 G. Picotin, J. Roziere, J. Potier, A. Potier, *Adv. Mol. Relax. Proc.* **1975**, *7*, 177.

Part III
Hydrogen Transfer in Polar Environments

This section turns to reactions involving the transfer of a proton in (primarily) polar environments, opening with a theoretical presentation and finishing with three experimental chapters all exploiting the greatly enhanced acidity of aromatic photoacids (compared to ground electronic state acids) in probing the details of proton transfer.

Kiefer and Hynes start off in Ch. 10 with a recounting of the theoretical description of the microscopic mechanisms and rate constants of such processes, here focused on acid-base type reactions, developed over the years in the latter's group. The crucial quantum mechanical character of the proton's nuclear motion even when there is no tunneling is emphasized, as are the rate-governing roles of the rearrangement of the polar environment and the vibrational motion of the atoms or groups between which the proton transfer occurs, such that the proton is *not* the reaction coordinate. Instead the proton follows these slower motions. Since this perspective differs strongly from many traditional views, special attention is given to experimental measures supporting this description, including free energy relations and kinetic isotope effects. Electronic rearrangements involved in the proton transfer are also discussed, for both the ground and excited electronic states; these also differ from many traditional views.

Next, Ch. 11 by Lochbrunner, Schriever and Riedle deals with excited electronic state intramolecular tautomerization proton transfers in nonpolar, rather than polar, solvents. But there is a connection to the previous chapter: the ultrafast optical experiments discussed here emphasize evidence that the proton is not the reaction coordinate. The proton transfer is controlled by low vibrational modes of the photo-acids, rather than by the proton motion itself, an interpretation supported by separate vibrational spectroscopic studies and theoretical calculations The key role of modes reducing the donor-acceptor distance for proton transfer is highlighted, and for the featured molecule of this chapter, the proton adiabatically follows the low frequency modes, and no tunneling or barrier for the proton occurs. (See also Ch. 15 by Elsaesser for direct ultrafast vibrational studies on these issues).

In Ch. 12, Pines and Pines return to proton transfer in polar solvents and investigate the factors affecting the photoacidity of weak organic acids such as phenols

Hydrogen-Transfer Reactions. Edited by J. T. Hynes, J. P. Klinman, H. H. Limbach, and R. L. Schowen

ISBN: 978-3-527-30777-7

by examining the use of the Förster cycle connecting the pKas of the excited and ground state acids. Their detailed analyses, which also involve free energy relations, demonstrate the usefulness of these thermodynamic tools in identifying the reasons for the enhancement of the photoacidity over that of the ground electronic state acid (typically six orders of magnitude in the rate constant). One emphasis is on the importance of the product side of the reaction, an aspect whose prediction is discussed at the conclusion of Ch. 10. The authors' procedure is shown to be general, remarkably also being applicable for quantifying photoacidity even for systems which remain unreactive during the finite excited state lifetime.

Tolbert and Solntsev conclude this Section as well as Volume 1 with a continuation of the theme of photoacidity and proton transfer, focusing special attention on the various factors crucial for producing very strong, or 'super' photoacids. These authors consider the functioning of these photoacids not only in polar environments such as water, alcohols, and water-alcohol mixtures—whose differences are revealing for understanding the proton transfer details—but also some biological environments such as the green fluorescent protein.

10
Theoretical Aspects of Proton Transfer Reactions in a Polar Environment

Philip M. Kiefer and James T. Hynes

Overview

This chapter reviews some nontraditional theoretical views developed in this group on acid–base proton transfer reactions. Key ingredients in this picture are a completely quantum character for the proton motion, the identification of a solvent coordinate as the reaction coordinate, and attention to the hydrogen (H-) bond vibrational mode in the acid–base complex. Attention is also given to the electronic structure rearrangements associated with proton transfer. A general overview is presented for the rate constants, including the activation free energy, as well as for the associated primary kinetic isotope effects, for both the proton adiabatic (quantum but nontunneling) and nonadiabatic (tunneling) regimes. While the focus is on ground electronic state proton transfers, some aspects of excited state proton transfers are also discussed.

10.1
Introduction

In this chapter, we give an overview of some of the theoretical developments for proton transfer (PT) reactions, focusing on the efforts in the Hynes group [1–7]. PT is of obvious importance in both chemistry and biology [8], and consequently there has been extensive study of PT in solution [8–10] and other polar environments, e.g. enzymes [9c, 11]. Of particular importance in characterizing and comprehending PT reactions is understanding which properties of the reaction partners and surrounding environment affect the PT rate constant and related experimental observables. Among these observables, the most frequently employed by experimentalists are the reaction free energy ($\Delta G^{\ddagger}$) and its relation to the reaction free energy (ΔG_{RXN}) [8–10], and kinetic isotope effects (KIEs) [8–12]. In the description of PT provided in this chapter, these important aspects are given special focus and are analyzed with a picture that significantly differs from widely employed 'standard' pictures for PT. We will begin this Introduction with a brief synopsis of these standard pictures and then highlight some of the key features of

Hydrogen-Transfer Reactions. Edited by J. T. Hynes, J. P. Klinman, H. H. Limbach, and R. L. Schowen

ISBN: 978-3-527-30777-7

the nontraditional perspective. Here and throughout this chapter we make no pretence of being complete in our discussion or referencing, and refer the interested reader to the cited articles from the group for more extensive discussion and references.

Standard descriptions for PT have both classical 'over the barrier' (nontunneling) and tunneling components, with the latter regarded as a correction to the former [8, 12–17]. The standard nontunneling view for KIEs traces back to Westheimer and Melander (W-M) [14]. Actually, most of the W-M discussion was developed for, and is more properly applied to, hydrogen atom transfers rather than to PT. Since however the W-M perspective has remained the cornerstone for most PT discussions, we present it here, making some changes to make it more appropriate for PT. In its simplest version, the W-M picture uses a collinear 3-center molecular system for PT, illustrated here for an acid–base reaction within a hydrogen-bonded (H-bonded) complex, e.g.

$$\mathrm{AH\cdots B \Rightarrow A^{-}\cdots HB^{+}} \tag{10.1}$$

While the basic formalism we present is more general, we will typically refer to the PT reaction Eq. (10.1) in a polar solvent throughout this chapter to illustrate the importance of the reactant and product charge distributions and the polarity of the environment. The reaction potential surface is a function of two coordinates, the A–H proton distance and the distance between the two heavy moieties, A···B. Along the minimum energy path, the reaction begins at a large heavy atom separation with the A–H proton distance constant, proceeds through a transition state (TS) A··H··B, and goes on to products with a large A–B separation and constant H–B distance to produce $A^{-}\cdot\cdot HB^{+}$. In the limiting symmetric reaction case, the reaction coordinate at the TS is the unstable asymmetric stretch motion, which for heavy A and B moieties, is essentially the proton coordinate. Passage through the TS region is thus classical motion of the proton over the barrier in the proton coordinate. For asymmetric reactions, the TS location will shift (see below), but the proton's motion across the TS always remains classical in the W-M description.

Further aspects of the standard W-M approach are most usefully discussed via its treatment of KIEs for the rate constant resulting from isotopic substitution – most often replacement of H with D – which is widely used in assessing the character of the TS and the proton's role therein. With the use of TS theory, the KIE arises from the exponentiated activation energies [8, 12–14]. For H versus D transfer, the KIE is given by

$$\begin{aligned} k_{\mathrm{H}}/k_{\mathrm{D}} &\approx \exp[-(\Delta G_{\mathrm{H}}^{\ddagger} - \Delta G_{\mathrm{D}}^{\ddagger})/RT; \\ \Delta G_{\mathrm{H}}^{\ddagger} - \Delta G_{\mathrm{D}}^{\ddagger} &\approx \mathrm{ZPE}_{\mathrm{H}}^{\ddagger} - \mathrm{ZPE}_{\mathrm{H}}^{\mathrm{R}} - \mathrm{ZPE}_{\mathrm{D}}^{\ddagger} + \mathrm{ZPE}_{\mathrm{D}}^{\mathrm{R}} \end{aligned} \tag{10.2}$$

As is of course well-known, the KIEs originate, in this framework, from isotopic zero point energy (ZPE) differences between the reactant and the TS [8, 12–14, 18].

The relevant reactant region ZPE is just that of the A–H vibration in the reactant A–H···B complex, the motion transverse to the reaction coordinate in this region. The reactant ZPE is larger for H than for D, $ZPE_H^R > ZPE_D^R$, due to the lower reactant AH stretch vibration frequency for the more massive D. For a thermodynamically symmetric reaction ($\Delta G_{RXN} = 0$, a very important reference situation), the reaction path through the TS consists solely of the proton's *classical* motion over the barrier (as noted above), so that there is no proton ZPE associated with this motion at the TS. Rather, the TS ZPE is associated with the transverse motion at the TS (just as for the reactant), which in the collinear model is a symmetric stretch, the heavy particle A–B vibration. Hence, at the TS, $ZPE^{\ddagger}_H = ZPE^{\ddagger}_D$ for such a symmetric reaction. The net result is the complete "loss" of the ZPE for the proton stretching mode on going from reactant to TS for a symmetric reaction. Typical reactant proton stretch frequencies (ω_{CH}~3000 cm^{-1}) and a simple mass correlation between ZPEs (i.e. $ZPE_D \approx ZPE_H/\sqrt{2}$) give an H vs. D KIE of about 7 at room temperature [8, 12–14].

For an asymmetric reaction, the reaction coordinate at the TS in the traditional view includes both proton and heavy particle classical motions: consistent with the Hammond postulate [15], the TS becomes more geometrically similar to the product as the reaction becomes more endothermic, and more similar to the reactant as it becomes more exothermic. Thus the *transverse* vibration at the TS – whose ZPE is relevant for the rate – more and more involves the proton motion, and in either limit approaches the bound proton stretch vibration of the product BH or the reactant AH. This effect decreases the KIE, resulting in a k_H/k_D vs. ΔG_{RXN} trend which is maximal at ΔG_{RXN}=0 and drops off as the reaction becomes more endo- or exo-thermic. A description of this behavior in terms of the well-known Marcus relation for the activation free energy [19] will be discussed in Sections 10.2.1–10.2.3. (With apologies, we do not use the proper terms 'exergonic' and 'endergonic' in connection with the sign of the reaction free energy).

The standard picture for classical proton motion over the barrier at the TS just described is sometimes supplemented with a quantum contribution via a tunneling correction [13, 16, 17, 20, 21]. Addition of tunneling corrections to the standard PT rate will obviously affect the KIE, including its reaction asymmetry dependence [13, 17]. Indeed, it has been argued that variation of the tunneling contribution versus reaction asymmetry is primarily responsible for the broad range in magnitude of observed KIE versus reaction asymmetry plots, instead of the variation of ZPE at the TS described above [17]. Various departures from experimental observations, such as non-Arrhenius rate behavior or KIEs much in excess of 10, typically are taken as indicating tunneling [8, 11–13, 17, 24]. We also remark that it is typical in such assessments of tunneling that it is regarded as a partial contribution to the rate, with the other contribution being classical over the barrier motion.

Despite the success of the standard picture just described, it can be argued that a different picture would be more plausible. First, the standard description has a certain logical inconsistency in the TS description. In the symmetric case, proton motion is viewed as completely classical over the proton barrier. For any finite reaction asymmetry, however, the proton motion's quantum character as a bound quantum vibration becomes extremely important, since it is that character that influences the frequency, and thus the ZPE, of the transverse TS motion. It seems difficult to maintain that proton motion within an H-bond can be both classical and quantum. Second, the standard picture presented above makes no reference to the solvent (while we focus here on the case of a solvent environment, very similar ideas will apply to more general polar environments such as enzyme active sites). To the degree that the solvent is included in standard descriptions, it is imagined to alter the rate via a differential equilibrium solvation of the TS and the reactant, again all within the standard framework recounted above [21, 25]. But the equilibrium solvation assumption – which requires the solvent motion to be fast compared to the relevant motion of the reacting solutes in the TS region – is completely implausible for the high frequency quantum proton motion; indeed, the opposite situation is more appropriate: the solvent is generally slow compared to the proton motion [1, 2, 26–28].

In the alternate, nontraditional picture of PT reactions [1–7, 26–28] employed herein, the reaction is driven by configurational changes in the surrounding polar environment, a feature of much modern work on PT reactions [1–7, 26–28], and the reaction activation free energy is largely determined by the environmental reorganization. In this picture, the rapidly vibrating proton follows the environment's slower rearrangement, thereby producing a perspective in terms of the instantaneous proton potential for different environmental arrangements. (We will expand upon this statement in the two different regimes of PT described in succeeding sections).

In order to clarify the essential features of the nontraditional picture, Fig. 10.1 displays the key features for a model overall symmetric PT reaction in a linear H-bonded complex. Three proton potential curves versus the proton coordinate are displayed, for solvent configurations appropriate to that of the reactant pair, the TS, and the product pair. These different states of solvation – or more simply the solvent's nuclear electrical polarization state – distort the potential from being initially asymmetric favoring the proton residing on the acid, through an intermediate situation where a proton symmetric double well is established, and on to an asymmetric potential now favoring the proton residing on the base. The solvent motion is critical, due to the strong coupling of the reacting pair's evolving charge distribution to the polar solvent polarization field. Furthermore, the high frequency quantum proton vibration adiabatically adjusts to the reorganizing solvent in the reactant and product regions; the reaction coordinate is a solvent coordinate, rather than the proton coordinate, and there is a free energy change up to the TS activation free energy as the solvent rearranges.

For each of the three proton potentials in Fig. 10.1, the quantized ground proton vibrational energy, i.e. the ZPE, is indicated. For the TS solvent configuration,

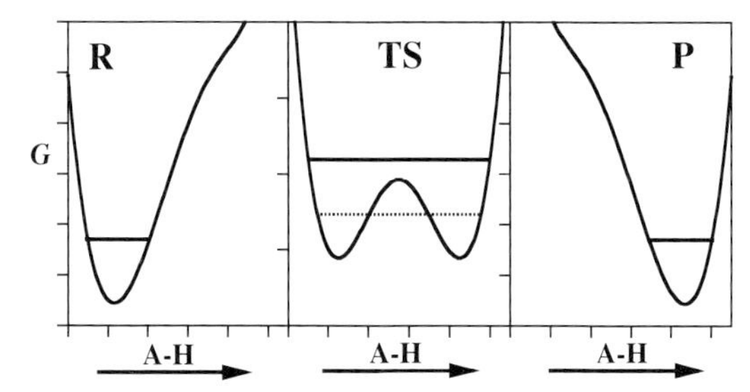

Figure 10.1 Free energy curves versus proton position at the reactant R, transition state TS, and product P solvent configurations for a symmetric reaction. For both R and P, the ground state proton vibrational energy level (solid line) is indicated. For the transition state, the ground state proton vibrational energy level is indicated for a small (solid line) and a large (dotted line) H-bond separation, which correspond to adiabatic and nonadiabatic PT, respectively.

there are two possible cases for the zero point level: if it is above the proton barrier, the system is in what we term the quantum *adiabatic* PT limit, while if instead the level is below the barrier top, the reaction would involve proton tunneling [29], which we term the quantum *nonadiabatic* limit. In either case, the proton motion is a bound quantum vibration; there is *no* classical barrier crossing of the proton, in contrast to a conventional TS theory for the standard description.

The distinction between the two regimes, whether the ground proton vibrational level is below or above the barrier in the proton coordinate, is critically determined by another important coordinate, the acid–base H-bond A–B separation Q. For weak H-bonds with large equilibrium A–B separations, the barrier in the proton coordinate will be large, requiring the proton to tunnel through the barrier, while for stronger H-bonds and smaller equilibrium A–B separations, the barrier is reduced such that nontunneling adiabatic PT occurs. Furthermore, the A–B vibrational frequency will also determine the thermal accessibility of smaller A–B separations where either tunneling is decreased with reduced barriers or, for nontunneling PT, the quantum vibrational proton mode adiabatically passes from R to P.

Before proceeding, we should discuss some electronic structure aspects of PT. In much of the modeling discussed within for Eq. (10.1), we have employed a 2 valence bond (VB) state description, with the VB states corresponding to the limiting reactant (R) and product (P) forms of Eq. (10.1) for any geometry, i.e. the electronic wavefunction is

$$|\Psi_{\mathrm{el}}\rangle = c_{\mathrm{R}}|\Psi_{\mathrm{R}}\rangle + c_{\mathrm{P}}|\Psi_{\mathrm{P}}\rangle \tag{10.3}$$

The two VB wavefunctions depend on the solute reacting pair coordinates; the coefficients also depends on these coordinates but in addition depend on a solvent coordinate characterizing the electric polarization state of the solvent, as further discussed in Section 10.2. In this description, the solute's electronic wavefunction is polarized over the VB states by the polar environment [6, 30–32].

The wavefunction description Eq. (10.3) corresponds closely to the Mulliken charge transfer picture of PT [33] shown in Fig. 10.2, in which a non-bonding (n_B) electron of the base is transferred into an antibonding (σ^*_{AH}) orbital of the acid. A naive interpretation of this would be that the transfer into the antibonding orbital of the acid weakens the AH bond and assists in allowing the transfer of the H species to the base; since the base is now positively charged, this would be transfer of an hydrogen atom H, not a proton H^+. Such a view ignores the fact that the electronic coupling between the two VB states is very strong, ~1 eV [6, 28, 34–36] and the charge transfer is electronically adiabatic. (We stress that this is true for both the proton adiabatic and nonadiabatic regimes as we have defined them.) A more correct picture then is that electron transfer and H atom transfer are concerted and the charge carried by the hydrogen moiety in the PT would be something close to +0.5. The validity of the Mulliken picture has been tested in molecular orbital calculations, making no reference to the two VB state approximation, for the first PT of the acid dissociation of hydrofluoric acid HF in water [2c]. It was found that indeed during the PT there is electronic charge transfer out of the H_2O base nonbonding orbital into the antibonding orbital of the HF acid, and that the charge on the hydrogenic species is approximately +0.5. This appears to be the first demonstration of the validity of the Mulliken charge transfer picture for PT. (There are further electronic structure issues for excited electronic state reactions, addressed in Section 10.4).

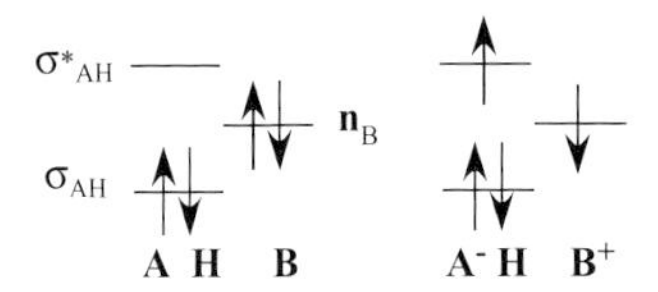

Figure 10.2 Qualitative orbital energy diagram of the Mulliken charge transfer picture for the two VB states of the H-bond PT complex AH···B.

Since both involve charge transfer, PT reactions share with (outer sphere) electron transfer (ET) reactions a key role for the polar environment. There are, however, a number of essential differences between PT and ET reactions. The feature just discussed, that the electronic coupling is large – as is typical for bond breaking and making reactions (See e.g. Ref. [31]) – is in strong contrast to the weak to very weak electronic coupling which applies for ET [37]. Another key difference is a strong dependence for PT on the H-bond A–B separation Q which must be accounted for in the dynamics. This is especially important for nonadiabatic tunneling PT reactions where the tunneling probability is exponentially sensitive to Q; the corresponding dependence for tunneling ET is much weaker, a feature arising from the more extreme quantum character of the electron compared to the proton [1]. As a consequence of these differences, one cannot directly take over equations valid for ET and apply them to PT, as is sometimes done.

In succeeding sections, we recount the essential features of the adiabatic and nonadiabatic PT regimes. Section 10.2 describes the adiabatic PT picture as well as the resulting rate constant, activation free energy and KIEs, while Section 10.3 presents the nonadiabatic PT picture rate constant and KIEs. We focus throughout on acid–base PT reactions in the ground electronic state, e.g. Eq. (10.1). Some aspects of excited electronic state PT reactions are referred to at various locations in the chapter. Concluding remarks are offered in Section 10.4.

10.2 Adiabatic Proton Transfer

10.2.1 General Picture

In the adiabatic PT limit, proton motion is quantum but is not tunneling, the reaction coordinate is the solvent, and the internal H-bond coordinate plays several important roles, to be described presently. These features have been examined in the Hynes group in a number of detailed realistic combined electronic structure/dynamics simulations of specific systems such as hydrochloric and hydrofluoric acids in water [2a–c]. While we will occasionally refer to results of these studies, for present purposes it is most useful to couch most of the discussion in terms of analytic studies by the present authors [3, 4] of the primary acid–base PT reaction event in a H-bonded complex, cf. Eq. (10.1), in a polar solvent, with the acid and base modeled to represent typical oxygen and nitrogen acid–base systems. The solvent is treated as a nonequilibrium dielectric continuum and the electronic structure aspects necessary to account for the bond breaking and making features of the PT are treated in terms of a two valence bond (VB) state description Eq. (10.3) involving VB wavefunctions corresponding to the limiting reactant and product forms, at every geometry, of Eq. (10.1) [6], as was mentioned in the Introduction in connection with the Mulliken charge transfer picture of PT. Details can be found in Ref. [3]. Finally, for the adiabatic PT regime, the A–B H-bond strength is to be sufficiently strong that the equilibrium reactant pair A–B separation is sufficiently small that, when the solvent rearrangement gives a symmetric double well in the proton coordinate, the proton potential barrier is sufficiently low such that the ZPE level of the proton lies above this barrier.

Figure 10.3 displays the key features for a model overall symmetric PT reaction; for the moment, we consider first the simplified case where the A–B H-bond length Q is held fixed. Figure 10.3(a) shows the proton potential curve versus the proton coordinate and its ground vibrational wavefunctions and level, for solvent configurations appropriate to that of the reactant pair R, the transition state TS, and the product pair P. The proton vibrational level in the symmetric proton potential situation is indeed above the barrier, as required in the adiabatic PT limit. Inspection of the displayed proton wavefunction for the TS solvent configuration shows that the quantum delocalization of the proton is so extensive that any classical description is completely inappropriate. Thus the proton extends over a significant portion of the double well; in a classical, conventional TST description, one would consider instead the classical motion of the always localized proton over the top of this proton potential, taking the proton coordinate as the reaction coordinate, with an imaginary frequency in the limited region of the top of the proton potential. In the correct quantum description, the reaction coordinate is instead the solvent. Examples illustrating the above points involving a microscopic description of the solvent and a more detailed electronic structure description are available [2, 7] (see also Ref. [36]); Fig. 10.4 illustrates the proton level pattern for the first PT step in the acid dissociation of HCl in water [2b].

The physical picture of the adiabatic limit is that as the solvent reorganizes, the high frequency quantum proton vibration adiabatically adjusts. Indeed, this is one source of the terminology 'adiabatic PT'. Thus, the reaction coordinate is a solvent coordinate, rather than the proton coordinate, and there is a free energy change up to the TS activation free energy as the solvent rearranges, as shown in Fig. 10.3(b). The total free energy versus the solvent coordinate can be usefully decomposed into two basic contributions, shown in Fig. 10.3(b). These are, respectively, a 'bare' free energy, G_{min}, which is that corresponding to the situation where the

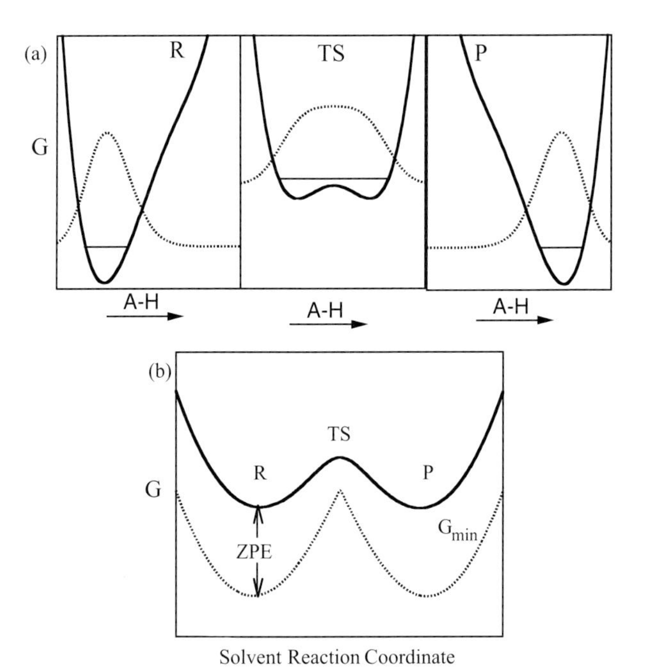

Figure 10.3 (a) Proton potential energy curves versus the proton coordinate at the reactant R, transition state TS, and product state P solvent configurations for a symmetric reaction. (The ordinate is labeled as a free energy G, since the potential energy curves at each solvent configuration point shown have a constant free energy contribution related to the solvent interaction). In each case, the ground state proton vibrational energy level (solid line) and wavefunction (dotted line) are indicated. (b) Free energy curve for the symmetric PT system displayed in (a) with the proton quantized in its vibrational ground state versus solvent reaction coordinate (solid line). The solvent coordinate critical points corresponding to the proton potentials in the upper panel are indicated. The free energy at the minimum of the proton potential along the solvent coordinate G_{min} (dotted line) is also shown. The difference between G_{min} and G is the zero point energy.

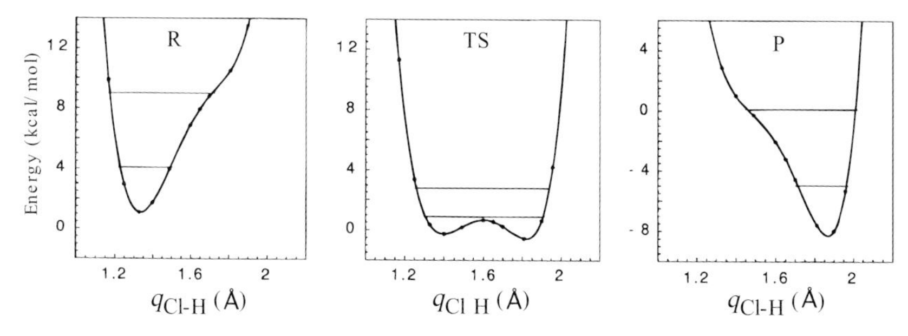

Figure 10.4 Proton transfer potentials characteristic of the reactant (left panel), TS (middle panel, and product (right panel) regions for the acid ionization PT of HCl in water [2b]. The Cl–O distance is 2.91 Å. The ground and first excited proton vibrational levels are indicated.

proton is located at its classical minimum free energy position for any given solvent coordinate, and the proton's vibrational ZPE, measured from the latter potential minimum energy. The ZPE decreases as the TS in the solvent is approached, since the proton potential becomes more symmetric and the proton is delocalized in a larger potential region. Finally, as a technical point, the solvent coordinate in this figure is an alter ego of the solvent polarization, related to a certain energy gap ΔE defined such that for a thermodynamically symmetric PT reaction, it equals zero at the solvent TS where the proton potential is symmetric [2, 3, 7].

In order to extend the above description to include the H-bond coordinate Q, we display a contour plot in Fig. 10.5 of the free energy of the PT system with the proton motion already quantized in its ground vibrational, zero point level, as a function of Q and the solvent coordinate ΔE. Reactant and product wells are

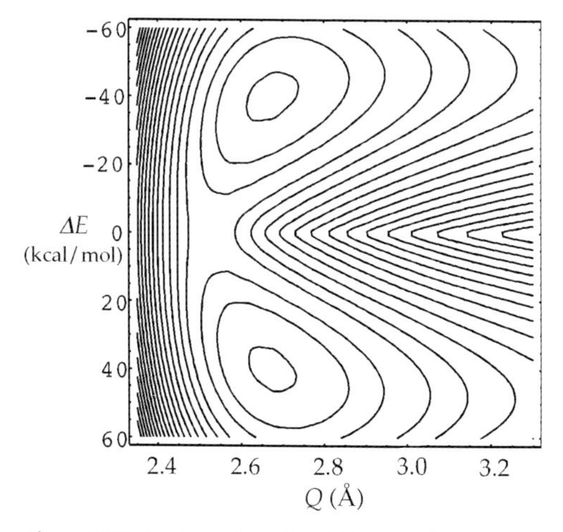

Figure 10.5 Contour plot of the PT system free energy versus the solvent coordinate, ΔE, and the H-bond coordinate separation Q for a symmetric reaction. Contour spacings are 1 kcal mol^{-1}.

clearly evident, with the passage between them involving an H-bond coordinate compression, i.e., a stronger H-bond. Thus, the H-bond frequency will increase near the saddle point. For this model system, the Q frequencies are ~290 cm^{-1} near the reactant and product configurations and 550 cm^{-1} near the saddle point [3b]. Thus, the Q motion should *also* be quantized, and this was done in the approximation that the H-bond vibration is fast compared to the solvent motion.

Figure 10.6 shows the resulting free energy curves versus the solvent coordinate for the first several H-bond vibrational states. While the excited vibrational state curves can be examined to deduce the impact of H-bond excitation on the rate [2d, 3b], under ordinary room temperature conditions, the rate will be dominated by the ground state $l = 0$ level curve, and we restrict further discussion to that case. Note that now there is a further contribution to the ZPE to G at the solvent TS and reactant (R) configuration arising from the quantized H-bond vibration. For the model symmetric reaction, the total change $\Delta ZPE^{\ddagger}$ from R to TS is –2.1 kcal mol^{-1}, of which –2.5 kcal mol^{-1} arises from the quantized proton motion, which is more delocalized at TS than in R (see Fig. 10.1(a)), while +0.4 kcal mol^{-1} arises from the H-bond vibration, reflecting the (stronger) H-bond higher frequency at TS compared to R (cf. Fig. 10.5). Since the ZPE change enters the activation free energy, the H-bond ZPE exerts a non-negligible effect on the rate constant.

Transition state theory (TST) can be applied in the solvent coordinate for the $l = 0$ curve in Fig. 10.6 and for the corresponding curves for different thermodynamic reaction asymmetry ΔG_{RXN} [2d, 3b], and the rate constant for the adiabatic PT reaction is then given by [2–4, 6, 7]

$$k_{TST} = (\omega_S/2\pi)\exp[-\Delta G^{\ddagger}/RT] \tag{10.4}$$

in which the prefactor involves a solvent frequency related to the free energy-solvent coordinate curvature of the solvent reactant well shown in Fig. 10.6 ($l = 0$), and $\Delta G^{\ddagger}$ is the activation free energy evident in that same figure. There are several important points to make here. First, while the quantum proton motion is accounted for, this is a classical TST result; the TST assumption of no barrier recrossing has been applied to the classical *solvent* reaction coordinate. Second,

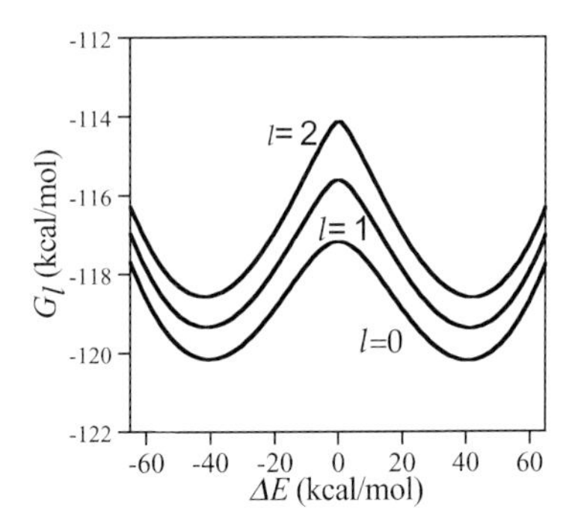

Figure 10.6 Free energy curves versus the solvent coordinate ΔE for a symmetric PT reaction with the quantized proton in its ground vibrational state and the quantized H-bond vibrational mode in the ground ($l = 0$), first excited ($l = 1$), and second excited ($l = 2$) energy states.

there can be barrier re-crossing effects that reduce the rate constant compared to the TST value. Since TS passage means changing the solvent polarization, there is a time dependent friction related to such polarization changes, leading to barrier recrossing. Such a correction has been calculated for a model adiabatic PT between phenol and trimethylamine in a molecular level solvent and found [2d, 38] to be in good agreement with Grote–Hynes theory [39], but in that case and in most cases such corrections should be minor; indeed Grote–Hynes theory shows why TST is generally an excellent approximation to the rate constant.

Third, and most important, while a symmetric reaction has been discussed here, we have analytically derived [3] expressions for the individual components of a $\Delta G^{\ddagger}$ vs. ΔG_{RXN} quadratic free energy relation (FER)

$$\Delta G^{\ddagger} = \Delta G_{o}^{\ddagger} + \frac{1}{2}\Delta G_{RXN} + \frac{1}{2}\bar{a}'_{o}\Delta G_{RXN}^{2} \tag{10.5}$$

(The over-bar notation for the Brønsted coefficient and its derivative, e.g. $\bar{a}'_{o}$, is introduced to distinguish it from the inverse coupling length a_L used in Section 10.3) This result is closely related to, but more fundamentally based than the well-known and widely employed (cf. Refs. [9c, 12]) Marcus relation [19]

$$\Delta G^{\ddagger} = \Delta G_{o}^{\ddagger} + \frac{1}{2}\Delta G_{RXN} + \frac{(\Delta G_{RXN})^{2}}{16\Delta G_{o}^{\ddagger}} \tag{10.6}$$

Actually Eq. (10.6) was never derived for PT reactions and thus has to be regarded as empirical for them. It was instead derived [19, 37] for two separate cases: (i) for outer sphere electron transfer reactions, where there is strong coupling to the solvent but there is no bond breaking and forming, and thus no strong electronic coupling between the R and P VB states, and (ii) (more approximately) for gas phase atom transfer reactions, where there is strong electronic coupling but no strong electrostatic coupling to a polar environment. As we have emphasized earlier in this chapter, PT reactions are characterized simultaneously by strong electrostatic coupling to the environment and strong electronic coupling. While we refer the reader to Ref. [3] for a more detailed discussion, our Eq. (10.5) can be regarded as the appropriate FER for proton adiabatic PT reactions; the coefficient of its quadratic term differs in principle from that of Eq. (10.6) (although the numerical magnitude of the coefficients can sometimes be similar [3]). We make a few further remarks on this in Sections 10.2.2 and 10.2.3.

Another FER exists that is based on an underlying electronic diabatic perspective (although aspects of the electronic coupling are included); the FER in the electronically adiabatic ET limit for a curve-crossing picture is [40]

$$\Delta G^{\ddagger} \approx \frac{(\lambda + \Delta G_{RXN})^{2}}{4\lambda} - \beta^{\ddagger} + \frac{(\beta^{R})^{2}}{(\lambda + \Delta G_{RXN})} \tag{10.7}$$

Here, λ is an electronically diabatic reorganization energy, now including both solvent and nuclear rearrangements, and the last two terms in Eq. (10.7) are leading order corrections to the barrier height with finite electronic resonance coupling at TS ($\beta^{\ddagger}$) and R (β^{R}). This equation has only been tested under conditions such that a linear FER predominates [40]. It differs considerably from our Eq. (10.5). A first issue for Eq. (10.7) is that the proton is not quantized, a crucial feature in our perspective; even when some quantum corrections are incorporated [22 b–d], there is no proton ZPE contribution at the TS for a symmetric reaction, in contrast to the adiabatic PT result. Other issues concern the Brønsted coefficient variation and the associated Hammond postulate, e.g. Eq. (10.7) does not give a Brønsted coefficient of 0.5, which it should. The reader is referred to Ref. [3b] for a more detailed discussion of Eq. (10.7).

We have emphasized above that the free energy contains a proton ZPE contribution, as seen in Fig. 10.3(b) and implicit in Fig. 10.5, which also contains an H-bond ZPE contribution. For the activation free energy $\Delta G^{\ddagger}$, this will contribute the difference of the H and D ZPEs at the solvent TS and for the reactant. This leads [4] to quite a different picture for kinetic isotope effects (KIEs) than the standard one [8, 12–17], sketched in Section 10.1 and to be further discussed in Section 10.2.3.

In the above description, the solvent coordinate ΔE has not been given a molecular visualization, other than our initial characterization in terms of the degree of the solvent electrical polarization. An explicit characterization in terms of solvent molecular motion can be given, although this requires simulation analysis. For example, Fig. 10.7(a) shows the key motions of the water molecules in the PT reaction of an excess proton in aqueous solution [2a, b]. Briefly, the coordination number of the proton–accepting water molecule is 4, while that of the hydronium ion H_3O^+ is 3 [2, 41] – and all these coordinations involve H-bonds. Thus, passage of the solvent coordinate through its TS involves the reduction of this coordination number, effected via a rearrangement of the water initially H-bonded to the proton-accepting water oxygen of the H_2O to break that initially existing H-bond, while a similar but opposite rearrangement occurs to form an H-bond with the proton-donating H_3O^+ appropriate to the coordination that this H_2O will become. This explicitly identified reaction coordinate motion involving H-bond rearrangement [2a, b], is closely related to the original suggestion of Newton [41], and has subsequently become widely accepted and discussed [42, 43]. Figure 10.7(b) shows a related picture of an acid ionization in aqueous solution $HCl \cdots H_2O \rightarrow Cl^- \cdots H_3O^+$ [2a, b], and similar results have been found of the ionization of HF in water [2c, d]). Again the key requirement of 3 H-bond coordination for H_3O^+ and 4 H-bond coordination for H_2O is involved: an H-bond must be broken for the proton-accepting H_2O to be appropriate to the coordination of the H_3O^+ that this H_2O will become.

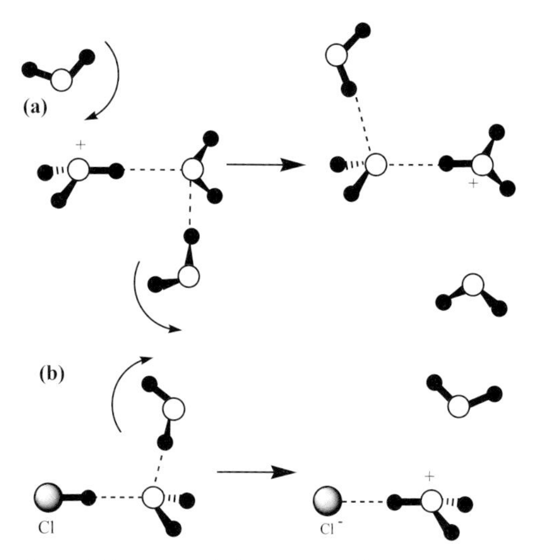

Figure 10.7 Schematic illustration of the reaction-promoting solvent motions for (a) the first PT from H_3O^+ to H_2O and (b) the first PT step in the acid ionization proton transfer reaction of HCl in aqueous solution.

10.2.2
Adiabatic Proton Transfer Free Energy Relationship (FER)

We now turn to a detailed discussion of the activation free energy $\Delta G^\ddagger$ which determines the adiabatic PT TST rate constant Eq. (10.4). In particular, we discuss the derived activation free energy–reaction free energy relation Eq. (10.5) and its components, in a molecular description. We focus on transfer of a proton, but include some aspects for deuteron transfer in preparation for a discussion of KIEs in Section 10.2.3.

We begin with an overview of the total free energy of the reacting system versus the solvent reaction coordinate, which can be usefully decomposed into two basic contributions [3, 7], as shown in Fig. 10.3(b),

$$G = G_{\min} + \mathrm{ZPE} \tag{10.8}$$

These are respectively, a 'bare' free energy, $G_{\min}$, corresponding to the situation where the proton and H-bond Q coordinate are located at their respective *classical* minimum position for any given solvent coordinate value, and the vibrational ZPE contributions of the proton and the H-bond mode, measured from the latter potential energy minimum. Figure 10.8 displays the H and D free energy curves for a symmetric and an asymmetric reaction for a model OH···O system [3, 4], explicitly showing the decomposition of the total free energy G according to Eq. (10.8). The ZPE contributions for both H and D from Fig. 10.8(a) and (b) are dis-

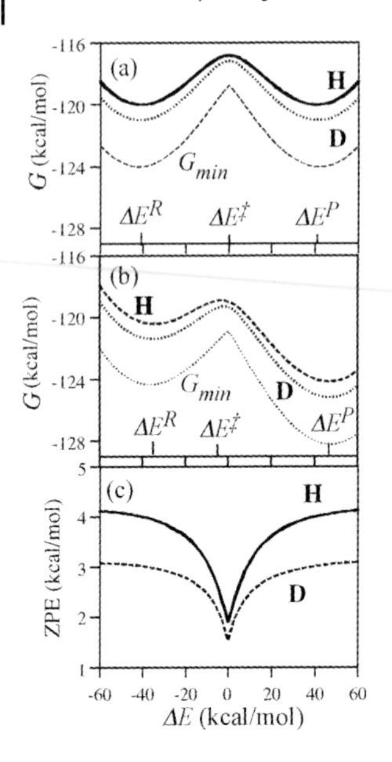

Figure 10.8 Ground state free energy curves for adiabatic PT (solid lines H and dotted lines D) with both the proton/deuteron and H-bond vibrations quantized: (a) symmetric reaction and (b) exothermic reaction. Dashed lines show the free energy curves G_{min} excluding the ZPE. (c) The ZPE for the proton (solid line) and deuteron (dashed line), including H-bond vibration, vs. ΔE. The dashed curves in (a) and (b) plus the ZPE in (c) give the full free energy G, the solid curves in (a) and (b). ΔE^R, ΔE^P, and $\Delta E^{\ddagger}$ denote the reactant, product, and TS solvent coordinate values, respectively.

played as Fig. 10.8(c), and contain both the ZPE of the H or D vibration and that of the H-bond mode. The ZPE decreases as the TS in the solvent coordinate is approached, mainly due to the proton/deuteron potential becoming more symmetric and the proton/deuteron is delocalized in a larger region. The reaction barrier increases starting from an exothermic case (Fig. 10.8(b)) to an endothermic case (the reverse of Fig. 10.8(b)). From Fig. 10.8(a) and (b), it is seen that both the reactant well frequency ω_S and barrier height $\Delta G^{\ddagger}$ are isotope-dependent. However, the contribution of ω_S to KIEs is minimal since it is largely governed by the solvent [4]; accordingly we regard ω_S as isotope-independent in all that follows.

Figures 10.8(a) and (b) also indicate that the TS position $\Delta E^{\ddagger}$ along the solvent coordinate shifts with reaction asymmetry. This is due to the fact that the addition of the ZPE to an asymmetric G_{min} (cf. Fig. 10.8(b)) shifts the maximum of G away from the maximum of G_{min} at $\Delta E = 0$, in the direction consistent with the Hammond postulate [15], e.g. later for endothermic reactions $\Delta G_{RXN}>0$. This indicates that the ZPE contribution at $\Delta E^{\ddagger}$ to the free energy barrier $\Delta G^{\ddagger}$ in the solvent coordinate will increase with increasing reaction asymmetry, a crucial qualitative characteristic [3] to which we will return. Also visible in Fig. 10.8(b) is the isotope dependence of the shift $\Delta E^{\ddagger}$ and the associated increase in ZPE at $\Delta E^{\ddagger}$ with increasing reaction asymmetry; the latter leads to a KIE reduction, since the ZPE contribution at $\Delta E^{\ddagger}$ will become more and more similar to that of the reactant. Here one can see that the *variation* in ZPE along the reaction coordinate and its isotopic difference plays a significant role in the reaction free energy barrier variation, and hence the KIE as well.

We now turn to the individual components of the FER in Eq. (10.5). These are $\Delta G^{\ddagger}_o$, the 'intrinsic' reaction barrier $\Delta G^{\ddagger}_o = \Delta G^{\ddagger}(\Delta G_{RXN} = 0)$ for the thermodynamically symmetric reaction, a term linear in ΔG_{RXN} whose coefficient is the Brønsted coefficient

$$\bar{a} = \frac{\partial \Delta G^{\ddagger}}{\partial \Delta G_{\mathrm{RXN}}} \tag{10.9}$$

evaluated for the symmetric reaction, $\bar{a}_{\mathrm{o}} = 1/2$, and a quadratic term in ΔG_{RXN} with a coefficient $\bar{a}'_{\mathrm{o}}$, which is the Brønsted coefficient slope evaluated at $\Delta G_{\mathrm{RXN}} = 0$ [44]. The intrinsic free energy barrier $\Delta G^{\ddagger}_{\mathrm{o}}$, which provides a reference barrier height, can be analytically expressed as [3]

$$\Delta G^{\ddagger}_{\mathrm{o}} = \Delta G^{\ddagger}_{\mathrm{m,o}} + \mathrm{ZPE}^{\ddagger}_{\mathrm{o}} - \mathrm{ZPE}^{\mathrm{R}}_{\mathrm{o}} \tag{10.10}$$

which has a contribution from a certain solvent reorganization free energy, here called $\Delta G^{\ddagger}_{\mathrm{m,o}}$, to go from solvent arrangements appropriate to the reactant complex to those TS solvent arrangements appropriate for a symmetric double well proton potential, and from the change $Z^{\ddagger}_{\mathrm{o}} - Z^{\mathrm{R}}_{\mathrm{o}} = \Delta \mathrm{ZPE}^{\ddagger}_{\mathrm{o}}$ of the ZPE in the proton and H-bond coordinates on going from the reactant to the TS for the symmetric reaction. Note that in the adiabatic PT picture, $Z^{\ddagger}_{\mathrm{o}}$ refers to a proton vibration at the symmetric reaction TS; the proton coordinate is a transverse, nonreactive coordinate.

The Brønsted coefficient Eq. (10.9) has played an important role in the FER. To discuss it, we need to specify the underlying electronic structure description we have employed for the PT reaction Eq. (10.1). This is a two state valence bond model [3, 6] in which the electronic wavefunction is given in terms of the limiting reactant and product electronic configurations for Eq. (10.1), i.e. Eq. (10.2) which we repeat here for convenience

$$|\Psi_{\mathrm{el}}\rangle = c_{\mathrm{R}}|\Psi_{\mathrm{R}}\rangle + c_{P}|\Psi_{\mathrm{P}}\rangle \tag{10.11}$$

The validity of this description was discussed in the Introduction. The Bronsted coefficient can be quantitatively described in the adiabatic PT picture by any of three differences between the TS and the reactant: of the separation in the solvent coordinate, of the electronic structure, and of the quantum-averaged nuclear structure [3, 4]

$$\bar{a} = \frac{\Delta E^{\ddagger} - \Delta E^{\mathrm{R}}}{\Delta E^{\mathrm{P}} - \Delta E^{\mathrm{R}}} = \frac{\langle c_{\mathrm{P}}^2\rangle^{\ddagger} - \langle c_{\mathrm{P}}^2\rangle^{\mathrm{R}}}{\langle c_{\mathrm{P}}^2\rangle^{\mathrm{P}} - \langle c_{\mathrm{P}}^2\rangle^{\mathrm{R}}} = \frac{\langle q - Q/2\rangle^{\ddagger} - \langle q - Q/2\rangle^{\mathrm{R}}}{\langle q - Q/2\rangle^{\mathrm{P}} - \langle q - Q/2\rangle^{\mathrm{R}}} \tag{10.12}$$

Here $\Delta E^{\ddagger} - \Delta E^{\mathrm{R}}$ is the solvent reaction coordinate distance between the TS and reactant, and $\Delta E^{\mathrm{P}} - \Delta E^{\mathrm{R}}$ is the corresponding distance between the product and reactant. $\langle c_{\mathrm{P}}^2\rangle$ is the *quantum average* over the proton and H-bond vibrations of $c_{\mathrm{P}}{}^2$, the limiting product contribution to the electronic structure. The electronic structure for each critical point (c = R, P, and ‡) is evaluated at the respective critical point position ΔE_c ($\Delta E_c = \Delta E_{\mathrm{R}}$, ΔE_{P}, or $\Delta E^{\ddagger}$) along the reaction coordinate. The structural element $\langle q - Q/2\rangle$ is the quantum-averaged proton distance (over both proton and H-bond coordinates) from the H-bond's center. As noted above, the TS structure (and the TS position along the reaction coordinate) for a symmetric reaction is halfway between that of the reactant and product, so that $\bar{\alpha}_o = 1/2$, independent of isotope.

We pause to remark that the Brønsted coefficient $\bar{\alpha}$ has often been used to describe TS structure via the Hammond postulate [15] or the Evans–Polanyi relation [45], where $\bar{\alpha}$ is viewed as a measure of the relative TS structure along the reaction coordinate, usually a bond order or bond length. The important point is that, although adiabatic PT has a quite different, environmental, coordinate as the reaction coordinate, Eq. (10.12) is consistent with that general picture, with a proper recognition that quantum averages are involved.

The TS structure's variation with reaction asymmetry is described by the Brønsted coefficient slope $\bar{\alpha}'_o$, the derivative of Eq. (10.12) with respect to ΔG_{RXN} evaluated for the symmetric reaction. In this manner, $\bar{\alpha}$for the FER in Eq. (10.12) is linearly related to the reaction asymmetry

$$\bar{\alpha} = \frac{1}{2} + \bar{\alpha}'_o \Delta G_{RXN} \tag{10.13}$$

Expressions for $\bar{\alpha}'_o$ have been explicitly derived in Ref. [3]. A convenient expression is in terms of the free energy's force constants along the reaction coordinate, k_R and $k^\ddagger$, at the reactant and TS positions, and the reaction coordinate distance between the reactant and product $\Delta\Delta E = \Delta E^P - \Delta E^R$ (cf. Eq. (1.5) of Ref. [3b])

$$\bar{\alpha}'_o = \frac{1}{\Delta\Delta E^2}\left(\frac{1}{k^\ddagger} + \frac{1}{k_R}\right) \tag{10.14}$$

Figure 10.9 displays a FER for a model system [3, 4], as well as our FER Eq. (10.5) using Eq. (10.14) for $\bar{\alpha}_o$, and it shows that the analytical description for the FER gives a good account of the activation free energy–reaction free energy relation for the rate constant.

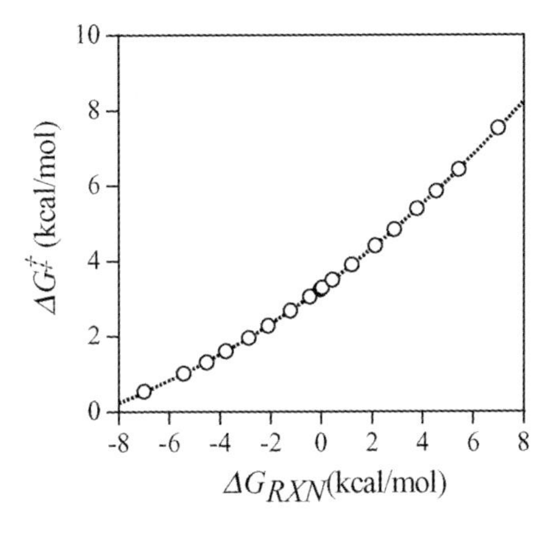

Figure 10.9 Free energy relationship $\Delta G^\ddagger$ versus ΔG_{RXN} for proton transfer for a model O···O system (o). Dotted line is Eq. (10.5) using Eq. (10.14) to evaluate $\bar{\alpha}'_o$ ($\Delta G^\ddagger_{oH} = 3.27$ kcal mol^{-1} and $\bar{\alpha}'_{oH} = 0.03$ mol kcal^{-1}).

As seen from Fig. 10.8, a key component of the TS structure variation is reflected in the variation of the ZPE along the reaction coordinate. This feature is incorporated in Eq. (10.14) since the force constant is the sum (via Eq. (10.6)) of the ZPE and G_{min} variation [3, 4]. (We pause to note that Eq. (10.4) shows that the coefficients in the FERs Eqs. (10.5) and (10.6) are not the same [3, 4].) Further, since $\bar{\alpha}$ is also directly related to the relative difference in structure between R and TS, i.e. the last expression in Eq. (10.12), the variation of ZPE versus ΔE directly correlates with structural variation along a 'reaction path'. A comparison between a 'reaction path' described with quantum averages via the adiabatic PT picture and those with a classical description is presented in Section 10.2.3.5.

We now turn to the isotope dependence of the FER, which will be important for the discussion of KIEs in Section 10.2.3. This dependence arises from the components $\Delta G^{\ddagger}_{o}$ and $\bar{\alpha}_{o}$of the FER. The isotope dependence of the intrinsic free energy barrier $\Delta G^{\ddagger}_{o}$ given by Eq. (10.4) is, as is apparent in Fig. 10.8(a), due solely to the difference in the H and D ZPEs

$$\Delta G^{\ddagger}_{oD} - \Delta G^{\ddagger}_{oH} = Z^{\ddagger}_{oD} - Z^{R}_{oD} - Z^{\ddagger}_{oH} + Z^{R}_{oH} = \Delta ZPE^{\ddagger}_{oD} - \Delta ZPE^{\ddagger}_{oH} \tag{10.15}$$

Recall that the ZPE contains both that of isotope L and that of the H-bond vibrational mode. The latter's contribution is, however, smaller in magnitude than the negative ZPE difference associated with the proton vibrational mode (–2.5 kcal mol^{-1} in Fig. 10.8). Thus, $\Delta ZPE^{\ddagger}_{o}$ is overall negative (e.g –2.1 kcal mol^{-1} from Fig. 10.8). Furthermore, this ZPE difference decreases as the mass of the transferring particle L increases, as one would expect from a ZPE $\propto 1/\sqrt{m_L}$ mass dependence. This ZPE mass dependence is the key ingredient for adiabatic PT KIEs.

The isotope dependence of the Brønsted slope $\bar{\alpha}'_{o}$ is most conveniently discussed in terms of the derivative of the expression involving force constants Eq. (10.14). These force constants certainly depend on the variation of the ZPE along the solvent reaction coordinate via Eq. (10.8). Accordingly, $\bar{\alpha}'_{o}$ can be cast in terms of these slopes plus the variation in the ZPE value at the reactant and TS positions with reaction asymmetry [4]. Since the ZPE variation is largest in the TS region, the first term in Eq. (10.14) is the most significant, and thus, the essential point is that the isotope difference $\bar{\alpha}'_{oH} - \bar{\alpha}'_{oD}$ is approximately proportional to the difference in the rate of increase of ZPE‡ with increasing reaction asymmetry between H and D.

Further analysis of the isotope dependence of the intrinsic barrier $\Delta G^{\ddagger}_{o}$ is useful for the KIE discussion. The intrinsic barrier's isotope dependence in Eq. (10.15), arising only from the difference in ZPEs, illustrates a key common point of connection between the present and standard perspectives: in both cases, the difference in intrinsic barrier heights is related to the difference in a ZPE between the reactant and TS between both isotopes, resulting in a KIE which is maximal for $\Delta G_{RXN} = 0$ and falls off with increasing asymmetry.

In standard treatments, the isotope mass scaling for the L contribution is $ZPE_L \propto 1/\sqrt{m_L}$, which assumes harmonic potentials [12–14]. The proton potentials in Fig. 10.3 are not harmonic, especially *at* the TS, where there is a double well, and thus it is not obvious that the simple ZPE mass scaling holds. However, the following relation based on Eq. (10.15) and assuming that all ZPEs scale according to $ZPE_L \propto 1/\sqrt{m_L}$

$$\Delta G^{\ddagger}_{oL_2} - \Delta G^{\ddagger}_{oL_1} = \Delta ZPE^{\ddagger}_{oL_2}\left(1 - \sqrt{\frac{m_{L_2}}{m_{L_1}}}\right) = \Delta ZPE^{\ddagger}_{oH}\left(\sqrt{\frac{m_H}{m_{L_2}}} - \sqrt{\frac{m_H}{m_{L_1}}}\right) \quad (10.16)$$

was shown to hold quite well numerically [4], to within 10%. (All differences in $\Delta G^{\ddagger}_{o}$ have been scaled to the ZPE difference $\Delta ZPE^{\ddagger}_{0}$ for H in the last member). The significance of the numerical validity of Eq. (10.16) is that the adiabatic PT picture *also* generates the mass scaling of standard KIE theory.

10.2.3
Adiabatic Proton Transfer Kinetic Isotope Effects

The KIE for adiabatic PT is the ratio of individual rate constants, where each of these is of the form in Eq. (10.4), e.g. H versus D transfer

$$\frac{k_H}{k_D} \approx \exp\left(-(\Delta G^{\ddagger}_{H} - \Delta G^{\ddagger}_{D})\big/ RT\right) \quad (10.17)$$

Here, the reactant reaction coordinate frequencies $\omega_S{}^{H,D}$ in Eq. (10.4) have been assumed equal [4]. From the FER analysis of Eq. (10.5), the explicit form for the KIE dependence on ΔG_{RXN} is

$$\frac{k_H}{k_D} = \exp\left(-(\Delta G^{\ddagger}_{oH} - \Delta G^{\ddagger}_{oD})/RT\right)\exp\left(-(\bar{a}'_{oH} - \bar{a}'_{oD})\Delta G^2_{RXN}/2RT\right) \quad (10.18)$$

Further, an equivalent form re-expresses the first part of this in terms of the KIE for the symmetric reaction:

$$\frac{k_H}{k_D} = \frac{k_{Ho}}{k_{Do}}\exp\left(-(\bar{a}'_{oH} - \bar{a}'_{oD})\Delta G^2_{RXN}/2RT\right) \quad (10.19)$$

involving the isotopic difference of the symmetric reaction Bronsted slope Eq. (10.14).

Before proceeding with the KIE analysis for adiabatic PT, it is worth stressing, for comparison with the standard picture, that there are four common experimental observations which are consistent with the standard picture for nontunneling PT KIEs, and which are thus viewed as supporting that picture: (i) the Arrhenius temperature dependence of the KIE (as well as of the individual isotope rate constants); (ii) the KIE – ΔG_{RXN} behavior described in Section 10.1 (i.e. maximal for the symmetric case); (iii) the KIE range is ~2–10; and (iv) the wide applicability of the Swain–Schaad relationship [13, 46] connecting KIE ratios (e.g. $k_H/k_T = (k_D/k_T)^{3.3}$). These observations have done much to maintain the standard picture as a widespread perspective for KIEs. It is therefore important that, as now discussed, these also follow from the present adiabatic PT picture [3, 4].

10.2.3.1 KIE Arrhenius Behavior

The Arrhenius form for the adiabatic PT KIE in Eqs. (10.17)–(10.19) is consistent with the first set of experimental results (i), and the general form for the KIE is identical to that of the standard picture (i.e. the adiabatic PT Eq. (10.17) is similar to the standard Eq. (10.2)), despite significant differences in ingredients between the two pictures. The adiabatic PT rate constant's Arrhenius temperature dependence follows from a temperature-independent $\Delta G^{\ddagger}$. Additional temperature dependence is in principle present in both $\Delta G^{\ddagger}$ (and the prefactor) in the above KIE expressions, but these effects are, with some exceptions, negligible for highly polar solvents [47]. The temperature dependence of $\Delta G^{\ddagger}$ will be discussed further in Section 10. 2.4.

10.2.3.2 KIE Magnitude and Variation with Reaction Asymmetry

The KIE behavior versus reaction asymmetry for adiabatic PT follows directly [4] from insertion of the isotopic difference between the FER curves described in Eqs. (10.18) and (10.19). The general feature that the KIE is maximal for $\Delta G_{RXN} = 0$ follows from a Brønsted coefficient for a symmetric reaction that is isotope-independent, $\bar{a}_o = 1/2$, which reflects the symmetric nature of the electronic structure of the reacting pair at the TS (cf. Eq. (10.12)) [48]. The decrease from the maximum, characterized by a gaussian fall-off with increasing reaction asymmetry, is due to the isotope dependence $\bar{a}'_{oH} > \bar{a}'_{oD}$. As discussed in Section 10.2.2, this isotope dependence is primarily due to the differential rate of change of ZPE‡ versus reaction asymmetry between H and D.

Figure 10.10(a), which displays the H versus D KIE ($T = 300$ K) for the Fig. 10.8 PT system, makes these points concrete. The calculated KIE is maximum at $\Delta G_{RXN} = 0$ and drops off symmetrically as the reaction asymmetry is increased. The maximum KIE for the symmetric reaction and the KIE magnitude throughout the whole range are both consistent with experimental observations, (ii) and (iii), respectively. The origin of this last aspect is as follows. The intrinsic KIE magnitude in the adiabatic PT view is directly related to the isotopic difference TS–R ZPE difference $\Delta ZPE^{\ddagger}_o = Z^{\ddagger}_o - Z^R_o$ (see Eqs. (10.15) and (10.17)), whose special feature is the presence of the ZPE for the bound proton vibration at the solvent coordinate TS. Together with the $\Delta ZPE^{\ddagger}_o$ mass dependence following from the ZPE mass-scaling discussed in Section 10.2.2, the maximum KIE magnitude will automatically fall in the same general range as in the standard view. Further, the KIE will fall off due to the increase in TS ZPE with increasing reaction asymmetry, also similar to the standard view.

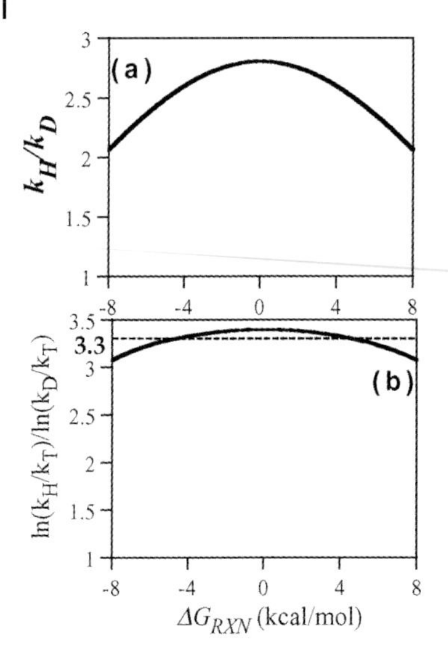

Figure 10.10 (a) KIE k_H/k_D versus ΔG_{RXN} ($T = 300$ K) for an adiabatic PT system (see the text for details). (b) Swain–Schaad slope $\ln(k_H/k_T)/\ln(k_D/k_T)$ versus reaction asymmetry calculated for the PT system in (a).

The reaction free energy dependence on the KIE has often been modeled with an expression based on the empirical Marcus FER for PT Eq. (10.6) [9c, 19].

$$k_H/k_D = k_{Ho}/k_{Do} \exp\left[-\left(\frac{\Delta G^2_{RXN}}{16\Delta G^{\ddagger}_{oH}\Delta G^{\ddagger}_{oD}}\right)\ln\left(k_{Ho}/k_{Do}\right)\right]$$
$$\approx k_{Ho}/k_{Do} \exp\left[-\left(\frac{\Delta G_{RXN}}{4\Delta G^{\ddagger}_{oH}}\right)^2 \ln\left(k_{Ho}/k_{Do}\right)\right] \quad (10.20)$$

Equation (10.20) is similar in form as well as numerically to Eqs. (10.18) and (10.19), but it relates the rate of fall-off of the KIE to the magnitude of the intrinsic reaction free energy barrier $\Delta G^{\ddagger}_{o}$, and not with the change of TS structure (i.e. variation in $\bar{a}$) described above, a feature that is common between the present picture and the W-M picture described in Section 10.1. To further elucidate this point, consider $\Delta G^{\ddagger}$, for the standard picture which is a function of a classical activation energy for the MEP $\Delta V^{\ddagger}$ and a difference in ZPE between R and TS (analogous to Eq. (10.10))[4]

$$\Delta G^{\ddagger} = \Delta V^{\ddagger} + \mathrm{ZPE}^{\ddagger} - \mathrm{ZPE}^{R} \quad (10.21)$$

The isotopic difference $\bar{a}'_{oH} - \bar{a}'_{oD}$ is thus

$$\bar{a}'_{oH} - \bar{a}'_{oD} \approx \frac{\partial^2 \mathrm{ZPE}^{\ddagger}_{H}}{\partial \Delta G^2_{RXN}} - \frac{\partial^2 \mathrm{ZPE}^{\ddagger}_{D}}{\partial \Delta G^2_{RXN}} \quad (10.22)$$

where we have used the fact that the second derivative of $\Delta V^{\ddagger}$ is isotope independent. The variation of reactant ZPE is much less than that in the TS, so that the latter dominates Eq. (10.22). With Eq. (10.22), one can clearly see the connection between ZPE variation and the fall-off of the KIE with increased reaction asymmetry for the W-M picture. This is not the case for Eq. (10.20), however, where $\bar{a}'_{oH} - \bar{a}'_{oD}$ depends only on the magnitude of $\Delta G^{\ddagger}_{o}$ and the ZPEs, and does not depend on any variation in ZPE. A detailed discussion and comparison with the Marcus relation is given in Ref. [3, 4].

10.2.3.3 Swain–Schaad Relationship

The Swain–Schaad relationship has been an important experimental probe for PT reaction KIEs [11, 24, 46]. We have used [4] one of its forms for illustration

$$\ln(k_H/k_T) = 3.3 \ln(k_D/k_T) \tag{10.23}$$

which assumes the ZPE mass correlation $ZPE \propto 1/\sqrt{m}$, discussed in Section 10.2.2, to relate the H, D, and T ZPEs in Eq. (10.2). Figure 10.10(b) displays the calculated adiabatic PT $\ln(k_H/k_T)/\ln(k_D/k_T)$ versus reaction asymmetry for the same PT systems as in Fig. 10.10(a), and shows little variation from Eq. (10.23). Thus, conventional Swain–Schaad behavior also follows from the adiabatic PT picture. We now recount the reasons for this [4].

From the ΔG_{RXN}-dependent form in Eq. (10.18) for the KIE, the ratio of natural logarithms needed for the Swain–Schaad relation in Eq. (10.23) can be written as

$$\ln\left(\frac{k_H}{k_T}\right) \Big/ \ln\left(\frac{k_D}{k_T}\right) = \frac{\Delta ZPE^{\ddagger}_{oT} - \Delta ZPE^{\ddagger}_{oH} - \Delta G^2_{RXN}(\bar{a}'_{oH} - \bar{a}'_{oT})/2}{\Delta ZPE^{\ddagger}_{oT} - \Delta ZPE^{\ddagger}_{oD} - \Delta G^2_{RXN}(\bar{a}'_{oD} - \bar{a}'_{oT})/2} \tag{10.24}$$

A first significant point is that the adiabatic PT form in Eq. (10.24) has the same important feature as the standard picture, via Eq. (10.2): the Swain–Schaad relation is independent of temperature. We first examine the symmetric case $\Delta G_{RXN} = 0$, for which the adiabatic PT expression via Eq. (10.15) shows that the magnitude is related solely to the reactant and TS ZPE difference. These ZPE differences were shown to obey the same mass scaling used to derive the Swain–Schaad relations, cf. Eq. (10.16); hence the Fig. 10.10(b) plot maximum is close to the traditionally expected value. While Fig. 10.10(b) also shows that there is a small variation with reaction asymmetry, in the adiabatic PT perspective, of the Swain–Schaad slope. This has a minimal net effect, however, as discussed in Ref. [4].

10.2.3.4 Further Discussion of Nontunneling Kinetic Isotope Effects

We have already repeatedly emphasized several important fundamental distinctions between the adiabatic PT and the standard view. Despite these distinct differences in physical perspective between adiabatic PT and the standard Westheimer-Melander (W-M) picture, we have emphasized [4] that a remarkable *general* similarity exists between the two perspectives. For adiabatic PT, the symmetric reaction KIE depends on the difference in magnitude of the TS–reactant difference $\Delta ZPE^{\ddagger}_{o}$, and the KIE variation with reaction asymmetry is due to the variation of TS ZPE (and structure). These two points, in fact, are shared with the W-M picture (cf. Eq. (10.2)). We now enumerate the numerical and physical differences between the two perspectives [4].

The adiabatic PT maximum KIE in Fig. 10.10(a) is in the range of KIEs commonly expected in the standard W-M picture, item (iii), but it is somewhat smaller than the higher KIEs 5–10 that one would expect with the standard view. (The argument for this latter range is given in Refs. [12–14].) From Eqs. (10.16) and (10.17), the maximum H/D KIE is that of the symmetric reaction

$$\frac{k_{\mathrm{Ho}}}{k_{\mathrm{Do}}} = \exp\left[-\left(\Delta G^{\ddagger}_{\mathrm{oH}} - \Delta G^{\ddagger}_{\mathrm{oD}}\right)\Big/ RT\right]$$
$$= \exp\left[-\Delta \mathrm{ZPE}^{\ddagger}_{\mathrm{oH}}\left(1 - \sqrt{\frac{1}{2}}\right)\Big/ RT\right] \tag{10.25}$$

with the second line following from the mass scaling of the ZPEs. Equation (10.25) is also used [8, 12, 13] as an estimate for the KIE in the standard W-M picture (cf. Eq. (10.2)). The different symmetric reaction KIE limits for the adiabatic PT and W-M pictures is entirely due to their different views of the TS reaction and transverse coordinates: for a symmetric reaction, there is always a finite proton TS ZPE contribution for adiabatic PT (~1 kcal mol^{-1} for H and ~ 0.7 kcal mol^{-1} for D [4]), whereas the proton TS ZPE is zero ($Z^{\ddagger}_{\mathrm{oH}} = 0$) in the W-M description. The maximum KIE is thus always smaller in the adiabatic PT view; using ω_{R} ~3200 cm^{-1} as the maximum reactant frequency, the maximum KIE without tunneling is ~6 at 300 K. A reduced ω_{R} reduces the minimum value to less than 3 (cf. Fig. 10.10(a)).

10.2.3.5 Transition State Geometric Structure in the Adiabatic PT Picture

We have emphasized, throughout, the quite different perspectives of the standard PT and the adiabatic PT pictures, for the reaction coordinate and the relevant barriers, as well as for the rate constant and KIEs. While we have argued for the validity of the adiabatic PT picture, it is useful to pause here and add an important remark. One of the most widespread and important uses of KIEs is in making inferences, via the standard PT picture, about the geometrical structure of the TS of the PT reaction [14, 15, 45]. Indeed, images of the entire reaction path are generated in this fashion. This obviously involves a classical perception of the coordinates, and it is important to ask whether such assessments can legitimately be made when the coordinates of the acid–base PT system are treated classically.

We have addressed this issue in Ref. [3b], where, for a model PT system in solution treated in the adiabatic PT fashion, we have generated a certain reaction path in the following fashion. At each value of the solvent coordinate ΔE, we have calculated the *quantum averaged* values $\langle q \rangle$ and $\langle Q \rangle$ of the proton and H-bond coor-

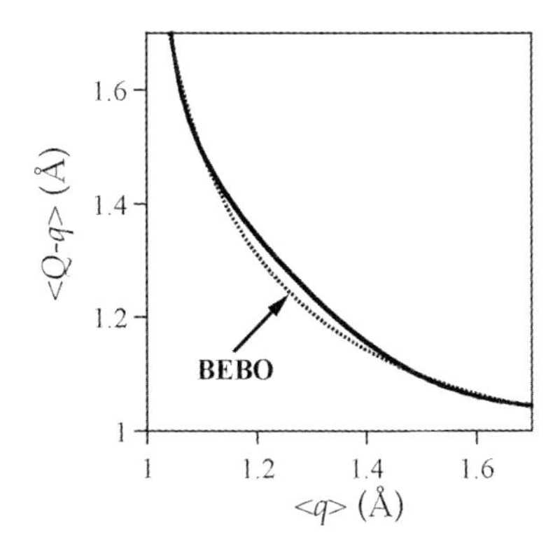

Figure 10.11 Calculated quantum averages $<Q-q>$ vs. $<q>$ for an O···O system (solid line) [3b] and the BEBO curve (dotted line).

dinates, respectively. The relation between $<q>$ and $<Q-q>$ is displayed in Fig. 10.11[3b] and is compared with a reaction path generated from a bond energy–bond order (BEBO) model [49], which is often used in the standard picture of PT and which completely ignores the solvent. It is seen that the two curves are quite close despite distinct differences between the two methods, most notably that in one case the proton and H-bond coordinates are treated fully quantum, while in the other case they are treated completely classically. This comparison demonstrates that one, in a certain sense, can retain the picture of a path in terms of quantum averaged coordinates and the connection between 'transition state structure' and reaction asymmetry [3, 4].

We pause to note that we have determined that a recently proposed empirical quantum correction [50] improves the agreement with the quantum averaged solid line in Fig. 10.11, especially in the TS region. This correction was based on solid state NMR studies of H/D isotope effects on the geometries of strong NHN and OHN hydrogen bonded solids. The success of the model implies that the effects of different solvent configurations on H-bond geometries are similar to those produced by a combination of molecular acidity, basicity and local crystal fields.

10.2.4 Temperature Solvent Polarity Effects

The above discussion has assumed a polar environment in which the polarity does not significantly change upon temperature (T) variation. However, over a sufficiently *large* temperature range (larger than those we have considered), the static dielectric constants ε_0 of liquids are known to change significantly with T, the solvent polarity decreasing with increasing T[51, 52]. For example, ε_0 for water is ~88 at 273 K and decreases to ~55 at 373 K[51]. This type of solvent polarity change will drastically affect H-bonding and PT for systems such as acid ionization PT reactions Eq. (10.1) where the charge character of the reactant differs significantly from that of the product. For the reaction class in Eq. (10.1), the magnitude of the product solvation free energy will increase relative to the reactant with increasing solvent polarity, and thus the reaction asymmetry changes with ε_0 variation, as dis-

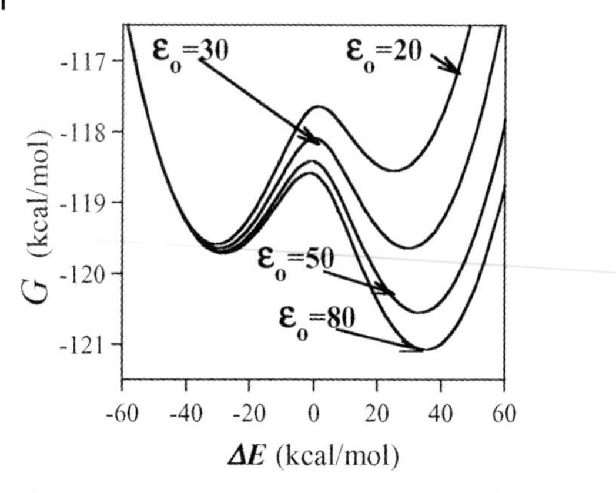

Figure 10.12 Reaction free energy curves for solvent dielectric ε_0 values = 20, 30, 50, and 80. See Ref. [47] for details of calculation.

played in Fig. 10.12 [47]. This effect has been demonstrated experimentally [52] and theoretically [6]. One can also see in Fig. 10.12 that the PT reaction free energy barrier also changes, increasing with increasing *T*. The resulting PT rate constant *k* has a reduced effective activation energy in an Arrhenius plot due to the *T*-dependent solvent polarity effect [47]. It should be noted though that the Arrhenius behavior for the KIE is largely unchanged by solvent polarity variation via a cancellation of effects between isotopes [47].

10.3 Nonadiabatic 'Tunneling' Proton Transfer

We now turn to the proton nonadiabatic, or tunneling, regime. We first briefly review the PT tunneling rate constant formalism [1], including the role of the H-bond mode, and then summarize the resulting KIE behaviors, focusing on adherence (or not) with the same KIE trends (i–iv) discussed in Section 10.2.3 for nontunneling PT. We here restrict the temperature to be close to room temperature and above where the H-bond mode with frequency $\hbar\omega_Q$ is significantly populated, i.e. $\hbar\omega_Q$~RT and $\hbar\omega_Q$<<RT. Other regimes exist and the reader is referred to Ref. [1] for details.

Before proceeding, we first place our basic treatment [1] in perspective with other descriptions, of which there are several, although we make no attempt to be exhaustive. First, there are efforts within the framework of corrections to a Westheimer-Melander (W-M approach) [13, 17]. However, these picture tunneling as occurring through the potential energy along a miminum energy path – a perspective which has long since been replaced by 'corner-cutting' paths [20] – and the role of the solvent is not taken into account. More recent approaches [21, 22] are based on a reference classical path, sometimes with the solvent equilibrated, with quantum features subsequently added. In our view, these suffer from some difficulties (for discussion, see Refs. [3–5]). The Russian school [26, 53, 54] pioneered in the late '60s a perspective where the solvent played a key role in tunneling PT. While the approach of Refs. [26, 53, 54] shares some features with our perspective [1], there are several importance differences. First, the H-bond vibration is always treated classically, rather than by the general quantum mechanical treatment of Ref. 1. Most importantly, it is assumed in the Russian school approach that PT is electronically diabatic, in very strong contrast to the present assumption of electronic adiabaticity, a difference which leads to significant differences in experimental predictions [5b]. Our view is that most tunneling regime PT reactions are electronically adiabatic, with a rather strong electronic coupling – an aspect related to the fact that chemical bonds are broken and made – and that an electronically diabatic description is inappropriate [5b]. Finally, we should note that so-called proton -coupled electron transfer is a quite different reaction class, involving transfer of an electron over larger distances, where different considerations apply [34, 35].

10.3.1 General Nonadiabatic Proton Transfer Perspective and Rate Constant

Figure 10.13 displays the physical picture for nonadiabatic PT, with a fixed H-bond separation, a constraint later relaxed. The system free energy as a function of the proton coordinate – involving the electronically adiabatic proton potential – is displayed with the reactant and product *diabatic* proton vibrational states indicated, for three values of the solvent coordinate characterizing different environmental configurations: reactant state **R** (Fig. 10.13(a)), transition state **TS** (Fig. 10.13(b)), and product state **P** (Fig. 10.13(c)). The **R** and **P** proton diabatic levels are found by solving the nuclear Schrödinger equation for the proton in each of the reactant and product wells, respectively. (Proton *adiabatic* levels are found by solving the Schrödinger equation for the entire proton potential.) The evolving diabatic ground proton vibrational states define free energies as a function of the environment rearrangement, shown in Fig. 10.13(d). At the thermally activated TS position, the proton reactant diabatic vibrational state is in resonance with the corresponding ground proton product state, and the proton can thus tunnel.

The rate constant expressions quoted in this section were originally derived [1b,1c] from a dynamic perspective, i.e. from the time integral of the time correlation function (tcf) of the probability flux associated with the tunneling PT, and subsequently via other approaches [1a, d, e] including a curve crossing perspective [1a]. Beyond the rate constant results themselves, the dynamic tcf approach is revealing in connection with the passage from coherent tunneling – which leads to spectroscopic tunneling splitting but no kinetics – to the incoherent, kinetic limit where a tunneling PT rate constant exists; interestingly only a very small coupling of the PT system to the environment is required for this transition [1b, c].

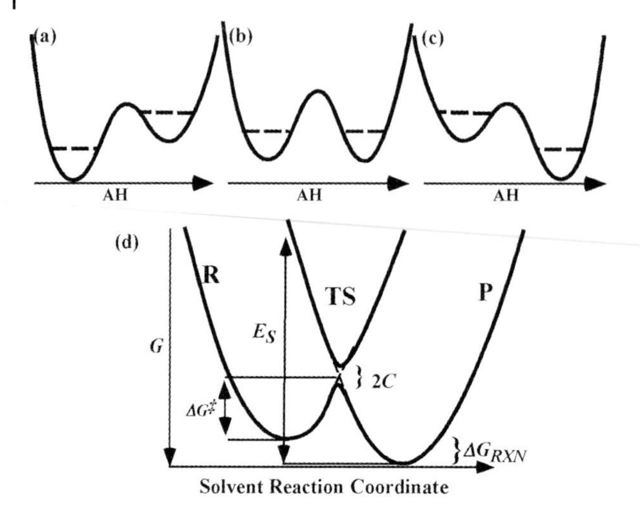

Figure 10.13 Free energy curves versus the proton position at (a) the reactant R, (b) transition state TS, and (c) product state P solvent configurations for nonadiabatic PT. In each case, the ground diabatic proton vibrational energy levels are indicated for both the reactant and product proton wells. (d) Free energy curves versus the solvent reaction coordinate for both diabatic proton levels displayed in (a–c).

The rate constant for nonadiabatic PT between reactant and product proton ground vibrational states with the H-bond separation Q fixed is [1, 26]

$$k = \frac{C^2}{\hbar}\sqrt{\frac{\pi}{E_S RT}}\exp\left[-\frac{\Delta G^{\ddagger}}{RT}\right] \tag{10.26}$$

where the free energy barrier $\Delta G^{\ddagger}$ is

$$\Delta G^{\ddagger} = \frac{(\Delta G_{RXN} + E_S)^2}{4E_S} \tag{10.27}$$

and E_S is the solvent reorganization (free) energy [55]. The tunneling probability is governed by the square of the proton coupling C, defined as one-half the resonance proton vibrational splitting [55], cf. Fig. 10.13(d). C increases as the H-bond separation decreases due to the increased tunneling probability for a smaller proton barrier; its Q dependence is predominantly linear exponential [1]

$$C_L(Q) = C_{eqL}\exp[-a_L(Q - Q_{eq})]\,; \quad C_{eqL} = C_L\,(Q_{eq}) \tag{10.28}$$

where Q_{eq} is the equilibrium H-bond separation in the reactant state and a_L is the exponent characterizing the exponential dependence(L=H, D, and T). The mass dependence in Eq. (10.28) is contained within a_L and C_{eqL}. In particular, a_L is expected to be of the form $a_L \propto \sqrt{m_L}$ (e.g. $a_D \approx \sqrt{2}a_H$, and $a_T \approx \sqrt{3}a_H$), with typical values α_H~25–35 Å^{-1} [1]. We limit our discussion to the single most important mode modulating the proton barrier – the H-bond mode; but other modes that regulate the barrier through which the proton must tunnel, e.g. H-bond bending modes [4, 28, 56], can be treated in a similar fashion. For simplicity, a harmonic H-bond vibration $U_Q(Q) = U_{Q,eq} + \frac{1}{2} m_Q \omega_Q^2 (Q - Q_{eq})^2$ has been assumed, with an effective mass m_Q and vibrational frequency ω_Q, and for the moment, we retain the restriction to PT between ground proton vibrational levels in the reactant and product.

For extremely low temperatures $\hbar\omega_Q \gg RT$, the Q vibrational mode resides primarily in its ground state, and the PT rate expression is [1]

$$k_L = \frac{C_{00}^2}{\hbar}\sqrt{\frac{\pi}{E_S RT}} \exp\left[-\frac{(\Delta G_{RXN} + E_S)^2}{RT4E_S}\right] \tag{10.29}$$

which is similar to Eq. (10.26) except that the proton coupling C is replaced by its ground Q-vibrational state quantum average

$$C_{00}^2 = |\langle 0|C(Q)|0\rangle|^2 = C_{eqL}^2 \exp\left[\alpha_L \Delta Q + \frac{(E_{\alpha L} - E_Q)}{\hbar\omega_Q}\right] \tag{10.30}$$

Here $\Delta Q = Q_{P,eq} - Q_{R,eq}$ is the difference in the P and R equilibrium Q positions, and $E_Q = \frac{1}{2} m_Q \omega_Q^2 \Delta Q^2$ is the associated reorganization energy. $E_{\alpha L}$ is a quantum energy associated with the tunneling probability's variation with the Q vibration

$$E_{\alpha L} = \hbar^2 \alpha_L^2 / 2m_Q \tag{10.31}$$

Even with $\Delta Q = 0$ ($E_Q = 0$), C is increased from its fixed value $C(Q_{eq})$ by $\exp(E_{\alpha L}/\hbar\omega_Q)$: there is a finite probability of smaller H-bond separations even at low T due to Q's the zero point motion. The ratio $E_{\alpha L}/\hbar\omega_Q$ identifies $E_{\alpha L}$ as a quantum energy scale for the localization of the Q wavefunction [1, 5]. When $E_{\alpha L}/\hbar\omega_Q \ll 1$, the coupling C is essentially that for fixed $Q = Q_{eq}$. As $E_{\alpha L}/\hbar\omega_Q$ increases, C increases, corresponding to increased quantum accessibility of smaller Q values.

As T is increased, contributions from excited Q vibrational states become more significant. For moderate to high temperatures $\hbar\omega_Q$~RT and $\hbar\omega_Q \ll RT$, the focus here, many Q vibrational states are thermally excited. In this regime, the PT rate expression [1] for $\Delta Q = 0$ is

$$k_L = \frac{\langle C^2 \rangle}{\hbar}\sqrt{\frac{\pi}{(E_S + \tilde{E}_{\alpha L})RT}} \exp\left[-\frac{\Delta G_L^\ddagger}{RT}\right] \tag{10.32}$$

where the reaction activation free energy is given by

$$\Delta G_L^\ddagger = \frac{(\Delta G_{RXN} + E_S + E_{\alpha L})^2}{4(E_S + \tilde{E}_{\alpha L})} \tag{10.33}$$

$$\tilde{E}_{\alpha \mathrm{L}} = E_{\alpha \mathrm{L}} \tfrac{1}{2} \beta \hbar \omega_Q \coth(\tfrac{1}{2} \beta \hbar \omega_Q) \tag{10.34}$$

The square proton coupling factor in Eq. (10.32) is the thermal average over the Q vibrational states [1]

$$\langle C^2 \rangle = C_{\mathrm{L}}^2(Q_{\mathrm{eq}}) \exp\left(2 \frac{E_{\alpha \mathrm{L}}}{\hbar \omega_Q} \coth(\tfrac{1}{2} \beta \hbar \omega_Q) \right) \tag{10.35}$$

The inverse length scale of the coupling, α_{L}, contributes significantly (via $E_{\alpha \mathrm{L}}$, Eq. (10.31)) to this average square coupling. In particular, the sensitivity of the coupling of C to Q dynamics is characterized by the ratio $E_{\alpha \mathrm{L}}/\hbar \omega_Q$ so that $\langle C^2 \rangle$ increases as this ratio increases.

The energy factor $E_{\alpha \mathrm{L}}$ also appears in the reaction barrier in Eq. (10.33), as an energetic contribution to that barrier due to thermal activation of the H-bond mode. The isotopic dependence $E_{\alpha \mathrm{L}} \propto m_{\mathrm{L}}$ in the barrier plays a key role in isotope and temperature effects, but before recounting these effects, we describe the inclusion of excited proton vibrational levels [1, 5].

In the discussion above, PT has been assumed to occur from the reactant ground proton diabatic vibrational state to the corresponding state in the product. However, for very exothermic or endothermic reactions ($|\Delta G_{\mathrm{RXN}}| \geq E_{\mathrm{S}} + \hbar \omega_Q$), excited proton vibrational states can become important: the proton can be transferred into an excited proton product vibrational state for an exothermic case and from a thermally excited reactant proton vibration for an endothermic case. Accordingly, free energy curves corresponding to excited diabatic proton vibrational states are added to the ground diabatic proton vibrational states in Fig. 10.13(d), displayed as Fig. 10.14. Each free energy curve corresponds to a diabatic proton vibrational level. There are now several transitions possible, with a specific tunneling resonance situation associated with each TS or intersection of the proton diabatic free energy curves. We now discuss a number of such transitions.

Figure 10.15 displays the TS proton potentials for the four transitions in Fig. 10.14 [57]. Figure 10.15(b) shows the symmetric proton potential for the ground-state to ground-state (0–0) transition and the corresponding first excited state transition (1–1). The 1–1 transition will have a higher transition probability (larger C) because the excited proton level is closer to the proton barrier top. But this

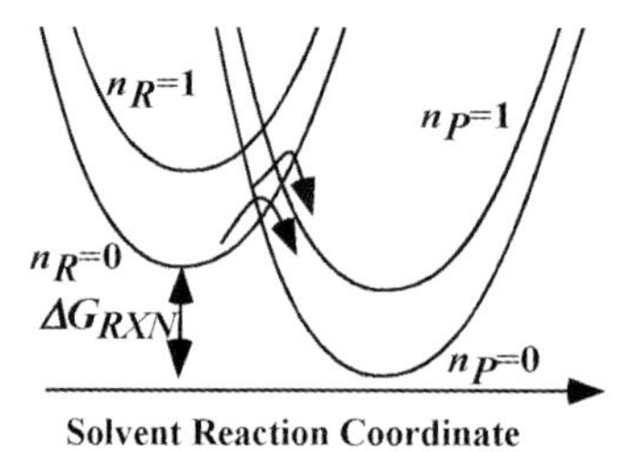

Figure 10.14 Proton diabatic free energy curves versus the solvent reaction coordinate for individual reactant (n_{R}) and product (n_{P}) proton vibrational states. Several transitions are qualitatively indicated.

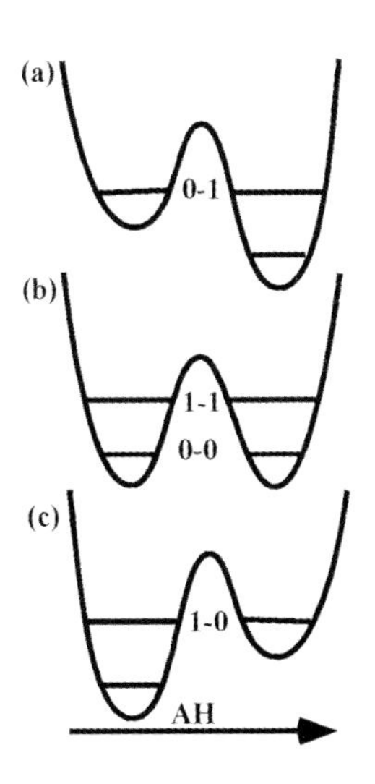

Figure 10.15 Proton potentials for the solvent coordinate TS for four proton vibrational transitions (n_R–n_P): (a) 0–1, (b) 0–0 and 1–1, and (c) 1–0. The lines indicate diabatic proton vibrational levels.

increased 1–1 transition tunneling probability comes at a cost of 1 quantum of proton vibration excitation, which is added to the activation energy. Figures 10.15(a) and (c) display the proton potentials with 0–1 and 1–0 transitions, respectively. Both will have an increased tunneling probability compared with the 0–0 transition due to a smaller proton barrier through which to tunnel. Reactant proton vibrational mode thermal excitation leads to the 1–0 transition, assisting endothermic reactions, while extra solvent activation leads to the 0–1 transition, assisting exothermic reactions. The interplay between the cost of thermal excitation and the gain from increased tunneling probability, and their isotope dependence, plays a significant role in KIEs.

In a general formulation, excited proton vibrational states are included in the PT rate as a sum over all state-to-state PT rates $k_{n_R \to n_P}$ from a proton reactant state n_R to a product state n_P

$$k_L = \sum_{n_R} \sum_{n_P} P_{n_R} k_{n_R \to n_P} \tag{10.36}$$

where each state-to-state rate is weighted by the reactant state thermal occupation $P_{n_R} = \exp(-\beta E_{n_R}) / \sum_{n_R} \exp(-\beta E_{n_R})$, and $E_{n_R} = \hbar\omega_R(n_R + 1/2)$. $k_{n_R \to n_P}$ is a modified version of Eq. (10.32).

$$k_{n_R \to n_P} = \frac{\left\langle C^2_{n_R,n_P} \right\rangle}{\hbar} \sqrt{\frac{\pi}{(E_S + \tilde{E}_{\alpha L})RT}} \exp\left[-\frac{\Delta G^{\ddagger}_{n_R,n_P}}{RT} \right] \tag{10.37}$$

where the reaction free energy barrier $\Delta G^{\ddagger}_{nR,nP}$ is transition-dependent

$$\Delta G^{\ddagger}_{n_R,n_P} = \frac{(\Delta G_{RXN} + n_P\hbar\omega_P - n_R\hbar\omega_R + E_S + E_{aL})^2}{4(E_S + \tilde{E}_{aL})} \tag{10.38}$$

The proton coupling $C_{eqL}(n_R>n_P)$ is also transition-dependent, and increases as the quantum numbers n_R and n_P increase (more properly, the difference) because the proton coordinate barrier's width and height are smaller as the proton level sits higher in either well [5]. The Q dependence of the coupling can still be approximated by the same form in Eq. (10.28) [5]

$$C_{n_R,n_P}(Q) = C_{n_R,n_P}(Q_{eq})\exp[-a_L(Q - Q_{eq})] \tag{10.39}$$

such that the thermal average of C^2 for Eq. (10.37) is accordingly

$$\left\langle C^2_{n_R,n_P}\right\rangle = C^2_{n_R,n_P}(Q_{eq})\exp\left(2\frac{E_{aL}}{\hbar\omega_Q}\coth(\tfrac{1}{2}\beta\hbar\omega_Q)\right) \tag{10.40}$$

Figure 10.16 displays the logarithm of the rate constant versus ΔG_{RXN} behavior using Eqs. (10.36) and (10.37) for an example PT reaction (T = 300 K; $\hbar\omega_Q$ = 300 cm^{-1}, $V^{\ddagger}$ = 25 kcal mol^{-1}, E_S = 8 kcal mol^{-1}, m_Q = 20 amu, $\hbar\omega_H$ = 3200 cm^{-1}, $\hbar\omega_H^{\ddagger}$ = 2700 cm^{-1}, and a_H = 28 Å^{-1}). Contributions from individual transitions ($n_R - n_P$) are also indicated. In particular, Fig. 10.16 describes the dominance of the 0–0 transitions near ΔG_{RXN} = 0 and the increased contributions from the 0–1 transition for exothermic reactions and from the 1–0 transition for endothermic reactions, as discussed above. Indeed, the rate constants for excited proton vibrational transitions will dominate for more asymmetric reactions. These aspects have an important influence on activation free energy behavior: the full rate constant continuously increases going from endo- to exothermic reactions due to the increased contributions of 0–n_P transitions as the reaction becomes more exothermic, while, the drop in rate constant with increased reaction endothermicity is decreased with contributions from n_R–0 transitions.

In the next subsection, we will discuss an analytic FER for nonadiabatic PT, mainly for use in discussing KIEs. Figure 10.17 shows an application to the PT rate constants themselves and indicates that such a description captures the rate

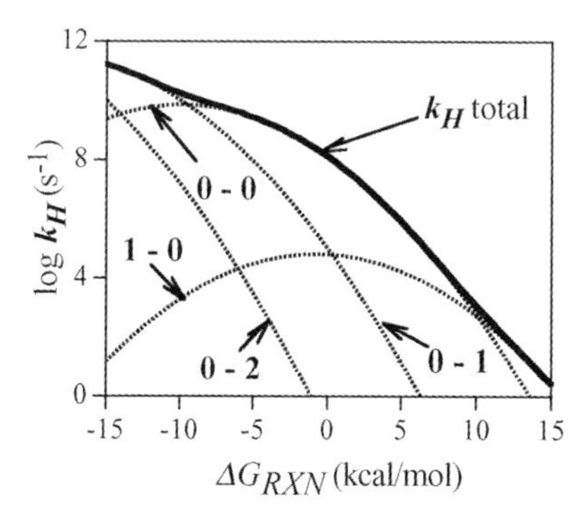

Figure 10.16 Log k versus ΔG_{RXN} (T = 300 K) for H including excited proton vibrational states (solid lines). Dotted lines indicate individual contributions from 0–0, 0–1, 1–0, and 0–2 transitions. Rate constants were calculated with Eqs. (10.36) and (10.37). (T = 300 K; $\hbar\omega_Q$ = 300 cm^{-1}, $V^{\ddagger}$ = 25 kcal mol^{-1}, E_S = 8 kcal mol^{-1}, m_Q = 20 amu, $\hbar\omega_H$ = 3200 cm^{-1}, $\hbar\omega_H^{\ddagger}$ = 2700 cm^{-1}, and a_H = 28 Å^{-1}).

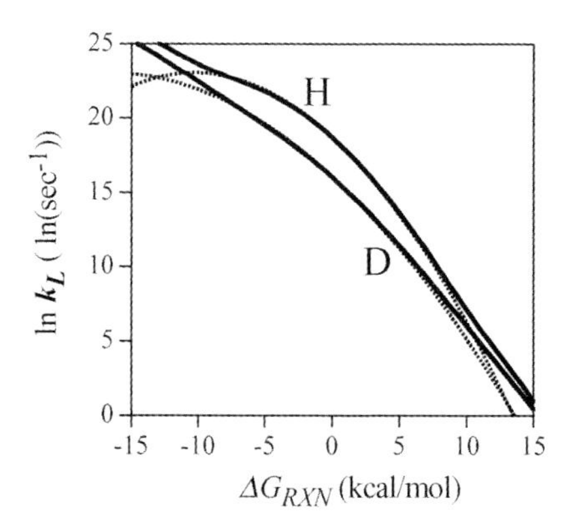

Figure 10.17 ln k_H and ln k_D versus ΔG_{RXN} (bold lines) for the same system in Fig. 10.16 (parameters for D are appropriately mass scaled). Dotted lines are Eq. (10.36) using Eq. (10.37) to evaluate $\bar{\alpha}'_{oL}$.

constant behavior with reaction asymmetry as long as the asymmetry is not too large; we return to this issue below.

10.3.2
Nonadiabatic Proton Transfer Kinetic Isotope Effects

We now review the KIE behaviors that follow from this nonadiabatic PT formalism, focusing on the four KIE observables (i)–(iv) analyzed in Section 10.2.3 [5]. The KIE magnitude and its variation with reaction asymmetry are first summarized, which serves to demonstrate the importance of excited proton and H-bond vibrational states, and the temperature dependence is then reviewed. We conclude with discussion of the Swain–Schaad relations. As will be seen, the present perspective gives several KIE results which are similar to that of traditional perspectives not including tunneling; there are however important exceptions to this statement which we will discuss. We pause to emphasize that the same basic nonadiabatic PT formalism can be applied to *H atom* transfers (for which there will generally be only weak coupling to the environment). For H atom transfers, the predicted KIEs can be very large ~10^4 [1e] and non-Arrhenius T dependence is marked – as opposed to the modest KIE values and Arhennius T behavior discussed here for PT; in such cases, the signatures of tunneling are obvious, in marked contrast to the situation for PT.

10.3.2.1 Kinetic Isotope Effect Magnitude and Variation with Reaction Asymmetry

Traditional treatments of KIEs, mentioned at the beginning of Section 10.3, including those invoking tunneling along a minimum energy path, predict that the KIE is maximal for a symmetric reaction $\Delta G_{RXN} = 0$ [17]. As now recounted, a similar behavior results in the present perspective. We will also pay attention to the tunneling PT KIE magnitude (focusing on tunneling PT systems that give smaller KIE magnitudes which might be confused with *non*tunneling PT).

Before presenting the KIE variation with reaction asymmetry for nonadiabatic PT, it will prove useful to first discuss the individual isotope PT rate constant Eq. (10.36)'s variation with reaction asymmetry, which must include tunneling prefactor terms as well as the activation free energy. This behavior was analyzed up through quadratic terms in ΔG_{RXN} [5] to find

$$\ln k_L = \ln k_L^o - \frac{\bar{a}_{oL}\Delta G_{RXN}}{RT} - \frac{\bar{a}'_{oL}\Delta G_{RXN}^2}{2RT} \tag{10.41}$$

where $k^o{}_L$ is the symmetric reaction $\Delta G_{RXN} = 0$ rate constant and $\bar{a}_{oL}$ and $\bar{a}'_{oL}$ are respectively the familiar Brønsted coefficient [13–16] and its derivative evaluated for the symmetric reaction

$$\bar{a}_{oL} = -RT\frac{\partial \ln k_L}{\partial \Delta G_{RXN}}\bigg|_o \; ; \quad \bar{a}'_{oL} = -RT\frac{\partial^2 \ln k_L}{\partial \Delta G_{RXN}^2}\bigg|_o \tag{10.42}$$

This analysis anticipates that the 0–0 rate k_{L00} will have a significant contribution near $\Delta G_{RXN} = 0$, and thus the rate expression in Eq. (10.36) can be written

$$k_L = k_{L00}\rho_L \; ; \quad \rho_L = \sum_{n_R}\sum_{n_P} F_{n_R,n_P} \tag{10.43}$$

in terms of k_{L00}

$$k_{L00} = \frac{\langle C_{0,0}^2 \rangle}{\hbar}\sqrt{\frac{\pi}{(E_S + \tilde{E}_{aL})RT}}\exp\left[-\frac{\Delta G^{\ddagger}_{L0,0}}{RT}\right] \tag{10.44}$$

and the rate enhancement ρ_L due to excited proton vibrational states. The coefficient of each transition $F_{nR,nP}$ is [5]

$$F_{n_R,n_P} = P_{n_R}\exp\left[\frac{\pi(\hbar\omega_R n_R + \hbar\omega_P n_P)}{\hbar\omega^{\ddagger}}\right]\exp\left[-\frac{\Delta\Delta G^{\ddagger}_{n_R,n_P}}{RT}\right] \tag{10.45}$$

$\Delta\Delta G^{\ddagger}_{n_R,n_P}$ is the difference between the general reaction barrier $\Delta G^{\ddagger}_{n_R,n_P}$ Eq. (10.38) and that, $\Delta G^{\ddagger}_{L0,0}$, for the 0–0 case

$$\begin{aligned}\Delta\Delta G^{\ddagger}_{n_R,n_P} &= \Delta G^{\ddagger}_{n_R,n_P} - \Delta G^{\ddagger}_{L0,0} \\ &= \frac{(n_P\hbar\omega_P - n_R\hbar\omega_R)(2(\Delta G_{RXN} + E_S + E_{aL}) + n_P\hbar\omega_P - n_R\hbar\omega_R)}{4(E_S + \tilde{E}_{aL})}\end{aligned} \tag{10.46}$$

Here, properties from the TS proton potential (cf. Fig. 10.1) are included, i.e. the curvatures in the wells, ω_R and ω_P, and at the top of the barrier $\omega^{\ddagger}$, as a means to relate $C_{nR,nP}$ to $C_{0,0}$ [5].

The Brønsted coefficient $\bar{\alpha}_{\mathrm{oL}}$ and its derivative $\bar{\alpha}'_{\mathrm{oL}}$ in Eq. (10.41) can thus be written as a sum of the 0–0 case plus a correction due to the contribution of excited proton vibrational transitions:

$$\bar{\alpha}_{\mathrm{oL}} = \frac{1}{2}\frac{E_{\mathrm{S}} + E_{\alpha\mathrm{L}}}{\left(E_{\mathrm{S}} + \tilde{E}_{\alpha\mathrm{L}}\right)} + \left\langle \frac{\partial \Delta\Delta G^{\ddagger}_{n_{\mathrm{R}},n_{\mathrm{P}}}}{\partial \Delta G_{\mathrm{RXN}}} \right\rangle_{\mathrm{F}} \tag{10.47}$$

$$\bar{\alpha}'_{\mathrm{oL}} = \frac{1}{2\left(E_{\mathrm{S}} + \tilde{E}_{\alpha\mathrm{L}}\right)} - \left[\left\langle \left(\frac{\partial \Delta\Delta G^{\ddagger}_{n_{\mathrm{R}},n_{\mathrm{P}}}}{\partial \Delta G_{\mathrm{RXN}}}\right)^{2} \right\rangle_{\mathrm{F}} - \left\langle \frac{\partial \Delta\Delta G^{\ddagger}_{n_{\mathrm{R}},n_{\mathrm{P}}}}{\partial \Delta G_{\mathrm{RXN}}} \right\rangle_{\mathrm{F}}^{2}\right] \tag{10.48}$$

where $<\ldots>_{\mathrm{F}}$ denotes a certain average over the vibrationally excited proton states for the symmetric reaction (cf. Eqs. (10.42)–(10.44)) [5]; Eqs. (10.47) and (10.48) reduce to their respective quantities from the FER in Eq. (10.33) if these excitations are not taken into account. (As discussed in Ref. [5], the symmetric reaction Brønsted coefficient $\bar{\alpha}_{\mathrm{oL}}$ deviates slightly from the value of 1/2, an effect which vanishes almost entirely in a more refined treatment.) One can explicitly relate $\bar{\alpha}_{\mathrm{oL}}$ and $\bar{\alpha}'_{\mathrm{oL}}$ to the proton excitations by considering the quantum number dependence in Eq. (10.46) yielding

$$\bar{\alpha}_{\mathrm{oL}} = \frac{1}{2}\frac{\left\{E_{\mathrm{S}} + E_{\alpha\mathrm{L}} + \langle n_{\mathrm{P}}\hbar\omega_{\mathrm{P}} - n_{\mathrm{R}}\hbar\omega_{\mathrm{R}}\rangle_{\mathrm{F}}\right\}}{\left(E_{\mathrm{S}} + \tilde{E}_{\alpha\mathrm{L}}\right)} \tag{10.49}$$

$$\begin{aligned}\bar{\alpha}'_{\mathrm{oL}} = {} & \frac{1}{2\left(E_{\mathrm{S}} + \tilde{E}_{\alpha\mathrm{L}}\right)} \\ & \times \left\{1 - \frac{1}{2RT\left(E_{\mathrm{S}} + \tilde{E}_{\alpha\mathrm{L}}\right)}\left[\left\langle \left(n_{\mathrm{P}}\hbar\omega_{\mathrm{P}} - n_{\mathrm{R}}\hbar\omega_{\mathrm{R}}\right)^{2} \right\rangle_{\mathrm{F}} - \langle (n_{\mathrm{P}}\hbar\omega_{\mathrm{P}} - n_{\mathrm{R}}\hbar\omega_{\mathrm{R}})\rangle_{\mathrm{F}}^{2}\right]\right\}\end{aligned} \tag{10.50}$$

Finally, from development of the isotopic rate constant up through quadratic order in ΔG_{RXN} Eq. (10.41), the logarithmic KIE is [5]

$$\ln(k_{\mathrm{H}}/k_{\mathrm{D}}) = \ln(k^{\mathrm{o}}_{\mathrm{H}}/k^{\mathrm{o}}_{\mathrm{D}}) - \frac{(\bar{\alpha}'_{\mathrm{oH}} - \bar{\alpha}'_{\mathrm{oD}})\Delta G^{2}_{\mathrm{RXN}}}{2RT} \tag{10.51}$$

where the position of the maximum in a KIE versus reaction asymmetry plot (H vs. D) occurs for a symmetric reaction $\Delta G_{\mathrm{RXN}} = 0$, a direct result of the isotope independence of $\bar{\alpha}_{\mathrm{oL}} \approx 1/2$. The asymmetry variation is governed by the isotopic difference $\bar{\alpha}'_{\mathrm{oH}} - \bar{\alpha}'_{\mathrm{oD}}$ in the Brønsted slope derivative Eq. (10.48), which is positive so that the KIE diminishes with increasing reaction asymmetry. We will illustrate the usefulness of this analytic result in a moment.

The maximal KIE behavior for tunneling PT is illustrated numerically in Fig. 10.18, which employs the full rate constant expression (10.36), with Eq. (10.37), for the H and D isotopes, evaluated for the same PT system used to generate Fig. 10.16, (D parameters are appropriately mass scaled) [5]. Also displayed in Fig. 10.16 is the analytical behavior Eq. (10.49), utilizing Eq. (10.48) for $\bar{a}'_{\mathrm{oL}}$ (dotted lines). The agreement between Eq. (10.51) and the actual behavior is quite close, although a breakdown ($|\Delta G_{\mathrm{RXN}}| > 10$ kcal mol^{-1}) is apparent (see also Fig. 10.17). The KIE is maximal for $\Delta G_{\mathrm{RXN}} = 0$ and falls off with increasing reaction asymmetry. This maximal KIE behavior is due to increased excitation in both the proton and H-bond modes, excitations that become more facile with increased reaction asymmetry [5]. H-bond excitation benefits D more than H because the D tunneling probability is more sensitive to changes in Q: see the $E_{a\mathrm{L}}/\hbar\omega_Q$ ratio in Eq. (10.40) with $E_{a\mathrm{H}} < E_{a\mathrm{D}}$. The contribution from the excited states (cf. the second term in Eq. (10.48)) contributes about 20% to the coefficient difference for the model reactions examined. $\bar{a}_{oL}$ is larger for D than for H, since excited states are more easily accessed due to the smaller quanta $\hbar\omega_{\mathrm{D}} < \hbar\omega_{\mathrm{H}}$[5].

The KIE magnitude in Fig. 10.18 is actually fairly small compared to expectations for a PT tunneling reaction. In fact, the KIE magnitude for fairly mildly asymmetric reactions might be considered consistent with *nontunneling* PT. To emphasize this important point, the KIE with a slightly lower H-bond vibrational frequency $\hbar\omega_Q$=275 cm^{-1} is also included, where the KIE magnitude decreases by a factor of 3, emphasizing the sensitivity of the KIE to the donor–acceptor frequency. Even for the symmetric reaction, the KIE is far smaller than traditionally expected for a tunneling reaction. Indeed, the KIE behavior versus ΔG_{RXN} for this case cannot be distinguished from that for nontunneling PT.

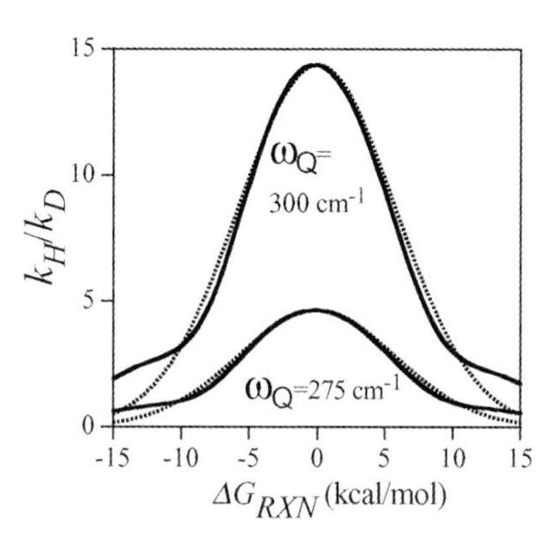

Figure 10.18 $k_{\mathrm{H}}/k_{\mathrm{D}}$ for a nonadiabatic PT system with $\hbar\omega_Q$= 300 cm^{-1} and $\hbar\omega_Q$= 275 cm^{-1} (solid lines). Dotted lines use the analytical form for the KIE versus ΔG_{RXN} behavior in Eq. (10.51), using Eq. (10.48) to evaluate $\bar{a}_{\mathrm{oL}}$.

10.3.2.2 Temperature Behavior

We now turn to the T dependence of the rate in Eq. (10.36) [5], which in general is certainly not Arrhenius. However, one must bear in mind that most experiments are conducted over a reasonably restricted temperature range where the behavior can appear to be Arrhenius, even though the PT is completely tunneling in character.

We first consider the individual transition rates Eq. (10.37) which are weighted to give the overall rate constant Eq. (10.36). There are two contributions to the T dependence of these individual transition rates which dominate in Eq. (10.37). The first is contained within the exponential containing the reaction free energy barrier, which gives Arrhenius behavior if the components of the reaction barrier, i.e. $\tilde{E}_{\alpha L}$ (see Eq. (10.34)) and E_S [5, 47], have only a minor T dependence. An important point in this connection is that the impact of any such T dependence is suppressed if the reorganization energy is significant ($E_S > E_{\alpha L}$), as it often will be for PT in a polar environment. The second contribution comes from the thermally averaged square proton coupling Eq. (10.40), and in principle is not Arrhenius. In addition to these T dependences for the individual transition rate constants, the thermal sum over excited proton transitions for the full rate in Eq. (10.36) is clearly also not in principle Arrhenius. Altogether, these contributions give rise to a nonlinear T dependence in an Arrhenius plot, as expected for tunneling PT [8, 11, 13, 14, 58]. (We immediately stress that this is *not* a non-Arrhenius behavior associated with a transition from high temperature, classical "over the barrier" PT to tunneling PT at lower temperatures; the entire description here is in the tunneling regime.) Nonetheless, the T dependence of the tunneling rate constant Eq. (10.36) was shown [5] to be effectively linear in an Arrhenius plot for a limited but non-negligible temperature range [59]. This is now discussed.

In the analysis [5], the PT rate in proximity to a specific temperature T_o is written in an Arrhenius form

$$k_L = k_L(T_o)\exp[-(\beta - \beta_o)E_{A_L}] \tag{10.52}$$

where the Arrhenius intercept is just the extrapolation from the rate at $T = T_o$ to infinite temperature: $A_L = k_L(T_o)\exp[\beta_o E_{A_L}]$, and E_{A_L} is determined by the slope in an Arrhenius plot.

For illustrative purposes, the same system as in Fig. 10.16 was taken, and T was varied (T = 300–350 K), while keeping the reaction asymmetry constant, $\Delta G_{RXN} = 0$. The apparent Arrhenius rate and KIE behavior obtained in this limited T range are displayed in Fig. 10.19. The apparent activation energies for H and D differ considerably, with E_{AD} almost twice E_{AH}: E_{AH} = 5.7 kcal mol^{-1} and E_{AD} = 10.6 kcal mol^{-1}; this results in a significant effective activation energy for the KIE $E_{AD} - E_{AH}$ = 5.0 kcal mol^{-1}, displayed in Fig. 10.19(b). These slopes can be quantitatively analyzed [5] to determine the contributions from the H-bond and proton vibration excitations.

For this determination, the expansion in Eq. (10.43) of the rate constant in terms of the 0–0 transition and the contribution from excited proton vibrational

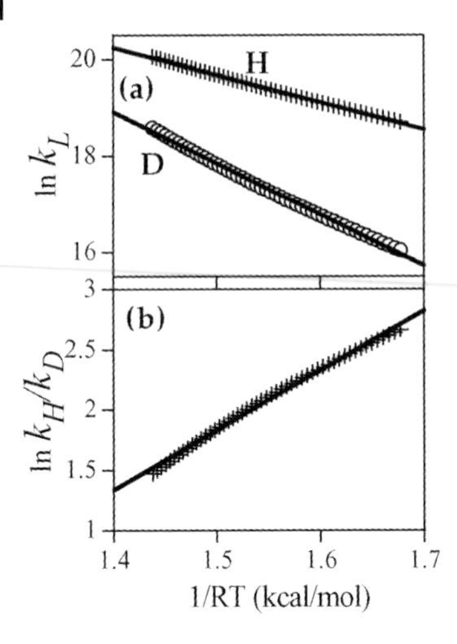

Figure 10.19 (a) lnk_H (+) and lnk_D (o) versus $1/RT$ (T = 300–350 K) for the PT system in Fig. 10.16 with $\Delta G_{RXN} = 0$. (b) $\ln(k_H/k_D)$ (+) for rate constants in (a). The lines are linear fits to the points. The slopes of the lines give the activation energies (a) $E_{AH} = 5.7$ kcal mol^{-1}; $E_{AD} = 10.6$ kcal mol^{-1} and (b) KIE $E_A = 5.0$ kcal mol^{-1}.

states is also used. Here, the 0–0 rate k_{L00} and the excited state factor ρ_L are evaluated at the mid-range temperature T_o $k_{oL} = k_{L00}(T_o)$ (e.g. $T_o = 325$ K in Fig. 10.19), such that $k_L(T_o) = \rho_L k_{oL}$. The parameters in the resulting Arrhenius form Eq. (10.52) for the rate constant in this limited T region have the following forms [5]

$$A_L = k_{oL}\rho_L(T_o)\exp(\beta_o E_{A_L});$$

$$E_{AL} = E_{aL}[\coth^2(\beta_o\hbar\omega_Q/2) - 1] + \Delta G^{\ddagger}_{L0,0} + \left\langle \Delta\Delta G^{\ddagger}_{n_R,n_P}\right\rangle_L \quad (10.53)$$

Here $\Delta G^{\ddagger}_{L0,0}$ is the 0–0 reaction free energy barrier Eq. (10.33) and $\left\langle \Delta\Delta G^{\ddagger}_{n_R,n_P}\right\rangle_L$ is the activation free energy barrier contribution from excited proton states

$$\left\langle \Delta\Delta G^{\ddagger}_{n_R,n_P}\right\rangle_L = \frac{\sum_{n_R}\sum_{n_P} F_{o_{R,P}}\left(\hbar\omega_R n_R + \Delta\Delta G^{\ddagger}_{n_R,n_P}\right)}{\sum_{n_R}\sum_{n_P} F_{o_{R,P}}} \quad (10.54)$$

where the symmetric reaction transition coefficient is $F_{oR,P} = F_{nR,nP}(T = T_o)$ (cf. Eq. (10.45)). When compared to the numerical results in Fig. 10.19, Eq. (10.53) gives reasonable estimates for E_{AH} and E_{AD}, $E_{AH} = 6.1$ kcal mol^{-1} and $E_{AD} = 11.2$ kcal mol^{-1}, which differ by less than 10% from the obtained numerical values. The decomposition of these apparent activation energies via Eq. (10.46) is useful [5] in determining which contributions are most important and how these contributions change with T, $\hbar\omega_Q$, reaction asymmetry, and solvent reorganization energy E_S, as now reviewed.

The first term in Eq. (10.53) is the activation energy contribution from the thermally averaged square coupling $\langle C^2\rangle$ Eq. (10.40), and as such is extremely sensitive to parameters affecting the H-bond mode-tunneling coupling, namely T, $\hbar\omega_Q$, and E_{aL}. For the present system, this term dominates the activation energy for both H (60%) and D (66%). Furthermore, since $E_{aL} \propto m_L$ is mass sensitive, the predominant contribution to the activation energy difference determining the Arrhenius activitation energy factor for the KIE will be dominated by this first term. The coefficient $\{\coth^2(\beta_o\hbar\omega_Q/2) - 1\}$ in this term is *very* sensitive to T_o and $\hbar\omega_Q$, increasing drastically as T_o is increased or $\hbar\omega_Q$ decreases, and the ratio $\hbar\omega_Q/RT_o$ determines the relative contribution for this first term.

The second term in Eq. (10.53), the activation free energy barrier $\Delta G^{\ddagger}_{L0,0}$, is for the present system also significant for both H (39%) and D (25%). Of course, the magnitude of this term changes with reaction asymmetry, decreasing as the reaction goes from endo to exo-thermic (cf. Eq. (10.33)).

Finally, the last term in Eq. (10.53) for E_{AL} is the least important for the present example system, <1% for H and 9% for D. Its lack of importance correlates with the significance of the 0–0 transition in the overall rate, described here by ρ_L~1, $\rho_H = 1.004$ and $\rho_D = 1.25$. More generally, however, ρ_L will obviously change as the reaction becomes more asymmetric as well as with increasing T.

With the above individual isotope Arrhenius parameter results in Eq. (10.53), the Arrhenius parameters for the KIE can be examined. The KIE Arrhenius slope is thus determined by the isotopic difference of the apparent activation energies E_{AL}

$$E_{AD} - E_{AH} = E_{\alpha H}[\coth^2(\beta\hbar\omega_Q/2) - 1] + \left[\Delta G^{\ddagger}_{Do} - \Delta G^{\ddagger}_{Ho}\right] + \left[\langle\Delta\Delta G_{n_R,n_P}\rangle_D - \langle\Delta\Delta G_{n_R,n_P}\rangle_H\right] \tag{10.55}$$

For the chosen example, the first term arising from thermal H-bond excitation effects on the average squared coupling contributes, as predicted above, the most, 72%, the final term is next in significance at 20%, and the middle term contributes only 8%. The minimal significance of the (middle term) difference in 0–0 reaction barriers reflects the disparity $E_S > E_{\alpha L}$ that we have noted above should be expected for PT in a sufficiently polar environment. The increased contribution (third term) of the excited proton state contribution is due to the differential contribution of the 0–0 transition to the total rate, $\rho_H < \rho_D$ [60].

The Arrhenius intercept A_L in Eq. (10.53) is the extrapolation from the rate at $T = T_o$ $k_L(T_o) = k_{oL}\rho_L$ to infinite T, and thus the ratio of intercepts (H vs. D) is the extrapolation of the H vs. D KIE at T_o to infinite temperature

$$A_H/A_D = \left(k_H/k_D\right)_o \exp[-\beta_o(E_{AD} - E_{AH})] \tag{10.56}$$

where

$$\left(k_H/k_D\right)_o = \left(k_{oH}\rho_H/k_{oD}\rho_D\right) \tag{10.57}$$

The significant isotopic difference of Arrhenius intercepts, i.e. $A_H \neq A_D$, is a signature for a tunneling process [11]. For the Fig. 10.19(a) system, $A_H < A_D$, which is the case where the right-hand side of Eq. (10.56) is less than 1. If, however,

$E_{AD} - E_{AH}$ were ~1 for the system (and not ~5 kcal mol^{-1} as in Fig. 10.19), then one would instead have $A_H > A_D$; thus, tunneling itself imposes no relative size for the Arrhenius prefactor ratio. Clearly, the interplay between the $(k_H/k_D)_o$ magnitude and the difference $E_{AD} - E_{AH}$ in Eq. (10.56) determines whether $A_H > A_D$ or vice versa. This suggests that an alternate yet equivalent isotope analysis of Arrhenius plots would be to analyze $(k_H/k_D)_o$ and $E_{AD} - E_{AH}$ rather than A_H/A_D and $E_D - E_H$ [5]. The advantage is that there is a direct connection between the KIE magnitude and Arrhenius slope and the H-mode characteristics. Specifically, such an analysis [5] clearly demonstrates that larger KIE magnitudes result from longer (large acid–base separations) and more rigidly H-bonded complexes (large $\hbar\omega_Q/RT$), and small Arrhenius slopes arise with less probability of H-bond and proton mode excitation (i.e. low T and especially higher $\hbar\omega_Q/RT$ ratios).

10.3.2.3 **Swain–Schaad Relationship**

This final category of KIE behavior concerns the relative KIEs between the three isotopes H, D, and T via the Swain–Schaad relationship, cf. Eq. (10.23). *Deviation* from this relationship is regarded as a clear indication of tunneling [11, 61]. We noted in Section 10.2.3.3 that for the adiabatic PT regime, close adherence to the Swain–Schaad relations results from the fact that the isotope dependence arises from ZPE differences in the reaction free energy barrier, cf. Eq. (10.15). Further, these ZPE differences obey the same mass scaling used to derive the Swain–Schaad relations, cf. Eq. (10.16). In the nonadiabatic PT regime, the significant isotope dependencies lie in the tunneling *prefactor* (e.g $\langle C^2 \rangle$ in Eq. (10.32)), not in ZPEs, and thus lead to deviations from Swain–Schaad behavior as a clear signature of tunneling. However, in the remaining discussion, we show that the situation need not be so straightforward.

Traditionally, the Swain–Schaad relationship has been experimentally assessed by varying system parameters and plotting $\ln(k_H/k_T)$ versus $\ln(k_D/k_T)$ and determining whether this produces a line which goes through the origin and has a slope ~3.3 [8, 13, 14, 46]. However, Eq. (10.23) has also been assessed by plotting the ratio $\ln(k_H/k_T)/\ln(k_D/k_T)$ versus a system parameter, such as temperature [11]. If the ratio deviates significantly from the value of ~3.3, the PT system is said to be tunneling. Here we recount our examination [5] of whether or not a nonadiabatic tunneling PT system can exhibit the Swain–Schaad behavior in Eq. (10.23) and simultaneously have a KIE magnitude that is normally consistent with adiabatic nontunneling PT, i.e. $k_H/k_D \leq 6$ at $T = 300$ K [4, 8, 13, 14].

Figure 10.20(a) displays the ratio $\ln(k_H/k_T)/\ln(k_D/k_T)$ for the same system in Fig. 10.16 except that the H-mode frequency has been increased to $\hbar\omega_Q = 375$ cm^{-1} ($T = 300$ K) [5]. The Swain–Schaad ratio is at the expected traditional value, *but* the H versus D KIE in Fig. 10.20(b) is clearly large enough to indicate tunneling PT. Beyond this, the T-variation of the Swain–Schaad ratio for this system shown in Fig. 10.20(c) displays a distinct deviation from Eq. (10.23) for part of the temperature range, and thus also allows confirmation of tunneling PT [62]. The lesson of this example is that while the traditionally appropriate Swain–Schaad ratio can be

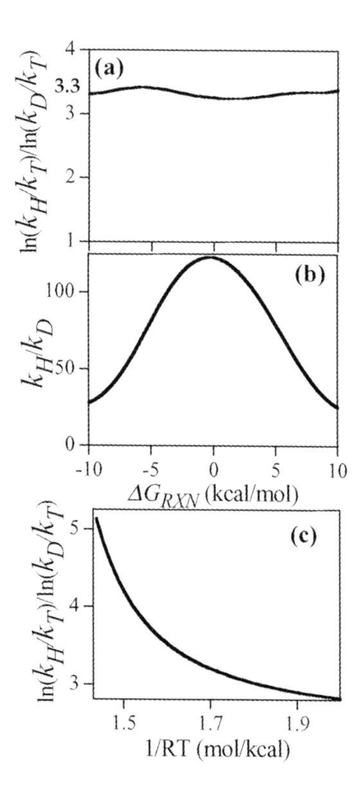

Figure 10.20 (a) Swain–Schaad ratio $\ln(k_H/k_T)/\ln(k_D/k_T)$ versus reaction asymmetry for a non-adiabatic PT system with $\hbar\omega_Q = 375\ \text{cm}^{-1}$. (b) k_H/k_D for system in (a). (c) $\ln(k_H/k_T)/\ln(k_D/k_T)$ versus $1/RT$ ($T = 250$–350 K) for the symmetric reaction in (a).

obtained – implying no tunneling – even for a tunneling PT system, it is difficult to find a tunneling PT reaction with *both* the Swain–Schaad ratio and KIE magnitude of the traditionally appropriate, i.e. nontunneling, values. This analysis indicates that the Schwain–Schaad ratio can be a clear indicator of tunneling [11], but examination over a parameter (e.g. *T*) range and simultaneous examination of the KIE magnitude can be necessary to establish this.

10.4 Concluding Remarks

In concluding this brief review, we content ourselves with a few remarks concerning perspectives for, and some limitations and extensions of, the theoretical treatments described.

As we have discussed within, many of the experimental measures thought to support traditional descriptions of proton transfer (PT) also follow from the nontraditional perspective that we have presented. It is therefore important to ask which experimental signatures can be employed to distinguish between the different perspectives. We have noted several of these throughout the chapter. For example, in the proton adiabatic regime, the KIE magnitude for a symmetric reaction for the standard picture is larger (≥5) than that for the nontraditional view (≤6); this point is discussed in Section 10.2.3.4 and in more detail elsewhere [4]. Indeed, experimental observations of PT KIEs exist [9c, 9e, 63, 64] which are consistent with the adiabatic PT predictions, but not with the standard picture. But we focus here instead on what we will call 'smoking guns', by which we mean qualitative experimental signatures which would follow from the nontraditional perspective but which would definitely not be predicted from traditional points of view.

A possible smoking gun would be infrared (IR)-induced PT, e.g. the driving of the acid–base PT reaction $AH + B \rightarrow A^- + HB^+$, Eq. (10.1), by the IR excitation of the AH proton vibration. In a traditional perspective, the only role of the solvent would be to thwart the IR-induced reaction by vibrational de-excitation of the AH vibration. In the nontraditional perspective however, the nonequilibrium solvent can *assist* the IR-induced reaction [65]. This is currently under study via detailed simulations.

In the nonadiabatic, tunneling PT regime, large KIE magnitudes and strong deviation from Swain–Schaad behavior for kinetic isotope effects are clear signatures of tunneling, as discussed in Section 10.3.2. However, these can follow not only from the nonadiabatic PT perspective critically involving nonequilibrium solvation reviewed here, but also from e.g. treatments where equilibrium solvation is assumed and some type of tunneling prefactor for the PT rate is included. A possible smoking gun for the nontraditional tunneling perspective where nonequilibrium solvation is central would be the observation of an inverted regime for PT. In analogy to electron transfer reactions [37], inverted regime behavior would be the eventual decrease of the rate constant with increasingly negative reaction free energy ΔG_{RXN}. Such behavior has been reported experimentally [66, 67]; indeed agreement has been achieved [66] between the experimental findings and certain of the tunneling rate constant equations given in Sec. 10.3, for both the transfer of a proton and a deuteron. This is quite encouraging, but the situation is not so straightforward. It is likely that special characteristics of the experimental system studied in Ref [66] are involved in allowing inverted regime behavior to be observed. Indeed, Fig. 10.16 illustrates that in the two internal solute coordinate description used throughout, inverted regime behavior is observed only if the 0–0 transition between ground proton vibrational states in the reactant and product is accounted for. However, Fig. 10.16 also shows that this behavior disappears if excited proton vibrational state contributions to the rate are included. In brief, before the 0–0 rate constant has significantly declined, the 0–1 rate contribution has increased to avoid the decline of the overall rate constant, and so on for the various 0–n_P transition contributions to higher product proton excited states. The special characteristics of the experimental system of Ref. [66] permitting inverted regime behavior – most likely related to the curvilinear character of the PT – and are still under study.

There are assorted limitations to the theoretical descriptions we have described within. We refer the interested reader to the original papers for discussion of these and here concern ourselves solely with electronic structure issues. In the analytical treatments described in this chapter, the electronic structure aspects were described in terms of a two valence bond (VB) basis, whose support was recounted in the Introduction. However, this kind of description is unlikely to apply to certain proton transfers, notably those involving carbon acids (cf. Ref. 42 in Ref. [3b]), where a third VB state will be necessary. Further electronic structure considerations are also necessary when treating excited electronic state proton transfer (ESPT) [68–71]. Among these is the origin of the well-known enhancement of the acidity in the excited state, i.e. a pK_a reduction of ~ 6 units, for aromatic acids ArOH. Traditionally this has been explained in terms of a charge transfer (CT), upon electronic excitation, from the oxygen (O) of the acidic OH moiety into the aromatic portion of the acid; this would create a partial positive charge on O, thereby repelling the proton. However, it has been shown [70, 71] that there is very little such CT produced in the electronic excitation of the ground state acid; instead the acidity enhancement arises from a strong CT on the *product* side of the excited state reaction, i.e. from the O of the conjugate base of the acid into the aromatic ring system. Beyond this key aspect of ESPT, a number of issues concerning the proper description of the electronic state during ESPT [68–71] remain to be clarified in the future, an important task given that much current experimental investigation of PT is via ultrafast spectroscopy of excited state systems[43b–d, 72].

Acknowledgments

This work was supported in part by NSF grants CHE-9700419, CHE-0108314, and CHE-0417570. The work of PK at ENS was supported by a grant from the French ANR.

References

1 (a) Borgis, D.; Hynes, J. T. *J. Phys. Chem.* **1996**, *100*, 1118–1128; (b) Borgis, D.; Hynes, J. T. *Chem. Phys.* **1993**, *170*, 315–346; (c) Borgis, D.; Lee, S.; Hynes, J. T. *Chem. Phys. Lett.* **1989**, *162*, 19–26; (d) Borgis, D.; Hynes, J. T. *J. Chem. Phys.* **1991**, *94*, 3619–3628; (e) Lee, S.; Hynes, J. T. *J. Chim. Phys.* **1996**, *99*, 7557–7567.

2 (a) Ando K.; Hynes, J. T. *J. Mol. Liq.* **1995**, *64*, 25–34; (b) Ando, K.; Hynes, J. T. *J. Phys. Chem. B* **1997**, 101, 10464–10478; (c) Ando, K.; Hynes, J. T. *J Phys Chem A* **1999**, *103*, 10398–10408; (d) Staib, A.; Borgis, D.; Hynes, J. T. *J. Chem. Phys.* **1995**, *102*, 2487–2505; (e) Ando, K.; Hynes, J. T. *Faraday Discuss.* **1995**, *102*, 435–441; (f) Ando, K.; Hynes, J. T. *Adv. Chem. Phys.* **1999**, *110*, 381–430.

3 (a) Kiefer, P. M.; Hynes, J. T. *J. Phys. Chem. A* **2002**, *106*,1834–1849; (b) Kiefer, P. M.; Hynes, J. T. *J. Phys. Chem. A* **2002**, *106*, 1850–1861.

4 Kiefer, P. M.; Hynes, J. T. *J. Phys. Chem. A* **2003**, *107*, 9022–9039.

5 (a) Kiefer, P. M.; Hynes, J. T. *J. Phys. Chem. A*, **2004**, *108*, 11793–11808; (b) Kiefer, P. M.; Hynes, J. T. *J. Phys. Chem. A* **2004**, *108*, 11809–11818.

6 Timoneda, J. J.; Hynes J. T. *J. Phys. Chem.* **1991**, *95*, 10431–10442.

7 Gertner, B. J.; Hynes, J. T. *Faraday Discuss.* **1998**, *110*, 301–322.

8 (a) Bell, R. P. *The Proton in Chemistry.* 2nd edn. Cornell University Press, Ithaca, NY 1973; (b) Caldin, E.; Gold, V., *Proton Transfer Reactions*, Chapman and Hall, London 1975; (c) Kresge, A. J. *Acc. Chem. Res.* **1975**, *8*, 354–360; (d) Hibbert, F. *Adv. Phys. Org. Chem.* **1986**, *22*, 113–212; (e) Hibbert F. *Adv. Phys. Org. Chem.* **1990**, *26*, 255–379.

9 (a) Kreevoy, M. M.; Konasewich, D. E. *Adv. Chem. Phys.* **1972**, *21*, 243–252; (b) Kreevoy, M. M.; Oh, S-w. *J. Am. Chem. Soc.* **1973**, *95*, 4805–4810; (c) Kresge, A. J.; Silverman, D. N. *Meth. Enz.* **1999**, *308*, 276–297; (d) Kresge, A.J.; Chen, H. J.; Chiang, Y. *J. Am. Chem. Soc.* **1977**, 99, 802–805; (e) Kresge, A.J.; Sagatys, D. S.; Chen, H. L. *J. Am. Chem. Soc.* **1977**, *99*, 7228–7233; (f) McLennan, D. J.; Wong, R. J. *Aust. J. Chem.* **1976**, *29*, 787–798; (g) Lee, I.-S.; Jeoung, E.H.; Kreevoy, M. M. *J. Am. Chem. Soc.* **2001**, *123*, 7492–7496.

10 (a) Bell, R. P., Goodall, D. M. *Proc. R. Soc. London Ser. A* **1966**, *294*, 273–297; (b) Dixon, J. E.; Bruice, T. C. *J. Am. Chem. Soc.* **1970**, *92*, 905 –909; (c) Pryor, W. A.; Kneipp, K. G. *J. Am. Chem. Soc.* **1971**, 93 5584 –5586; (d) Bell, R. P.; Cox, B. G. *J. Chem. Soc. B* **1971**, 783–785; (e) Bordwell, F. G.; Boyle, W. J. *J. Am. Chem. Soc.* **1975**, *97*, 3447–3452.

11 (a) Cha, Y.; Murray, C. J.; Klinman, J. P. *Science* **1989**, *243*, 1325–1330; (b) Kohen, A.; Klinman, J. P. *Acc. Chem. Res.* **1998**, 31, 397–404; (c) Kohen, A.; Klinman, J. P. *Chem. Biol.* **1999**, *6*, R191–R198; (d) Kohen, A.; Cannio, R.; Bartolucci, S.; Klinman, J. P. *Nature* **1999**, *399*, 496–499.

12 Kresge, A. J., in *Isotope Effects on Enzyme-Catalyzed Reactions*, Cleland, W. W.; O'Leary, M. H.; Northrop, D. B. (Eds.), University Park Press, Baltimore, MD 1977, pp. 37–63.

13 Melander, L.; Saunders, W. H. *Reaction Rates of Isotopic Molecules*, Wiley, New York 1980.

14 (a) Westheimer, F. H. *Chem. Rev.* **1961**, *61*, 265–273; (b) Melander, L. *Isotope Effects on Reaction Rates*, The Ronald Press Co., New York 1960.

15 (a) Hammond, G. S. *J. Am. Chem. Soc.* **1955**, *77*, 33–338; (b) Lowry, T. H.; Richardson, K. S. *Mechanism and Theory in Organic Chemistry.* 3rd edn. Harper Collins, New York 1987.

16 More O'Ferral, R. A. in *Proton Transfer Reactions*, Caldin, E.; Gold, V. (Eds.), Chapman and Hall, London 1975, pp. 201–261.

17 (a) Bell, R. P., *The Tunnel Effect in Chemistry*, Chapman and Hall, NewYork 1980; (b) Bell, R. P.; Sachs, W.H.; Tranter, R. L. *Trans. Faraday Soc.* **1971**, *67*, 1995–2003.

18 A KIE expression of the form in Eq. (10.2) neglects any mass dependence in

the prefactor, the contribution of which is usually negligible, except in the case of tunneling [8, 11, 17], as well as assumes that the proton primarily resides in it ground (stretching and bending)vibrational state.

19 (a) Marcus, R. A. *J. Phys. Chem.* **1968**, *72*, 891–899; (b) Cohen, A. O.; Marcus, R. A. *J. Phys. Chem.* **1968**, *72*, 4249–4255; (c) Marcus, R.A. *J. Am. Chem. Soc.* **1969**, *91*, 7224–7225; (d) Marcus, R. A. *Faraday Symp. Chem. Soc.* **1975**, *10*, 60–68.

20 (a) Babamov, V. K.; Marcus, R. A. *J. Chem. Phys.* **1981**, *74*, 1790–1798; (b) Hiller, C.; Manz, J.; Miller, W. H.; Römelt, J. *J. Chem. Phys.* **1983**, *78*, 3850–3856; (c) Miller, W. H. *J. Am. Chem. Soc.* **1979**, *101*, 6810–6814; (d) Skodje, R. T.; Truhlar, D. G.; Garrett, B. C. *J. Chem. Phys.* **1982**, *77*, 5955–5976; (e) Marcus, R. A.; Coltrin M. E. *J. Chem. Phys.* **1977**, *67*, 2609–2613; (f) Garrett, B. C.; Truhlar, D. G. *J. Chem. Phys.* **1983**, *79*, 4931–4938; (g) Kim, Y.; Kreevoy, M. M. *J. Am. Chem. Soc.* **1992**, *114*, 7116–7123.

21 Calculations of KIEs derived from a classical reaction path (e.g. the MEP) in the presence of a solvent or polar environment typically add quantum corrections to that path [22]. Such a reaction path, however, includes *classical* motion of the proton, especially near the TS, and thus this technique exhibits no difference in quantum corrections between H and D at the TS for a symmetric reaction (ΔG_{RXN}=0) [22b], in contrast to the present picture. In variational TS theory for gas phase H atom transfer, the TS significantly deviates from the MEP TS and is isotope-dependent [23]. This feature has been calculated for PT in an enzyme, where the KIE has been diminished because the TS position significantly differs between H and D even in a symmetric case [22e].

22 For numerical calculations consistent with the standard view, see (a) Alhambra, C.; Corchado, J. C.; Sánchez, M. L.; Gao, J.; Truhlar, D. G. *J. Am. Chem. Soc.* **2000**, *122*, 8197–8203; (b) Hwang, J.-K.; Warshel A. *J. Phys Chem.* **1993**, *97*, 10053–10058; (c) Hwang, J.-K.; Chu, Z. T.; Yadav, A.; Warshel, A. *J. Phys. Chem.* **1991**, *95*, 8445–8448; (d) Hwang, J.-K.; Warshel, A. *J. Am. Chem. Soc.* **1996**, *118*, 11745–11751; (e) Alhambra, C.; Gao J.; Corchado, J. C.; Villa, J.; Truhlar, D. G. *J. Am. Chem. Soc.* **1999**, *121*, 2253–2258; (f) Cui, Q.; Karplus, M. *J. Am. Chem. Soc.* **2002**, *124*, 3093–3124.

23 Garrett, B.C.; Truhlar, D. G. *J. Am. Chem. Soc.* **1979**, *101*, 4534–4547.

24 Saunders, W. H. *J. Am. Chem. Soc.* **1985**, *107*, 164–169.

25 For examples of the equilibrium solvation picture for PT in a complex system, see (a) Bash, P. A.; Field, M. J.; Davenport, R. C.; Petsko, G. A.; Ringe, D.; Karplus, M. *Biochemistry* **1991**, *30*, 5826–5832; (b) Cui, Q.; Karplus, M. *J. Am. Chem. Soc.* **2001**, *123*, 2284–2290; (c) Alhambra, C.; Corchado, J.; Sánchez, M. L.; Garcia-Viloca, M.; Gao, J.; Truhlar, D.G. *J. Phys. Chem. B* **2001**, *105*, 11326–11340; (d) Cui, Q.; Elstner, M.; Karplus M. *J. Phys. Chem. B* 2002, 106, 2721–2740.

26 (a) Dogonadze, R. R.; Kuznetzov, A. M.; Levich, V. G. *Electrochim. Acta* **1968**, *13*, 1025–1044; (b) German, E. D.; Kuznetzov, A. M.; Dogonadze, R. R. *J. Chem. Soc., Faraday Trans. II* **1980**, *76*,1128–1146; (c) Kuznetzov, A. M. *Charge Transfer in Physics, Chemistry and Biology: Physical Mechanisms of Elementary Processes and an Introduction to the Theory.* Gordon and Breach, Amsterdam 1995; (d) Kuznetzov, A.M.; Ulstrup, J. *Can. J. Chem.* **1999**, *77*, 1085–1096; (e) Sühnel, J.; Gustav, K. *Chem. Phys.* **1984**, *87*, 179–187.

27 (a) Basilevsky, M. V.; Soudackov, A.; Vener, M. V. *Chem. Phys.* **1995**, *200*, 87–106; (b) Basilevsky, M. V.; Vener, M. V.; Davidovich, G. V., Soudackov, A. *Chem. Phys.* **1996**, *208*, 267–282; (c) Vener, M.V.; Rostov, I. V.; Soudackov, A.; Basilevsky, M. V. *Chem. Phys.* **2000**, *254*, 249–265.

28 (a) Agarwal, P. K.; Billeter, S. R.; Hammes-Schiffer S. *J. Phys. Chem. B* **2002**, *106*, 3283–3293; (b) Agarwal, P. K.; Billeter, S. R.; Rajagopalan, P. T.; Benkovic, S. J.; Hammes-Schiffer, S. *P. N. A S.* **2002**, *99*, 2794–2799; (c) Hammes-Schiffer, S. *Chem Phys Chem*

2002, *3*, 33–42; (d) Hammes-Schiffer, S.; Billeter, S. R.; *Int. Rev. Phys. Chem.* 2001, 20, 591–616; (e) Billeter, S. R.; Webb, S. P.; Agarwal, P. K.; Iordanov T, Hammes-Schiffer S. *J. Am. Chem. Soc.* **2001**, *123*, 11262–11272; (f) Billeter, S. R.; Webb, S. P.; Iordanov, T.; Agarwal, P. K.; Hammes-Schiffer, S. *J. Chem. Phys.* **2001**, *114*, 6925–6936.

29 While these two regimes are distinctly separated, there may exist real systems where different isotopes will be in different regimes. For example, the proton potential barrier at the solvent TS configuration may be small enough such that the proton ground vibration state may be above the barrier, but still large enough such that the deuteron ground vibration state, with a smaller ZPE, may be below the barrier. The present discussion assumes that all isotopes transfer in the same regime.

30 Kim, H. J.; Hynes, J. T. *J. Chem. Phys.* **1992**, *96*, 5088–5110 .

31 (a) Kim, H. J.; Hynes, J. T. *J. Am. Chem. Soc.* **1992**, *114*, 10508–10528; (b) Kim, H. J.; Hynes, J. T. *J. Amer. Chem. Soc.* **1992**, *114*, 10528–10537 .

32 Fonseca, T.; Kim, H. J.; Hynes, J. T. *J. Photochem. Photobiol. A: Chem.* **1994**, *82*, 67–79.

33 (a) Mulliken, R. S. *J. Phys. Chem.* **1952**, *56*, 801–822; (b) Mulliken, R. S. *J. Chim. Phys.* **1964**, *61*, 20–38; (c) Mulliken, R. S.; Person, W. B. *Molecular Complexes* , Wiley, New York, 1969.

34 (a) Cukier, R. I. *J. Phys. Chem.* **1996**, *100*, 15428–15443; (b) Cukier, R. I.; Nocera, D. *Annu. Rev. Phys. Chem.* **1998**, *49*, 337–369.

35 (a) Soudackov, A.; Hammes-Schiffer, S. *J. Chem. Phys.* **2000**, 113, 2385–2396; (b) Iordanova, N.; Decornez, H.; Hammes-Schiffer, S. *J. Am. Chem. Soc.* **2001**, 123, 3723–3733; (c) Iordanova, N.; Hammes-Schiffer, S. *J. Am. Chem. Soc.* **2002**, *124*, 4848–4856; (d) Hammes-Schiffer, S. *Acc. Chem. Res.* **2001**, *34*, 273–281.

36 Thompson, W. H.; Hynes J. T. *J. Phys. Chem. A* **2001**, *105*, 2582–2590.

37 (a) Marcus, R. A. *J. Chem. Phys.* **1956**, *24*, 966–978; (b) Marcus, R. A. *J. Chem. Phys.* **1956**, *24*, 979–988; (c) Marcus, R. A.; Sutin, N. *Biochim. Biophys. Acta* **1985**, *811*, 265–322; (d) Sutin, N. *Prog. Inorg. Chem.* **1983**, *30*, 441–498.

38 The recrossing correction was calculated for both ground and excited H-bond vibrational states at the transition state in Ref. [2d]. The reaction is in the adiabatic PT limit, as opposed to an earlier, more approximate treatment (Azzouz, H.; Borgis, D. *J. Chem. Phys.* **1993**, *98*, 7361–7374; Azzouz, H.; Borgis, D. *J. Mol. Liq.* **1994**, *61*, 17–36), an important feature to emphasize here since this earlier treatment of the reaction has been subsequently widely used as a reference model for various calculation methods.

39 Grote, R. F.; Hynes, J.T. *J. Chem. Phys.* **1980**, *73*, 2715–2732.

40 (a) Warshel, A. *Computer Modeling of Chemical Reactions in Enzymes and Solutions*, John Wiley & Sons, New York, 1991; (b) Kong, Y. S.; Warshel, A. *J. Am. Chem. Soc.* **1995**, *117*, 6234–6242; (c) Warshel, A.; Schweins, T.; Fothergil, M. *J. Am. Chem. Soc.* **1994**, *116*, 8437–8442; (d) Schweins, T.; Warshel, A. *Biochemistry* **1996**, *35*, 14232–14243; (e) Åqvist, J.; Warshel, A. *Chem. Rev.* **1993**, *93*, 2523–2544; (f) Hwang, J.-K.; King, G.; Creighton, S.; Warshel, A. *J. Am. Chem. Soc.* 1988, *110*, 5297–5311; (g) Warshel, A.; Hwang, J.-K. *Faraday Discuss.* **1992**, *93*, 225–238.

41 (a) Newton, M.D.; Ehrenson, S. *J. Am. Chem. Soc.* **1971**, *93*, 4971–4990; (b) Newton, M.D. *J. Chem. Phys.* **1977**, *67*, 5535–5546.

42 (a) Agmon, N. *Chem. Phys. Lett.* **1995**, *244*, 456–462; (b) Marx, D.; Tuckerman, M. E.; Hutter, J.; Parrinello, M. *Nature* **1999**, *397*, 601–604; (c) Geissler, P. L.; Dellago, C.; Chandler, D.; Hutter, J.; Parrinello, M. *Science* **2001**, *291*, 2121–2124; (d) Schmitt, U.W.; Voth, G. A. *J. Phys. Chem. B* **1998**, 102, 5547–5551; (e) Schmitt, U.W.; Voth, G. A. *J. Chem. Phys.* **1999**, *111*, 9361–9381; (f) Vuilleumier, R.; Borgis, D. *J. Phys. Chem. B* **1998**, 102, 4261–4264; (g) Vuilleumier, R.; Borgis, D. *J. Chem. Phys.* **1999**, *111*, 4251–4266.

43 (a) For a perspective on this, we refer the reader to (b). In this connection, see also (c) and (d) concerning the issue of a hydronium ion in the neighborhood of anion produced by PT. (b) Pines, E.; Pines D. in *Ultrafast Hydrogen Bonding Dynamics and Proton Transfer Processes in the Condensed Phase*, Elsaesser, T.; Bakker, H.J. (Eds.), Kluwer Academic, Amsterdam, 2002, 155–184; (c) Rini, M.; Magnes, B.-Z.; Pines, E.; Nibbering, E. T. J. *Science* **2003**, *301*, 349–352; (d) Mohammed, O. F.; Pines, D.; Dreyer, J.; Pines, E.; Nibbering, E. T. J. *Science* **2005**, *310*, 83–86.

44 ΔG_{RXN} is defined throughout as the free energy difference between reactant and product H-bond complexes, a free energy difference that is rarely experimentally determined. The full connection to experimentally determined quantities is discussed in Ref. 43 of Ref. [4].

45 (a) Evans, M. G.; Polanyi, M. *Trans. Faraday Soc.* **1936**, *32*, 1333–1360; (b) Evans, M. G.; Polanyi, M. *Trans. Faraday Soc.* **1938**, *34*, 11–24. The Evans–Polanyi relations were developed mainly for gas phase reactions but are also useful in solution in the context of the standard picture.

46 Swain, G. G., Stivers, E. C.; Reuwer, J. F.; Schaad, L. J. *J. Am. Chem. Soc.* **1958**, *80*, 5885–5893.

47 Kiefer, P. M.; Hynes J. T. *Israel J. Chem.* **2004**, *44*, 171–183.

48 For real systems, $\bar{a}_0$ is not expected to be exactly 0.5 due to 'intrinsic' asymmetry, but the deviation from 0.5 for either H or D is not expected to be significant. For further discussion, see Ref. 45 of Ref. [4].

49 Pauling, L. *J. Am. Chem. Soc.* **1947**, *69*, 542–553.

50 (a) Limbach,H.-H.; Pietrzak, M.; Benedict, H.; Tolstoy, P. M.; Golubev, N. S.; Denisov, G. S. *J. Mol. Structure* **2004**, *706*, 115–119; (b) Limbach, H.-H.; Pietrzak, M.; Sharif, S.; Tolstoy, P. M.; Shenderovich, I. G.; Smirnov, S. N.; Golubev, N. S.; Denisov, G. S. *Chem. Eur. J.* **2004** *10*, 5195–5204; (c) Limbach, H.-H.; Denisov, G. S.; Golubev, N. S. in *Isotope Effects In Chemistry and Biology*, Kohen, A.; Limbach,H.-H. (Eds.), Taylor & Francis, Boca Raton FL, 2005, Ch. 7, pp. 193–230.

51 (a) Eisenberg, D.; Kauzman, W. *The Structure and Properties of Water*, Oxford University Press, New York, 1969; (b) Riddick, J. A.; Bunger, W. B. *Organic Solvents: Physical Properties and Methods of Purification*, 3rd edn., Wiley-Interscience, New York, 1970, pp. 67–88.

52 Shenderovich, I. G.; Burtsev, A. P.; Denisov, G. S.; Golubev, N. S.; Limbach, H. H. *Magn. Reson. Chem.* **2001**, *39*, S91–S99.

53 (a) Siebrand, W.; Wildman, T. A; Zgierski, M. Z. *J. Am. Chem. Soc.* **1984**, *106*, 4083–4089; (b) Siebrand, W.; Wildman, T. A; Zgierski, M. Z. *J. Am. Chem. Soc.* **1984**, *106*, 4089–4096.

54 (a) Cukier, R. I. *J. Phys. Chem. B* **2002**, *106*, 1746–1757; (b) Cukier, R. I.; Zhu, J. *J Phys. Chem. B* **1997**, *101*, 7180–7190.

55 The expression in Eqs. (10.26) and (10.27) is similar in form to that for weakly electronically coupled (electronically nonadiabatic) electron transfer [37], in that the proton coupling *C* is analogous to the electronic resonance coupling and the reorganization energy E_S is analogous to the electronic *diabatic* solvent reorganization energy. Even though the reorganization energies and couplings are analogous, the physical picture behind the two reaction types is quite different. The reorganization energy for proton tunneling is the free energy difference associated with a Franck–Condon-like excitation (all nuclear and solvent modes other than the proton mode are held fixed) of the ground *diabatic* proton vibrational state at the equilibrium reactant solvent position to the ground product *diabatic* proton vibrational state, followed by relaxation along the solvent coordinate to the equilibrium solvent product position (See Fig. 10.13(d)).

56 Discussion of the influence of bending on these results can be found in Refs. [4] and [5a].

57 The H-bond vibrational mode is assumed to remain significantly unchanged while the reaction coordinate fluctuates from the 0–0 TS to either the 0–1 or 1–0 TS.

58 This is particularly true for H atom transfer reactions because they are weakly coupled to a polar environment, i.e. small reorganization energies (cf. H atom transfer reaction in Ref. [1e]).

59 We note in passing that even for data over a broad temperature range where nonlinear behavior is observed, the analysis is useful to analyze different subregions in the nonlinear plot where the behavior is effectively linear, i.e. rate and KIE expressions for a given T_o and the local slope in an Arrhenius plot at T_o are obtained.

60 Excited state contributions described by ρ_L have a key characteristic in that they increase with increased reaction asymmetry ΔG_{RXN} and decreased reorganization energy E_S [5]. Furthermore, the isotopic disparity $\rho_H < \rho_D$ also increases with these trends, resulting in an increase in significance of the third difference in Eq. (10.55) with increased ΔG_{RXN} and decreased E_S [5].

61 (a) Antoniou, D.; Schwartz, S. D. *P. N. A. S.* **1997**, *94*, 12360–12365; (b) Antoniou, D.; Schwartz, S. D. *J. Chem. Phys.* **1999**, *110*, 465–472; (c) Karmacharya, R.; Schwartz, S. D. *J Chem. Phys.* **1999**, *110*, 7376–7381.

62 Further remarks on the T dependence of the Swain–Schaad ratio can be found in Section 3c of Ref. [5].

63 E. Pines, in *Isotope Effects in Chemistry and Biology*, Kohen, A.; Limbach, H.-H. (Eds.), Marcel Decker, Inc., New York, 2005, Ch.16, pp. 451–464.

64 The observed KIEs [9c, 9e, 63] were measured by changing the solvent from H_2O to D_2O, and while this change in solvent introduces other possible solvent isotope effects (i.e. viscosity), the rate limiting step in each case has been shown to be a PT step (or a series of PT steps), and thus the measured KIE corresponds to P.

65 Kim, H. J.; Staib, A.; Hynes, J. T. in *Ultrafast Reaction Dynamics at Atomic-Scale Resolution Femtochemistry and Femtobiology*, Nobel Symposium 101, Villy Sundstrom (Ed.), Imperial College Press, London, 1998, pp. 510–527.

66 (a) Peters, K. S; Cashin, A.; Timbers, P. *J. Am. Chem. Soc.* **2000**, *122*, 107–113; (b) Peters, K. S; Cashin, A. *J. Phys. Chem. A* **2000**, *104*, 4833–4838; (c) Peters, K. S.; Kim, G. *J. Phys. Chem. A* 2001, 105, 4177–4181.

67 Andrieux, C. P.; Gamby, J.; Hapiot, P.; Saveant, J.-M. *J. Am. Chem. Soc.* **2003**, *125*, 10119–10124.

68 Tran-Thi, T.-H; Gustavsson, T.; Prayer, C.; Pommeret, S.; Hynes, J. T. *Chem. Phys. Lett.* **2000**, *329*, 421–430 .

69 Tran-Thi, T.-H; Prayer, C.; Millie, P.; Uznanski, P.; Hynes, J. T. *J. Phys. Chem. A* **2002**, *106*, 2244–2255.

70 Hynes, J. T.; Tran-Thi, T.-H; Granucci, G. *J. Photochem. Photobiol.; A. Photochemistry* **2002**, *154*, 3–11.

71 Granucci, G.; Hynes, J. T.; Millie, P.; Tran-Thi, T.-H. *J. Am. Chem. Soc.* **2000**, *122*, 12235–12245.

72 Elsaesser, T. in *Ultrafast Hydrogen Bonding Dynamics and Proton Transfer Processes in the Condensed Phase*, Elsaesser, T.; Bakker, H.J. (Eds.), Kluwer Academic, Amsterdam, 2002, pp. 119–153.

11
Direct Observation of Nuclear Motion during Ultrafast Intramolecular Proton Transfer

Stefan Lochbrunner, Christian Schriever, and Eberhard Riedle

11.1
Introduction

Hydrogen dynamics is a fundamental issue in chemistry and molecular biology as the contributions in this handbook show. All organic compounds contain a large fraction of hydrogen atoms and the phenomenon of acidity as well as the sensitivity of many reactions on the pH-value mirror the ability of molecules to exchange protons [1, 2]. Hydrogen bridges are, in particular, important for the structure and function of proteins and their interaction with the environment [3]. Proton pumps for example are an important class of natural micromachines responsible for the energy management in halobacteria [4]. To understand hydrogen dynamics is therefore of fundamental importance [5].

In this chapter we describe the progress in understanding hydrogen dynamics via ultrafast spectroscopy of excited state intramolecular proton transfer (ESIPT). Femtosecond pump–probe spectroscopy provides a time resolution of the order of periods of high frequency vibrations and allows one to observe in real time the molecular motions during rearrangements and chemical reactions [6]. This results in a detailed picture of the reaction course and of the relevant mechanisms. ESIPT is particularly well suited to apply this technique to hydrogen dynamics. Contrary to ground state proton transfer, the process can be initiated by ultrashort light pulses, and the starting geometry is very well defined due to the intramolecular character. As we will see, ESIPT is associated with a wavepacket motion which allows one to reconstruct the reaction path in the multidimensional space of the relevant molecular coordinates.

The chapter focuses on the ultrafast ESIPT found in molecules containing an H-chelate ring (see Fig. 11.1). We will discuss the spectral features associated with the ESIPT which are observed in experiments with a time resolution down to 30 fs. They are interpreted in terms of a wavepacket motion on an adiabatic potential energy surface connecting the Franck–Condon region of the enol-form with the minimum of the electronically excited keto configuration. The reaction coordinate is reconstructed from the coherently excited vibrational product modes observed in the experiment. Strong evidence is provided that skeletal modes deter-

Hydrogen-Transfer Reactions. Edited by J. T. Hynes, J. P. Klinman, H. H. Limbach, and R. L. Schowen

ISBN: 978-3-527-30777-7

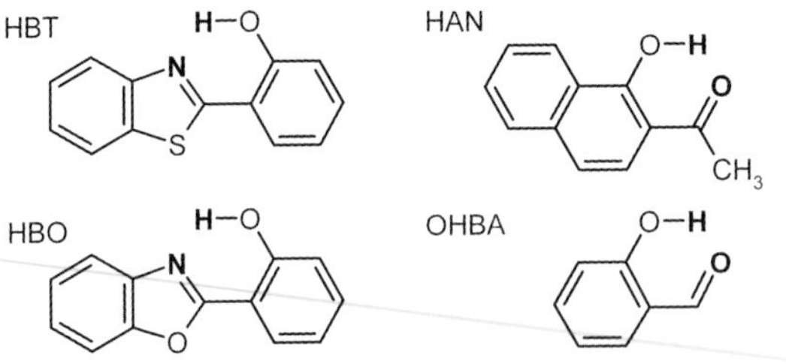

HBT HAN HBO OHBA

Figure 11.1 Four ESIPT compounds considered in this study: 2-(2′-hydroxyphenyl)benzothiazole (HBT), 1-hydroxy-2-acetonaphtone (HAN), 2-(2′-hydroxyphenyl)benzoxazole (HBO), and *ortho*-hydroxybenzaldehyde (OHBA).

mine the dynamics whereas the proton itself plays a passive role and tunneling does not contribute significantly.

The chapter is organized as follows. In the remainder of the introduction the investigated molecular systems are introduced. The experimental section describes a typical pump–probe setup for experiments with a very high time resolution. Then the transient absorption and the contributing processes are discussed. In the fourth section a model of the reaction mechanism is developed from the experimental findings. Then we discuss briefly the situation of parallel reaction channels, and finally the results and conclusion are summarized.

The H-chelate ring is the common motif of many molecules exhibiting ESIPT (see Fig. 11.2). In the electronic ground state the enol-form with the hydrogen atom bound to the donor oxygen of the H-chelate ring is the stable tautomer. If the molecule is promoted to its first electronically excited state the hydrogen atom of the hydroxy group is transferred to a proton acceptor at the opposite site of the ring. The acceptor is typically a nitrogen or a second oxygen atom. Associated with the transfer is a shift of double bonds in the ring and a variation of the aro-

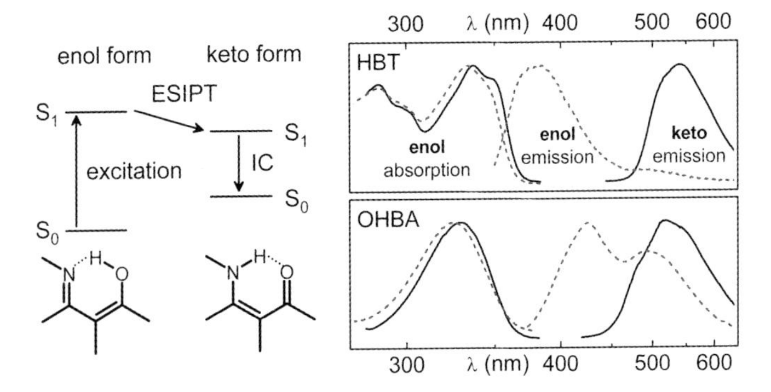

Figure 11.2 ESIPT scheme and steady state absorption and fluorescence spectra of HBT in cyclohexane (solid lines) and in ethanol (dashed lines) and of OHBA in cyclohexane (solid lines) and in methanol (dashed lines).

maticity of the neighboring groups [7]. The ESIPT leads to a strong and characteristic Stokes shift between the UV absorption of the original enol-form and the visible fluorescence (see Fig. 11.2). Already in the seminal work of Weller the fluorescence is attributed to the reaction product, the keto-form [8]. The breaking of the OH-bond and the formation of a keto bond were unequivocally proven for the ESIPT of 2-(2′-hydroxyphenyl)benzothiazole (HBT) by time-resolved infrared spectroscopy [9, 10]. It was found that the absorption band of the OH stretching vibration disappears upon UV excitation of HBT, and the C=O stretching band arises due to the ESIPT. After the ESIPT the molecules return by radiative decay or internal conversion (IC) to the electronic ground state of the keto-form (see Fig. 11.2). A very fast ground state proton transfer brings the molecule back to its enol-form [11].

In aprotic solvents, there is an intramolecular hydrogen bond between the hydroxy group of the H-chelate ring and the proton acceptor in the electronic ground state [12]. As long as this hydrogen bond exists, efficient proton transfer is found independent of the solvent. In polar solvents the hydrogen bond is broken for a high percentage of ESIPT molecules. The hydroxy group then forms an intermolecular hydrogen bond to solvent molecules and in many cases a cis–trans isomerisation takes place in the electronic ground state [13, 14]. In these molecules no ESIPT is possible and the fluorescence of the enol-form is observed, like e.g. for HBT in ethanol (see Fig. 11.2). In the case of *ortho*-hydroxybenzaldehyde (OHBA) dissolved in methanol the fluorescence exhibits two bands, one due to molecules which can undergo ESIPT and one at shorter wavelengths due to those which cannot. However, if ESIPT occurs, very similar time scales are observed for the process in gas phase and in different solvents. This indicates that the solvent has only a minor impact on the transfer dynamics itself [15]. Currently, investigations in our laboratory are in progress to clarify this point. In the following we assume that the influence of the solvent on the ESIPT can be neglected due to the intramolecular character of the reaction.

In early investigations, transfer times of about 100 fs or less were observed for HBT in tetrachloroethylene [16], 2-(2′-hydroxyphenyl)benzoxazole (HBO) in cyclohexane [17], and for methyl salicylate (MS) [18] and OHBA [19] in gas phase. For a number of ESIPT molecules very fast transfer times were found, even at cryogenic temperatures [20]. To resolve the evolution of the transfer and to learn about the mechanism the experimental time resolution had to be improved to better than 50 fs. Such experiments were only possible with the advent of Ti:sapphire laser systems and novel nonlinear sources for ultrashort tunable light pulses. Within the last ten years several experiments with extremely high time resolution have been performed. They revealed rich spectroscopic dynamics and resulted in a detailed picture of the ESIPT and the underlying mechanisms. Their comparative discussion is the central issue of this article.

11.2
Time-resolved Absorption Measurements

The transient absorption setup used in our experiments is a versatile tool providing both the necessary time resolution and the tunability to investigate different compounds. The pump–probe spectrometer is based on two noncollinearly phase matched optical parametric amplifiers (NOPAs) (see Fig. 11.3) [21, 22]. The NOPAs are pumped by a regenerative Ti:sapphire laser amplifier delivering approximately 100 fs NIR pulses at around 800 nm with a repetition rate of 1 kHz.

The NOPAs provide pulses tunable throughout the visible spectral region that are compressed by a fused silica prism sequence to below 20 fs. The output of NOPA 1 is frequency doubled to provide pump pulses which are suitable to excite the UV absorbing enol-form of the ESIPT molecules. The frequency doubling is typically done with a 100 μm thick BBO crystal cut for type I phase matching, and the dispersion of the resulting UV pulses is compensated with a second fused silica prism sequence. After passing a motorized delay stage the UV pulses are focused onto the sample to a spot size of 120 μm. A chopper blocks every second pump pulse to measure the pump-induced absorption changes with two consecutive probe pulses. As probe beam the output of NOPA 2 is used. It is focused to a size of 80 μm at the sample and crosses the pump beam with a small angle of 3°. The sample transmission is measured by detecting the energy of the probe pulses with a photodiode behind the sample. To account for intensity fluctuations of the probe beam, a reference beam is split off before the sample and measured with a second photodiode. On a time scale of a few picoseconds orientational relaxation has no influence on the signal. For measurements covering longer delay times the polarizations of pump and probe beam can be set to the magic angle (54.7°) relative to each other with a half wave plate in the pump beam path in order to avoid the influence of orientational relaxation. The sample is a 70 μm thick, free flowing liquid jet or a flow cell with a channel thickness of 120 μm. In both cases the excited volume of the sample solution is replaced by a fresh one between successive laser shots. The ESIPT molecules are dissolved in cyclohexane or other UV

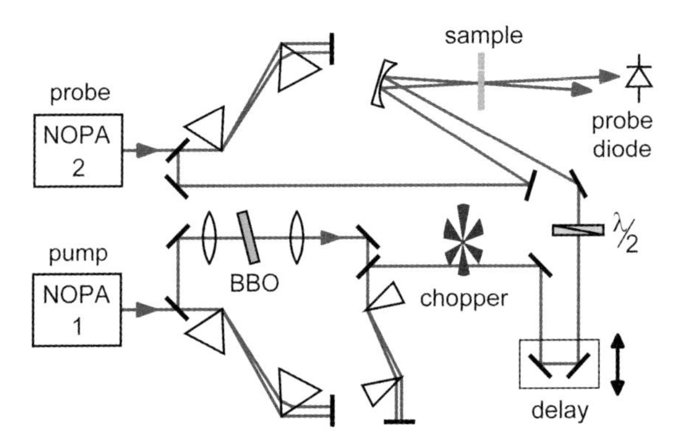

Figure 11.3 Pump–probe absorption setup based on two NOPAs.

compatible solvents with a concentration of 10^{-2}–10^{-3} M resulting in an absorption of about 50% of the pump light.

The detailed analysis of the time dependent absorption changes given below is only possible because the transmitted energy of the probe pulses is integrally detected and measurements at various spectral positions are performed by tuning the NOPA: In principle almost the complete visible region can be covered by the spectrum of a sub-5 fs pulse. If such pulses are applied they are dispersed before detection in order to obtain spectral selectivity [23, 24]. However, this results in strong coherent artifacts around time zero due to cross phase modulation with the pump pulses [25]. In such experiments the early time window of the experimental data, of the order of 100 fs, is usually excluded from the analysis of the molecular dynamics. Since the cross phase modulation does not change the energy of the probe pulses the integral detection is not sensitive to these artifacts and allows one to observe the molecular dynamics also in the direct vicinity of time zero.

11.3 Spectral Signatures of Ultrafast ESIPT

In the following we discuss the spectral signatures observed by ultrafast absorption spectroscopy of ESIPT molecules with a time resolution sufficient to temporally resolve the nuclear motions during the transfer. Pioneering work was performed on 2-(2′-hydroxy-5′-methylphenyl)benzotriazole (TINUVIN P), and for the first time oscillatory signal contributions associated with the ESIPT were observed [26].

Unfortunately, the extremely short lifetime of the electronically excited state of 150 fs [11] makes the analyses particular challenging. In order to uncover the characteristic features of the ESIPT dynamics and to develop a universal model we investigated a series of molecules with a longer S_1 lifetime [27–29]. Two of them, HBT and HBO have a nitrogen atom as proton acceptor while in the other two, OHBA and 1-hydroxy-2-acetonaphtone (HAN) the proton switches between two oxygen atoms (see Fig. 11.1). The IC of these molecules varies markedly, but is much slower than in the case of TINUVIN P and does not mask the ESIPT dynamics. In the following the emphasis is put on HBT due to the extensive analyses available for this molecule. Meanwhile, however, similar dynamics have also been observed in 1,8-dihydroxyanthraquinone (DHAQ) [30], 2,5-bis(2′-benzaoxazolyl)-hydroquinone (BBXHQ) [31], and 10-hydroxybenzo[*h*]quinoline (10-HBQ) [32]. The general features of ESIPT will be elaborated by comparison between HBT and the other compounds.

11.3.1
Characteristic Features of the Transient Absorption

To initiate the proton transfer, ESIPT molecules are promoted to the S_1 state by exciting them with an ultrashort UV pulse at their first absorption band, e.g. at 347 nm for HBT. Figure 11.4 gives an overview of the transient transmission after excitation of HBT, HBO, and HAN in cyclohexane, measured with a white light continuum in magic angle configuration. The spectra are dominated by excited state absorption (ESA), i.e. the absorption from the first electronically excited state to higher electronic states. Only in the UV and at about 500 to 600 nm are positive signals observed. In the UV, the transmission increase results from the bleaching of the electronic ground state. In the visible it is due to stimulated emission (SE). Comparing its spectral position and shape with the steady state fluorescence indicates that the stimulated emission stems from the electronically excited keto-form (compare with Fig. 11.2). This shows that the ESIPT and the emergence of the keto-form occurs well below the picosecond time scale. The changes observed on the picosecond time scale can be attributed to vibrational redistribution processes and internal conversion (see below).

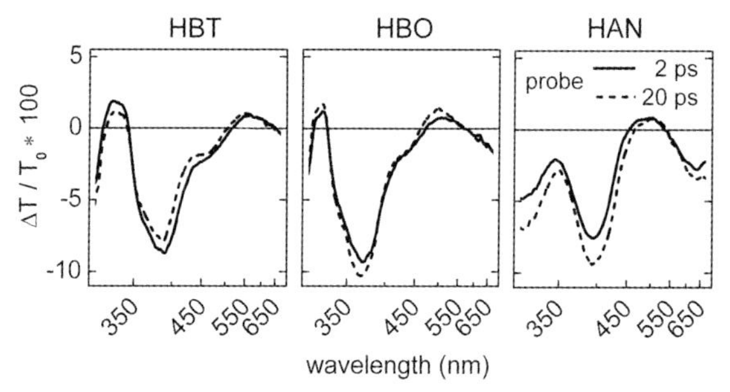

Figure 11.4 Transient spectra after optical excitation of HBT, HBO, and HAN in cyclohexane measured with a white light continuum.

To characterize the proton transfer and the subsequent dynamics with 30 fs time resolution, the setup described in Section 11.2 is used. The transmission change, and thereby the appearance and evolution of the keto emission, is measured at various probe wavelengths covering the entire fluorescence band. The pump and probe polarization are set parallel since the transition dipole moment of the stimulated emission is parallel to the ground state absorption. The transmission change in dependence on the probe wavelength and the time delay between the pump and the probe pulses is shown in Fig. 11.5 for HBT [33]. Traces measured at probe wavelengths of 505 nm and 597 nm are depicted in Fig. 11.6(a). Figure 11.6(b) shows the time-dependent transmission change at 500 nm and 560 nm due to excitation of HBO at 340 nm in cyclohexane.

The initial transmission decrease is caused by excited state absorption (ESA) which appears as soon as the molecule is promoted to the S_1 state. The subse-

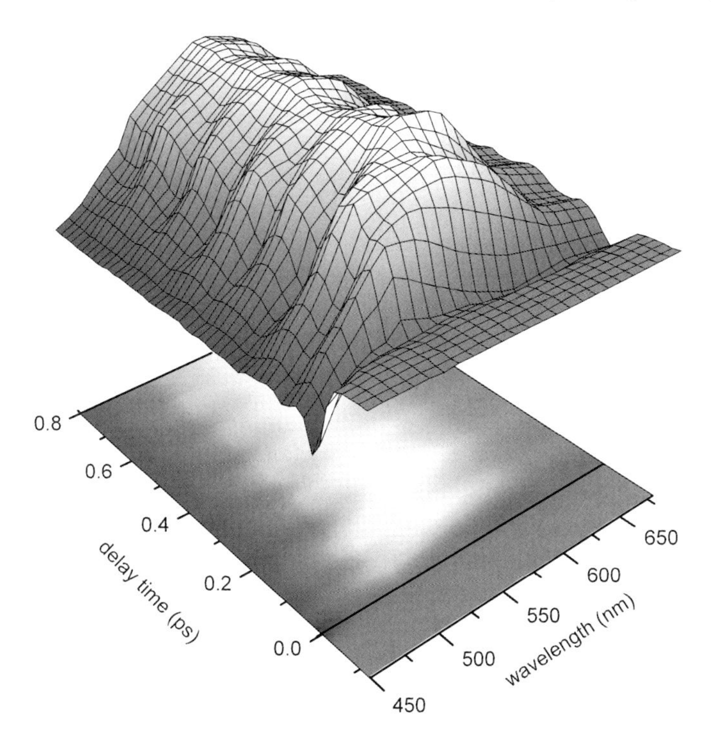

Figure 11.5 Time-dependent transmission change for HBT dissolved in cyclohexane induced by optical excitation at 347 nm. The spectral evolution is reconstructed from time scans taken at various probe wavelengths.

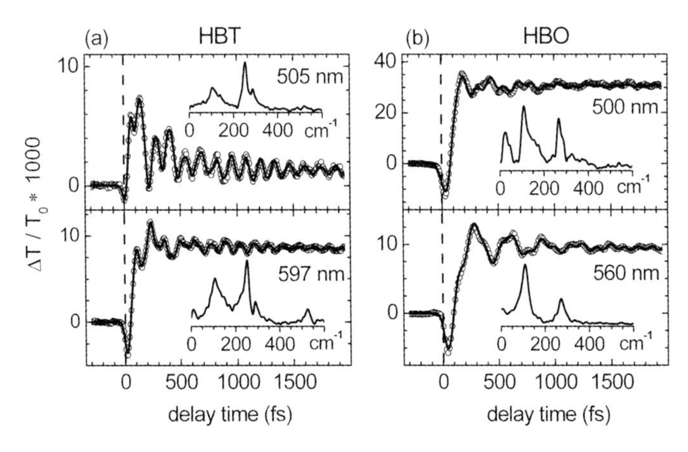

Figure 11.6 (a) Time-dependent transmission change due to excitation of HBT at 347 nm probed at 505 nm and 597 nm. (b) Time-dependent transmission change in case of HBO excited at 340 nm and probed at 500 nm and 560 nm. The open circles are the data and the solid lines the fitted model functions. The insets show the Fourier transforms of the oscillatory signal components.

quent positive transmission change is due to stimulated emission from the electronically excited keto-form, i.e. the product of the proton transfer [27]. In principle it can also result from the bleaching of ground state molecules. However, this possibility can be ruled out since ESIPT molecules in the electronic ground state absorb only in the UV and the positive transmission change is observed in the visible. The same holds for the oscillatory signal contributions which have to originate from the dynamics in the electronically excited state and are not due to ground state wavepackets. If the excitation and probe wavelengths are similar, ground state wavepackets prepared by impulsive stimulated Raman scattering contribute to the signal and usually mask oscillatory excited state contributions [34, 23]. In the case of ESIPT we can be sure that only dynamics in the electronically excited state contribute to the observed transmission changes, and this unambiguous assignment greatly simplifies the interpretation of the experiments.

We obtained very similar experimental results for HBO [29], OHBA [35], and HAN [36] (see Fig. 11.6 and 11.7). They all exhibit a delayed transmission increase and pronounced oscillatory signal contributions with a couple of frequencies in the range up to 700 cm^{-1}, as can be seen from the Fourier transforms of the oscillatory contributions depicted in Fig. 11.6 and 11.7 as insets. This observation was also made for a number of other ESIPT compounds [30–32].

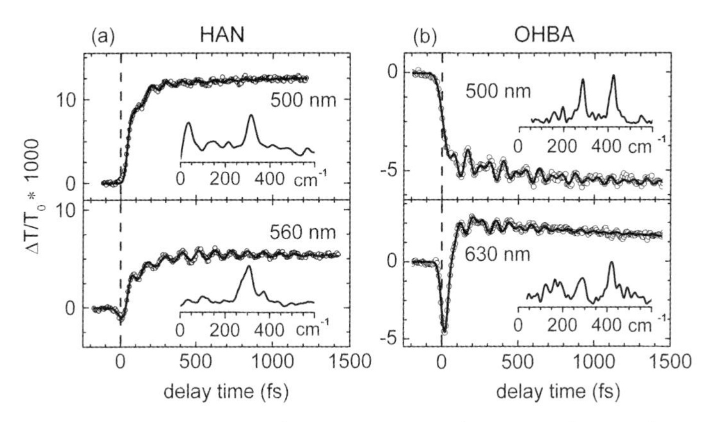

Figure 11.7 (a) Transmission change at 500 nm and 560 nm in the case of HAN excited at 340 nm. (b) Transmission change measured at 500 nm and 630 nm in the case of OHBA excited at 340 nm.

11.3.2
Analysis

The analysis of the experimental results is performed by fitting a model function $S(t)$ to the time traces of all applied probe wavelengths [33]. The function consists of a transmission decrease $S_{\mathrm{ESA}}(t)$ at time zero due to ESA and a transmission increase $S_{\mathrm{SE}}(t)$ which rises step-like with a small delay and reflects the onset of the product emission. The total transient signal is convoluted with the Gaussian cross correlation CC(t).

$$S(t) = \mathrm{CC}(t) \otimes \left\{ S_{\mathrm{ESA}}(t) + S_{\mathrm{SE}}(t) + \sum_i S_{\mathrm{OSC}}^i(t) \right\}$$

$$= \mathrm{CC}(t) \otimes \left\{ A_{\mathrm{ESA}} \mathrm{e}^{-\frac{t}{\tau_{\mathrm{slow}}}} \Theta(t) + \begin{bmatrix} A_{\mathrm{SE}}^{\mathrm{fast}} \mathrm{e}^{-\frac{t-t_{\mathrm{SE}}}{\tau_{\mathrm{fast}}}} + A_{\mathrm{SE}}^{\mathrm{slow}} \mathrm{e}^{-\frac{(t-t_{\mathrm{SE}})}{\tau_{\mathrm{slow}}}} \\ + \sum_i A_i \cos(2\pi v_i t + \varphi_i) \mathrm{e}^{-\frac{(t-t_{\mathrm{SE}})}{\tau_i}} \end{bmatrix} \Theta(t - t_{\mathrm{SE}}) \right\}$$

The excited state absorption $S_{\mathrm{ESA}}(t)$ sets in with a step at time zero modeled by the heavyside function $\Theta(t)$ and a negative valued amplitude A_{ESA}. The non-oscillating part of the stimulated emission $S_{\mathrm{SE}}(t)$ is delayed by t_{SE}. It consists of a fast exponential component with amplitude $A_{\mathrm{SE}}^{\mathrm{fast}}$ and a decay time τ_{fast} of about 250 fs and a slow component with amplitude $A_{\mathrm{SE}}^{\mathrm{slow}}$ and a decay time τ_{slow}. The fast contribution accounts for a red shift of the emission spectrum due to intramolecular vibrational redistribution and the slow one corresponds to the emission from the equilibrated keto-form in the electronically excited state [33]. The decay of the emission is due to the limited S_1 lifetime, and therefore, the decay of the ESA is modeled with the same time constant τ_{slow}. The convolution with the cross correlation CC(t) can be expressed as a product with an appropriate error function. The oscillating signal contributions are modeled by damped cosine functions with frequencies v_i, amplitudes A_i, phases φ_i and damping times τ_i. Their emergence is delayed by the same amount t_{SE} as the emission since they are attributed to the ESIPT product. With appropriately fitted parameters the model function can accurately reproduce the experimental data. This is illustrated by the fits shown in Fig. 11.6 and 11.7 as solid lines. The consistency of the parameters found is checked by their dependence on the probe wavelength [33].

In the following we concentrate on the delay of the emission rise and the oscillatory contributions. For HBT we find a delay of 33 fs in the blue wing and of 55 fs in the red wing of the emission spectrum [33]. In the case of HBO, OHBA and HAN the delay is determined to be 80 fs, 45 fs and 30 fs. The fitted frequencies of the oscillatory contributions match the Fourier transformations depicted in Fig. 11.6 and 11.7. Comprehensive listings of the fit results are given in Refs. [33, 35, 36].

11.3.3
Ballistic Wavepacket Motion

As discussed above, a similar delay of the emission rise of about 50 fs is found for all investigated molecules. The wavelength dependence of the emission amplitude follows the cw emission spectrum [33] which is attributed to the keto-form, the product of the ESIPT. The transmission increase is therefore identified as the delayed rise of the emission from the keto-form. This assignment is also in agreement with recent fluorescent up-conversion experiments which measure the time evolution of the fluorescence [37]. The delay is the time passed until the molecule adopts the keto-form and represents the duration of the ESIPT. The observed delays are in agreement with the timescales determined in experiments on BBXHQ and MS even though these experiments assumed rate behavior for the ESIPT process [18, 31].

A second important observation is that the temporal shape of the emission rise follows a step function. This is demonstrated in Fig. 11.8. It shows the results of fitting an exponential increase convoluted with the crosscorrelation (a), a delayed step like rise convoluted with the crosscorrelation (b) and the complete model function (c) to an experimental trace of HBT at a probe wavelength of 564 nm where the oscillatory contributions are quite weak. The exponential increase and the delayed step function give almost the same ESIPT time [33]. However, the exponential increase deviates significantly from the data at short delay times whereas the step function matches quite accurately the essential shape of the trace.

An exponential signal dependence is expected if the dynamics can be described as a rate governed population transfer between two states. This seems to be an inadequate model for the ESIPT. The step function on the other hand points to an almost classical ballistic motion along the potential energy surface (PES) [27]. The wavepacket prepared by the optical excitation moves completely to the product state without pronounced spreading or splitting. According to Ehrenfest's theorem the center of gravity of the wavepacket then follows a classical trajectory along the PES. The population appears delayed but within a very short time interval in the product state. Such a ballistic wavepacket motion implies that the wavepacket stays confined during the process, in agreement with the observation of a coherent excitation of specific vibrations due to the ESIPT (see below). A ballistic wavepacket motion was experimentally demonstrated for the direct photoinduced dissociation of ICN [38] and iodobenzene [39]. Only recently, indications for a ballis-

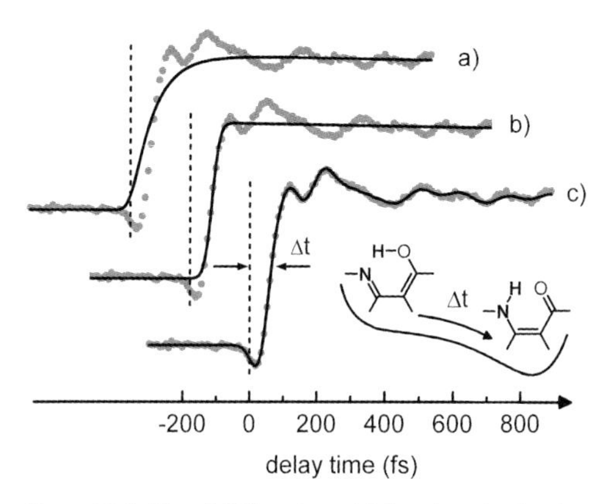

Figure 11.8 Fits of different model functions to the time resolved transmission change of HBT at 564 nm. (a) Exponential rise, (b) delayed step, and (c) complete model function. All functions are convoluted with the cross correlation.

tic behavior in a more complex situation were found in the case of the ring opening of 1,3-cyclohexadiene [40].

A ballistic wavepacket motion is incompatible with a tunneling process of the proton from the enol to the keto site. The transition probability of a single attempt is much smaller than 1 and many tunnel events are necessary for an efficient population transfer leading to a gradual population rise in the product state. However, if the proton itself would move from the enol to the keto site via a barrierless path, the ESIPT would take less than 10 fs because of the small proton mass [18]. This is a first indication that slower motions of the molecular skeleton are the speed determining factors and that the proton mode is not the relevant reaction coordinate [27].

11.3.4 Coherently Excited Vibrations in Product Modes

The oscillatory signal contributions are due to vibrational wavepacket motions in the electronically excited product state and allow one to identify the participating modes. A vibrational wavepacket can be described as a phase coherent superposition of vibronic eigenstates belonging to the same electronic state. As long as wavepacket motion is observed the vibrational phase is still quite well defined. The electronic phase, i.e. the phase of the electronic wavefunction relative to other electronic states, is scrambled in solution within some 10 fs due to the interaction with the solvent [41], but this does not affect the wavepacket motion on a single PES. The vibrational wavepacket motion leads to a periodic variation of the optical transition energy and to an oscillatory spectral shift of the S_1 emission and the transient S_n–S_1 absorption spectra as depicted in Fig. 11.9. It causes the characteristic oscillation of the HBT emission spectrum shown in Fig. 11.5. The amplitude of the oscillatory signal contributions has a dependence on the probe wavelength which reflects directly the slope of the emission spectrum [31, 33].

The Fourier transforms as well as the fitting procedure reveal, depending on the molecule, two to four relevant frequencies below 700 cm^{-1} in the oscillatory signal contributions (see Fig. 11.6 and Fig. 11.7). The oscillations extend to some picoseconds. Obviously the vibrational dephasing occurs on a picosecond time scale

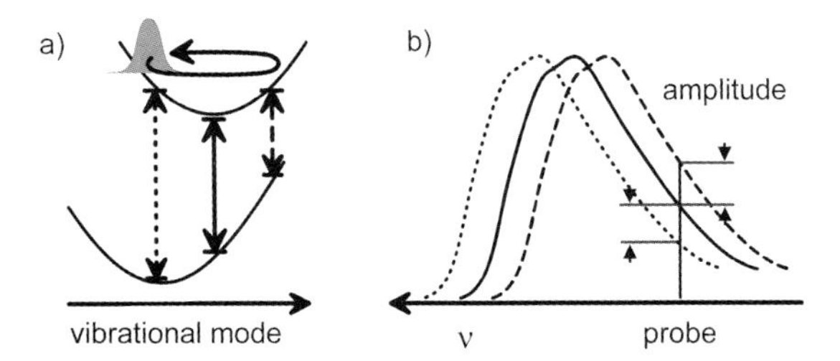

Figure 11.9 An oscillating wavepacket leads to periodic variation in the mean transition energy between two electronic states (a) and thereby to an oscillatory shift of the spectrum (b). The amplitude of the oscillating signal is proportional to the slope of the spectrum.

although the wavepacket propagates on a reactive PES and the ESIPT molecule interacts with solvent molecules. Only modes with very low frequencies are damped within shorter times [33]. A faster damping for the coherently excited mode with the lowest frequency was also observed for 10-HBQ [32]. It was argued that this mode has a particularly large projection onto the reaction coordinate and exhibits therefore a larger anharmonicity, resulting in a faster spreading of the wavepacket along this coordinate. Preliminary experiments in our laboratory indicate that, contrary to intuition, the influence of the solvent on the damping time is very minor and the vibrational dephasing is an intrinsic property of the molecular PES.

As argued above the observed frequencies are those of the vibrational eigenmodes of the electronically excited keto-form. To identify the coherently excited modes the frequencies are compared to steady state vibrational and Raman spectra as well as to results of *ab initio* calculations. The calculations are particularly helpful for the interpretation because they reveal not only the frequencies but also the nuclear movements and deformations of the molecule associated with a vibrational normal mode. *Ab initio* calculations of electronically excited states in molecules exhibiting reactive dynamics are very challenging. High level PESs for the complete ESIPT pathway from the Franck–Condon point to the keto minimum are up to now only available for HBT [7] and very few other compounds [42, 43]. It turns out that the frequencies of the skeletal modes vary only little from the electronically excited to the ground state. Therefore, we usually calculate the vibrational eigenmodes in the electronic ground state [35, 36] by means of density functional theory, as implemented in the Gaussian98 program package [44].

Figure 11.10 shows the calculated vibrational modes of HBT which appear in the transient absorption measurements [33]. The coherent excitation of the 113 cm^{-1} mode was also observed by time resolved vibrational spectroscopy [10]. In the frequency range up to 700 cm^{-1} several other eigenmodes exist but their

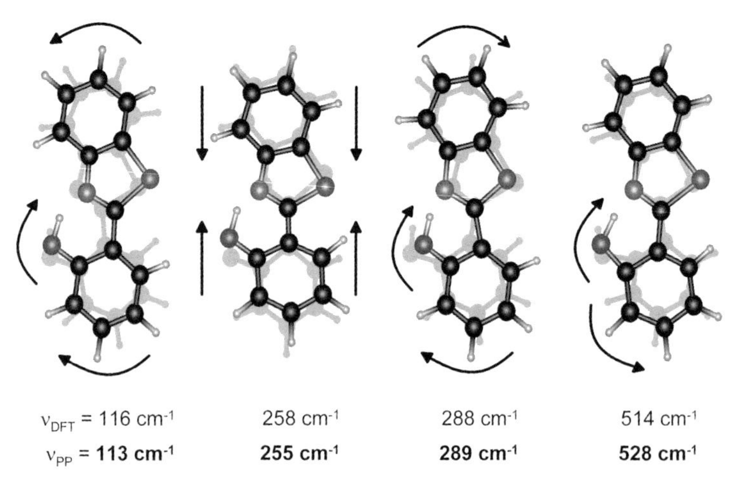

Figure 11.10 Coherently excited vibrational modes of HBT. ν_{DFT} are the frequencies calculated by density functional theory [7] and ν_{PP} the frequencies observed in pump–probe experiments [33].

frequencies are not found in the experimental data and they obviously do not take part in the observed dynamics. The vibrational modes contributing to the wave-packet motion are in-plane deformations of the H-chelate ring [33]. They change strongly the distance between the proton donor and acceptor, in the case of HBT the oxygen and nitrogen atom. In addition, bond angles and lengths within the chelate ring are varied whereas moieties adjacent to the chelate ring are moved as a whole and are barely deformed. For the other molecules investigated we encounter a very similar situation [29, 35]. As a second example the coherently excited modes of HAN are shown in Fig. 11.11.

The specific mechanisms responsible for the coherent excitation of vibrational modes can be determined from the phase of the signal oscillations. At the time a vibration starts, the molecule is at a turning point of its harmonic motion and the argument in the corresponding cosine function can be set to zero. According to this consideration the starting time of an oscillation can be calculated modulo half of a vibrational period from the fitting result of the phase. This analysis was performed in the case of HBT and is discussed in [33]. The three vibrations of higher frequency start at about 30 fs to 50 fs after the optical excitation when the product emission is observed for the first time. It can be concluded that these vibrational motions are excited by the electronic configuration change associated with the change from the enol to the keto-form.

For the vibration with the lowest frequency (113 cm^{-1}) the phase corresponds to a starting time shifted by a quarter of a vibrational period relative to the configuration change. Apparently, the molecule is initially accelerated along this coordinate. The mode is a bending motion of the entire molecule which reduces primarily the proton donor–acceptor distance and introduces only slight changes in other parameters (see Fig. 11.10). A coherent excitation of very similar vibrational modes was found in all ESIPT molecules we investigated and also in DHAQ [30], BBXHQ [31], and 10-HBQ [32]. In HBT and HBO this mode also modulates the

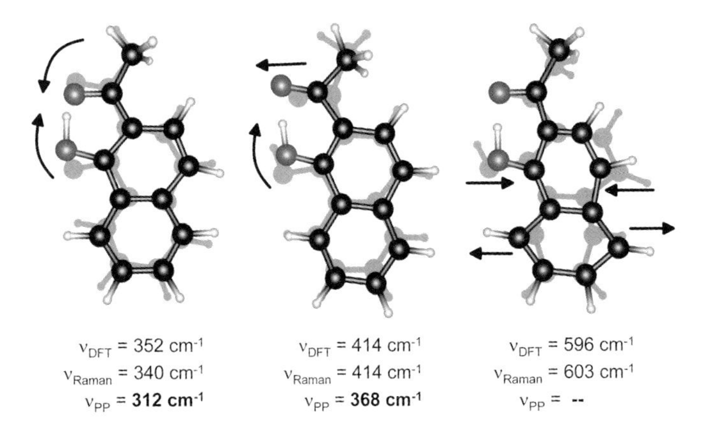

Figure 11.11 Coherently excited vibrational modes of HAN. The frequencies determined by *ab initio* calculations (ν_{DFT}) and found in resonance Raman spectra (ν_{Raman}) are compared to pump–probe measurements (ν_{PP}) [36].

shape of the absorption spectrum which supports the conclusion that the initial motion after optical excitation has a large component along the in-plane bending coordinate of the entire molecular skeleton [17, 45]. For HBT and 10-HBQ it was found that the coherent oscillations in this mode show a faster damping than in higher frequency modes [32, 33]. It was argued that this is an indication for a pronounced anharmonicity reflecting again the intimate connection between this mode and the reaction coordinate [32].

11.4 Reaction Mechanism

11.4.1 Reduction of Donor–Acceptor Distance by Skeletal Motions

The observed proton transfer times of the order of 50 fs have already been discussed in earlier work with respect to the importance of skeletal vibrations [16–18, 46]. It was proposed that a reduction in the distance between the proton donor and the acceptor results in a decrease in the energetic barrier between the enol- and the keto-form. At times when the barrier is suppressed the proton can tunnel or jump from its enol position to the keto site. In the case of HBO it was suggested that, in particular, the in-plane bending vibration modulates the donor–acceptor distance and thereby enables the proton movement [17]. This model was then applied to MS and to 2-(2′-hydroxyphenyl)-5-phenyloxazole [18, 46]. However, due to insufficient time resolution of these experiments it was not possible to give experimental evidence for this model.

Only when TINUVIN P was investigated with 25 fs laser pulses were oscillatory signal contributions detected [26]. They were attributed to coherently excited vibrations in two low frequency modes at 250 cm^{-1} and 470 cm^{-1}. One is a translational motion of the triazole and the phenyl moieties against each other and the other a geared in-plane rotation of the two subunits. The authors interpreted the results in the framework of the model presented above, but attributed the modulation of the proton transfer barrier to a geared in-plane rotation.

We and others found in many ESIPT molecules a strong coherent excitation of a skeletal in-plane bending mode and concluded that predominantly this motion results in the reduction of the donor–acceptor distance as it is discussed in Section 11.3.4. For TINUVIN P no direct evidence for the coherent excitation of the in-plane bending mode was found [26]. However, in this case the short S_1 lifetime of 150 fs does not allow the observation of oscillations with a period of about 300 fs which would result from the coherent excitation of the bending mode. Therefore it can be assumed that also in this case the in-plane bending motion of the molecular skeleton provides the primary contribution to the reduction of the donor–acceptor distance. In HBT, HBO, and TINUVIN P the skeletal stretch vibration is very strongly coherently excited [26, 27, 29] and contributes probably also quite strongly to the initial motion.

11.4.2
Multidimensional ESIPT Model

The irreversibility and efficiency of the ESIPT cannot be understood by a model consisting of only one low frequency mode and one high frequency proton mode. During every period of the low frequency vibration the ON-separation adopts a suitable distance where the energy barrier is suppressed and a back transfer of the proton should be possible. As discussed in Section 11.4.3 this inconsistency can be resolved by assuming that several low frequency modes are directly involved in the ESIPT.

Our investigations lead to a multidimensional model of the ESIPT which can provide a comprehensive understanding of the experimental findings presented above [28, 29, 33]. The proposed evolution of the reaction is sketched in Fig. 11.12. The ballistic wavepacket motion demonstrates that no tunneling is involved and no significant energy barrier is encountered along the reaction path [27]. After optical excitation the molecule accelerates out of the Franck–Condon region, predominantly along the vibrational coordinate which causes an in-plane bending of the entire molecular skeleton and reduces the proton donor–acceptor distance. When it is shortened by a sufficient amount an electronic configuration change from the enol to the keto configuration takes place. The configuration change manifests itself in an occupation redistribution of the molecular orbitals and results in the breakage of the OH bond and the formation of the hydrogen acceptor bond and also in the shift of the double bonds in the chelate ring. Due to the small mass of the electrons, the configuration change itself can be regarded as instantaneous on the time scale of nuclear motions. In HBT it happens 30 to 50 fs after the excitation into the S_1 state [27, 33]. Time-resolved IR experiments found the emergence of the C=O stretching vibration exactly on this time scale [10]

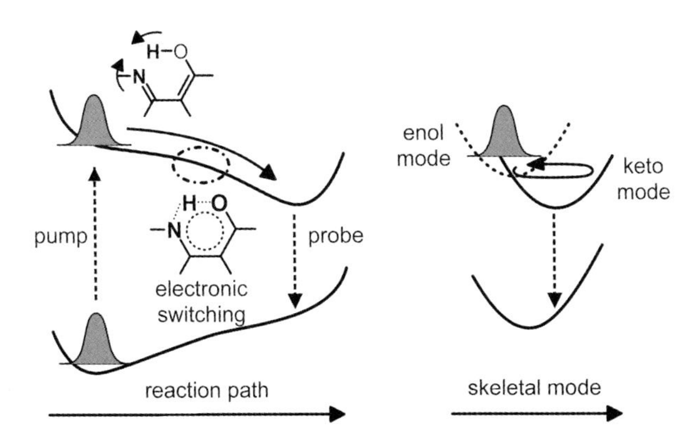

Figure 11.12 ESIPT model: directly after the optical excitation the H-chelate ring contracts and the donor–acceptor distance is reduced. Then an electronic configuration change (electronic switching) occurs and the keto bonds are formed. Subsequently, the molecule is accelerated towards the keto minimum and starts to oscillate around the equilibrium geometry in coherently excited modes.

strongly supporting this interpretation. The new bonds are associated with a changed equilibrium geometry corresponding to the product minimum of the PES. At this moment the skeleton of the molecule has not yet adopted the new equilibrium geometry and is displaced from the product minimum (see right-hand panel of Fig. 11.12). It accelerates towards this minimum and vibrational modes are coherently excited which reflect these changes in geometry. The phases of the coherently excited vibrations indicate that in this stage of the process the higher frequency skeletal modes are more involved than in the initial motion [33].

In previous models the proton position was treated as the key parameter. However, we find that the role of the proton is a very passive one [29, 33]. During the initial motion the proton is shifted by the movement of the oxygen towards the acceptor atom. Then the electronic configuration change breaks the bond between the proton and the donor and forms a new bond to the acceptor. During the subsequent wavepacket motion and the associated geometrical changes the proton is already fixed to the acceptor atom. At any time the proton is at its local potential minimum and is shifted in an adiabatic fashion from the enol to the keto site [27]. Several findings support this point of view: If the ESIPT is mainly the direct motion of the proton from the donor site to the acceptor site then it should be much faster because of the small proton mass [18] or must be hindered by a barrier. Tunnelling through the barrier as well as thermally activated crossing over the barrier should not lead to a ballistic wavepacket motion of the whole system as is observed (see Section 11.3.3). Furthermore, if the proton does not stay in its local minimum an excitation of its local vibrational mode should occur. However, high frequency vibrational modes associated with the N–H and C=O bond of the keto-form are not excited, as was experimentally demonstrated for HBT by time-resolved IR experiments [47]. This is in agreement with the following energy considerations [33]: Overall, the proton transfer releases only an energy of about 2500 cm^{-1}. However, a couple of coherently excited skeletal modes is observed. Since all these modes must contain a few vibrational quanta after the transfer there is not enough energy left to significantly excite a high frequency mode. Recent picosecond time-resolved resonance Raman experiments were able to verify this consideration in the case of HBO [48]. After the ESIPT, anti-Stokes signals were found for several vibrational modes below 1000 cm^{-1} whereas high-frequency modes did not show indications for a vibrational excitation. It should be possible to prove the passive role of the proton with experiments on deuterated compounds. As soon as the motion or tunneling of the proton governs the transfer time a drastic increase in this time due to deuteration is expected. However, no variation of the ESIPT dynamics with deuteration has been observed so far [18, 19, 49, 50] although the precision of those experiments might not have been sufficient for a final decision. The passive role of the proton is further supported by resonance Raman experiments on HAN and 2-hydroxy-acetophenone in which no resonance enhancement for the OH stretching vibration was found [36, 51, 52].

In *ab initio* calculations a transition state for the proton transfer was found which is indeed characterized by a reduced donor–acceptor distance [53] and it was concluded that a strong coupling to the skeletal in-plane bending mode exists

[54, 55]. Recently, the S_1-PES of HBT was characterized by extensive *ab initio* calculations [7]. The complete minimum energy path from the Franck–Condon point to the keto S_1 minimum was calculated. It was found that first the oxygen nitrogen distance is reduced, then an electronic configuration change takes place, and subsequently the molecular skeleton relaxes to the equilibrium geometry of the keto-form. The analysis of the path in terms of normal modes reproduced the coherent excitation of exactly those modes which have been found in the experiment. Whether there is an energy barrier along the reaction path or not is difficult to decide on the basis of *ab initio* calculations. With increasing quality of the applied method almost vanishing barriers were calculated for HBT and OHBA [43]. The reason for the sensitivity to the method is that two electronic configurations determine the shape of the PES and electronic correlation effects have to be handled very accurately [42, 56].

In the case of HBT and HBO the proton acceptor is a nitrogen atom whereas in the case of OHBA and HAN it is a second oxygen atom. For the latter two the chelate ring is fairly symmetric. The degree of symmetry is much less for HBT and HBO. Nevertheless very similar evolutions of the ESIPT are found [28, 29, 36]. Thus the degree of symmetry seems not to be important for the mechanism. Likewise, the ESIPT is insensitive to subtleties of the energetics and the electronic wavefunction. The dynamics of the process is rather determined by motions of the nuclei along in-plane skeletal coordinates. Due to the insensitivity of the ESIPT to details of the reaction center we conclude that the proposed mechanism should be quite general for compounds exhibiting an ultrafast ESIPT within a H-chelate ring. And indeed, the experimental results on DHAQ [30], BBXHQ [31], and 10-HBQ [32] are in agreement with the model.

11.4.3
Micro-irreversibility

The multidimensional character of the process, i.e. the significant participation of several nuclear coordinates is also responsible for its irreversibility [28, 29]. For a reversal of the reaction a full recurrence in all coordinates would be necessary (see Fig. 11.13).

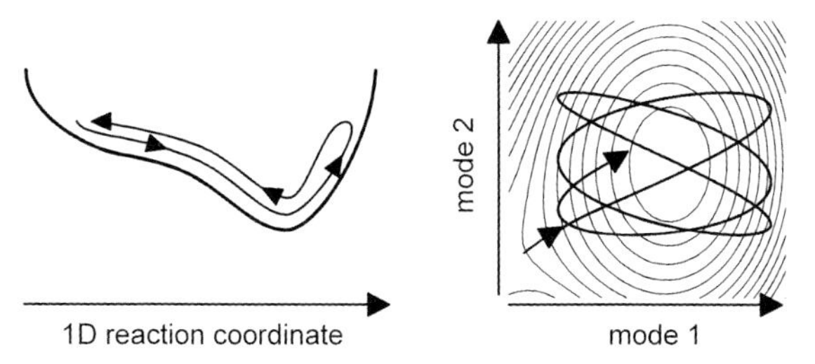

Figure 11.13 (a) In a one-dimensional system the proton transfer would be fully reversible as long as dissipation can be neglected. (b) In a multidimensional situation the wavepacket can not directly return to its origin because a simultaneous recurrence in all involved coordinates is necessary.

This would need much longer than the period of a representative vibration. On this strongly extended time scale the energy flow into other degrees of freedom becomes significant and inhibits the back reaction. This leads to irreversibility, even though the coherence and the energy are lost within a duration many times longer than the actual ESIPT. Out-of-plane twisting vibrations might play a role as energy accepting modes. The coherent excitation of the twisting mode with the lowest frequency (about 60 cm^{-1}) was recently observed in HBT [10] and in *o*-hydroxyacetophenone [57] and points to such vibrational redistribution processes. The example of ESIPT shows that vibrational coherences can exist for picoseconds, even in the situation of reactive dynamics. Therefore we think that the multidimensional character of ultrafast reactions is a prerequisite for their irreversibility.

11.4.4 Topology of the PES and Turns in the Reaction Path

After the electronic configuration change the donor–acceptor distance increases again to adopt the equilibrium geometry of the keto-form resulting in a turn in the reaction path. This is indicated in Fig. 11.14 as an ellipse on top of the reaction path at the transition from the enol to the keto dominated region of the S_1 PES. The path thereby evades an energy barrier which separates the Franck–Condon region from the keto minimum, and which inhibits the direct transfer or jump of the proton to the keto site. This topology of the PES is in good agreement with previous suggestions for the case of TINUVIN P [26, 58] and was also found by the *ab initio* calculations of the ESIPT in HBT [7].

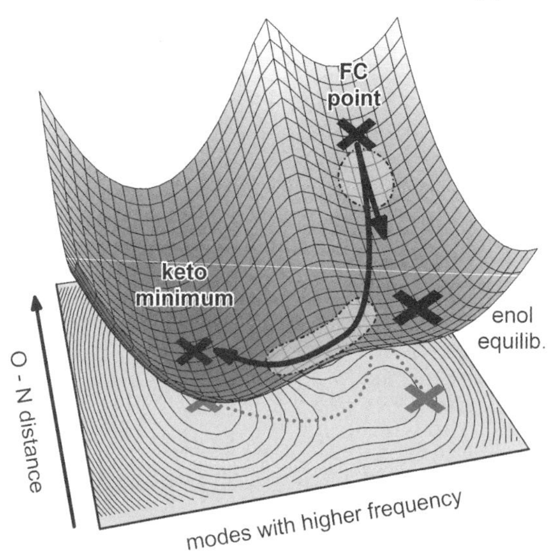

Figure 11.14 Sketch of the S_1 PES and the proton transfer path. The reaction path from the Franck–Condon point (FC-point) to the keto-minimum exhibits two turns (ellipses). The slope at the Franck–Condon point is directed towards the equilibrium geometry of the enol configuration (enol-equilib.).

At the Franck–Condon point, the separation between the donor and acceptor is large, preventing a mixing between the enol and the keto configuration. The PES in this region is of almost pure enol character, resulting in the barrier for the OH-stretch motion. Accordingly, Raman experiments did not find a resonance amplification for the OH-stretch vibration [36, 51, 52]. However, resonance Raman spectra of ESIPT compounds indicate that an additional bend in the reaction path exists directly after the Franck–Condon region. The resonance Raman cross section is large for coordinates with a strong slope of the potential energy at the Franck–Condon point and reveals the initial acceleration of a wavepacket launched by the optical excitation [59]. The intensity distributions in the resonance Raman spectra of HBT [60, 61] are different from the relative vibrational amplitudes in the time traces [33]. The slope at the Franck–Condon point does not reflect the excitation of the modes found after the proton transfer, and the major contributions to their excitation seems to occur after the wavepacket has left the Franck–Condon region. To clarify this point we investigated HAN with resonance Raman and transient absorption spectroscopy [36]. After the proton transfer a strong coherent excitation was found for the bending mode at 312 cm^{-1} and weaker excitation of the modes at 280 cm^{-1} and 368 cm^{-1} (see Fig. 11.7(a)). However, in the low frequency range of the resonance Raman spectrum a mode at 603 cm^{-1} gives the strongest signal which is completely absent in the time-resolved absorption data. Figure 11.11 shows that modes at 312 cm^{-1} and 368 cm^{-1} are dominated by in-plane deformations of the H-chelate ring, while the mode at 603 cm^{-1} is associated with a strong deformation of the naphthalene chromophore. The resonance Raman spectrum of HAN also exhibits strong lines in the spectral region of CC-stretch vibrations which indicates that the optical transition changes the electron density along the system of conjugated bonds [52]. Because of the limited amount of energy released by the proton transfer it can be excluded that these modes are significantly excited by the transfer. For HBO, time-resolved Raman experiments found that high frequency modes are indeed not excited after the ESIPT [48]. These observations show that the slope of the PES at the Franck–Condon point has projections on several coordinates which contribute negligibly to the transfer path. The slope points from the Franck–Condon point to the equilibrium geometry of the electronically excited enol-form (see Fig. 11.14). This direction reflects the geometric relaxation of the electronically excited chromophore and of the π-bonds. However, as soon as the launched wavepacket leaves the Franck–Condon region, the interaction with the keto configuration gains importance and the direction of the slope changes towards the reduction of the donor–acceptor distance. In conclusion, the resonance Raman spectrum provides evidence that the reaction path makes a first bend shortly after the Franck–Condon point.

These considerations show also that the vibrations are dominantly excited by the reactive dynamics itself and the optical transition contributes only little. This is in agreement with the analysis of the vibrational phases. There it was found that in some of the modes the molecule starts to oscillate with a significant delay, which is not expected for a pure optical excitation [33].

11.4.5
Comparison with Ground State Hydrogen Transfer Dynamics

Proton transfer in the electronic ground state is responsible for various phenomena in chemistry like e.g. the acidity of substances. The relative concentrations of involved molecules are given by the free enthalpies if a stationary equilibrium is established. However, dynamic aspects and reaction rates depend on the actual evolution of the transfer. Since proton transfer in the electronically excited state takes place on an adiabatic PES some of the conclusions can be transferred to ground state reactions. They also proceed on an adiabatic surface, the ground state PES, to which several electronic configurations associated with the different molecular arrangements and bond configurations contribute. Our results on ESIPT show that the transfer itself is governed by skeletal vibrational motions and the exchange of the proton represents an electronic configuration change occurring at specific deformations of the involved molecules. Therefore, the transfer rates in the electronic ground state should also be strongly affected by skeletal vibrations of the participating molecules. Consequently, recent theoretical work treating the proton dynamics focuses on the influence of skeletal vibrational coordinates and on solvent configurations [62–64]. The same holds for photoinduced intermolecular proton transfer between donor molecules and surrounding solvent molecules where several studies include low frequency coordinates in the description of the intermolecular transfer process [65, 66].

There are even deeper parallels. A comparison of hydrogen bonded systems with differing bond strength revealed the following correlation [67, 68]. With increasing hydrogen bond strength, the separation between the hydrogen atom and the donor atom increases. This increase is strictly related to a shortening of the donor–acceptor distance. The systems with the shortest donor–acceptor distances are always those in which the hydrogen atom is just in the middle between the donor and the acceptor atom. Shifting the hydrogen atom even more towards the acceptor just exchanges the role of donor and acceptor and their separation increases again. For hydrogen bonded systems, the dependence of the separation between hydrogen and donor atom on the donor–acceptor distance is sampled by different compounds. In the case of ESIPT a very similar correlation is fulfilled during the course of the reaction. The ESIPT path in the space spanned by the donor–acceptor distance and the hydrogen position resembles the correlation diagrams found in hydrogen bonded systems for these two quantities.

11.4.6
Internal Conversion

Most compounds exhibiting ESIPT experience subsequently a rather fast IC (see Section 11.1), which brings the molecule back to the electronic ground state. As discussed above, the ESIPT proceeds in all these molecules in a very similar manner. On the contrary, the time-resolved absorption measurements reveal that the lifetime of the first electronically excited state depends strongly on the molecule.

In solution it varies by three orders of magnitude from 300 ps in the case of HBT [16] down to 150 fs for TINUVIN P [11]. For OHBA we found that the S_1 lifetime after excitation at 340 nm is 55 ps when the molecule is dissolved in cyclohexane, while in the gas phase the lifetime is 13 times shorter [19, 35]. The S_1 lifetime exhibits a strong dependence on the excess energy which obeys an Arrhenius law with an energy barrier of about 200 meV. This points to a statistical behavior and one can apply the concept of an internal molecular temperature [35, 69]. The IC of OHBA is not sensitive on deuteration, indicating that the proton plays no important role in the IC process [19]. After the ESIPT, the dominant portion of the excess energy is stored in the vibrational modes, reflecting the geometry change between the enol and the keto-form (see above). If the IC were sensitive to the amount of vibrational energy in these modes it would be quite fast in the first few picoseconds and then it would slow down when vibrational redistribution leads to an energy flow into other modes. Since the IC does not show any indications of such behavior, we conclude that the significant coordinates for both processes are orthogonal to each other [35]. In this case the IC cannot directly profit from the energy content in the modes participating in the ESIPT and a vibrational redistribution is necessary, leading to a statistical energy distribution and behavior.

A realistic model for the IC process must be able to explain the energy barrier, the statistical behavior, the efficiency and the variation with the molecule. *Ab initio* calculations on malonaldehyde indicate that a $\pi\sigma^*$ state interacts with the S_1 state, which has $\pi\pi^*$ character, resulting in an avoided crossing and an energy barrier [70]. With increasing deformation along the promoting coordinate, the $\pi\sigma^*$ state also crosses the electronic ground state. A conical intersection is formed by the two PESs allowing a very fast and efficient transition to the ground state. A particularly attractive aspect of this model is that the barrier results from an avoided crossing between two electronically excited states. The barrier height depends sensitively on the relative energetic location of the two states, and slight variations in the energetics enter via the barrier height exponentially in the S_1 lifetime. This would nicely explain why the lifetime varies that strongly with the molecule [28]. However, the suggested promoting coordinate is associated with a hydrogen detachment from the H-chelate ring [70]. In this case the process should be sensitive to deuteration, in contradiction to the experimental findings. Therefore we think that skeletal vibrations are responsible for the coupling between the electronic states [28].

By means of calculations it is also discussed that a torsion around the central carbon bond might be responsible for the IC. In the case of 2-(2′-hydroxyphenyl)-triazole a conical intersection between the electronically excited state and the ground state was calculated for a twist angle of nearly 90° [71]. However, the S_1 lifetime of 10-HBQ, which cannot perform twisting motions due to its rigid geometry, is 260 ps [32] indicating that other modes are involved in the IC. At the moment, what are the important coordinates for the IC and whether a common mechanism for all ESIPT molecules exists are completely open questions.

11.5
Reaction Path Specific Wavepacket Dynamics in Double Proton Transfer Molecules

For systems with parallel and branching reaction channels the question arises if, even in this situation, coherent wavepacket dynamics can be observed and, if so, can it be used to characterize the different channels and get an improved understanding of the mechanisms. A promising system to answer these questions is [2,2′-bipyridyl]-3,3′-diol ($BP(OH)_2$) which contains two H-chelate rings (see Fig. 11.15). In aprotic solvents it exhibits both single and concerted double intramolecular proton transfer after it is promoted to the electronically excited state. Glasbeek and coworkers found that the single proton transfer occurs within less than 100 fs and leads to an intermediate mono-keto isomer which subsequently transforms with a time constant of 10 ps to the final di-keto-form [72, 73]. In addition, a second reaction channel exists which leads to the final di-keto product within less than 100 fs by a simultaneous transfer of both protons. The branching ratio between both channels varies strongly with the excess energy [72, 73].

We performed time-resolved absorption studies on $BP(OH)_2$ solvated in cyclohexane applying various excitation and probe wavelengths. The time resolution of 30 fs allows one to observe both transfer processes and the associated coherent wavepacket dynamics in real time [50]. Figure 11.15 shows the time-resolved transmission change of $BP(OH)_2$ in cyclohexane excited at 350 nm and probed at 480 nm. It exhibits pronounced oscillatory signal contributions typical for the ESIPT process. The two most dominant contributions at 196 cm^{-1} and 295 cm^{-1} (see inset of Fig. 11.15) are identified by comparison with *ab initio* calculations. The mode at 196 cm^{-1} is an antisymmetric in-plane bending vibration and is attributed to the single proton transfer. The mode at 295 cm^{-1} is a symmetric stretch vibration and participates in the double proton transfer. If $BP(OH)_2$ is excited at 375 nm the oscillatory contributions at 196 cm^{-1} are strongly suppressed and much weaker than those at 295 cm^{-1}. Since the mono-keto yield at 350 nm is 30% and only 16% at 375 nm [73] this observation confirms our assignment of the two modes to the two different reaction channels. The coherent excitation of the vibra-

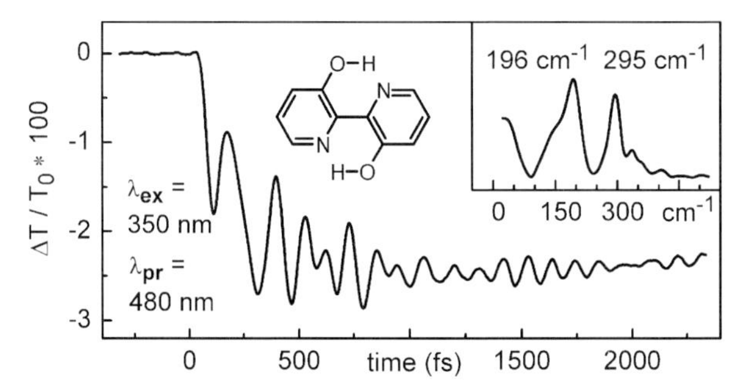

Figure 11.15 Transmission change after optical excitation of $BP(OH)_2$ at 350 nm probed at 480 nm. The Fourier transformation (inset) shows that two modes are dominating the oscillatory signal contributions.

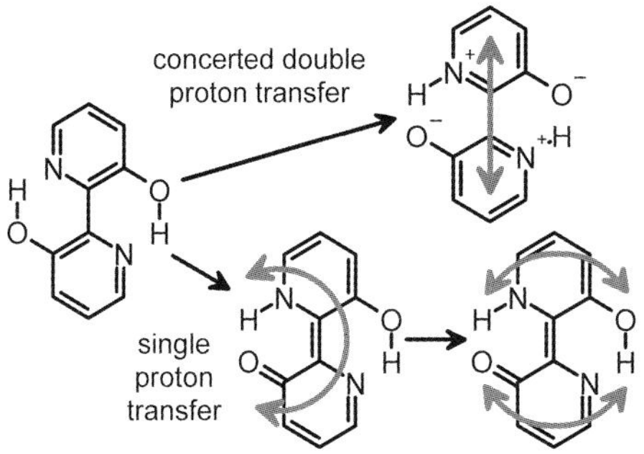

Figure 11.16 Reaction scheme and coherently excited vibrations in $BP(OH)_2$. For the concerted double proton transfer both donor–acceptor distances have to be reduced simultaneously, leading to a symmetric contraction of the molecule and to the coherent excitation of the symmetric stretch vibration. For the single proton transfer the donor–acceptor distance in only one of the two H-chelate rings is compressed by an antisymmetric bending motion.

tional modes can be understood in the following way (see Fig. 11.16) [50]: In the case of the concerted double proton transfer the donor–acceptor distances in both H-chelate rings have to be reduced simultaneously by a symmetric contraction of the molecule. This results in the coherent excitation of the symmetric stretch vibration. For the single proton transfer the donor–acceptor distance in only one of the two H-chelate rings has to be compressed. This is most efficiently accomplished by an antisymmetric bending motion. The example demonstrates that different reaction channels result in different coherent wavepacket dynamics and these wavepacket motions can be distinguished from each other by measurements with different excitation wavelengths. Thereby the mechanisms responsible for both reaction channels can be uncovered.

$BP(OH)_2$ exhibits inversion symmetry. Because of the selection rules for electronic dipole transitions, the symmetric vibrational modes show up in the Raman spectrum and the antisymmetric modes in the infrared absorption spectrum [74]. The wavepacket motion in the antisymmetric bending vibration observed in the time-resolved experiments cannot be directly excited by the optical excitation. Therefore it has to be excited by the symmetry breaking single proton transfer itself. This shows unambiguously that the coherent wavepacket motion results from the ultrafast ESIPT and the associated electronic configuration change and does not reflect the direct optical excitation of the contributing vibronic levels. It answers positively the long-standing question whether a coherent wavepacket motion can be induced by an ultrafast reaction.

11.6 Conclusions

Transient absorption experiments with a time resolution sufficient to resolve the motion of nuclei showed that the ultrafast ESIPT proceeds as a ballistic nuclear wavepacket motion. The wavepacket stays confined during the whole process and moves from the Franck–Condon region to the product minimum within about 50 fs. This time is given by the inertia of the involved nuclei. The subsequent ringing of the molecule in specific modes reflects the structural changes during the reaction. These signatures have been observed for a large number of ESIPT compounds and provide strong evidence for a common mechanism.

In the Franck–Condon region the initial slope of the PES corresponds to a relaxation of the chromophore. However, the further evolution is dominated by an in-plane bending motion of the molecular skeleton, resulting in a reduction in the donor–acceptor distance. An electronic configuration change occurs when this distance is sufficiently shortened. Thereby the bonds are altered from the enol to the keto configuration. Then the donor–acceptor distance increases again and the molecular geometry relaxes along several skeletal modes towards the keto minimum of the S_1 state. The proton itself stays at its local potential minimum and is passively shifted from the enol to the keto site by the skeletal motions during the transfer. This model accounts for the observation of a ballistic wavepacket motion, the coherent excitation of skeletal in-plane vibrations and the lack of excitation in high frequency modes. The topology of the S_1 PES can be understood in terms of an enol / keto configuration mixing which is extremely sensitive to the donor–acceptor distance resulting in a high barrier for the OH stretch motion at the ground state equilibrium distance.

Besides the electronic degrees of freedom several vibrational modes contribute and the essential features of the dynamics can only be understood by a multidimensional model. The multidimensional character causes an irreversible course of the ESIPT even though the transfer itself takes only 50 fs and vibrational dephasing occurs on a picosecond time scale. Nevertheless, most of the energetically accessible vibrations do not play a significant role and a realistic description of the ESIPT has to consider only a restricted number of vibrational degrees of freedom. As discussed in Ref. [22] we think that these are general features for many ultrafast molecular processes. The observation of vibrational wavepacket dynamics in a number of systems [6, 75, 76], which exhibit other ultrafast processes, supports this conclusion.

A comparison between ESIPT and ground state hydrogen dynamics can be drawn in two ways. It can be concluded that the kinetics of ground state reactions are determined by skeletal modes and intermolecular motions which modulate the proton transfer barrier in such a way that at certain configurations it almost disappears. Second, the correlation found between the strength of a hydrogen bond and the donor–acceptor distance exhibits a very similar topology to the S_1 PESs of ESIPT compounds and reflects the importance of configuration mixing.

In the case of intramolecular double proton transfer a wavepacket motion is found which depends, via the excess energy, on the branching ratio between concerted double and single proton transfer. It demonstrates that the coherent wavepacket dynamics in ESIPT molecules is driven by the ESIPT itself and is specific for the reaction path.

Acknowledgement

We thank Regina de Vivie-Riedle, Kai Stock, and Alexander Wurzer for most valuable contributions and acknowledge gratefully the financial support of the German Science Foundation.

References

1 H.-H. Limbach, J. Manz (Eds.), special issue on Hydrogen Transfer: Experiment and Theory, *Ber. Bunsen-Ges. Phys. Chem.*, **1998**, *102*, 289–592.

2 A. J. Barnes, H.-H. Limbach (Eds.), special issue Horizons in Hydrogen Bond Research 2003, *J. Mol. Struct.* **2004**, *700*, 1–254.

3 T. Bountis (Ed.), *Proton Transfer in Hydrogen-Bonded Systems*, Plenum Press, New York, **1992**.

4 U. Haupts, J. Tittor, D. Oesterhelt, *Annu. Rev. Biophys. Biomol. Struct.* **1999**, *28*, 367–399.

5 P. F. Barbara, H. P. Trommsdorff (Eds.), special issue on Spectroscopy and Dynamics of Elementary Proton Transfer in Polyatomic Systems, *Chem. Phys.* **1989**, *136*, 153–360.

6 A. H. Zewail, *J. Phys. Chem. A* **2000**, *104*, 5660–5694.

7 R. de Vivie-Riedle, V. De Waele, L. Kurtz, E. Riedle, *J. Phys. Chem. A* **2003**, *107*, 10591–10599.

8 A. Weller, *Z. Elecktrochem.* **1956**, *60*, 1144–1147.

9 T. Elsaesser, W. Kaiser, *Chem. Phys. Lett.* **1986**, *128*, 231–237.

10 M. Rini, A. Kummrow, J. Dreyer, E. T. J. Nibbering, T. Elsaesser, *Faraday Discuss.* **2002**, *122*, 27–40.

11 C. Chudoba, S. Lutgen, T. Jentzsch, E. Riedle, M. Woerner, T. Elsaesser, *Chem. Phys. Lett.* **1995**, *240*, 35–41.

12 A. L. Huston, G. W. Scott, A. Gupta, *J. Chem. Phys.* **1982**, *76*, 4978–4985.

13 M. Mosquera, J. C. Penedo, M. C. Ríos Rodríguez, F. Rodríguez-Prieto, *J. Phys. Chem.* **1996**, *100*, 5398–5407.

14 J. A. Organero, I. García-Ochoa, M. Moreno, J. M. Lluch, L. Santos, A. Douhal, *Chem. Phys. Lett.* **328** (2000) 83–89.

15 O. K. Abou-Zied, R. Jimenez, E. H. Z. Thompson, D. P. Millar, F. E. Romesberg, *J. Phys. Chem. A* **2002**, *106*, 3665–3672

16 F. Laermer, T. Elsaesser, W. Kaiser, *Chem. Phys. Lett.* **1988**, *148*, 119–124.

17 T. Arthen-Engeland, T. Bultmann, N. P. Ernsting, M. A. Rodriguez, W. Thiel, *Chem. Phys.* **1992**, *163*, 43–53.

18 J. L. Herek, S. Pedersen, L. Bañares, A. H. Zewail, *J. Chem. Phys.* **1992**, *97*, 9046–9061.

19 S. Lochbrunner, T. Schultz, M. Schmitt, J. P. Shaffer, M. Z. Zgierski, A. Stolow, *J. Chem. Phys.* **2001**, *114*, 2519–2522

20 P. F. Barbara, P. K. Walsh, L. E. Brus, *J. Phys. Chem.* **1989**, *93*, 29–34.

21 E. Riedle, M. Beutter, S. Lochbrunner, J. Piel, S. Schenkl, S. Spörlein, W. Zinth, *Appl. Phys. B* **71** (2000) 457–465.

22 A. J. Wurzer, S. Lochbrunner, E. Riedle, *Appl. Phys. B* **2000**, *71*, 405–409.

23 G. Cerullo, G. Lanzani, M. Muccini, C. Taliani, S. De Silvestri, *Phys. Rev. Lett.* **1999**, *83*, 231–234.

24 S. Adachi, V. M. Kobryanskii, T. Kobayashi, *Phys. Rev. Lett.* **2002**, *89*, 027401.
25 M. Lorenc, M. Ziolek, R. Naskrecki, J. Karolczak, J. Kubicki, A. Maciejewski, *Appl. Phys. B* **2002**, *4*, 19–27.
26 C. Chudoba, E. Riedle, M. Pfeiffer, T. Elsaesser, *Chem. Phys. Lett.* **1996**, *263*, 622–628.
27 S. Lochbrunner, A. J. Wurzer, E. Riedle, *J. Chem. Phys.* **2000**, *112*, 10699–10702.
28 S. Lochbrunner, E. Riedle, *Recent Res. Devel. Chem. Phys.* **2003**, *4*, 31–61.
29 S. Lochbrunner, K. Stock, E. Riedle, *J. Mol. Struct.* **2004**, *700*, 13–18.
30 J. Jethwa, D. Ouw, K. Winkler, N. Hartmann, P. Vöhringer, *Z. Phys. Chem.* **2000**, *214*, 1367–1381.
31 N. P. Ernsting, S. A. Kovalenko, T. Senyushkina, J. Saam, V. Farztdinov, *J. Phys. Chem. A* **2001**, *105*, 3443–3453.
32 S. Takeuchi, T. Tahara, *J. Phys. Chem. A* **2005**, *109*, 10199–10207.
33 S. Lochbrunner, A. J. Wurzer, E. Riedle, *J. Phys. Chem. A* **2003**, *107*, 10580–10590.
34 S. L. Dexheimer, Q. Wang, L. A. Peteanu, W. T. Pollard, R. A. Mathies, C. V. Shank, *Chem. Phys. Lett.* **1992**, *188*, 61–66.
35 K. Stock, T. Bizjak, S. Lochbrunner, *Chem. Phys. Lett.* **2002**, *354*, 409–416.
36 S. Lochbrunner, A. Szeghalmi, K. Stock, M. Schmitt, *J. Chem. Phys.* **2005**, *122*, Art. No. 244315-1–244315-9.
37 M. Sanz, A. Douhal, *Chem. Phys. Lett.* **2005**, *401*, 435–439.
38 M. J. Rosker, M. Dantus, A. H. Zewail, *Science* **1988**, *241*, 1200–1202.
39 P. Y. Cheng, D. Zhong, A. H. Zewail, *Chem. Phys. Lett.* **1995**, *237*, 399–405.
40 W. Fuß, W. E. Schmid, S. A. Trushin, *J. Chem. Phys.* **2000**, *112*, 8347–8362.
41 C. J. Bardeen, C. V. Shank, *Chem. Phys. Lett.* **1994**, *226*, 310–316.
42 A. L. Sobolewski, W. Domcke, *Phys. Chem. Chem. Phys.* **1999**, *1*, 3065–3072.
43 A. J. A. Aquino, H. Lischka, C. Hättig, *J. Phys. Chem. A* **2005**, *109*, 3201–3208
44 M. J. Frisch et al., Gaussian 98, Revision A.7.
45 M. Wiechmann, H. Port, *J. Lumin.* **1991**, *48 & 49*, 217–223
46 A. Douhal, F. Lahmani, A. H. Zewail, *Chem. Phys.* **1996**, *207*, 477–498
47 M. Rini, A. Kummrow, J. Dreyer, E. T. J. Nibbering, T. Elsaesser, *Chem. Phys. Lett.* **2003**, *374*, 13–19
48 V. Kozich, J. Dreyer, A. Vodchits, W. Werncke, *Chem. Phys. Lett.* **2005**, *415*, 121–125.
49 W. Frey, F. Laermer, T. Elsaesser, *J. Phys. Chem.* **1991**, *95*, 10391–10395.
50 S. Lochbrunner, K. Stock, C. Schriever, E. Riedle, in *Ultrafast Phenomena XIV*, T. Kobayashi, T. Okada, T. Kobayashi, K. Nelson, S. De Silvestri (Eds.,) Springer-Verlag, Berlin, **2005**, pp. 491–495.
51 L. A. Peteanu, R. A. Mathies, *J. Phys. Chem.* **1992**, *96*, 6910–6916.
52 A. V. Szeghalmi, V. Engel, M. Z. Zgierski, J. Popp, M. Schmitt, *J. Raman Spectrosc.* **2006**, *37*, 148–160.
53 M. A. Rios, M. C. Rios, *J. Phys. Chem.* **1995**, *99*, 12456–12460.
54 M. A. Rios, M. C. Rios, *J. Phys. Chem. A* **1998**, *102*, 1560–1567.
55 V. Gualler, V. S. Batista, W. H. Miller, *J. Chem. Phys.* **2000**, *113*, 9510–9522.
56 S. Scheiner, *J. Phys. Chem. A* **2000**, *104*, 5898–5909.
57 C. Su, J.-Y. Lin, R.-M. R. Hsieh, P.-Y. Cheng, *J. Phys. Chem. A* **2002**, *106*, 11997-12001.
58 M. Pfeiffer, A. Lau, K. Lenz, T. Elsaesser, *Chem. Phys. Lett.* **1997**, *268*, 258-264.
59 E. J. Heller, R. L. Sundberg, D. Tannor, *J. Phys. Chem.* **1982**, *86*, 1822-1833.
60 M. Pfeiffer, K. Lenz, A. Lau, T. Elsaesser, *J. Raman Spectrosc.* **1995**, *26*, 607–615.
61 M. Pfeiffer, K. Lenz, A. Lau, T. Elsaesser, T. Steinke, *J. Raman Spectrosc.* **1997**, *28*, 61–72.
62 K. Ando, J. T. Hynes, *J. Phys. Chem. B* **1997**, *101*, 10464–10478.
63 O. Kühn, *J. Phys. Chem. A* **2002**, *106*, 7671–7679.
64 P. M. Kiefer, J. T. Hynes, *J. Phys. Chem. A* **2003**, *107*, 9022–9039.
65 D. Borgis, J. T. Hynes, *J. Phys. Chem.* **1996**, *100*, 1118–1128.
66 R. I. Cukier, J. J. Zhu, *J. Chem. Phys.* **1999**, *110*, 9587–9597.
67 H. Benedict, H.-H. Limbach, M. Wehlan, W.-P. Fehlhammer, N. S. Golubev, R. Janoschek, *J. Am. Chem. Soc.* **1998**, *129*, 2949–2950.

68 P. M. Tolstoy, S. N. Smirnov, I. G. Shenderovich, N. S. Golubev, G. S. Denisov, H.-H. Limbach, *J. Mol. Struct.* **2004**, *700*, 19–27

69 F. Emmerling, M. Lettenberger, A. Laubereau, *J. Phys. Chem.* **1996**, *100*, 19251–19256.

70 A. L. Sobolewski, W. Domcke, *J. Phys. Chem. A* **1999**, *103*, 4494–4504

71 M. J. Paterson, M. A. Robb, L. Blancafort, A. D. DeBellis, *J. Am Chem. Soc.* **2004**, *126*, 2012–2922.

72 H. Zhang, P. van der Meulen, M. Glasbeek, *Chem. Phys. Lett.* **1996**, *253*, 97–102.

73 D. Marks, P. Prosposito, H. Zhang, M. Glasbeek, *Chem. Phys. Lett.* **1998**, *289*, 535–540.

74 P. Borowicz, O. Faurskov-Nielsen, D. H. Christensen, L. Adamowicz, A. Les, J. Waluk, *Spectrochim. Acta A* **1998**, *54*, 1291–1305.

75 A. J. Wurzer, T. Wilhelm, J. Piel, E. Riedle, *Chem. Phys. Lett.* **1999**, *299*, 296–302.

76 V. De Waele, M. Beutter, U. Schmidhammer, E. Riedle, J. Daub, *Chem. Phys. Lett.* **2004**, *390*, 328–334.

12 Solvent Assisted Photoacidity

Dina Pines and Ehud Pines

12.1 Introduction

Proton transfer [1–33] and electron transfer [33–38] are among the most common classes of chemical reactions in nature. Similar to electron transfer, the main outcome of a proton transfer reaction is the net transfer of a charge. However, unlike electron transfer, proton transfer reactions involving large organic molecules are usually localized between two donor and acceptor atoms. Proton transfer reactions are inherently reversible in the ground electronic state of the proton donor and proton acceptor molecules. The inherent reversibility of the proton transfer reaction is usually maintained in the electronic excited-state of photoacids and photobases in aqueous solutions. Most proton transfer reactions involve relatively small changes in the backbone structure of the proton donor and proton acceptor molecules. These ensuing changes fully reverse upon back-transfer of a proton, either by the back-recombination of the dissociated (geminate) proton or following recombination with a proton coming from the bulk solution [39–62].

Proton transfer is very sensitive to the environment, which usually affects both the yield and the rate of the proton transfer reaction. Aqueous solutions are the most common environment accommodating proton transfer reactions due to the high dielectric constant and extensive hydrogen-bond interactions of the aqueous medium which act both to stabilize charged products and to establish the reaction coordinate along which the proton is transferred. Following the pioneering work of Brønsted and Lowry it has been customary to define proton donors as Brønsted acids and proton acceptors as Brønsted bases [1, 2]:

$$\text{AH (acid)} + \text{B (base)} \leftrightarrows \text{A}^- \text{(conjugated base)} + \text{BH}^+ \text{(conjugated acid)} \tag{12.1}$$

where the acid and base molecules prior to proton transfer may either be charged or neutral species.

The pK_a and pK_b scales in aqueous solutions ($B{=}H_2O$, $BH^+{\equiv}H^+$ in Eq. (1)) serve to define the extent of acidity and basicity of the proton donor and the proton acceptor, respectively, Eqs. (12.2) and (12.3).

Hydrogen-Transfer Reactions. Edited by J. T. Hynes, J. P. Klinman, H. H. Limbach, and R. L. Schowen

ISBN: 978-3-527-30777-7

$$K_a = [A^-][H^+]/[AH] \quad (12.2)$$

$$K_b = [AH][OH^-]/[A^-] \quad (12.3)$$

with $K_a K_b = K_w$, $K_w = [H^+][OH^-] = 10^{-14}$ at room temperature. The self-concentration of water $[H_2O] = 55.3$ M at room temperature is usually included in the respective equilibrium constants.

12.2 Photoacids, Photoacidity and Förster Cycle

12.2.1 Photoacids and Photobases

Photoacids [5, 9, 17–20, 24–28] and photobases are organic dyes which become stronger acids or stronger bases in the electronic excited-state (Figs. 12.1–12.3). They have been used extensively in the past 50 years to study the kinetics and mechanism of proton-transfer reactions in aqueous solutions [3–28, 39–81] and as very efficient means of creating rapid change in the pH of a solution, the so called pH-jump [82–86]. Unlike excited-state electron transfer reactions, proton transfer reactions are usually reversible on the potential surface of the excited-state of the photoacid and the photobase [39–62]. For that reason excited-state proton transfer reactions are very useful for modeling ordinary ground state acid–base reactions.

Research into photoacids and photobases mainly originated with the seminal studies of Förster [3–6] and Weller [7–11]. Förster correctly assigned the large Stokes shift in the fluorescence emission of several hydroxy- and amine-substituted dyes to very fast excited-state proton transfer reaction to the solvent [3]. The unit charge change upon proton transfer results in the fluorescence emission originating from a new fluorescing chromophore, the conjugated photobase. It was Förster who suggested that the acidity of photoacids in the excited state may be estimated by using a thermodynamic cycle, the so-called "Förster cycle" [5]. Förster's assumption was that electronic excitation of a photoacid acts to shift its ground state equilibrium constant K_a to a new value K^*_a, where $K^*_a > K_a$ and may be defined thermodynamically similarly to K_a,

$$K^*_a = [(A^-)^*][H^+]/[A^*H] \quad (12.4)$$

Where an asterisk on a concentration symbol indicates that the species is in the electronic excited state. In Eq. (10.4) A^*H is the excited photoacid and $(A^-)^*$ is the excited (conjugated) photobase.

12.2.2 Use of the Förster Cycle to Estimate the Photoacidity of Photoacids

The K^*_a of a photoacid may be estimated by combining the Förster cycle (Fig. 12.4) [5] with the equilibrium constant of the photoacid in the ground state, which should be independently known to facilitate the calculation.

Arguably, of even greater importance than the ability to estimate the absolute photoacidity when the ground state acidity is accurately known, the Förster cycle provides a general thermodynamic cycle for estimating the relative change in the acidity of a chromophore upon electronic excitation, $\Delta pK^*_a = pK_a - pK^*_a$, independent of prior knowledge of the ground state pK_a. One may, thus, distinguish between the Förster photoacidity, ΔpK^*_a, and the absolute photoacidity of a molecule when in the excited state as defined by its K^*_a (or pK^*_a) value. Furthermore, it is often correct to assume that, while the absolute photoacidity depends on intramolecular properties of the photoacid as well as on solvent properties, Förster photoacidity depends mainly on intramolecular rearrangements in the electron density of the photoacid and its conjugated photobase upon electronic excitation.

The Förster acidity, ΔpK^*_a, may be found with the aid of the Förster cycle (Fig. 12.4) and is given by Eq. (12.5)

$$\Delta pK^*_a = pK_a - pK^*_a = N\,(\Delta G_{FEG} - \Delta G'_{FEG})\,/\,(RT \ln 10) \qquad (12.5)$$

where N is the Avogadro constant and ΔG_{FEG} and $\Delta G'_{FEG}$ are the two Förster energy gaps which are the free-energy gaps separating the ground state and the stable (thermodynamic) energy levels of the photoacid and photobase while in their electronic excited states, respectively.

Clearly, Förster cycle bears a sound physical meaning when excited-state proton transfer is inherently reversible so the two thermodynamically stable energy levels of the photoacid and the photobase in the excited state could have equilibrated, providing that they lived long enough in the excited state. Assuming inherent reversibility of the proton transfer reaction in the excited state, one may proceed and define the equilibrium constant of such an excited-state process independent of the system actually reaching equilibrium populations during the finite (ns-short) lifetime of the excited state. In the following discussion we refer to the thermodynamically stable states of the photoacid and photobase while in the electronically excited state as the "Förster levels" and the free energy gap that separates them from their respective ground-state (thermodynamic) energy levels as the Förster energy gap (FEG, ΔG_{FEG}, $\Delta G'_{FEG}$, respectively). It is worth pointing out that the Förster cycle does not constitute by itself proof for its physical validity. It rather defines a general thermodynamic cycle for estimating the change in the equilibrium constant, ΔpK^*_a, of a photoacid upon electronic excitation, assuming the acid–base equilibrium is shifted from the ground electronic state to the excited electronic state of the photoacid.

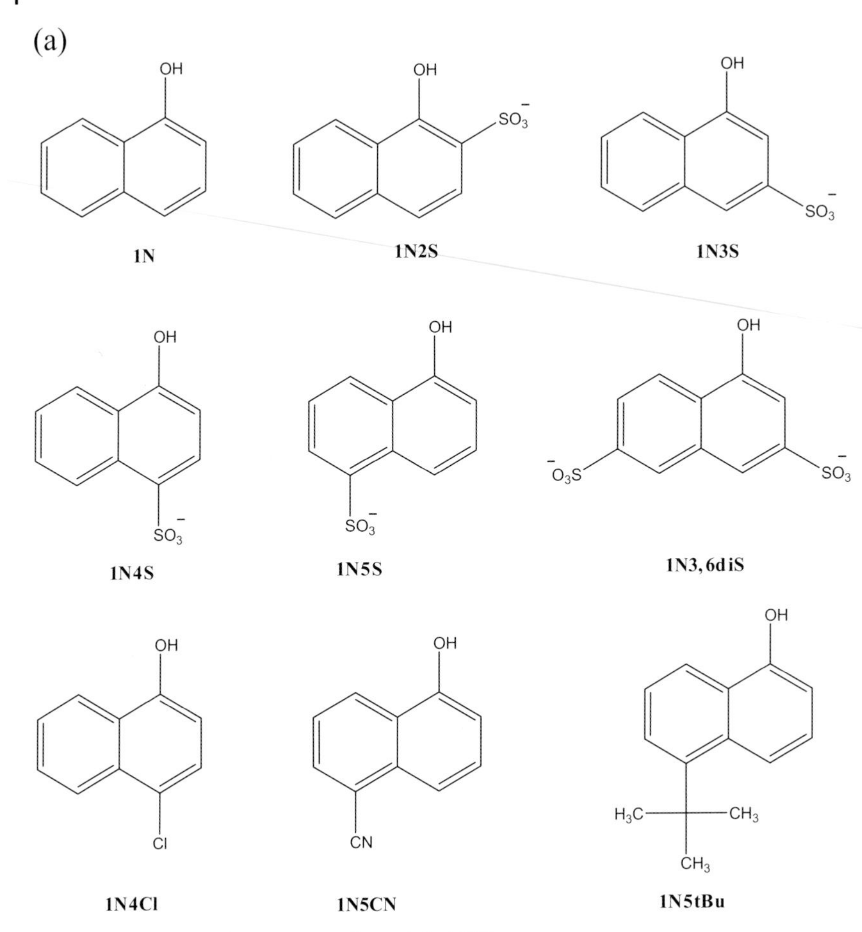

Figure 12.1 (a) Molecular structure of 1-naphthol (1N), 1-naphthol-2-sulfonate (1N2S), 1-naphthol-3-sulfonate (1N3S), 1-naphthol-4-sulfonate (1N4S), 1-naphthol-5-sulfonate (1N5S), 1-naphthol-3,6-disulfonate (1N3,6diS), 1-naphthol-4-chlorate (1N4Cl), 1-naphthol-5-cyano (1N5CN), and 1-naphthol-5-tetrabutyl (1N5tBu).

(b)

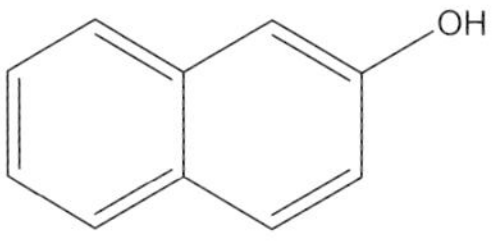

OH

2N

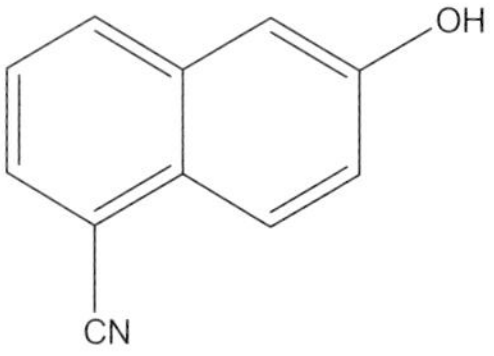

OH
CN

2N5CN

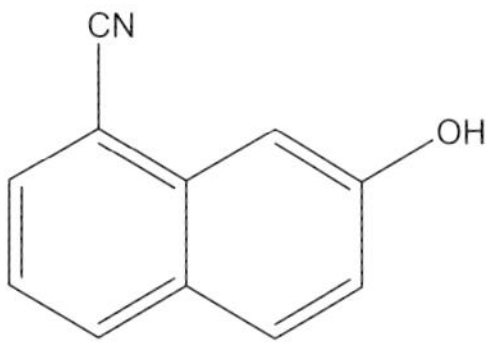

CN
OH

2N8CN

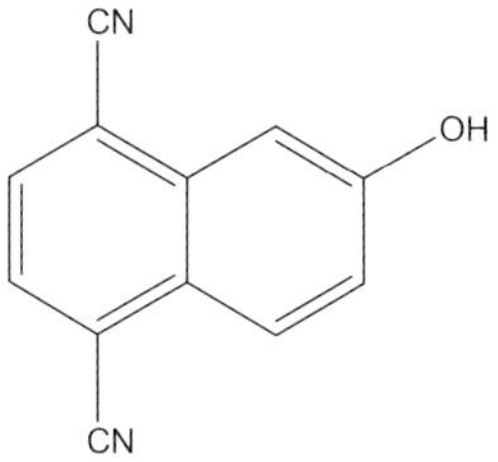

CN
OH
CN

2N5,8diCN

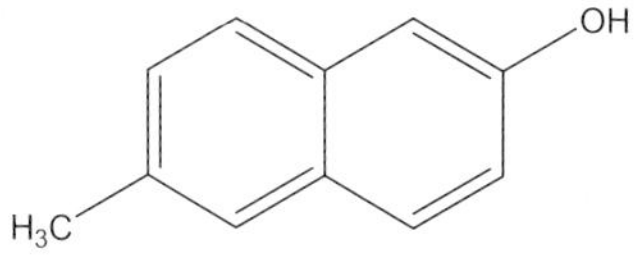

OH
H_3C

2N6Me

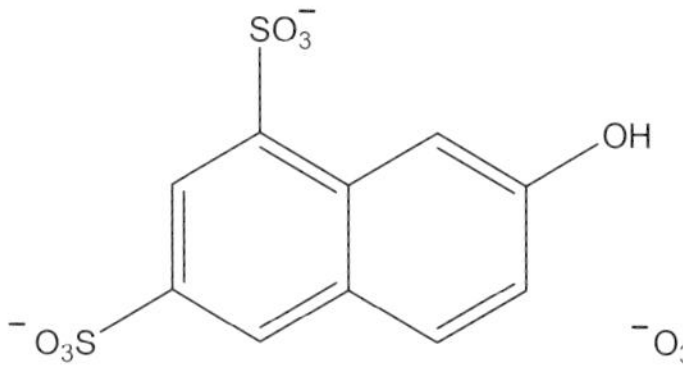

SO_3^-
OH
^-O_3S

2N6,8diS

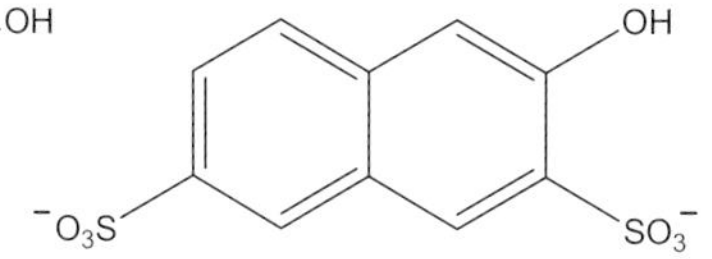

OH
^-O_3S
SO_3^-

2N3,6diS

Figure 12.1 (b) Molecular structure of 2-naphthol (2N), 2-naphthol-5-cyano- (2N5CN), 2-naphthol-8-cyano (2N8CN), 2-naphthol-6,8-disulfonate (2N6,8diS) 2-naphthol-6-methyl- (2N6Me), 2-naphthol-5,8-dicyano (2N5,8diCN) and 2-naphthol-3,6-disulfonate (2N3,6diS).

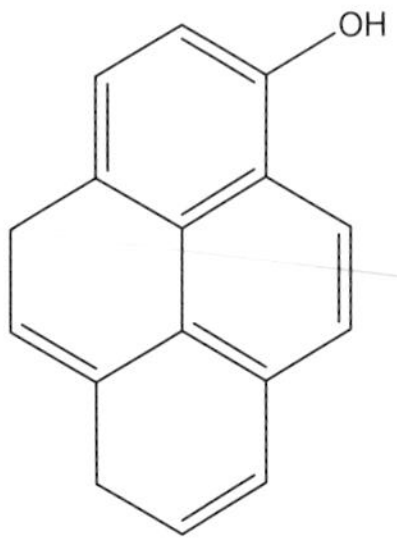

1HP

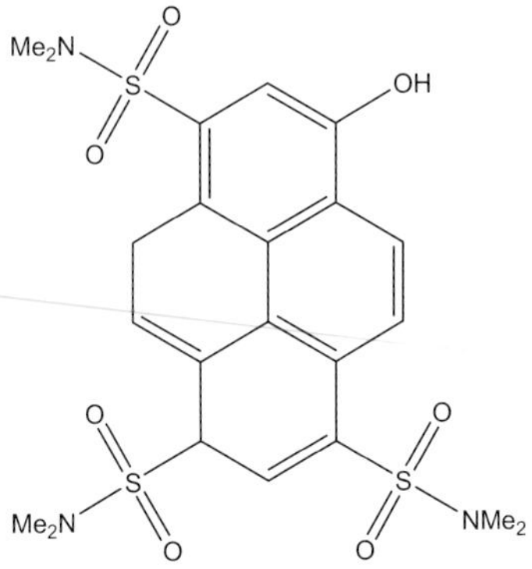

HPTA

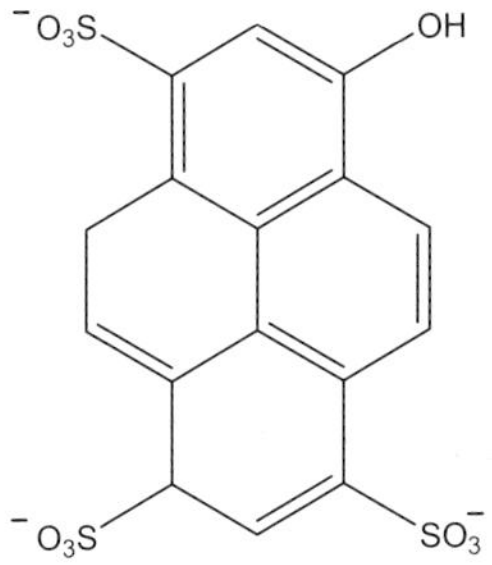

HPTS

Figure 12.2 Molecular structure of 1-hydroxypyrene (1HP), 8-hydroxypyrene-1,3,6-dimethylsulfamide (HPTA) and 8-hydroxypyrene 1,3,6-trisulfonate (HPTS).

AP

MAP

DMAP

APTS

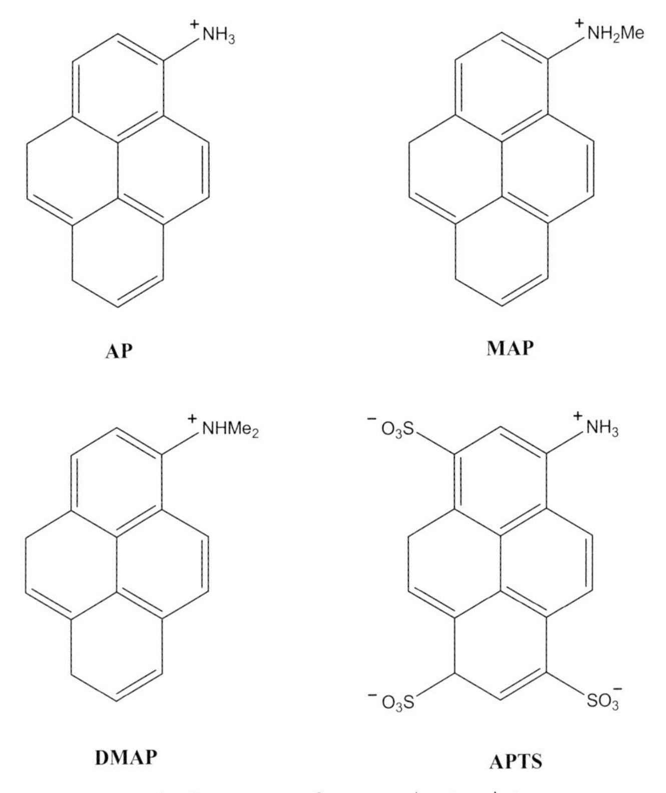

Figure 12.3 Molecular structure of protonated amine photoacids: 1-aminopyrene (1AP), *N*-methyl-1-aminopyrene (MAP), *N,N*-dimethyl-1-aminopyrene (DMAP), 8-aminopyrene-1,3,6-trisulfonate (APTS).

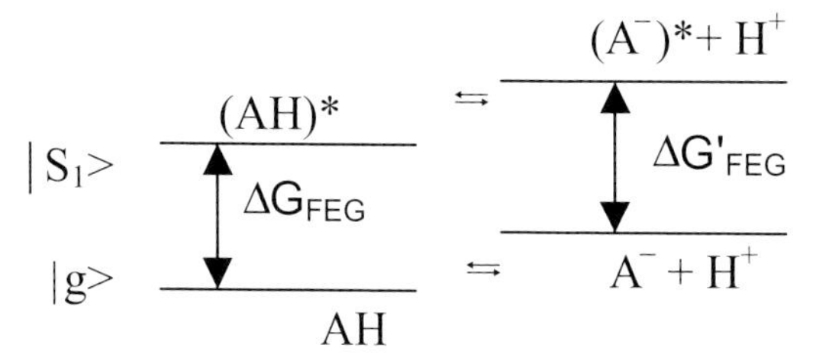

Figure 12.4 Förster cycle of photoacids in the gas phase. Energy levels are for a general photoacid $A^{*}H$ and its conjugated base $(A^{-})^{*}$. S_1> is the first singlet excited state and |g> is the ground-state. ΔG_{FEG} and $\Delta G'_{FEG}$ (the Förster energy gap) are the free energy gaps separating the vibronically relaxed ground state and the vibronically relaxed excited state energy levels of the photoacid and photobase, respectively.

The situation is not straightforward when the acid–base equilibria take place in solution. Unlike the schematic gas-phase situation (Fig. 12.4) vertical electronic transitions of large molecules in solution are followed by various intramolecular relaxation processes and by solvent relaxation around the excited photoacid and photobase. As a result, the location of the Förster levels with respect to their corresponding ground state levels may only be estimated from the corresponding vertical transition energies. There are several methods for determining the respective FEG energies of a photoacid and a photobase which have been reviewed by Grabowski and Grabowska [71]. The method which is usually recommended is to average the transition energies from and to the ground electronic state of the photoacid and the conjugated photobase [9, 19, 27, 71–76]. The transition energies are usually taken from the location of the peak absorption and peak fluorescence spectra of the photoacid and conjugated photobase, $h\nu_{Ab}$ and $h\nu_{Em}$ and $h\nu'_{Ab}$ and $h\nu'_{Em}$ for the photoacid and the photobase, respectively. The averages of the optical transition energies thus found are considered the FEG energies of the Förster cycle. Thus, the transition energies in the photoacid and conjugated photobase sides are calculated independently of each other, from their respective absorption and emission spectra. The procedure of averaging between the energies of the absorption transition, $h\nu_{Ab}$, and the fluorescence transition, $h\nu_{Em}$, is mainly aimed at minimizing the effect of solvent relaxation on determining the final (relaxed) location of the energy level of the acid (and base). The difference in free energy between the directly accessed electronic level and the thermodynamically relaxed Förster level generally depends on both intramolecular and intermolecular relaxation processes. For the electronic excited state one may denote the total free energy of relaxation following the absorption of a photon by ΔG_s and $\Delta G'_s$ for the photoacid and photobase, respectively. Similarly, ΔG_g and $\Delta G'_g$ denote the excess free energy of the vertically accessed level in the ground state of the photoacid and photobase, respectively, following the emission of a photon from the corresponding Förster levels of the photoacid and the photobase. The energy levels and averaging procedure appropriate for solutions of photoacids are depicted in Fig. 12.5.

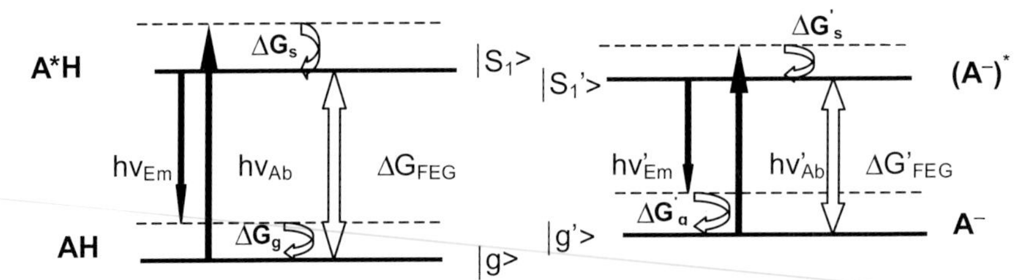

Figure 12.5 Förster cycle of photoacids in solution. Energy levels are for a general photoacid A^*H and its conjugated base $(A^-)^*$. $|S_1>$ is the excited-state of the acid and $|S'_1>$ of the base, $|g>$ is the ground-state of the acid and $|g'>$ of the anion, respectively. $h\nu_{Ab}$ and $h\nu'_{Ab}$ are the energy of the absorption transition. $h\nu_{Em}$ and $h\nu'_{Ab}$ are the energy of the fluorescence transition of the acid and base, respectively. ΔG_{FEG} and $\Delta G'_{FEG}$ (the Förster energy gaps) are the free energy gaps separating the thermodynamically relaxed ground state and the thermodynamically relaxed excited state energy levels of the photoacid and photobase, respectively. ΔG_g and ΔG_S are the relaxation free energies of the acid immediately following the electronic transition in the ground and excited state respectively and $\Delta G'_g$ and $\Delta G'_S$ are the corresponding free relaxation of the base.

Using the symbols of Fig. 12.5 one has for the average of the absorption and emission transitions of the photoacid (AH):

$$(h\nu_{Ab} + h\nu_{Em})/2 = (\Delta G_{FEG}+\Delta G_s + \Delta G_{FEG}-\Delta G_g)/2 = \Delta G_{FEG} + (\Delta G_s-\Delta G_g)/2 \qquad (12.6)$$

Similarly, averaging the energies of the optical transitions of the photobase (A^-) yields:

$$\begin{aligned}(h\nu'_{Ab} + h\nu'_{Em})/2 &= (\Delta G'_{FEG} + \Delta G'_s + \Delta G'_{FEG}-\Delta G'_g)/2 \\ &= \Delta G'_{FEG} + (\Delta G'_s-\Delta G'_g)/2\end{aligned} \qquad (12.7)$$

In Eqs. (12.6) and (12.7) ΔG_{FEG} and $\Delta G'_{FEG}$ are the Förster energy gaps (FEG) in the photoacid and photobase side, respectively. ΔG_s and ΔG_g are the total free energy of relaxation in the target electronic state following absorption and emission of a photon, respectively. It follows that averaging between the absorption and emission energies results in cancellation of errors, which usually results in better estimation of the FEG energies as compared to either using the absorption or emission energies alone. Furthermore, the total error in determining, ΔG_{FEG} and $\Delta G'_{FEG}$ is only half that of the difference in the relaxation energies in the ground and the excited state, thus reducing the residual error by half, Eqs. (12.6) and (12.7).

The total error in carrying out the Förster cycle with the averaged transition energies of the photoacid and photobase rather than with the thermodynamic ΔG_{FEG} and $\Delta G'_{FEG}$ values may be estimated by the following simple algebraic consideration, Eq. (12.8).

$$\Delta G_{FEG}-\Delta G'_{FEG} = (h\nu_{Ab}+h\nu_{Em})/2 - (h\nu'_{Ab}+h\nu'_{Em})/2 + (\Delta G_s-\Delta G_g-\Delta G'_s+\Delta G'_g)/2 \quad (12.8)$$

The difference between the ground-state and excited-state equilibrium constants of the photoacid as defined thermodynamically by the Förster cycle, ΔpK^*_a (therm), is given by:

$$\Delta pK^*_a(\text{therm}) = (\Delta G_{FEG}-\Delta G'_{FEG})/(RT\ln 10) = ((h\nu_{Ab}-\Delta G_s)-(h\nu'_A-\Delta G'_s))/(RT\ln 10) \quad (12.9)$$

So relaxations in the excited photoacid side to below the vertically accessed level act to decrease photoacidity while corresponding relaxations in the excited photobase side act to increase it.

The difference between the calculated value ΔpK^*_a (cal) when using averaged transition frequencies and the thermodynamic value, ΔpK^*_a (therm) may be found by the following procedure:

$$\Delta pK^*_a(\text{cal}) = [(h\nu_{Ab}+h\nu_{Em})/2 - (h\nu'_{Ab}+h\nu'_{Em})/2]/(RT\ln 10) \quad (12.10)$$

The error in ΔpK^*_a when applying Eq. (12.10) rather then Eq. (12.9) is given by:

$$\Delta pK^*_a(\text{therm}) - \Delta pK^*_a(\text{cal}) = [N(\Delta G_g - \Delta G_s + \Delta G'_s - \Delta G'_g)/2]/(RT\ln 10) \quad (12.11)$$

It is not immediately clear which of the relaxation free energy terms appearing in Eq. (12.11) are more important than the others. In conditions where there is little electronic rearrangement in the excited states of the photoacid and photobase ΔG_s and $\Delta G'_g$ are likely to be larger than $\Delta G'_s$ and ΔG_g. The former transitions involve solvent relaxation following vertical electronic transitions to the electronically excited photoacid which is the more polar form of the photoacid and transition to the ground-state photobase which is more charge localized than the excited-state photobase and hence undergoes stronger interactions with the solvent. This is the situation for electronic transitions which only involve modest changes in the electronic structure both in the photoacid and photobase sides. Photoacids where excitation to the relatively nonpolar ^1L_b state (see below) takes place may conform to this scenario. One such example is the S_0–S_1 transition of 2-naphthol. A useful classification of the electronic levels of aromatic molecules was given by Platt. The two lowest electronic levels common to all *cata*-condensed hydrocarbons are, according to Platt's notation [7, 88] the ^1L_a and ^1L_b levels. In Platt's notation the subscripts *a* and *b* refer to the direction of the electronic polarization. In general, *a* refers to an electronic state whose eigenfunction nodes pass through the carbon atoms forming the aromatic ring, and *b* refers to a state whose eigenfunction nodes pass through the carbon–carbon bonds. For the naphthalene ring system the *a* band is polarized along the short axis of the molecule in a transverse polarization and the *b* band is polarized along the long axis.

(a)

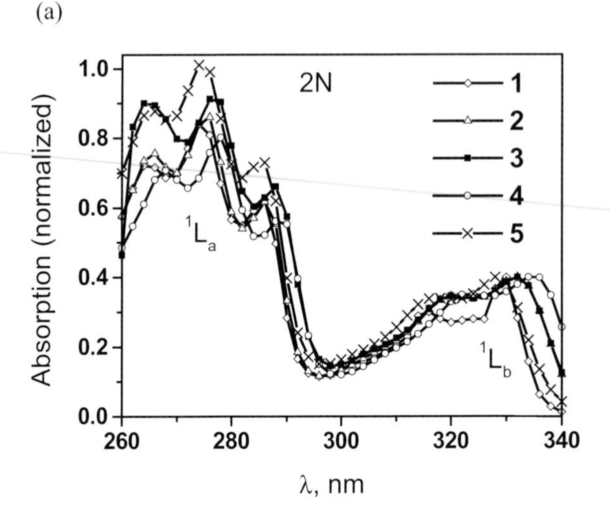

(b)

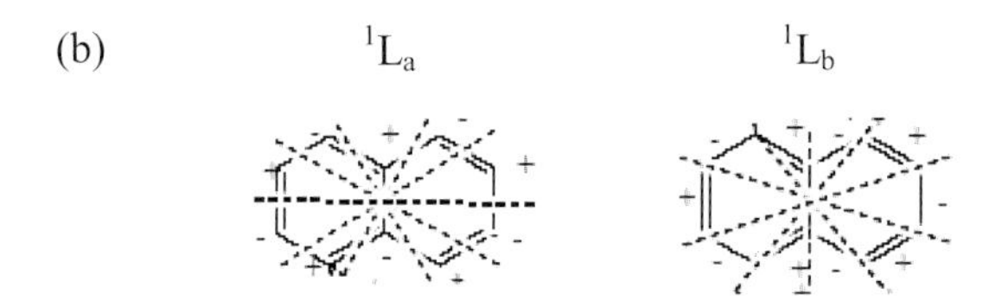

Figure 12.6 (a)Absorption spectra of 2-naphthol in several solvents of different polarity: 1- c-hexane, 2- ethanol, 3- formamide, 4-DMSO, 5-water. (b) L_a/L_b scheme for naphthalene. Adapted from Ref. [95].

The two lowest-energy electronic absorption bands of 2-naphthol in various solvents are shown in Fig. 12.6. These are assigned to transitions to the ^1L_b state (S_1) and to the ^1L_a state (S_2). In cases where the vertically accessed excited state level of the photoacid and the relaxed excited state level of the photoacid are both the relatively nonpolar ^1L_b state the difference between the calculated value and the thermodynamic value of pK^*_a is expected to be small. It has indeed been found that Förster cycle with average transition frequencies is a very good approximation for calculating ΔpK^*_a values of photoacids in the ^1L_b electronic excited state [75].

The situation is more complex when two different singlet states are involved in the photon-absorption and photon-emission processes of the photoacid [76, 89–96]. Following Baba and Suzuki [89–91], we have suggested that the blue-side in the absorption band of 1-naphthol belongs to the ^1L_a transition and the red-side of the absorption band of 1-naphthol belongs to the ^1L_b transition [92, 93] (see Fig. 12.7). The level structure of such a photoacid is congested and is portrayed in Fig. 12.8. Here the full Eq. (12.11) should be considered with potentially much larger deviations from the energies of the true Förster cycle transitions.

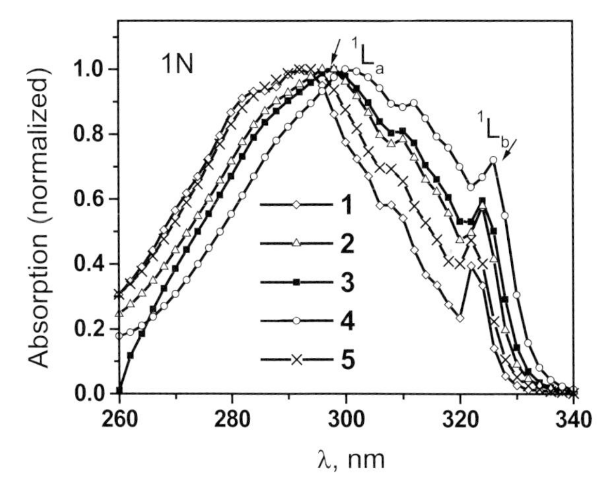

Figure 12.7 Absorption spectra of 1-naphthol in several solvents of different polarity: 1 – c-hexane, 2 – ethanol, 3 – formamide, 4 – DMSO, 5 – water. The spectral range is identical to that of Fig. 12.6. Adapted from Ref. [95].

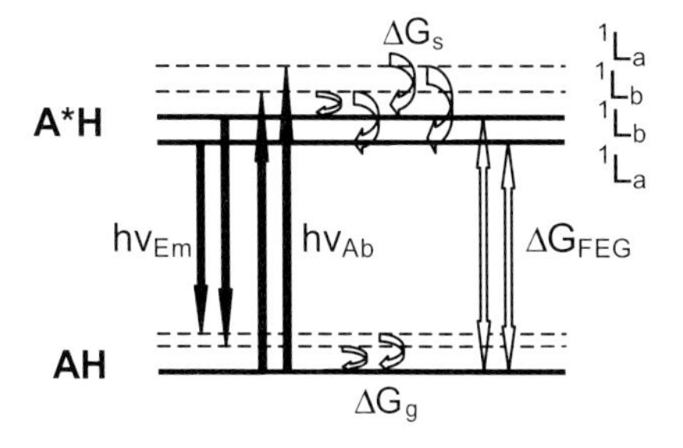

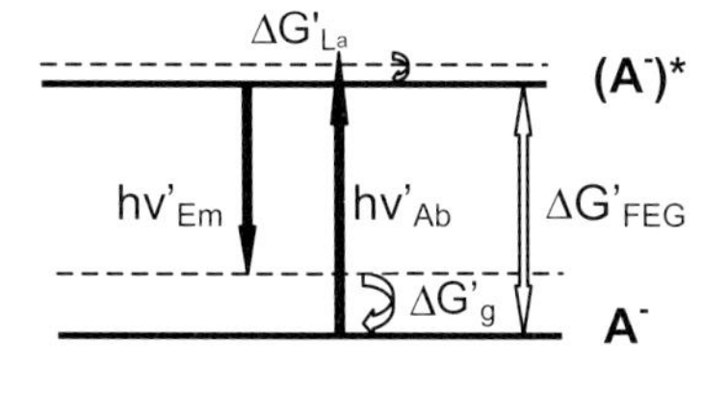

Figure 12.8 Förster cycle of photoacids in solution when two different singlet states are involved in the photon absorption and photon emission processes of the photoacid i.e., the ^1L_b and the ^1L_a states. The four possible excitation-emission cycle scenarios involving the ground state and the two excited state levels are: ^1L_a : ^1L_b, ^1L_a : ^1L_a, ^1L_b : ^1L_b, ^1L_b : ^1L_a (see text). All symbols are as defined in Fig. 12.5.

12.2.3
Direct Methods for Determining the Photoacidity of Photoacids

In certain cases of well-behaved photoacids, typically having pK^*_a values in the range of 0–3, Förster cycle predictions may be tested against pK^*_a values found by two direct experimental methods. The first is titration of the excited photoacid with a stronger mineral acid until an "end–point" is reached [9]. The titration of the excited photoacid is monitored by following the change in the relative quantum yield of the photoacid or the photobase as a function of the solution pH. Taking the steady-state fluorescence spectra of the photoacid at each titration point is sufficient for this purpose [7–11]. The value of K^*_a is found by analyzing the inflection point of the titration curve. This method is reliable when the acid–base equilibrium is not affected by proton quenching and when the photoacidity of the photoacid is not very large or not too small for the excited state dissociation to be appreciable.

The second independent method for finding K^*_a is by direct time-resolved measurements of the proton-dissociation and proton-recombination reaction rates of the excited photoacid. These measurements have traditionally utilized time-resolved fluorescence and absorption spectroscopy. They were originally developed by Weller [7–11] and Förster [3–6] and have been widely in use in photoacid research [17,18, 27].

Assuming homogenous proton recombination and unidirectional dissociation reaction one has, for the excited-state equilibrium constant:

$$pK^*_a = -\log(k^*_d/k^*_r) \tag{12.12}$$

where k^*_r and k^*_d are the bimolecular (homogeneous) proton recombination and the unimolecular proton dissociation rate constants of the excited photoacid, respectively.

Recently, the usefulness of fs-resolved mid-IR measurements of some vibrational markers of the photoacid and the photobase was demonstrated by Nibbering et al. [97–100]. Direct mid-IR absorption spectroscopy has thus proved to be an additional tool for directly monitoring the proton-transfer kinetics of photoacids while in the excited state.

The main disadvantage of the direct methods for the determination of K^*_a values is that these methods are usually cumbersome and are only feasible for a limited range of photoacidities of well behaved photoacids. This is due to limitations imposed by the short lifetime of the excited state and/or competing excited-state reactions and also due to either a very large or a very small K^*_a value of the photoacid. These complicating conditions are very common and put limits on the usefulness of the time resolved measurements of any observable that depends on monitoring the actual progress of the proton dissociation and proton recombination reactions of the photoacid. When the limitations on the time-resolved measurements are considerable, Förster cycle calculations do not usually agree well with the directly estimated pK^*_a value of the photoacid. Such discrepancies have been attributed many times to limitations inherent to the Förster cycle and have even have led to questioning of its general validity. In comparison, it has been much less often suggested that the source of the discrepancy lies in difficulties associated with the time resolved measurements. Below we review in detail the evidence for the general validity of the Förster cycle concept and the various experimental limitations imposed on its practical use.

12.3 Evidence for the General Validity of the Förster Cycle and the K^*_a Scale

Because of its general applicability, relative ease of the steady-state measurements, and the simplicity of the thermodynamic cycle, the Förster cycle has become the main tool for the initial determination of the extent of photoacidity [9, 27, 71–78]. However, this has not been done without considerable debate about the validity of a thermodynamic approach to short-lived excited-state species. Additional concern has been with the correct identification of the thermodynamically stable excited-state energy levels using conventional steady-state optical spectroscopy in solution [16, 71, 75, 76]. Chief among the arguments against the routine use of Förster cycle has been the short ns-lifetime of the singlet state of most photoacids which many times does not allow even a partial establishment of a chemical equilibrium in the straightforward thermodynamic sense. In particular, the situation becomes unclear when either one of the two excited-state proton transfer reactions (proton dissociation and proton recombination) is much slower than the fluorescence lifetime of the excited photoacid. In the extreme situation, the photoacid may appear to be completely nonreactive within the lifetime of the excited state. Such situations prevent the observation of the excited-state proton transfer process and render impossible the determination of the excited-state K^*_a by any of the direct methods of measurement.

The general validity of the Förster cycle approach is undoubtedly linked first of all with the reality of the assumed microscopic reversibility of excited-state proton transfer reactions. Secondly, the reliability of the K^*_a scale should be checked, when possible, against directly determined K^*_a values of well behaved photoacids. Furthermore, for the general validity of the K^*_a scale to hold as defined by the Förster cycle its validity should not depend on the photoacid actually reaching equilibrium conditions or even on observing at all an excited-state proton transfer reaction within the finite lifetime of the excited state. These assertions should be carefully tested and checked before establishing the general applicability of the Förster cycle.

12.3.1 Evidence for the General Validity of the Förster Cycle Based on Time-resolved and Steady State Measurements of Excited-state Proton Transfer of Photoacids

Arguably, the first evidence for the general validity of the pK^*_a scale came from steady-state fluorescence titrations of well behaved photoacids such as 2-naphthol [27]. As already indicated, this method was largely developed by Weller [9] and resulted in K^*_a values which were in general agreement with the Förster cycle predictions (see below).

More direct evidence for the inherent microscopic reversibility of an excited-state proton transfer reaction was found in ps-time-resolved measurements of a strongly reactive photoacid, namely HPTS (Fig. 12.2). With its conjugated-base, fourfold charged, the observation of the back (geminate) recombination of the proton following the photoacid dissociation has become feasible. Pines and Huppert were first to report on the geminate recombination reaction of an excited photoacid [39–42]. They found that, following the dissociation of excited HPTS (also commercially known as pyranine), proton recombination occurred reversibly so that the ultimate fate of the so-formed excited photoacid was to dissociate again. Over relatively long times of observation, the repeated cycles of dissociation–recombination were found to occur in the excited state without quenching and to cause the populations of the reacting species to converge into a pseudo-equilibrium situation while being in the excited state. The equilibrating system was found to be continuously perturbed by the diffusion of the two geminate reactants away from each other. This gradual separation of the ion pair by diffusion away has been found to monotonically decrease the average concentration of the dissociated proton with respect to its geminate photobase anion. Following the initial series of observations and their correct physical modeling by Pines and Huppert [39–43] Pines, Huppert and Agmon [43–48] arrived at an analytic expression describing the decaying amplitude of the photoacid at long times:

$$[A^*H]_t \propto K^*_a \, (4\pi Dt)^{-3/2} \tag{12.13}$$

where t is the time elapsed from the moment of the initial dissociation of the photoacid and D is the mutual diffusion coefficient between the proton and the conjugated photobase.

Equation (5.12) was verified over relatively long observation times in conditions where only diminishing small concentrations of the photoacid remained in the excited state, down to about 10^{-4} of the initial population. (Fig. 12.9). The experimental verification of Eq. (12.13) was carried out after normalizing the decaying photoacid population with the observed fluorescence lifetime of the conjugated photobase, τ'_0, Eq. (12.14):

$$[A^*H]_t \exp(t/\tau_0') \propto K^*_a \, (4\pi Dt)^{-3/2} \tag{12.14}$$

Equation (12.14) was found to be exact by Gopich and Agmon when the nonreactive lifetime of the photoacid equals that of the photobase. Reviews of the extensive kinetic analysis done over the past 20 years in order to refine the basic geminate-recombination model have been recently published by Pines and Pines [25] and by Agmon [48].

A further step to establish the validity of the K^*_a has been undertaken by Pines and Fleming [49] and was extended by Pines et al. [50, 51] and by Solntsev et al. [52, 53]. These authors have shown that the concept of an excited-state equilibrium constant holds in the more kinetically demanding (and more general) situation of proton quenching in parallel to reversible geminate rocombination. 1-Naphthol exemplifies such a situation when, in addition to undergoing reversible proton dissociation, the population of the excited photoacid has been shown to be self-quenched by the dissociated geminate proton as well as by bulk protons. In such cases irreversible recombination (quenching) of the proton competes with

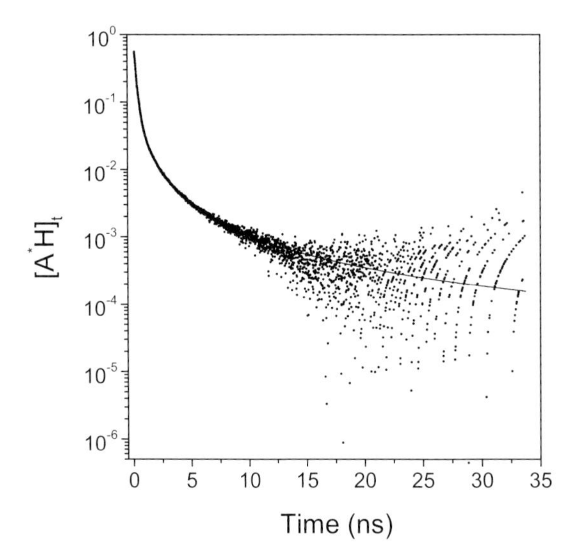

Figure 12.9 Semi-logarithmic plot of normalized fluorescence decay of HPTS. Points are experimental data ($\lambda_{ex} = 375$ nm, $\lambda_{em} = 420$ nm) after lifetime correction. Data taken in water at pH = 6.10 after background subtraction. The solid line is a numerical solution of the Debye–Smoluchovski equation (Ref. [44]) convoluted with the instrument response function. Parameters are $a = 6.30$ Å, $k_d = 125$ ps^{-1}, $k_r = 7.90$ Å ns^{-1}, $R_D = 28.3$ Å, $D = 930$ Å^2 ns^{-1} ($K^*_a = 26.5$). The asymptotic slope is –1.50. Adapted from Ref. [60].

$$A^*H \underset{k_r}{\overset{k_d}{\longleftrightarrow}} A^-* \cdots\cdots H^+ \xrightarrow{k_s} A^-* + H$$

$$\downarrow k_q$$

$$A^- + H^+$$

Scheme 1

reversible recombination of the proton at the original site of the dissociation, Scheme 12.1.

To account for the additional proton quenching reaction, the long time decay of the photoacid population, Eq. (12.14), should be corrected and take the form of Eq. (12.15) [50–52].

$$[A^*H]_t \exp(t/\tau') \propto K^*_a\, [A^{-*}]_\infty^{\ 2}\, (4\pi\, Dt)^{-3/2} \tag{12.15}$$

where $[A^{-*}]_\infty$ is the surviving fraction of the unquenched geminate pairs from the initial excited-state population. The normalized fraction of the surviving pairs is equal to the ultimate escape probability of the pair, while avoiding self-neutralization, Ω_∞. In the case of infinite lifetimes $\Omega_\infty = 1$ for reversible recombination reactions without quenching and is less than unity when a parallel quenching reaction takes place. Gopich and Agmon [54–57] have extended the above analysis even further to include conditions when the excited-state lifetimes of the photoacid differ from that of the conjugated photobase. The effect of unequal excited-state lifetimes of the photoacid and the photobase was originally considered by Weller in order to correct for pK^*_a values found by direct fluorescence-titration of the photoacid while in the excited state. He showed that the relative kinetic effectiveness of the proton dissociation reaction from the photoacid and the back-protonation reaction of the photobase depends on the ratio of their respective excited-state lifetimes [7, 9], Eq. (12.16).

$$\frac{\phi/\phi_0}{\phi'/\phi'_0} = \frac{1}{k_d \tau_0} + \frac{(k^*_r)\tau'_0}{k^*_d \tau_0} (\gamma_\pm)^2 [H^+] \tag{12.16}$$

where ϕ/ϕ_0 and ϕ'/ϕ'_0 are the relative fluorescence quantum yields of the photoacid and photobase, respectively, which change upon titration and serve to monitor its progress as a function of the concentration of the mineral acids. τ_0 and τ'_0 are the fluorescence lifetime of the photoacid and photobase, respectively, in the absence of proton transfer. $\gamma_\pm$ is the mean activity coefficient of the strong mineral acid used to titrate the photoacid. Equations (12.12) and (12.16) serve to demonstrate the macroscopic reversibility of the proton transfer reaction in the excited state while Eqs. (12.13)–(12.15) describe the microscopic reversibility of the same reaction. The observation that over long times the time dependence of the population of the photoacid followed Eq. (12.13) (or Eq. (12.15)) in the case of self-quenching [49–53, 58]) have demonstrated the general microscopic validity of K^*_a, even in a most demanding situation where the excited state is rapidly quenched back to the ground state.

The final stage of this yet unfinished saga has been to directly demonstrate the establishment of an actual excited-state (acid–base) equilibrium by performing time-resolved titration of the photoacid while in the excited-state (Fig. 12.10). This was done in conditions where proton dissociation was initiated by short laser pulse excitation in the presence of strong mineral acids [59–61]. Following the initial dissociation of the excited photoacid the population of the photoacid relaxed to its equilibrium concentration with the photobase. The relaxation-to-equilibrium process was carried out with the excited photobase reversibly reacting with both the dissociated proton and the large excess of bulk protons introduced by the mineral acid. At long times the reaction was predicted to follow Eq. (12.17) [62],

$$[A^*H]_t - [A^*H]_\infty \propto K^*_a (4\pi Dt)^{-3/2} / (1+cK_{eq})^3 \tag{12.17}$$

Equation (12.17) was verified by Pines and Pines [60] who were able to demonstrate the predicted analytic dependence of the relaxation kinetics on the bulk concentration of the mineral acid ($HClO_4$) used to titrate the excited photoacid.

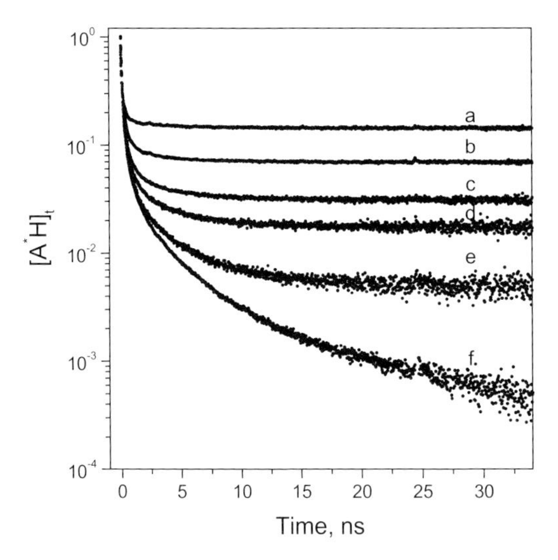

Figure 12.10 Semi-logarithmic plots of the normalized fluorescence decay of 2N6,8diS in the presence of a strong acid. Points are experimental data after lifetime correction; (a–e) in water acidified by $HClO_4$ (pH = 2.0, 2.52, 2.83, 3.11, 3.72, respectively), (f) in water at pH = 6.10, ($[HClO_4] = 0$). Adapted from Ref. [60].

The advantage of Eqs. (12.13)–(12.15) and (12.17) was that they allowed the direct determination of the excited-state equilibrium constant by a single kinetic measurement. The proton dissociation rate constant and hence also the proton recombination rate constant may also be found from the same measurement. Although this method has been applied successfully in only a few cases [60, 61], the K^*_a values thus found have been in very good agreement with K^*_a values independently estimated from the Förster cycle or by steady-state titrations.

12.3.2 Evidence Based on Free Energy Correlations

More subtle methods for verifying the validity of the Förster cycle for estimating the K^*_a of weak photoacids have been successfully used, even in cases of very weak photoacids where no apparent dissociation of the photoacid has been observed. These methods rely on measuring the reactivity of photoacids toward stronger bases than water. It was demonstrated that the reactivity of excited photoacids follows general structure–reactivity laws, photoacids having similar structural features but different excited-state acidities can be grouped and correlated [101–103] (Fig. 12.8). Once the reactivity of a photoacid has been correlated within a family of similar photoacids according to its K^*_a (either the Förster cycle value or a directly measured one), its reactivity toward strong bases could be estimated and then verified experimentally (Fig. 12.11). Such procedures using for example

Eq. (12.18) for the correlation between the proton transfer rate, $k_{p,}$ and K^*_a (the dependence on K^*_a enters through the free energy of activation term ΔG_a , see below) usually result in very good agreement between the observed reactivity of the photoacid and its Förster cycle K^*_a value.

$$k_p \propto k^*_o \exp(-\Delta G^*_a/kT) \tag{12.18}$$

where $(k^*_o)^{-1}$ is the frequency factor of the specific family of reactions, ΔG^*_a is the effective activation energy of the proton transfer reaction in the excited state which may be estimated using the Marcus BEBO equation [104], Eq. (12.19)

$$\Delta G_a = \Delta G^o/2 + \Delta G_o^{\#} + \Delta G_o^{\#} \cosh[\Delta G^o \ln 2/(2\Delta G_o^{\#})] / \ln 2 \tag{12.19}$$

or alternatively assuming the reaction takes place in the “normal” region of the celebrated Marcus charge-transfer theory (MCT) (as opposed to “inverted” reaction conditions when the activation energy increases although the reaction becomes increasingly favored thermodynamically) where the activation energy decreases when the reaction is more favorable thermodynamically. The MCT theory was originally developed for the activation free energy of electron transfer reactions in solution [34–36], Eq. (12.20)

$$\Delta G_a = (1+\Delta G^o/4\,\Delta G_o^{\#})^2\,\Delta G_o^{\#} \tag{12.20}$$

$\Delta G_o^{\#}$ is the solvent-dependent activation energy of the charge-exchange reaction when the total free energy change ($\Delta G^o = RT \log pK^*_a$) in the proton transfer is equal to zero. Eqs. (12.19) and (12.20) are practically equivalent in the photoacidity range that has been studied so far which seems to display only “normal” reaction behavior where the proton transfer rate increases monotonically as a function of the increase in the relative strength of the base compared to the acid (see Fig. 12.12).

A very convincing support for the existence of solvent controlled proton dissociation reactions in aqueous solutions has risen from the theoretical studies of Ando and Hynes [105–108] who have studied the proton dissociation of simple mineral acids HCl and HF in aqueous solutions. The two acids seem to follow a solvent-controlled proton transfer mechanism with a Marcus-like dependence of the activation energy on the acid strength. Recently, a free energy relationship for proton transfer reactions in a polar environment in which the proton is treated quantum mechanically was found by Kiefer and Hynes [109, 110]. Despite the quite different conceptual basis of the treatment the findings bear similarity to those resulting from the Marcus equation Eq. (12.19) which has been used to correlate the proton transfer rates of photoacids with their pK^*_a [101, 102].

The case of 1-hydroxypyrene is illuminating in this respect. Being one of the first photoacids studied by Weller [9], its $pK^*_a = -\log(K^*_a)$ was estimated by Weller using the Förster cycle, $pK^*_a = 3.7$, but the photoacid was not observed to dissociate in water [111]. Several explanations were offered for this apparent lack of con-

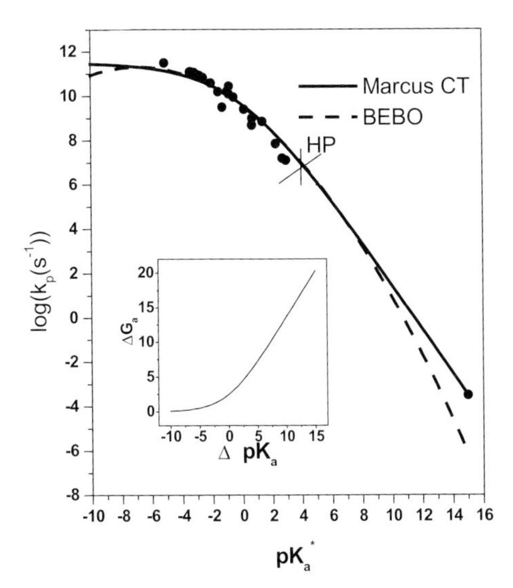

Figure 12.11 The free energy correlation found in photoacid dissociation reactions taken at room temperature. Solid line is the fit using Marcus BEBO equation (Eq. (12.19)) and dashed line the fit using Marcus CT equation (Eq. (12.20)). The free energy barrier for a symmetric proton transfer is 2.4 kcal mol^{-1} (BEBO equation) and 2.2 kcal mol^{-1} (Marcus CT equation), respectively. The dependence of the activation free energy on the pK_a^* of the photoacid as calculated using the BEBO equation (Eq. (12.19)) is shown as the inset.

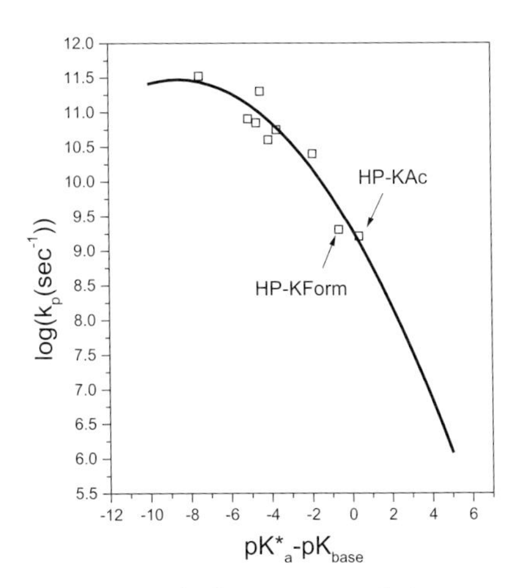

Figure 12.12 The free energy correlation found in photoacid dissociation to water and in the direct proton transfer reactions between photoacids and carboxylic bases (potassium acetate (KAc and potassium formate (KForm)) taken at room temperature. The free energy barrier, 2.9 kcal mol^{-1}, found for the total set of proton transfer reactions is 0.5 kcal higher than that found for the sub-set of the proton-dissociation-to-water reactions (Fig. 12.11).

sistency of the Förster cycle. These invoked the inadequacy of the Förster cycle for describing the photoacidity of very weak photoacids and the idea that weak photoacids may differ kinetically from strong photoacids by not dissociating so readily in the excited state. The reason given for the latter behavior invoked the electronic structure of weak photoacids which has been suggested to be inherently less polar and less reactive than that of strong photoacids [132–134]. The first singlet state of weak photoacids has been suggested to be the ^1L_b state while strong photoacids like HPTS have been thought to undergo solvent-influenced level crossing to a more polar ^1L_a state [135–137]. However, several reports have shown that 1-hydroxypyrene behaves as a proper photoacid in so far as the strengthening of its hydrogen-bonding interactions in the excited state and its ability to transfer a proton to stronger-than-water bases both in aqueous and nonaqueous solutions [9]. Using the correlation shown in Fig. 12.12 Pines et al. were able to show that in fact the reactivity of excited 1-hydroxypyrene toward acetate bases is consistent with a pK^*_a value of about 4.1, which agrees very well with the Förster cycle estimation of its pK^*_a, (Fig. 12.13). Finally, it has been demonstrated recently (Fig. 12.14) [112–114] that about 1% of the population of 1-hydroxypyrene does dissociate at room temperature ($k^*_p = 5 \times 10^6$ s^{-1}) which is in good agreement with predictions based on correlating the pK^*_a (Förster cycle) value by the BEBO model of Marcus (Fig. 12.11). It thus seems that the Förster cycle calculation does provide a reliable way of estimating K^*_a even when the proton transfer reaction is two orders of magnitude slower than the excited-state decay rate. One may also conclude from Fig. 12.12, which correlates both ^1L_b and ^1L_a, acids that the proton transfer rate within a family of photoacids is uniquely determined by the K^*_a value of the photoacid, regardless of whether it is in the ^1L_b or ^1L_a state. This means that internal changes in the electronic structure of excited photoacids leading to changes in the fluorescing level of the photoacid may not constitute a significant kinetic control for the rate of the proton transfer from weak photoacids. However, these electronic

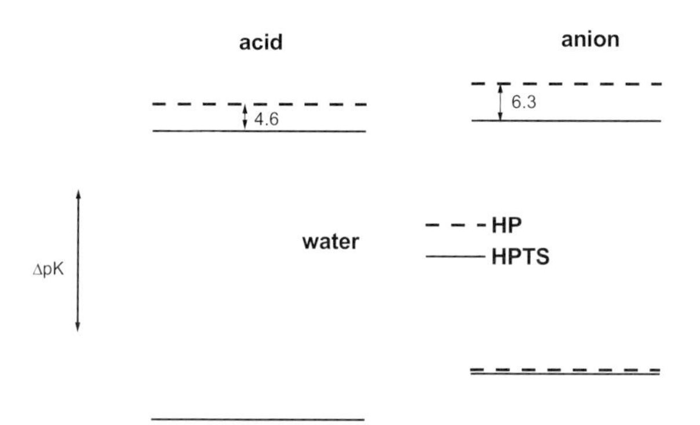

Figure 12.13 Förster cycle of HP and HPTS photoacids in water. Energy levels are for the photoacids A*H and their conjugated base $(A^-)^*$. The energy levels of both photoacids are normalized to the energy of the ground state level of the photoacids. The Förster cycle is plotted to scale in the units of p*K*, $pK = \Delta G / \log RT$.
HP: $pK_a = 8.7$, $\Delta pK^*_a = 5.1$, $pK^*_a = 3.6$. HPTS $pK_a = 8.0$, $\Delta pK^*_a = 6.1$, $pK^*_a = 1.9$.

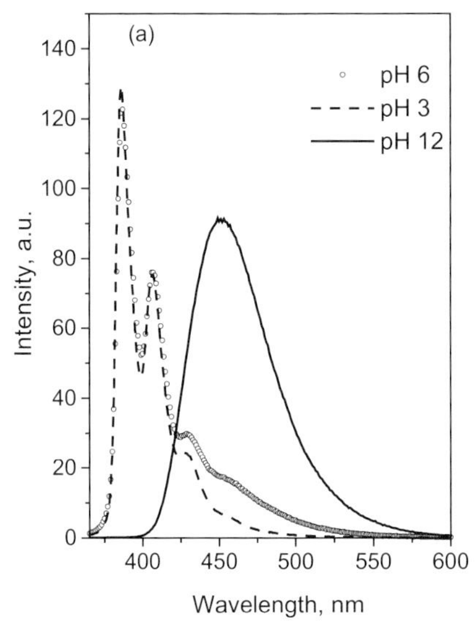

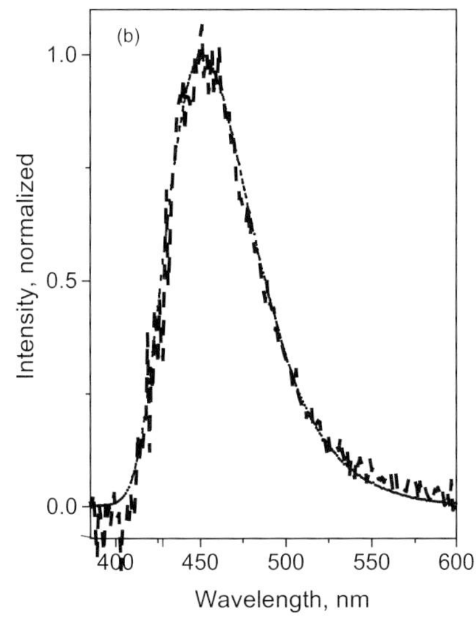

Figure 12.14 (a) Fluorescence spectra of 1-hydroxypyrene at 80 °C measured under the excitation of the photoacid at pH 6 (circles) and at pH 3 where negligible photoacid dissociation occurs due to much faster proton recombination process with bulk protons (dashes), and under direct excitation of the photobase at pH 12 (solid line). (b) Dashed line is the photobase fluorescence obtained by subtraction of fluorescence spectra of HP at pH 6 and pH 3 measured in H_2O at 20 °C. The photobase fluorescence obtained under direct excitation of the photobase at pH 12 is also plotted for comparison (solid line). Adapted from Ref. [114].

changes should result in a change in the K^*_a value of the photoacid, which in turn would affect the proton dissociation rate by means of Eqs. (12.18)–(12.20), see also the discussion in Section 12.5 below.

12.4 Factors Affecting Photoacidity

12.4.1 General Considerations

What are the important intramolecular and intermolecular factors affecting photoacidity? Clearly, photoacidity by itself is the result of some intramolecular changes in the electronic structure of the excited photoacid and its photobase enhancing its Brønsted acidity as compared to the situation existing in the electronic ground state. By the same token, acidity and basicity largely depend on the chemical and physical properties of the solvent in which the acid–base reactions take place. Judging by the available literature it seems that Förster cycle calculations are very useful in predicting the relative change in the K_a, i.e. ΔpK^*_a of most photoacids upon electronic excitation, independent of the photoacid being a strong or a weak acid in the ground state. However, the absolute value of $K^*_{a,}$ which determines the actual strength of the photoacid depends on the corresponding ground-state equilibrium constant which in turn is a solvent-dependent property. It follows that it is possible that the absolute acidity of an excited photoacid may be very low, depending on the solvent, while still exhibiting considerable Förster photoacidity.

12.4.2 Comparing the Solvent Effect on the Photoacidities of Neutral and Cationic Photoacids

Similar to the situation prevailing for uncharged ground-state acids, uncharged photoacids which exhibit strong acidity are most commonly found in water which is, arguably, the best overall solvent-medium for solvating free ions produced by the photoacid dissociation. This situation dramatically changes upon moving to less polar solvents where neutral photoacids become much weaker acids. This is essentially due to a decrease in their ground-state acidity and not because of a large decrease in their photoacidity. Some relevant data on the pK^*_a's is collected in Tables 12.1 and 12.2 where the Förster cycle photoacidity of 1-naphthol and 2-naphthol is listed in several solvents where the ground-state acidity of the photoacid is known. In comparison with the considerable decrease in the ground-state acidity on moving from aqueous solutions to the less polar solvents, the Förster acidities, as judged by the ΔpK^*_a values in the same solvents, $\Delta pK^*_a = pK_a - pK^*_a$, do not change by much and even seem to increase with decreasing solvent polarity. The extent of the marked decrease in the ground state acidity of neutral photoacids on moving from water to less polar solvents has been so overwhelming as to identify photoacidity with actually observing excited-state proton transfer which occurs almost exclusively in an aqueous environment. The term "enhanced photoacids" has been introduced by Tolbert [22–24] to describe neutral photoacids being strong enough to still appreciably dissociate in nonaqueous (albeit still polar) solvents. However, the situation becomes much more blurred when cationic photoacids (Fig. 12.3), which undergo an acid–base equilibrium of the form $AH^+ \leftrightarrows A + H^+$, are considered. In such cases proton dissociation does not cause the formation of an ion pair so the proton-dissociation reaction is isoelectric. Here the polarity of the solvent does not play a dominant role as with neutral-acid dissociation, while other factors such as solvent basicity toward the proton may become more important. Not surprisingly, cationic photoacids may even increase their acidity and their photoacidity in less polar but more basic solvents than pure water. An example is the proton dissociation reaction of protonated aminopyrenes $RN^*H_3^+$ (Fig. 12.3) [19, 25, 101, 115, 116]. The K^*_a values of this family of cationic photoacids are very large with some pK^*_a values approaching the acidity of strong mineral acids (Fig. 12.15). The proton transfer rate of these photoacids (and their pK_a values) increases in mixtures of water/organic solvent solutions until it reaches a maximum rate at water compositions of about 50–70% (M/M) [101]. Further decrease in the water content causes the proton transfer rate to decrease again until it reaches its value in the pure organic solvent. (Fig. 12.16). For tertiary amine photoacids, $RN(CH_3)_2H^+$ the maximum rate of proton transfer may be more than an order of magnitude greater than the corresponding rate in pure water. Such a complex dependence on solvent composition reveals the complex role that the solvent has in proton transfer reactions. Not only does the solvent need to stabilize the dissociating proton and its conjugated base, it also provides the hydrogen-bonding network necessary for the proton to transfer efficiently by the Grotthuss mechanism, i.e. along hydrogen-bonding networks of water [117–124]. The complex dependence of the proton transfer rate on the solvent composition is once more not a unique property of the excited photoacid–solvent system. It is rather similar in both ground-state and excited-state proton transfer reactions of cationic acids.

Tab. 12.1 pK_a, $pK_a{}^*$ and ΔpK^*_a values of 1-naphthol in water, methanol and DMSO.

Solvent	pK_a	ΔpK^*_a $^1L_a/^1L_b$	pK^*_a $^1L_a/^1L_b$ [a]
water	9.3	10.3/6.9	–1.0/2.4
methanol	13.9	11.1/8.4	2.8/5.5
DMSO	17.1[b]	12.3/9.7	4.8/7.4

a pK^*_a values when calculated by the Förster cycle using either the $S_0 \rightarrow S_2$ (^1L_a) absorption energy or the $S_0 \rightarrow S_1$ (^1L_b) absorption energy.
b Ref. 139.

Tab. 12.2 pK_a, ΔpK^*_a and pK^*_a values of 2-naphthol in water, methanol and DMSO from Förster cycle calculations.

Solvent	pK^0_a	ΔpK^*_a	pK^*_a
water	9.6	6.3	3.3
methanol	14.2	6.6	7.6
DMSO	16.2	10.0	6.2
pK(meth) – pK(water)	4.6	0.3	4.3

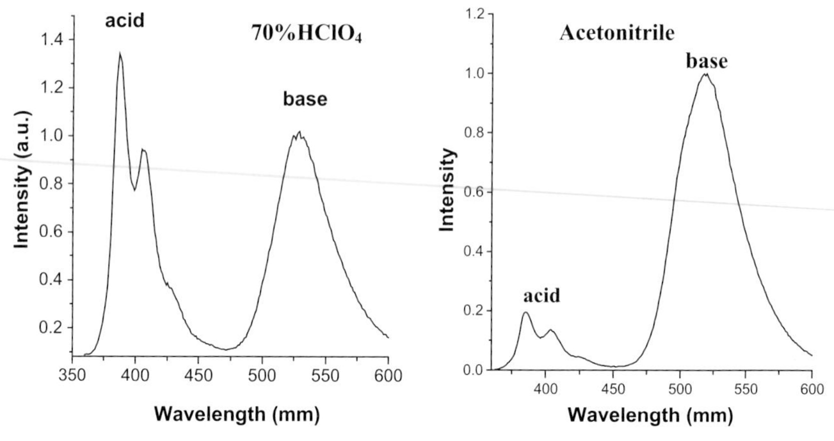

Figure 12.15 Dissociation of protonated APTS in acetonitrile and in 70% $HClO_4$. About 50% of the photoacid (fluorescence maximum at 395 nm) dissociates to form the conjugated photobase (fluorescence maximum at 530 nm) in $HClO_4$ and about 20% in acetonitrile within the excited-state lifetime of the photoacid. Adapted from Ref. [116].

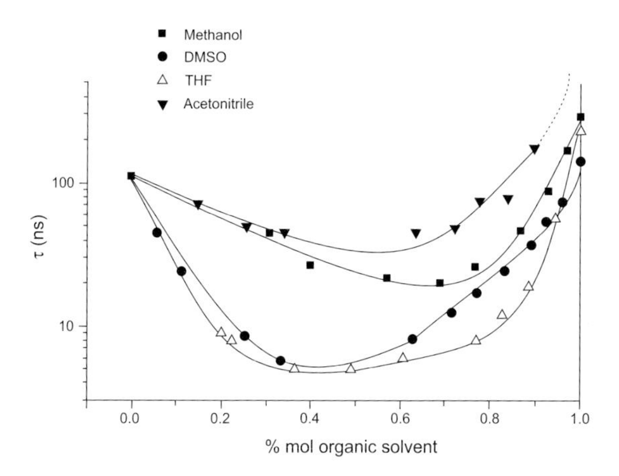

Figure 12.16 Proton dissociation lifetimes time of protonated 2-AP photoacid as a function of the molar fraction of organic co-solvent in water solutions. Ref. [116].

12.4.3
The Effect of Substituents on the Photoacidity of Aromatic Alcohols

The situation considered so far has been of the solvent affecting ground-state acidity while intramolecular changes in the charge density of the chromophore upon electronic excitation determine the extent of the photoacidity of the photoacid relative to its corresponding ground state acidity. The relative contribution of the

Förster photoacidity to the total (absolute) acidity of photoacids in their electronic excited state may be estimated directly from the corresponding pK_a and pK^*_a values. Additional questions are the extent to which photoacidity may be tuned by suitable substituents which also affect the ground state acidity of the photoacid and, alternatively, by the choice of the solvent. Table 12.3 compares the effect of several substituents on the ground- and excited-state acidities of several photoacids. The first conclusion that may be drawn from this table is that ring substituents cause the K_a and K^*_a of aromatic photoacids to change in the same direction. It follows that one may discuss the effect of various types of substituents on photoacidity using arguments and terminology that have been traditionally used for ground state acids. In particular, Hammett [127, 128] and Taft [129, 131] have contributed much to the discussion of the substituent effect on equilibrium and reactivity of aromatic acids in the ground electronic state. Their arguments seem to be valid also for the excited state of aromatic acids but with different scaling factors (i.e., different values in the Hammett Equation) [24].

Tab. 12.3 pK_a and pK_a* values of some common photoacids in water.

Photoacid	pK_a	Ref.	pK_a*	Ref.
1-naphthol	9.4	125	–0.2	49
1-naphthol 3,6-disulfonate	8.56	126	1.1	79
5-cyano-1-naphthol	8.5	103	–2.8	103
1-naphthol-4-sulfonate	8.27	74	–0.1	74
2-naphthol	9.6	125	–2.8	23
5,8-dicyano-2-naphthol	7.8	23	–4.5	23
5-cyano-2-naphthol	8.75	23	–0.3	23
8-cyano-2-naphthol	8.35	23	–0.4	23
2-naphthol-6,8-disulfonate	8.99	126	0.7	78
1-hydroxypyrene	8.7	27	3.6	27
HPTS	8.0	45	1.4	45
HPTA	5.6	116	–0.8	116

Finally, the observed net effect of substituents on either increasing or decreasing photoacidity is less than 3 pK^*_a units, even in the most extreme cases studied so far [24, 69]. This only constitutes about one third of the total acidity change upon electronic excitation. Apparently, in most cases, one may treat the substituent effect as a perturbation to the electronic structure of the unsubstituted chromophore even when the aromatic acid is in the electronic excited state [69].

Figure 12.13 compares the photoacidities of 1-hydroxypyrene and HPTS. The considerable effect of the 3 sulfonate groups that are located on the 3, 6, 8 positions of the pyrene system of HPTS is evident. The difference in the ground state acidity of the two photoacids is about 0.9 pK^*_a units, HPTS being the stronger ground-state acid. In the excited state the difference in acidity increases to about 2.3 pK^*_a units, again HPTS being the stronger photoacid. It follows that the Förster photoacidity of HPTS is larger than the Förster photoacidity of 1-hydroxypyrene by about 1.4 pK^*_a units. As the position of the OH group is the same for the two photoacids most of the change in photoacidity is likely to originate from the combined inductive effect of the 3-sulfonate groups. Apparently, the inductive effect is about three times larger in the excited state of HPTS than in the ground

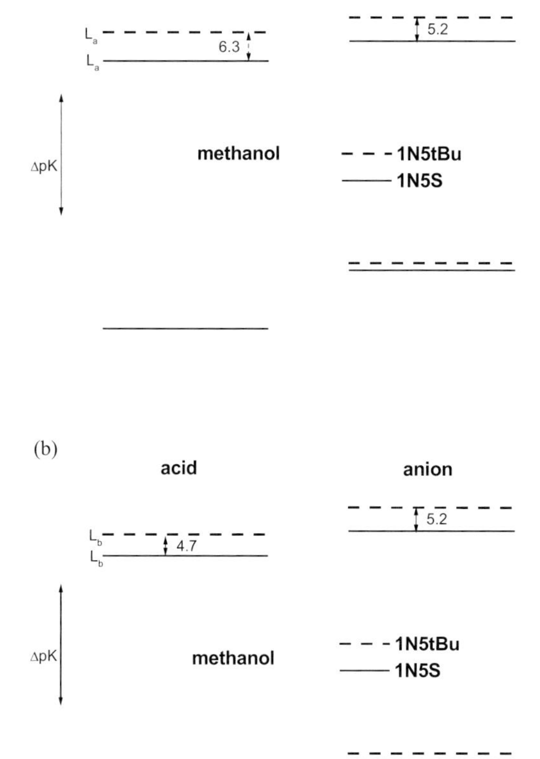

Figure 12.17 Förster cycle of 1N5tBu [150] and 1N5S photoacids in methanol. The energy levels are normalized to the energy of the ground state level of photoacids. (a) ^1L_a cycle: 1N5S: pK_a = 12.9 (estimated from pK_a in water), ΔpK^*_a = 8.5 and pK^*_a = 4.4. 1N5tBu: pK_a = 14.4 (estimated from pK_a in water), ΔpK^*_a =11.1 and pK^*_a = 3.3. (b) ^1L_b cycle: 1N5S: pK_a = 12.9 (estimated from pK_a in water), ΔpK^*_a = 7.4 and pK^*_a = 5.5. 1N5tBu: pK_a= 14.4 (estimated from pK_a in water), ΔpK^*_a = 8.4 and pK^*_a = 6.0.

state. Plotting the Förster cycle levels of the two photoacids on a normalized energy scale shows the increased photoacidity of HPTS to be the result of greater photobase stabilization, presumably because of the greater charge transfer away from the negatively charged oxygen atom to the substituted pyrene system. The enhanced charge transfer process from the OH group to the aromatic system of HPTS makes the photoacid considerably more acidic and its photobase a much weaker base as compared to the situation prevailing in the acid–base equilibrium of 1-hydroxypyrene. The net result of the two effects is to increase the acidity of HPTS over that of 1-hydroxypyrene, indicating larger sensitivity to the electronic excitation of the photobase side [69, 136, 137]. The effect of the three substituents in stabilizing the Förster level of the photoacid is nevertheless considerable. This may indicate a change in the nature of the Förster level of the photoacid, from ^1L_b in 1-hydroxypyrene to more polar S_1 level in HPTS which has some charge-transfer character [100].

As a final example we compare the photoacidities of 1-naphthol-5-sulfonate and 1-naphthol 5-tetrabutyl. The Förster cycle of the two photoacids was measured in methanol. The 5-position of the 1-naphthol system is considered the most sensitive to substituents (Table 12.4). As discussed in the HPTS case the sulfonate

Tab. 12.4 ΔpK^*_a values of some common hydroxyarene photoacids from Förster cycle calculations in methanol.

Photoacid	**ΔpK^*_a $^1L_a/^1L_b$** [a]
1-naphthol	11.1/8.4
1-naphthol-2-sulfonate	11.0/7.4
1-naphthol-3-sulfonate	11.3/8.0
1-naphthol-4-sulfonate	8.2/6.5
1-naphthol-5-sulfonate	8.5/7.4
1-naphthol-4-chlorate	9.4[b]
1-naphthol-5-tetrabutyl	11.1/8.4
1-naphthol 3,6-disulfonate	12.6/9.5
2-naphthol	6.6 [b]
2-naphthol-6,8-disulfonate	7.3 [b]
HPTS	6.9 [b]
HPTA	7.1 [b]

a ΔpK^*_a values when calculated the Förster cycle using either the $S_0 \rightarrow S_2$ (^1L_a) absorption energy or the $S_0 \rightarrow S_1$ (^1L_b) absorption energy.

b Calculated with the $S_0 \leftrightarrow S_1$ transition.

group enhances charge transfer from the oxygen atom to the aromatic system (Hammett σ_p value = 0.09). The situation is reversed with the tetrabutyl group which is an electron-withdrawing group (σ_p = − 0.197) [128]. Figure 12.17 is very illuminating in showing the large difference in the effect of the two types of substituents on the location of the Förster levels of the photoacids. Interestingly, in this case the photoacid side and the photobase side are almost equally affected by the change in substituents so the photoacidity of the two photoacids in methanol is almost equal. It follows that 5-sulfonate-1-naphthol is a much stronger photoacid, mainly because it is already a much stronger acid in the ground state having pK_a = 8.4 compared to the pK_a = 9.8 of 1-naphthol-5-tetrabutyl. Also shown in Fig. 12.17 is the substituent effect on the energy of the S_0–S_2 transition. Clearly, the more charge-transfer promoting substituent lowers the energy of the vertically accessed S_2 level more than the corresponding relative stabilization of the S_0–S_1 transition, presumably because the S_0–S_2 transition of 1-naphthol derivatives is to the polar ^1L_a state.

12.5
Solvent Assisted Photoacidity: The ^1L_a, ^1L_b Paradigm

A more complex situation may arise when polar interactions with the solvent not only stabilize the Förster levels of the photoacid and photobase but also select it. In such cases Förster acidity may depend to a large extent on the solvent. In particular, the possibility of the solvent determining the nature of the first singlet state of the photoacid has been of considerable interest and a matter of debate in recent years [132–137]. It was suggested that in polar solvents ^1L_a-type singlet states comprise the photoacidity states, while in less polar solvents less-polar ^1L_b-type singlet states make up the photoacidity states. This so-called "level inversion" in polar solvents has been suggested to be an additional, and sometimes the dominant mechanism which is responsible for neutral photoacids being so much less reactive in solvents other than water [132–137]. It has been argued that in water proton dissociation occurs from polar ^1L_a-type states which exhibit a negligible barrier for proton dissociation, while in less polar solvents proton dissociation occurs from relatively nonpolar ^1L_b-type states by an activated process which may involve a slow (solvent activated) internal conversion to the more reactive ^1L_a state. The difference in reactivity between the two close-lying singlet states of 1-naphthol was suggested to be so large that level crossing from the ^1L_b to the more polar ^1L_a state was considered the rate determining step for proton dissociation [132–134]. A support for this reactivity model was found in molecular dynamics simulations of the proton dissociation reaction of excited 1-naphthol in solvent clusters [131–134]. However, this model finds limited support in measurements of photoacids in liquid solutions.

More recently, a modified ^1L_b–^1L_a scenario was suggested to be a dominant factor in the photoacidity of HPTS in the liquid phase [135–138]. Based on sub-ps measurements of HPTS, proton dissociation was argued to consist of two activated processes. In the first stage, the local excited state of the photoacid converts to a charge transfer state from which proton dissociation occurs. However, the experimental evidence was not conclusive and this reactive model of HPTS remains under debate.

A recent time-resolved absorption study of HPTS carried out in the mid-IR range with sub-150 fs measurement did not find support for relatively slow internal conversion to a charge transferred state in the time range studied. It was observed, rather, that the patterns of the ring vibrations of excited HPTS change with the change in solvent but with no apparent time-resolved dynamics, thus suggesting solvent-influenced modification of the first singlet state of HPTS which occurs faster than the time resolution of the experiment (150 fs) [100].

As mentioned above, the extent by which the solvent influences the Förster photoacidity may be checked using the Förster cycle with the transition energies of the photoacid and the conjugated photobase measured independently in the solvent in question. However, the estimation of K^*_a values by Förster cycle in solvents where no proton transfer was observed has been a questionable practice over the more than 50 years of photoacid research. In such cases Förster cycle calculations may be carried out when it is possible to generate the conjugated photobase by deprotonating the photoacid in the ground state. This is usually done by introducing a strong proton-base to the photoacid solution. The gap between the ground-state and excited-state energy levels of the conjugated photobase may then be found by direct excitation of the ground-state photobase. Such measurements are independent of whether the photoacid dissociates appreciably in the excited-state to form the photobase. The energy gap between the two pairs of energy levels needed to complete the Förster cycle may thus be found and the value of ΔpK^*_a compiled.

Below we carry out a survey of the Förster cycle photoacidity of several well known photoacids in order to examine the $^1L_b - ^1L_a$ paradigm. Arguably, the best examples are 1-naphthol which is considered to be a ^1L_a photoacid, 2–naphthol (^1L_b acid), 1-hydroxypyrene (^1L_b) and HPTS (S_1).

Tables 12.1 and 12.2 compare the ΔpK^*_a values for 1 and 2-naphthols in water, methanol and DMSO solvents. It is clear that the Förster acidities of the photoacids do not diminish on moving from water to the less polar solvent but, rather, tend to increase in the less polar solvents. The observation that Förster cycle photoacidity increases on moving from water to less polar solvents was first made by a research group headed by Hynes [138]. The increase in the photoacidity in these solvents is more than compensated by the decrease in the ground-state acidity of the photoacid. Estimation of the ground state pK_a's in the same solvents show the considerable decrease in the overall acidity of the photoacids in nonaqueous solutions to be a ground-state effect (i.e., originating from a process also occurring in the ground state of the photoacid, so it cannot be uniquely associated with the excited electronic level of the photoacid). In addition, one may try to analyze the photoacidity behavior according to the nature of the first singlet state of the photoacid, being either ^1L_b or, presumably, the more polar ^1L_a state. Important conclusions may be drawn simply by comparing between photoacids having

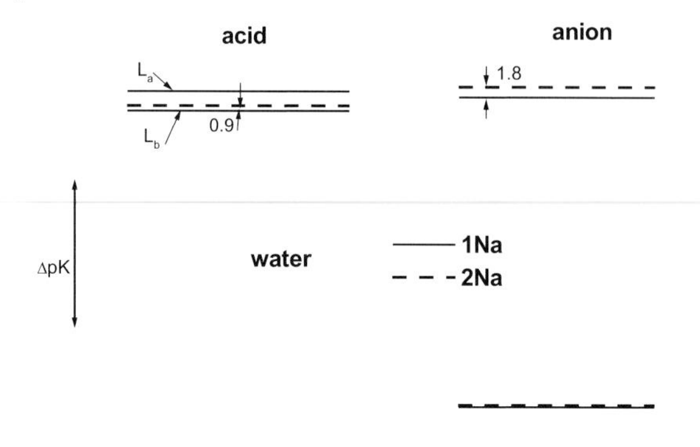

Figure 12.18 Förster cycle of 1N and 2N photoacids in water. The energy levels of both photoacids are normalized to the energy of the ground state level.1N: $pK_a = 9.3$, $\Delta pK^*_a = 10.3$ (^1L_a) and 6.9 (^1L_b), $pK^*_a = -1.0$ (^1L_a) and 2.4 (^1L_b). 2N: $pK_a = 9.6$, $\Delta pK^*_a = 6.3$ and $pK^*_a = 3.3$.

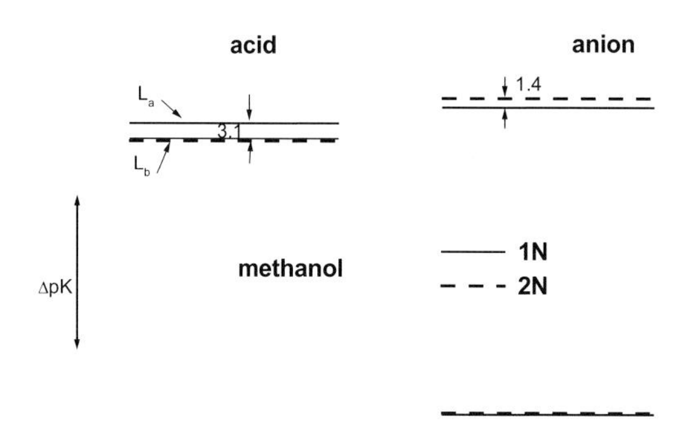

Figure 12.19 Förster cycle of 1N and 2N photoacids in methanol. The energy levels of both photoacids are normalized to the energy of the ground state level. 1N: $pK_a = 13.9$ (estimated from pK_a in water), $\Delta pK^*_a = 11.1$ (^1L_a) and 8.4 (^1L_b) $pK^*_a = 2.8$ (^1L_a) and 5.5 (^1L_b). 2N: $pK_a = 14.2$ (estimated from pK_a in water), $\Delta pK^*_a = 6.6$ and $pK^*_a = 7.6$

known excited-state level structure. Figures 12.13 and 12.17–12.21 portray the Förster cycle levels of several photoacids in different solvents as found by the spectra-averaging procedure discussed in this chapter. The figures are drawn to scale with the Förster energies of the different states shown with respect to the Förster energy of the photoacid in the ground state, which is considered the reference energy level. The energy separating between the ground state and the various excited state levels of the photoacid and the photobase is given in pK_a (pK^*_a) units.

The first general observation drawn from Tables 12.1 and 12.2 and Figs. 12.17–12.21 is that photoacids in ^1L_a-like levels are generally stronger than photoacids in ^1L_b-like levels by several pK^*_a units.

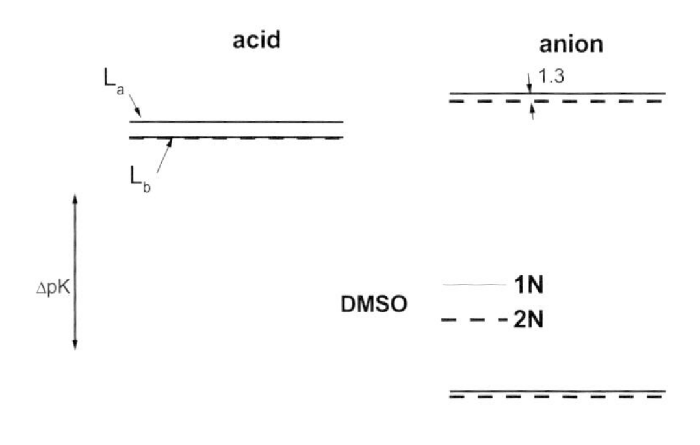

Figure 12.20 Förster cycle of 1N and 2N photoacids in DMSO. The energy levels of both photoacids are normalized to the energy of the ground state level. 1N: $pK_a = 17.1$ (estimated from pK_a in water), $\Delta pK^*_a = 12.3$ (^1L_a) and 9.7 (^1L_b), $pK^*_a = 4.8$ (^1L_a) and 7.4 (^1L_b). 2N: $pK_a = 16.2$ (estimated from pK_a in water), $\Delta pK^*_a = 10.0$, $pK^*_a = 6.2$.

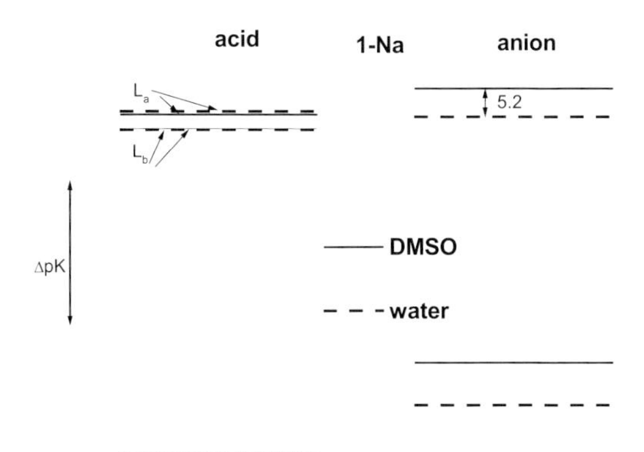

Figure 12.21 Förster cycle of 1N photoacid in DMSO and water. The energy levels of the photoacid in DMSO and water are normalized to the energy of the ground state level. DMSO: $pK_a = 17.1$ (estimated from pK_a in water), $\Delta pK^*_a = 12.3$ (^1L_a) and 9.7 (^1L_b), $pK^*_a = 4.8$ (^1L_a) and 7.4 (^1L_b). Water: $pK_a = 9.3$, $\Delta pK^*_a = 10.3$ (^1L_a) and 6.9 (^1L_b), $pK^*_a = -1.0$ (^1L_a) and 2.4 (^1L_b).

Figure 12.18 compares the locations of the Förster levels of the two isomers of naphthol, the 1-naphthol and 2-naphthol photoacids in water. The level structure of 1-naphthol is more congested than that of 2-naphthol which results in some uncertainty about the nature of the electronic level accessed from the ground state and the nature of the excited Förster level. Comparing the level structure of the two isomers reveals that the increased acidity of the 1-isomer over the 2-isomer is due to better stabilization of 1-naphtholate as compared to 2-naphtholate. This indicates a larger charge transfer process in the 1-isomer of the photobase than in the 2-naphtholate case.

When assuming the excitation to be directly to the ^1L_a state the calculated Förster photoacidity increases by 3.4 pK_a units. The observation that the S_0 to ^1L_b transition of 1-naphthol results in calculated Förster acidities considerably smaller than the experimental ones, while the S_0 to ^1L_a transition of the same molecule results in calculated Förster acidities being only slightly larger than the experimental photoacidity was made by Schulman [77] and discussed by Harris and Selinger [76]. This observation was put forward as an argument in favor of calculating the Förster cycle in cases of suspected ^1L_b to ^1L_a level crossing with the S_0–S_2 transition energy instead of the S_0–S_1 transition. However, level crossing in the acid side of 1-naphthol is unlikely to be the main reason for the Förster cycle failing to predict the extent of the photoacidity of the molecule so the better fit with experiment using the S_0–S_2 transition should be considered accidental. The improved correspondence between the Förster cycle and the experimental value of K^*_a is likely to be the result of cancellation of errors (Eq. (12.11)). The relative failure of the Förster cycle in the case of 1-naphthol is probably due to abnormally large reorganization energy in the excited photobase side of 1-naphthol and due to a large horizontal shift between S_1 and S_0 potential wells. This results in a vertical down-transition from the Förster level to high vibronic levels in the ground state. The resulting two large reorganization-energy terms apparently do not cancel out by the two other reorganization terms appearing in Eq. (12.11). The situation is different with 2-naphthol. In this case there is only a small Stokes shift between the absorption wavelength and the fluorescence wavelength in the photoacid side, pointing to an S_1 level resembling more the ground-state level than in the case of the 1-naphthol isomer.

Moving to less polar solvents usually increases the photoacidity of ^1L_a photoacids. This implies the destabilization of the Förster level of the photoacid with respect to the Förster level of the conjugated photobase on-going from aqueous to nonaqueous solutions, (Figs. 12.18–12.21). 1-Naphthol and 2-naphthol are again of particular interest. The two photoacid isomers differ by the position of the OH substituent and by the character of their first singlet state. Which of the two is more important in determining the extent of photoacidity? Comparison between the pK_a of 1-naphthol and 2-naphthol in water reveals that 1-naphthol is already the stronger acid in the ground state by about 0.3 pK_a units. In the excited state the difference in the acidity of the two naphthols isomers increases by about 150-fold, to about 2.5 pK_a units.

The increased acidity of 1-naphthol over 2-naphthol has been attributed to the first singlet state of 1-naphthol being in the more polar ^1L_a state. However, the effect of a photoacid crossing to the ^1L_a state ultimately being below a ^1L_b state is to stabilize the Förster level of the exited photoacid and so to decrease its photoacidity and not to increase it (Figs. 12.18–12.20). If level crossing in the acid side were the only process to occur then the photoacidity of 1-naphthol should have been less that that of 2-naphthol. To explain the overall increased photoacidity of 1-naphthol one has to invoke the anion side (base) of the photoacid, i.e. to assume much larger stabilization of the 1-naphtholate anion than that of the 2-naphtholate anion, which more than compensates for the changes occurring on the photoacid side. A conclusion may be drawn that the increased photoacidity of 1-naphthol over that of 2-naphthol must be due to much larger rearrangements in the electronic structure in the photobase side of 1-naphthol.

It is still a matter for further research to decide which process is dominant in making the basicity of the excited 1-naphtholate anion considerably smaller than the basicity of the 2-naphtholate anion. Is it because the α-position in the naphthalene ring is more susceptible to charge transfer than the β-position or is it because 1-naphtholate is in an altogether different (more polar) electronic state akin to the ^1L_a state of the photoacid side? Whatever the answer, it is clear that 1-naphthol is a stronger photoacid than 2-naphthol because of much greater intramolecular stabilization of the anionic charge of 1-naphtholate. These assertions are verified when the Förster cycle of the two isomers is compared. Figures 12.18–12.20 show the level structure of the two naphthol isomers both normalized to the energy of the ground state of the 2-naphthol molecule. Clearly evident is the effect of greater stabilization of the excited 1-naphtholate as compared to excited 2-naphtholate. The difference in the stabilities of the two anions increases in methanol, Fig. 12.19. The effect of assuming direct absorption to S_2 in the case of 1-naphthol is to bring the Förster level of 1-naphthol above that of 2-naphthol, which contradicts the starting point of this discussion, namely that the ^1L_a level may become the Förster level only in cases where solvent relaxation brings it down to below the ^1L_b level. Making the same assumption of direct excitation to ^1L_a in the less polar solvents DMSO and methanol only makes this inconsistency worse, Figs. 12.19 and 12.20).

Figure 12.21 compares the Förster cycle of 1-naphthol in water and in DMSO. DMSO is a polar aprotic solvent which is unable to stabilize the naphtholate anion by hydrogen-bonding interactions. This results in very large destabilization of the naphtholate anion in the ground state and a very large decrease in the ground-state acidity in DMSO, $pK_a = 17.1$, compared to 9.3 in water [139]. The situation is reversed in the excited state, where the photoacidity of 1-naphthol is much larger in DMSO than in water, ΔpK^*_a (Förster cycle) = 9.7 and 6.9 assuming ^1L_b transition, for DMSO and water, respectively. It is evident from Fig. 12.21 that the increased photoacidity in DMSO is due to a decrease in the destabilization of the excited naphtholate anion as compared to the situation in the ground state. The observation, that being in the excited state levels out the effect of the strong (specific) hydrogen-bonding interaction existing in the ground state of 1-naphtholate is important. It shows that 1-naphthol is a much weaker base in the excited state than in the ground state so it becomes insensitive in the excited state to whether the solvent is a strong hydrogen-bond donating one (water) or a very weak one (DMSO). A similar conclusion may be drawn by performing the Kamlet–Taft analysis [140–142] on the spectral shifts of 1-naphtholate in different solvents [143]. Indeed, such an analysis has shown that 1-naphtholate does not form hydrogen bonds in the excited state. Similar analysis on 2-naphtholate derivatives yielded the same conclusions [144].

As a final note we point out that proton transfer may involve a more complex mechanism than that implied by the Brønsted model for proton transfer between acid and base. One such model was recently discussed in connection with photoacidity in the gas phase. In this reaction model excited state hydrogen transfer (ESHT) occurs [145–149]. This model has been recently described by Sobolewski and Domcke [145, 146] and used in excited state dynamics studies of gas phase phenol clustered with ammonia or water molecules of the Jouvet group [147]. In this model it is argued that a level crossing between the initially excited $^1\pi\pi^*$ state and a $^1\pi\sigma^*$ state facilitates the migration of an electron from the photoacid to the solvent. The electron transfer reaction is followed by proton transfer with a net transfer of a hydrogen atom. Modeling photoacidity as ESHT has also been used as in an experiment where donor and acceptor groups are connected through a wire of ammonia molecules [148]. A conical intersection of the $^1\pi\sigma^*$ state with the S_0 state leads to an efficient internal conversion pathway for phenol–ammonia clusters. Net proton transfer on the other hand should involve at least one more step with an electron back transfer to the photoacid, producing the photobase and solvated proton as separate species. Conical intersections connecting the locally excited $^1\pi\pi^*$ state and the ground state with a $^1\pi\pi^*$ charge transfer state strongly stabilized by the transfer of a proton is the mechanism of the enhanced excited-state decay of hydrogen-bonded aromatic dimers in the experiments of Hertel group [149]. However, Brønsted photoacidity as described by the Förster cycle has been found to be the major mechanism of proton transfer from photoacids in aqueous solutions. However, parallel reactive routes which involve electron transfer apparently do exist in aqueous solutions of photoacids where proton and electrons are both detected following photoexcitation with extra energy over the S_1 excitation energies.

12.6 Summary

Photoacids have been consistently proven to be important research tools in solution chemistry over the past 50 years. Understanding photoacidity is a basic requirement for this field to develop further and to expand behind its traditional boundaries of acid–base research in aqueous solutions.

Photoacidity is affected by intramolecular charge rearrangement which is initiated by optical excitation and also by intermolecular interactions with the solvent. The Förster cycle is the primary tool for estimating the photoacidity of an aromatic dye. Important insight has been gained on the factors affecting photoacidity using direct time-resolved measurements and Förster cycle considerations:

1. Photoacidity may be defined regardless of proton dissociation occurring or not during the finite lifetime of the photoacid in the excited state. Photoacidity may also be defined in nonpolar solvents using the Förster cycle.
2. Generally speaking, photoacids behave similarly to ground-state acids as far as the solvent effect and the substituent effect on acidity and photoacidity is concerned.

3. Marked photoacidity is mainly the result of large rearrangements in the anion (photobase) side of the photoacid equilibrium.
4. Structural factors of an intramolecular process increasing the stability of only the photoacid side will decrease photoacidity. It follows that the reasons for the increased photoacidity of ^1L_a photoacids which undergo internal conversion from the locally excited, less polar ^1L_b state to the polar ^1L_a state should be found in the photobase side of the photoacid–photobase equilibrium rather than in the photoacid side.
5. Partial charge transfer to the aromatic ring may be important both in the photoacid and the photobase side. In the acid side this process is usually of a smaller magnitude than in the photobase side and will make the acidic proton more positively charged and more susceptible to hydrogen bonding. In the photobase side charge transfer to the aromatic ring is considerable and acts to increase the total photoacidity of the photoacid.

Acknowledgments

We cordially acknowledge the important contributions by our former group members Ben-Zion Magnes and Tamar Barak.

We are also grateful for the financial support of the German-Israeli Foundation for Scientific Research and Development (Project GIF 722/01), the Israel Science Foundation (Project ISF 562/04) and The James Franck German-Israeli Binational Program on Laser-Matter Interaction.

References

1 Brønsted, J. N., *Rev. Trav. Chim. 42* (1923), 718.
2 Brønsted, J. N., *Chem. Rev. 5* (1928), 231.
3 Förster, Th., *Natur. 36* (1949), 186.
4 Förster, Th., *Z. Electrochem. Angew. Phys. Chem. 42* (1952), 531.
5 Förster, Th., *Elektrochem. 54* (1950), 42.
6 Förster, Th., *Pure Appl. Chem. 4* (1970), 43.
7 Weller, A., *Z. Electrochem. 58* (1954), 849.
8 Weller, A., *Z., Phys. Chem. N. F. 17* (1958), 224.
9 Weller, A., *Prog. React. Kinet. 1* (1961), 187.
10 Urban, W., Weller, A., *Ber. Bunsen-Ges. Phys. Chem. 67* (1958), 787.
11 Weller, A. H., *Discuss. Faraday Soc. 27* (1959), 28.
12 Eigen, M., Kruse, W., De Maeyer, L., *Prog. React. Kinet. 2* (1964), 285.
13 Eigen, M., *Angew. Chem., Int. Ed. Engl. 3* (1964), 1.
14 Wei, Y., Wang, W., Yeh, J. M., Wang, B., Yang, D. C., Murray, J. K., *Adv. Mater. 6* (1994), 372.

15 Martinov, I. Y., Demyashkevich, A. B., Uzhinov B. M., Kuzmin, M. G., *Russ. Chem. Phys. 46* (1977), 3.

16 Politi, M. J., Brand, O., Fendler, J. H., *J. Phys. Chem. 89* (1985), 2345.

17 Vander Donckt, E., *Prog. React. Kinet. 5* (1970), 273.

18 Schulman, S. G., in *Modern Fluorescence Spectroscopy*, Wehry, E. L. (Ed.), Plenum Press, New York, 1976, Vol. 2, p. 239.

19 Shisuka, H., *Acc. Chem. Res. 18* (1985), 41.

20 Arnaut, L. G., Formosinho, S. J., *J. Photochem. Photobiol. A 75* (1993), 21.

21 Van, P., Shukla, D., *Chem. Rev. 93* (1993), 571.

22 Tolbert, L. M., Haubrich, J. E., *J. Am. Chem. Soc. 112* (1990), 863.

23 Tolbert, L. M., Haubrich, J. E., *J. Am. Chem. Soc. 116* (1994), 10593.

24 Tolbert, L. M., Solntsev, K. M., *Acc. Chem. Res. 35* (2002), 19.

25 Pines, E., Pines, D., in *Ultrafast Hydrogen Bonding Dynamics and Proton Transfer Processes in the Condensed Phase*, Elsaesser, T., Bakker, H. J. (Eds.), Kluwer Academic Publishers, Dordrecht, 2002, pp. 155–184.

26 Pines, E., in *Chemistry of Phenols*, Rappoport Z. (Ed.), Wiley, New York, 2003, pp. 491–529.

27 Ireland, J. F., Wyatt, P. A. H., *Adv. Phys. Org. Chem. 12* (1976), 131.

28 Agmon, N., *J. Phys. Chem. A 109* (2005), 13.

29 Luecke, H., Richter, H.T., Lanyi J. K., *Science 280* (1998), 1934.

30 Kühlbrandt, W., *Nature 406* (2000), 569.

31 DeCoursey, T. E., *Physiol. Rev. 83* (2003), 475.

32 Zimmer, M., *Chem. Rev. 102* (2002), 759.

33 Kosower, E. M., Huppert, D., *Annu. Rev. Phys. Chem. 37* (1986), 127.

34 Marcus, R. A., J. *Annu. Rev. Phys. Chem.* 15 (1964), 155.

35 Marcus, R. A., *J. Am. Chem. Soc.* 91 (1969), 7244.

36 Marcus, R. A., *Faraday Discuss. Chem. Soc.* 74 (1982), 7.

37 Kuznetov, M., Ustrup, J., *Electron Transfer in Chemistry and Biology*, Wiley- Interscience, Chichester, 1999.

38 *Electron Transfer in Chemistry*, Balzani, V. (Ed), Wiley-VCH, Weinheim, 2001.

39 Pines, E., Geminate Recombination in Excited State Proton Transfer, Ph.D. Thesis, Tel-Aviv University, 1989.

40 Pines, E., Huppert, D., *Chem. Phys. Lett. 126* (1986), 88.

41 Huppert, D., Pines, E., Intermolecular proton transfer, in *Advances in Chemical Reaction Dynamics*, NATO ASI Series Vol. 184 D, Riedel Publishing Company, 1986, pp. 171–178.

42 Pines, E,. Huppert, D., *J. Phys. Chem. 84* (1986), 3576.

43 Pines, E., Huppert, D., Agmon, N., *J. Chem. Phys. 88* (1988), 5620.

44 Agmon, N, Pines, E., Huppert, D., *J. Chem. Phys. 88* (1988), 5631.

45 Pines, E., Huppert, D., *J. Am. Chem. Soc. 111* (1989), 4096.

46 Huppert, D., Pines, E., Agmon, N., *J. Opt. Soc. Am. B 7* (1990), 1545.

47 Pines, E., and Huppert, D., Agmon, N., *J. Phys. Chem. 95* (1991), 666.

48 Agmon, N., *J. Phys. Chem. A 109* (2005), 13.

49 Pines, E., Fleming, G. R., *Chem. Phys. 183* (1994), 393.

50 Pines, E., Tepper, D., Magnes, B.-Z., Pines, Barak, T., *Ber. Bunsen-Ges. Phys. Chem. 102* (1998), 504.

51 Pines, E., Magnes, B-Z., Barak, T., *J. Phys. Chem. A 105* (2001), 9674.

52 Gopich, I.V., Solntsev, K. M., Agmon, N., *J. Chem. Phys. 110* (1999), 2164.

53 Solntsev, K. M., Abouou Al-Ainain, S., Il'ichev Y. V., Kuzmin M. G., *Photochem. Photobiol. A 175* (2005) 178.

54 Agmon, N., Gopich I. V., *Chem. Phys. Lett. 302* (1999), 399.

55 Gopich, I. V., Agmon, N., *J. Chem. Phys. 110* (1999), 10433.

56 Gopich, I. V., Agmon, N., *Phys. Rev. Lett. 84* (2000), 2730.

57 Agmon, N., Gopich, I. V., *J. Chem. Phys. 112* (2000), 2863.

58 Solntsev, K. M., Huppert, D., Agmon, N., *Phys. Rev. Lett. 15* (2001), 3427.

59 Huppert, D., Goldberg, S., Masad, Y., Agmon, N., *Phys. Rev. Lett. 68* (1992), 3932.

60 Pines, D., Pines, E., *J. Chem. Phys. 115* (2001), 951.

61 Solntsev, K. M., Huppert, D., Agmon, N., *J. Phys. Chem. A 105* (2001), 5868.
62 Gopich, I. V., Ovchinnikov, A. A., Szabo, A., *Phys. Rev. Lett. 86* (2001), 922.
63 Tolbert, L. M., Nesselroth, S. M., *J. Phys. Chem. 95* (1991), 1031.
64 Carmeli, I., Huppert, D., Tolbert, L. M., Haubrich, J. E., *Chem. Phys.,Lett. 260* (1996) 109.
65 Huppert, D., Tolbert, L., M., Linares-Samaniego, S. L., *J. Chem. Phys. A 101* (1997), 4602.
66 Solntsev, K. M., Huppert, D., Agmon, N., Tolbert, L. M., *J. Am. Chem. Soc. 120* (1998), 7981.
67 Solntsev, K. M., Huppert, D., Agmon, N., Tolbert, L. M., *J. Phys. Chem. A 104* (2000), 4658.
68 Clower, C., Solntsev, K.M., Kowalik, J., Tolbert, L. M., Huppert, D., *J. Phys. Chem. A 106* (2002), 3114.
69 Agmon, N., Rettig, W., Groth, C., *J. Am. Chem. Soc., 124* (2002), 1089.
70 Htun, M., Th., Suwaiyan, A., Klein, U. K. A., *Chem. Phys. Lett., 243* (1995), 506.
71 Grabowski, Z. R., Grabowska, A., *Z. Phys. Chem., 101* (1976), 197.
72 Mulder, W. H., *J. Photochem. Photobiol. A 161* (2003), 21.
73 Wehry, E. L., Rogers, L. B., *J. Am. Chem. Soc. 87* (1965), 4234.
74 Henson, M. C., Wyatt, A. H., *J. Chem. Soc., Faraday Trans. 2 71* (1974), 669.
75 Harris, C. M., Selinger, B. K., *J. Phys. Chem. 84* (1980), 891.
76 Harris, C. M., Selinger, B. K., *J. Phys. Chem. 84* (1980), 1366.
77 Schulman, S. G., *Spectrosc. Lett. 6* (1973), 197.
78 Schulman, S. G. Rosenberg, L. S.,Vincent, W. R. Jr., Underberg, W. J. M., *J. Am. Chem. Soc. 102* (1979),142.
79 Lee, J., Robinson, G. W., Webb, S. P., Philips, L.A., Clark, J. H., *J. Am. Chem. Soc. 108* (1986), 6538.
80 Rived, F., Roses, M., Bosch, E., *Anal. Chim. Acta 374* (1998), 309.
81 Roses, M., Rived, F., Bosch, E., *J. Chromatogr. A 867* (2000), 45.
82 Gutman, M., Huppert, D., Pines, E., *J. Am. Chem. Soc. 103* (1981), 3709.
83 Pines, E., Huppert, D., *J. Phys. Chem. 87* (1983), 4471.
84 Gutman, M., Nachliel, E., Gershon, E., Giniger, R., Pines, E., *J. Am. Chem. Soc. 105* (1983), 2210.
85 Gutman, M., *Methods Biochem. Anal. 30* (1984), 1.
86 Gutman, M., Nachliel, E., *Annu. Rev. Phys. Chem. 48* (1997), 3.
87 Platt, J. R., *J. Chem. Phys.* 1949, 17, 484.
88 Platt, J. R., *Systematics of the Electronic Spectra of Conjugated Molecules: A Source Book*, John Wiley & Sons, Chicago, 1964.
89 Baba, H., Suzuki S., *J. Chem. Phys. 35* (1118), 1961.
90 Suzuki, S., Baba, H., *Bull. Chem. Soc. Jpn. 40* (2199), 1967.
91 Suzuki, S., Fujii T., Sata, K., *Bull. Chem. Soc. Jpn. 45* (1937), 1972.
92 Sano, H., Tachiya, M., *J. Chem. Phys. 71* (1276), 1979.
93 Kasha, M., *Discuss. Faraday Soc. 9* (14), 1950.
94 Tramer, A., Zaborovska, M., *Acta Phys. Pol. 24* (1968), 821.
95 Magnes, B-Z., Strashnikova, N., Pines, E., *Isr. J. Chem. 39* (1999), 361.
96 Magnes, B.-Z., Pines, D., Strashnikova, N., Pines, E., *Solid State Ionics 168* (2004), 225.
97 Rini, M., Magnes, B.-Z., Pines, E., Nibbering, E. T., *Science 301* (2003), 349.
98 Rini, M., Mohammed, O. F., Magnes, B.-Z., Pines, E., Nibbering, E. T. J, Bimodal intermolecular proton transfer in water: photoacid-base pairs studied with ultrafast infrared spectroscopy, in "Ultrafast Molecular Events in Chemistry and Biology", J. T. Hynes, M. Martin (Eds.), Elsevier, Amsterdam, 2004, pp. 189–192.
99 Rini, M., Magnes, B.-Z., Pines, D., Pines, E., Nibbering, E. T. J., *J. Chem. Phys. 121* (2004), 959.
100 Mohammed, O. F., J. Dreyer, Magnes, B.-Z., Pines E., Nibbering, E.T. J., *ChemPhysChem 6* (2005), 625.
101 Pines, E., Fleming, G. R., *J. Phys. Chem. 95* (1991), 10448.
102 Pines, E., Magnes, B. Z., Lang, M. J., Fleming, G. R., *Chem. Phys. Lett. 281* (1997), 413.
103 Pines, E. Pines,D., Barak, T., Magnes, B.-Z., Tolbert, L. M., Haubrich, J. E.,

Ber. Bunsen-Ges. Phys. Chem. 102 (1998), 511.
104 Cohen, A. O., Marcus, R. A., *J. Phys. Chem. 72* (1968), 4249.
105 Ando, K., Hynes, J.T., in *Structure and Reactivity in Aqueous Solution: Characterization of Chemical and Biological Systems*, Cramer, C. J., Truhlar, D. G. (Eds.), American Chemical Society, Washington DC, 1994.
106 Ando, K., Hynes, J.T., *J. Mol. Liq. 64* (1995), 25.
107 Ando, K., Hynes, J. T., *J. Phys. Chem. B 101* (1997), 1046.
108 Ando, K., Hynes, J. T., *J. Phys. Chem. A 103* (1999), 10398.
109 Kiefer, P. M., Hynes, J. T., *J. Phys. Chem. A* 106 (2002), 1834.
110 Kiefer, P. M., Hynes, J. T., *J. Phys. Chem. A 106* (2002), 1850.
111 Thomas, J. K., Milosavljevic, B. H., *Photochem. Photobiol. Sci., 1* (2002), 1.
112 Pines, D. Barak, T. Magnes, B.-Z., Pines, E., The Photoacidity of 1-Hydroxypyrene from Förster Cycle to Marcus Theory, Summer School, *Molecular Basis of Fast and Ultrafast Processes*, Algarve (Portugal), June 12th-15th 2003.
113 Kombarova, S. V., Il'ichev, Y. V., *Langmuir 20* (2004), 6158.
114 Magnes, B. Z., Barak, T., Pines, D., Pines, E., The Photoacidity of 1-Hydroxypyrene from Förster Cycle to Marcus Theory, in preparation.
115 Shiobara, S., Tajima, S., Tobita, S., *Chem. Phys. Lett. 380* (2003), 673.
116 Barak, T., Dynamic and Thermodynamics Aspects of Proton Transfer from Cationic and Neutral Photoacids in Solutions, Ph.D. Thesis, Ben-Gurion University, 2005.
117 de Grotthuss, C. J. T., *Ann. Chim. Phys. LVIII* (1806), 54.
118 Danneel, *Z. Elektrochem. Angew. Phys. Chem. 11* (1905), 249.
119 Agmon, N., *Chem. Phys. Lett. 244* (1995), 456.
120 Hynes, J. T., *Nature 397* (1999), 565.
121 Marx, D., Tuckerman M. E., Hutter, J., Parrinello, M., *Nature 397* (1999), 601.
122 Vuilleumier, R., Borgis, D., *J. Chem. Phys. 111* (1999), 4251.
123 Schmitt U. W., Voth, G. A., *J. Chem. Phys. 111* (1999), 9361.
124 Lapid, H., Agmon, N., Petersen, M. K., Voth, G. A., *J. Chem. Phys. 122* (2005) in press.
125 Bryson, A., Matthews, R. W., *Aust. J. Chem. 16* (1963), 401.
126 *Dictionary of Organic Compounds*, 5th edn., Chapman and Hall, New York, 1982.
127 Hammett, L. P., *Trans. Faraday Soc., 34* (1938), 156.
128 Hammett, L. P., *Physical Organic Chemistry*, McGraw-Hill Book Co., New York, 1940, p. 186.
129 Taft, R. W., *J. Am. Chem. Soc. 74* (1952), 3120.
130 Taft, R. W., *J. Am. Chem. Soc. 86* (1964), 5189.
131 Taft, R. W., Lewis, I. C., *J. Am. Chem. Soc. 80* (1958), 2441.
132 Knochenmuss, R., Muino, P. L., Wickleder, C., *J. Phys. Chem. 100* (1996), 11218.
133 Knochenmuss, R., Leutwyler, S., *J. Chem. Phys. 91* (1989), 1268.
134 Knochenmuss, R., Fischer, I., *Int. J. Mass Spectrom.12137* (2002), 1.
135 Tran-Thi, T.-H., Gustavsson, T., Prayer, C., Pommeret, S., Hynes, J. T., *Chem. Phys. Lett. 329* (2000), 421.
136 Hynes, J. T., Tran-Thi, T.-H., Gustavsson, T., *J. Photochem. Photobiol. A 154* (2002), 3.
137 Tran-Thi, T.-H., Prayer C., Millie P., Uznansky, P., Hynes, J. T., *J. Phys. Chem. A 106* (2002), 2244.
138 Hynes, J. T., personal communication.
139 Bordwell, F. G., *Acc. Chem. Res. 21* (1988), 456.
140 Taft, R. W., Abboud, J.-L., Kamlet, M. J., *J. Org. Chem. 49* (1984), 2001.
141 Abraham, M. H., Buist, G., J., Grellier, P. L., Mcill, R. A., Ptior, D. V., Oliver, S., Turner, E., Morris, J. J., Taylor P.J., Nicolet, P., Maria, P.-C., Gal, J-F., Bboud, J.-L. M., Doherty, R. M., Kamlet, M. J., Shuely, W. J., Taft, R. W., *J. Phys. Org., Chem., 2* (1989), 540.
142 Reichardt, C., *Solvents and Solvent Effects in Organic Chemistry*, 2nd edn., VCH, Weinheim, 1990.
143 Magnes, B. Z., Photophysics and Solute–Solvent Interactions in the Excited

Electronic State of Photoacids. Ph.D. Thesis, Ben-Gurion University (2004).

144 Solntsev, K. M., Huppert, D., Agmon, N., *J. Phys. Chem. A 102* (1998), 9599.

145 Sobolewski, A. L., Domcke, W., Dedonder-Lardeux, C., Jouvet, C., *Phys. Chem. Chem. Phys. 4* (2002), 1093.

146 Domcke, W., Sobolewski, A. L., *Science 302* (2003), 1693.

147 David, O., Dedonder-Lardeux, C., Jouvet, C., *Int. Rev. Phys. Chem. 21* (2002), 499.

148 Tanner, C., Manca, C., Leutwyler, S., *Science 302* (2003), 1736.

149 Schultz, Th., Samoylova, E., Radloff, W., Hertel, I. V., Sobolewski, A. L., Domcke, W., *Science 306* (2004), 1765.

150 Synthesized by Prof. L. M. Tolbert's group, School of Chemistry and Biochemistry, Georgia Institute of Technology, Atlanta, GA 30332-0400.

13
Design and Implementation of "Super" Photoacids

Laren M. Tolbert and Kyril M. Solntsev

13.1
Introduction

Electron and proton transfer (see Eqs. (13.1) and (13.2), involve obvious similarities. When the starting materials are neutral, both result in the formation of charge and in the necessity for separation of charge to stabilize the product states. In principle, both can occur in the ground state, but transfer that is exoergic only in the excited state allows the use of time-resolved spectroscopic techniques to determine the details of solvation and structural reorganization. For electron transfer, the development of such techniques and the accompanying theoretical rationale, most especially the Marcus theory, has been one of the triumphs of modern mechanistic chemistry.

$$D + A \rightarrow D^{+\cdot} + A^{-\cdot} \tag{13.1}$$

$$AH + B \rightarrow A^- + BH^+ \tag{13.2}$$

The relationship between driving force and *proton* transfer has been much more elusive despite considerable evidence that the vast photosynthetic electron transfer machinery mainly exists to set up a charge gradient to drive proton transfer. This is due to a combination of two factors. First, there is a vast reservoir of readily available materials with which to examine electron transfer. Second, the relationship between rates and driving force for electron transfer, based upon the excitation energies and the relevant redox potentials (the Rehm–Weller equation [1]) is reasonably straightforward.

In contrast, the relationship between rates and driving force for excited-state

$$pK_a^* = pK_a - \Delta E_{0,0}/2.3RT \tag{13.3}$$

proton transfer (based upon relative pK_as and calculated through the Förster equation, Eq. (13.3) [2]) is less straightforward. For instance, the existence of a so-called "inverted" region for proton transfer has been the subject of much controversy

Hydrogen-Transfer Reactions. Edited by J. T. Hynes, J. P. Klinman, H. H. Limbach, and R. L. Schowen

ISBN: 978-3-527-30777-7

(see Chapter 10 by J. T. Hynes). This is due to a number of factors, of which the most important is the presence of at least one additional reaction coordinate, that of the proton transfer event itself. There are also a number of additional coordinates. For instance, because of the large solvent proton affinities, the proton is invariably accompanied by solvent ("chaperoned") on its trajectory from conjugate acid to base. Moreover, for hydroxylic molecules, there may be intervening solvent molecules between acid and base (defined by the Grotthus mechanism) [3] which may involve discrete protonated intermediates (stepwise mechanism) or a concerted mechanism with a single transition state (see Fig. 13.1). Finally, the solvation or protonation of the incipient conjugate base may contribute to the reaction mechanism. The kinetic method of pK_a^* determination is much more reliable, though not so widely used (see below).

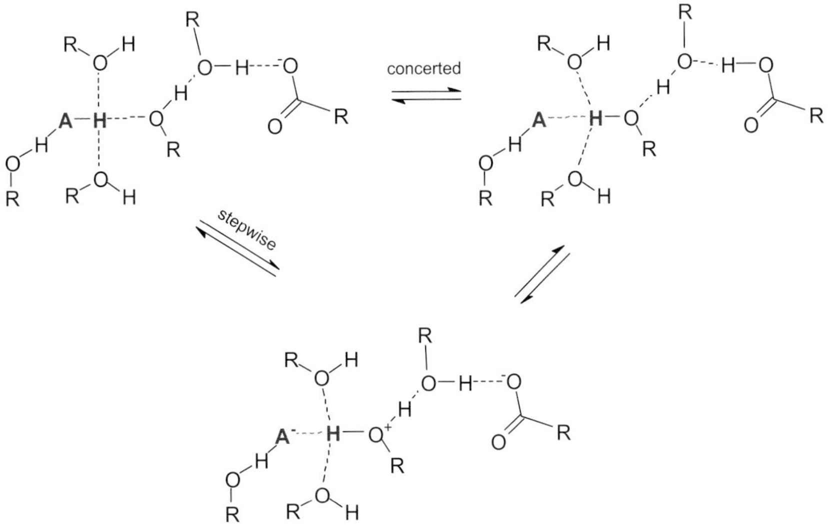

Figure 13.1 Stepwise or concerted proton transfer.

Since the vast photosynthetic and photoresponsive electron-transfer machinery exists largely to create a proton gradient, proton transfer is arguably a more important reaction. But because most redox systems, both natural and synthetic, can be optically pumped, electron transfer has been more extensively studied, leading to one of the triumphs of modern mechanistic chemistry, the Marcus theory. Progress on proton transfer has been more sluggish, due both to a paucity of natural and synthetic candidates for optically-pumped proton transfer and to the need for more sophisticated models for this complicated process. Moreover, since proton transfer in nature involves heterogeneous systems, understanding the *structural* and *energetic* requirements for proton transfer, including the initial solvation, nuclear motion within the transition structure, charge separation, and diffusional recombination becomes paramount. On a broader level, this will facilitate progress in a myriad of areas in which proton transfer is required, including "green" solvents such as supercritical fluids and "solventless" systems. This requires the development of systems which couple photoacids to biochemical reaction mechanisms.

The microscopic details of proton transfer are shown in Fig. 13.1. These include the initial hydrogen bonding step, solvation of the transferring proton, the Grotthus "relay" mechanism, and the nature of the proton acceptor. For the last several years, we and others have sought to elucidate details of each of these factors, ultimately weaving these into a comprehensive picture of proton transfer. Since many of the kinetic steps are extremely fast, this effort has required the synthesis of previously unavailable strong photoacids which result in ultrafast proton transfer upon photoexcitation, preparation of model compounds which incorporate "solvent" molecules, and rigorous photophysical analysis of the rates of proton transfer resulting from photoexcitation.

A number of molecules have been predicted to be strong photoacids [4, 5] An examination of the simple thermodynamic cycle represented by Eq. (13.3) for a proton-containing molecule and its conjugate base yields the simple yet deceptive prediction that in its excited state the molecule is a stronger acid than in the ground state if the absorption or emission spectrum of the conjugate base is characterized by a bathochromic shift relative to that of the conjugate acid. Thus molecules that undergo significant colorization upon deprotonation, e.g., triarylmethane dyes, should be powerful proton donors. Unlike the relatively simple Marcus theory for electron transfer, proton transfer rates cannot be correlated well with driving force. For instance, 9-phenylfluorene has a predicted pK_a^* of –13, yet it is photochemically inert, even in the presence of bases. Such thermodynamic acidity has not been evinced in prototropic behavior for photoexcited hydrocarbons to yield *observed* excited-state carbanions during our studies of carbanion photochemistry [6], although Wan has developed several possibilities for their intermediacy [7]. In contrast, hydroxyarenes have a long and rich history of use in excited-state proton transfer studies. The occurrence of proton transfer has been shown to be strongly correlated with the degree of ground-state hydrogen bonding, which provides a preorganization step for the proton transfer [8]. Thus oxygen-centered hydroxyarenes, which exhibit this phenomenon, have formed the basis for most such studies. However, their relatively high pK_a has limited most such studies to aqueous solvent systems.

A limited number of commercially available hydroxyarenes, including hydroxypyrenetrisulfonate ("pyranine", HPTS), have somewhat higher photoacidities than naphthols and have been used as photoacids for experiments requiring instant changes in acidity, e.g., "pH jump." Again, HPTS is limited to aqueous solvents. In order to improve photoacidity, therefore, strategic substitution by an electron-withdrawing group should result in proton transfer competitive with excited-state decay. The generic system is shown in Fig. 13.2. The hydroxyarenes have conjugate bases that are odd-alternant ions and thus yield to straightforward theoretical treatments. Moreover, odd-alternant ions have nonbonding molecular orbitals that are oxygen centered, while the excited states invariably produce charge distribu-

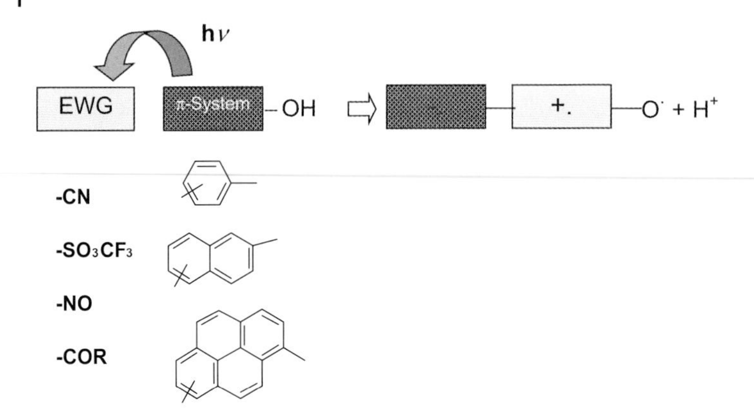

Figure 13.2 Generic photoacid.

tion at sites distal from oxygen, reducing the basicity of the excited-state anion and, by analogy, increasing the acidity of the conjugate base. This is equivalent to Weller's "intramolecular charge transfer" rationalization of the acidity in photoexcited hydroxyarenes [9]. Further compelling and instructive examples are provided by Lewis' studies of 3-hydroxy and 4-hydroxystilbene [10]. 3-Hydroxystilbene, for which the conjugate base does not allow ground-state delocalization into the distal aromatic ring, shows considerable excited-state basicity through population of a more delocalized excited state.

13.2 Excited-state Proton Transfer (ESPT)

13.2.1 1-Naphthol vs. 2-Naphthol

Naphthalene possesses nearly degenerate excited states, L_a and L_b, which are characterized by their polarization along the long axis ("through bond" or L_b) or along the short axis ("through atom" or L_a). The differences in symmetries can be understood by an appeal to the topologies of the HOMO and the two close lying LUMOs (see Fig. 13.3). Substitution by the hydroxy group apparently produces a reversal

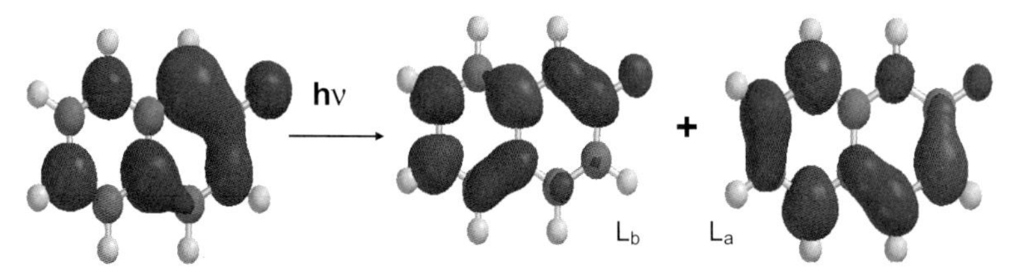

Figure 13.3 Relevant MOs for 2-naphtholate.

in the position of the lowest excited state, depending upon the site of substitution. For 1-naphthol the lowest energy state is L_a and for 2-naphthol L_b.

The photophysics of 1-naphthol is complicated by a pronounced proton quenching, which is manifested by the near absence of neutral fluorescence at all pHs and by a pK_a^* of 0.5 [11]. Years ago, we postulated that forward proton transfer to give 1-naphtholate was accompanied by reprotonation in the distal ring. The LUMO represented in Fig. 13.3 suggested that reprotonation should occur at C-5 or C-8. Indeed, we observed that 1-naphthol undergoes efficient H/D exchange at C-5 and C-8, with significant preference for the C-5 position [12]. In contrast, 2-naphthol exhibits little or no proton quenching and a "normal" pK_a^* of 2.8, a fact which has been attributed to the diffuse nature of the latter and a more localized L_a for the former. We propose that adiabatic protonation yields a highly delocalized tetraenone (see Fig. 13.4), which undergoes rapid internal conversion. In contrast, adiabatic protonation of 2-naphtholate is endothermic, yielding a higher-energy cross-conjugated enone. In support of this mechanism, we have synthesized 5-*tert*-butyl-1-naphthol. Indeed, this molecule undergoes an apparent excited-state (ipso) protonation at C-5 to regenerate 1-naphthol through a process analogous to the quenching of 1-naphthol itself (see Fig. 13.4).

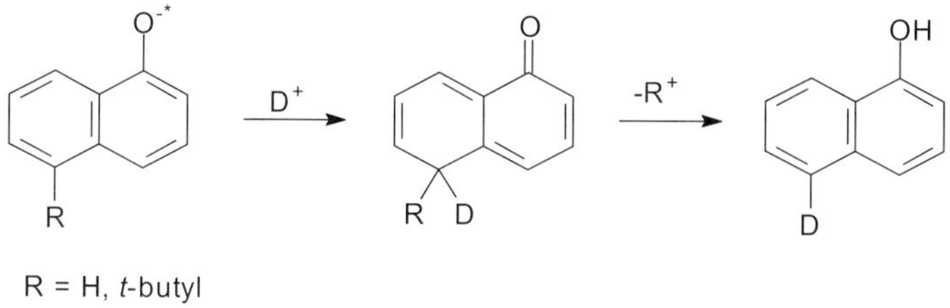

Figure 13.4 Exchange at C-5 in 1-naphthol.

Although simple hydroxyarenes such as 1- and 2-naphthol exhibit only modest gains in acidity upon photoexcitation, the quenching behavior of 1-naphthol, together with the theoretical underpinning, suggests that increases in acidity should occur by placement of electron-withdrawing groups at sites of large electron density in the excited state. Thus 5-cyano-, 8-cyano-, and 5,8-dicyano-2-naphthols show greatly increased photoacidity, and Förster acidities down to pK_a^* of –4.5 and lower have been obtained. These investigations have allowed the extension of proton transfer dynamics from aqueous to nonaqueous solvents and even to the gas phase, allowing a change of solvent from water (R = H in Fig. 13.1) to alcohol (R=alkyl). Interestingly, Agmon et al. predicted that 3,5,8-tricyano-2-naphthol might be the ultimate superphotoacid from this family [13].

13.2.2 "Super" Photoacids

The high acidity of 1-naphthol, coupled with the enhanced basicity at C-5 and C-8, suggested that the introduction of electron-withdrawing groups at these positions should lower the energy of the conjugate base and produce even higher acidities. Although the nitro group should be the most electron-withdrawing, nitroarenes are notoriously photoactive. Therefore, we concentrated our initial efforts on the more stable cyano and methanesulfonyl groups. Initially, 5-cyano-1-naphthol (5CN1) and 5,8-dicyano-1-naphthol (DCN1) were prepared but were discovered to undergo rapid quenching, a phenomenon we associate with a protonation (presumably on nitrogen) similar to the result observed in 1-naphthol. Also synthesized and studied were 5-, 6-, 7-, and 8-cyano-2-naphthols (5CN2, 6CN2, 7CN2, and 8CN2) and 5-methanesulfonyl-2-naphthol (5MSN2). With a Hammett σ-value for the methanesulfonyl group nearly identical to that for cyano, we did not see differences in photophysics between 5CN2 and 5MSN2 and abandoned further work with methanesulfonyl groups, although 5-methanesulfonyl-1-naphthol provided a first experimental verification of a theoretical prediction of a kinetic transition in reversible binding reactions, driven by the difference in effective lifetimes of the bound and unbound states [14] (see below). In contrast, 5,8-dicyano-2-naphthol (DCN2) [15], with a calculated Förster pK_a* of –4.5, has become the workhorse for our studies in nonaqueous solvents. For 2-naphthol, substitution in the 5 and 8 positions apparently adds additional vibronic modes that promote switching of the L_a and L_b states. This supposition has been confirmed by gas phase measurements [16].

The "true" acidity of a photoacid is a difficult-to-obtain quantity. Formally, since $pK_a^* = -\log(k_{pt}/k_{-pt})$, where k_{pt} and k_{-pt} are overall rates for forward and back proton transfer, respectively, the pK_a* obtained from the Förster calculation must be considered approximate.

$$R^*OH \underset{k_{rec}}{\overset{k_{dis}}{\rightleftharpoons}} [R^*O^{-*}...H^+]_{(r=a)} \overset{DSE}{\rightleftharpoons} R^*O^- + H^+ \tag{13.4}$$

Another approach is fluorescence titration, in which the emission from the conjugate base is examined as a function of pH. Again, the presence of competing processes may lead to erroneous results. In the scheme represented by Eq. (13.4), because of the electrostatic field generated by the counterion upon dissociation, an adiabatic recombination occurs to regenerate the excited acid (R*OH), which again dissociates and causing apparent nonexponential behavior for the excited state decay. In order to treat this problem, Huppert and Agmon used the Debye–Smoluchowski equation (DSE) to obtain the pK_a*s [17] This approach allows pK_a* determination from a single kinetic measurement at a neutral pH:

$$pK_a^* = -\log \frac{k_{dis} \exp(-R_D/a)}{k_{rec}} \tag{13.5}$$

In Eqs. (13.4) and (13.5) k_{dis} and k_{rec} are dissociation and recombination rate constants within a contact ion pair of a radius a, R_D is the Debye radius. For the systems with pronounced geminate recombination it is possible to fit nonlinear R*OH decay to a numerical solution of a system of DSE equations [18]. Equation (13.5) is a vivid example of the difference between excited-state *electron* and *proton* transfer studies. In the latter case both kinetic and thermodynamic parameters of the process could be determined directly from the same experiment, while for electron transfer ΔG of the reaction is usually estimated from electrochemical or other data.

In collaboration with Huppert, we were able to determine the pK_a*s by laser spectroscopy (Table 13.1). What emerges from this analysis is that (i) the measured pK_a*s are higher than the calculated ones, and (ii) substituents at C-5 and C-8 are more effective at lowering pK_a* for either 1- or 2-naphthol.

Table 13.1 Excited-state equilibrium constants for cyano-substituted 2-naphthols.

Compound	pK_a^* (DSE)	pK_a^* (Förster)	pK_a^*(fluor.)
DCN2		–4.5	
5CN2	–0.75	–1.2	1.7
6CN2	–0.37	0.2	0.5
7CN2	–0.21	–1.3	2.0
8CN2	–0.76	–0.4	0.7
N2		2.8	

In principle, 6-hydroxyquinoline (6HQ), which has a nitrogen atom at the position corresponding to C-5 in 2-naphthol, should exhibit enhanced acidity, especially if the nitrogen is converted into an electron-withdrawing group. This is also true for 7-hydroxyquinoline, by analogy with C-8 of 2-naphthol. Indeed, this is what Bardez and, more recently, Leutwyler report – an effect that involves a "hydrogen-bonding wire" [19]. In this case, the increased basicity at nitrogen leads to rapid nitrogen protonation (Fig. 13.5) [20]. Thus the photophysics of this species are characterized by weak emission from the tautomers rather than from the conjugate base, although one can estimate a pK_a* for the excited-state quinolinium species as approaching –13.

N-Methyl hydroxyquinolinium species, which are isoelectronic to the protonated form of hydroxyquinolines, have similar properties. These compounds demonstrate remarkable increased photoacidity – protolytic photodissociation in water

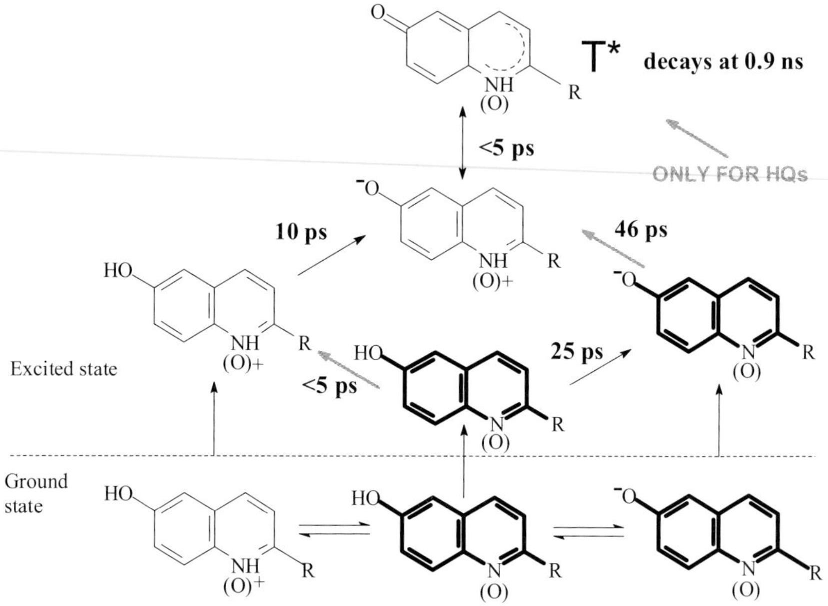

Figure 13.5 Tautomerization vs. proton transfer in hydroxy-quinolines and their *N*-Oxides.

occurs at 2 ps [21]. This rate for intermolecular ESPT to water is among the fastest reported to date. An ultrafast ESPT from this compound was also observed in the series of alcohols [22]. Several flavilium ions and anthocyanins [23], all carrying positive charge on the aromatic ring next to the phenolic one, are also reported to be superphotoacids with ultrafast ESPT to water.

An alternative approach, which avoids creating an organic salt, is to modify the acidity of 6HQ by converting the latter into 6-hydroxyquinoline-*N*-oxide (6HQNO), preserving the electronegativity of the molecule. 2-Methyl-6-hydroxy-quinoline-*N*-oxide (MeHQNO) was also prepared. Our studies with these molecules indicate a significant increase in acidity over the corresponding 5CN2 and the resulting photoproduct is an anion, not tautomer (Fig. 13.6) [24]. This happens due to oxidation of the nitrogen atom that diminishes its strong photobasicity, so in contrast to 6HQ no excited-state protonation is observed in neutral and basic media (Fig. 13.5). A complication of this chemistry is the slow but significant photodeoxygenation to yield 6HQ, as well as the known rearrangement to yield a quinolinone. Although the latter process can be inhibited by substitution at C-2 with methyl, the deoxygenation still proceeds slowly. Aromatic *N*-oxides, including a few hydroxyaromatic compounds [25], have a wide application in life sciences studies, so the combination of enhanced ESPT reactivity and efficient photodeoxygenation in combination with bright fluorescence makes hydroxyaromatic *N*-oxides possible candidates for the modulation of biological activity with simultaneous monitoring by fluorescence spectroscopy/microscopy methods.

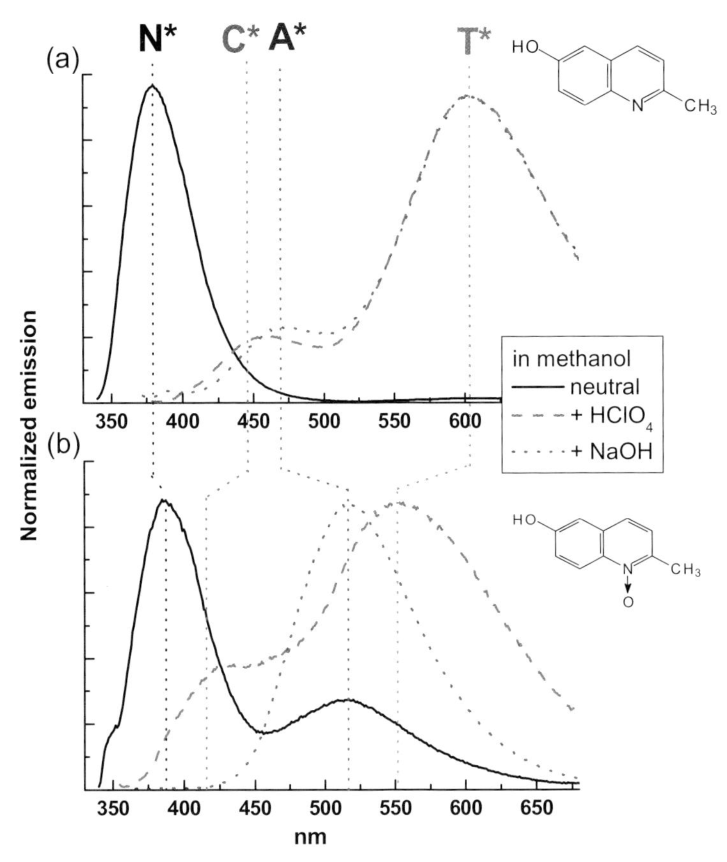

Figure 13.6 Emission from (a) MeHQ and (b) MeHQNO. N*, C*, A*, and T* correspond to neutral, cation, anion, and tautomer.

Notwithstanding the remarkable acidity of cyanated naphthols and hydroxyquinolines *N*-oxides, which allows them to transfer protons at rates competitive with excited-state decay to a number of organic acceptors, including sulfoxides and alcohols, the rates are still too sluggish to initiate bimolecular reactions (see below). We have synthesized the even more acidic perfluoroalkanesulfonylnaphthols, including 6-perfluorohexanesulfonyl-2-naphthol (6pFSN2) and 6-trifluoromethanesulfonyl-2-naphthol [26]. However, aggregation with these molecules has become a significant problem and has, up to now, defeated their use in proton transfer studies.

13.2.3
Fluorinated Phenols

Other phenols with electron-withdrawing groups have been reported to have behavior reminiscent of "super" photoacids. However, a complication has proved to be photohydrolysis of the carbon – fluorine bond, leading to carboxylic acid product. Thus trifluoromethylnaphthols undergo photohydrolysis to yield naphthalene carboxylic acids [27]. Boule has reported that fluorophenols undergo photodehalogenation to products of rearrangement and hydrolysis [28]. A recent report has appeared on the use of a fluorinated phenol [29] as a reversible photocatalyst for cationic polymerization of phenyl glycidyl ether. However, this report must be viewed with a great deal of skepticism, given the tendency for photohydrolysis of phenols. Thus irreversible formation of traces of acid which induce polymerization through a chemically amplified mechanism cannot be ruled out. Similarly, the polymerization of formaldehyde by low-temperature irradiation of a nitrophenol [30] may have similar mechanistic origins. Nevertheless, the latter reactions provide tantalizing support for the use of "super" photoacids in carrying out proton-initiated reaction.

13.3
Nature of the Solvent

13.3.1
Hydrogen Bonding and Solvatochromism in Super Photoacids

Proton transfer in both ground and excited states includes formation and breaking of hydrogen bonds. The degree of prior formation of hydrogen-bonded complexes and redistribution of the hydrogen bonds after excitation can be estimated using solvatochromic analysis of the absorption/excitation and the emission spectra. A Kamlet–Taft approach [31] allows separation of general (polar) and specific (H-bonding) solvation responsible for ground- and excited-state stabilization of super photoacids that have large excited-state dipole moments in the excited state. Using this method, we [8, 32] and others [33] have determined several types of hydrogen bonds of super photoacids with solvents (HS) and estimated their relative strengths. In amphoteric solvents, such as water and alcohols, two types of hydrogen bonds exist for neutral super photoacids (Fig. 13.7). Type 2 is much higher is energy than type 1 and makes a major contribution to the solvatochromism of ground- and excited-state species. The type 1 bond is totally cleaved in the excited state which supports the mechanism of intramolecular charge transfer after excitation. Stabilization of the excited state anion via a type 3 hydrogen bond is probably one of the major factors determining the relative reactivity of super photoacids. The time scale of type 2 bond strengthening in 8-hydroxypyrene-1,3,6-tris(dimethylsulfonamide), or, in other words, selective solvation due to H-bonding, is amazingly fast, about 55 fs [33].

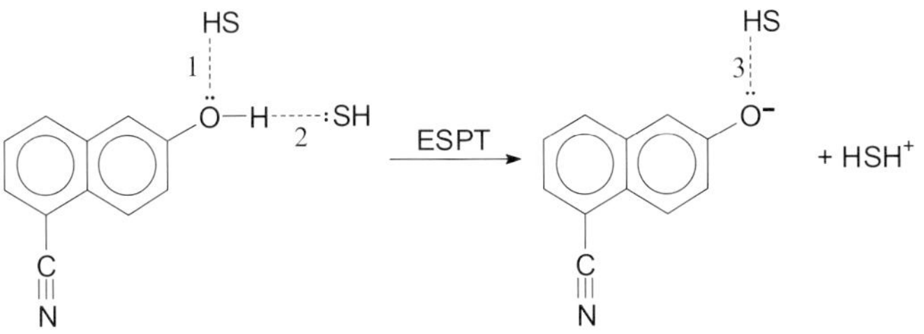

Figure 13.7 Hydrogen bonds in 5CN2.

13.3.2
Dynamics in Water and Mixed Solvents

As was mentioned before, in bulk water the pK_a* of super photoacids is negative, so protolytic photodissociation is exergonic and should not depend on the acid nature. Indeed, similar to electron-transfer reactions [1] a Bronsted-type plot k_{pt} (or k_{dis}) vs. pK_a* reaches a constant limiting value in the region of exergonic proton transfer. Several known characteristic ESPT dissociation times of super photoacids in water approach minimal values of a few picoseconds [21, 23, 34, 35]. These values are close to the longitudinal relaxation time around single-water-molecule charge in water that is probably the rate-limiting step of *intermolecular* ESPT in solution.

One of the earliest and most persistent questions concerning ESPT has been the anomalous solvent effect. The simple naphthols exhibit efficient proton transfer in water but not in alcohols. This observation is surprising, given the fact that the proton affinity of the simple alcohols is higher than that for water [36]. At intermediate water concentrations, the relationship between the rate of proton transfer and water concentration in methanol or ethanol solution is roughly fourth-order. The nonlinear response to water concentration has led to a series of papers on the structure of water (and other solvents) during proton transfer. Robinson [37] postulates that the proton transfer in aqueous solvent systems is the result of formation of a water cluster of order 4±1, that is, the generation of a tetrahedral coordination sphere for the proton that consists only of water molecules. Thus the underlying kinetics reflect the rate-limiting formation of a water cluster at the rate determining step. Huppert and Agmon have challenged Robinson's conclusions and the intervention of solvent clustering [38]. Rather, they postulate that the kinetics simply reflect the number of hydrogen bonds, both made and broken, required to facilitate the proton transfer at the transition state. The fact that water provides two hydrogen bonds, while methanol only one, is critical in that methanol alone must "give up" a hydrogen bond in order to form a new one, making the rate noncompetitive with excited-state decay. This distinction is subtle, but has important consequences for the kinetics. We have synthesized molecules in which an intramolecular solvating group, a hydroxyalkane, serves to substitute for one molecule of solvent, thus reducing the molecularity of the proton transfer. The resulting increase in proton transfer rate serves to demonstrate the importance of entropy in controlling the proton transfer transition state [39]. Recently, Hynes has used high-level calculations to make predictions about the trajectory of proton transfer and the role of solvent reorganization in the evolution of the reaction coordinate [40]. From these calculations emerges a picture that is consistent with the Huppert–Agmon model.

To study solvent effects in aqueous mixtures we have used the non-hydrogen-bond-donating solvent tetrahydrofuran. With 5CN2, the molecularity with respect to water is lowered to 3, while that for DCN2 is lowered to 2 (Fig. 13.8). Clearly there is a relationship between molecularity and driving force. It is also interesting that the molecularity, or order, with respect to water, is entirely contained in the hydrogen bonds donated by water, not alcohol, since there is no methanol present.

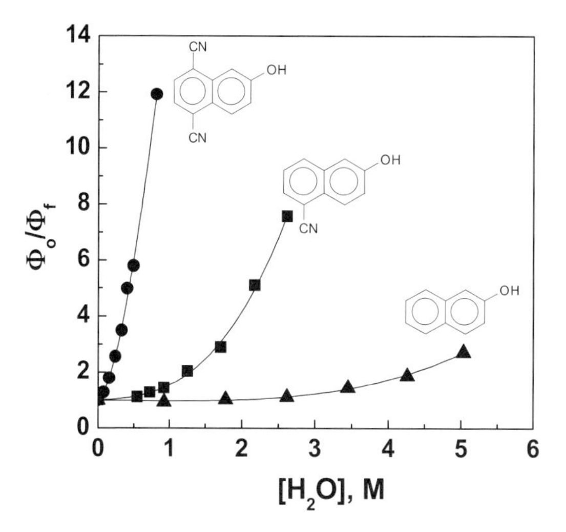

Figure 13.8 Effect of water on proton transfer from N2, 5CN2, and DCN2.

In methanol–water mixtures where ESPT is observed already in the absence of water, we separated the methanol- and water-dependent components of the protolytic dissociation rate constant for several photoacids [41]. As in THF–water mixtures, the water-dependent component has a power-law dependence on water concentration. The magnitude of the exponent decreased with the increase in photoacid strength.

13.3.3
Dynamics in Nonaqueous Solvents

A problem with 1- and 2-naphthol as substrates is that proton transfer is slower than solvent reorganization and, in any event, not observed in nonaqueous abasic solvents. In contrast, super photoacids exhibit anion fluorescence in a wide array of anhydrous solvents, including alcohols, as shown in the steady-state emission shown in Fig. 13.9. With superphotoacids, proton transfer still occurs efficiently, even at low temperatures in alcohols. The Arrhenius plot of proton transfer exhibits a two-stage process: (i) a nearly barrierless solvent-dependent process near room temperature and above and (ii) a ca. 3 kcal mol^{-1} barrier at low temperatures which is solvent independent (Fig. 13.10) [42]. The preexponential term at room temperature represents the proton transfer coordinate. At low temperature, solvent reorganization is rate-limiting. Thus we have for the first time observed proton transfer which is rate-limited by solvent relaxation, which supports the Hynes model in methanol [14]. Most curious is the presence of nearly identical kinetic deuterium isotope effects for both steps, an observation that is consistent with the hydrogen bonded network being involved both in solvent reorganization and in the proton transfer event itself. Similar effects were observed by Peon et al. in the symmetrical system, i.e. abstraction of proton from alcohols by super photobases (carbenes) was also rate limited by solvent relaxation [43].

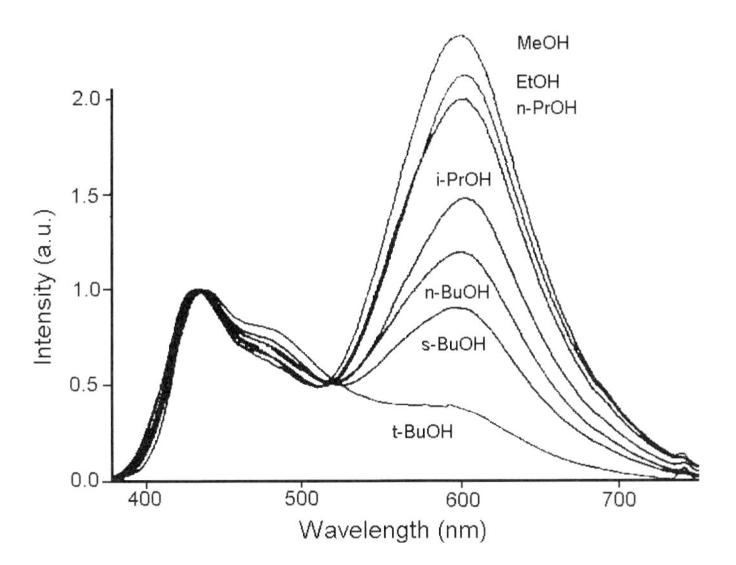

Figure 13.9 Steady-state emission of DCN2 in simple alcohols.

Nonlinear behavior is also observed in the wide-range (0.1–2.5 GPa) pressure dependence of the ESPT rate of DCN2 in alcohols [44]. At low pressure, the protolytic photodissociation rate slightly increases, reaching the maximum value. With further pressure increase this rate decreases below the initial value at atmospheric pressure (Fig. 13.11). To explain the unique nonexponential dependence of ESPT rate constants on pressure, as well as temperature, Huppert et al. have developed an approximate stepwise two coordinate proton-transfer model that bridges the high-temperature nonadiabatic proton tunneling limit with the rate constant $k_{\rm dis}^{\rm NA}$

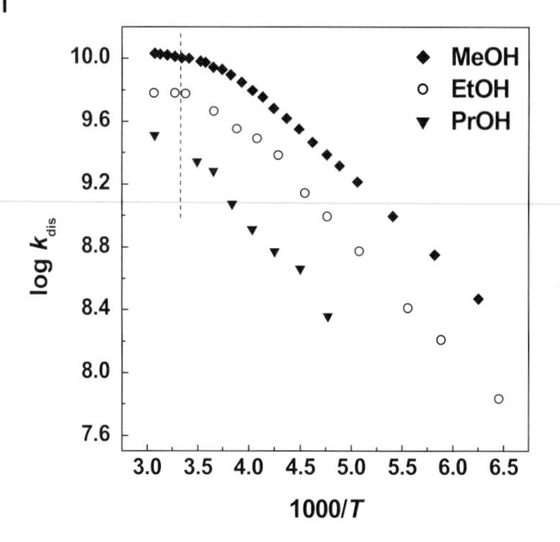

Figure 13.10 Arrhenius plot for photo-induced proton transfer from DCN2 to alcohols. Dashed line represents 25 °C.

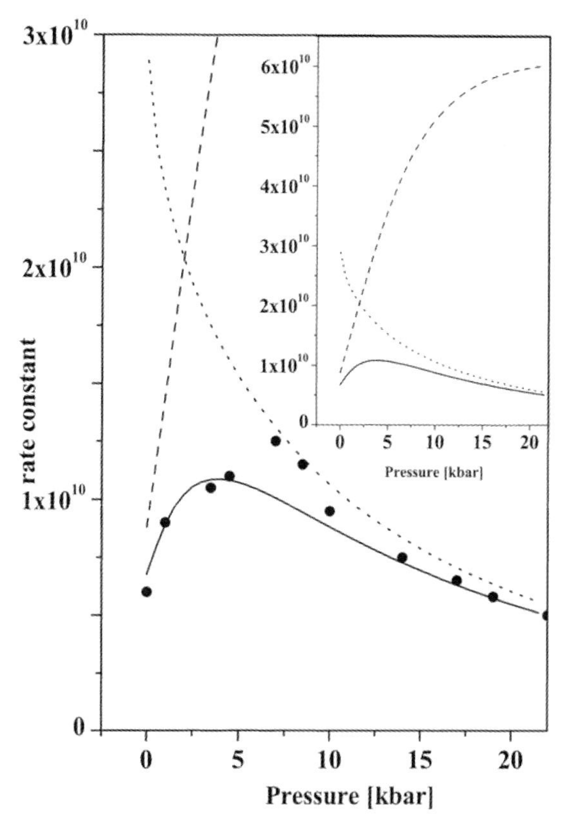

Figure 13.11 Fit to the stepwise two-coordinate model according to Eq. (13.6) as a function of pressure (solid line) along with the experimental data for DCN2 in ethanol. k_{dis}^{NA} and k_{dis}^{AD} are shown as dashed and dotted lines, respectively. Reproduced from Ref. [44a] by permission of the American Chemical Society.

and the low-temperature adiabatic solvent controlled limit with rate constant $k_{\mathrm{dis}}^{\mathrm{AD}}$ (Eq. (13.6)).

$$k_{\mathrm{dis}}(T, P) = \frac{k_{\mathrm{dis}}^{\mathrm{NA}}(T, P) k_{\mathrm{dis}}^{\mathrm{AD}}(T, P)}{k_{\mathrm{dis}}^{\mathrm{NA}}(T, P) + k_{\mathrm{dis}}^{\mathrm{AD}}(T, P)} \tag{13.6}$$

A decrease in the protolytic photodissociation rate of DCN2 with increasing pressure is also observed in supercritical CO_2/methanol mixtures with constant methanol molarity and molality [45]. This effect is currently under investigation.

Another interesting aspect of the excited-state proton transfer kinetics is the nonexponential long-time asymptotic behavior of R*OH and R*O⁻ fluorescence. This effect was initially observed for R*OH decay of pyranine in water and was quantitatively described by the diffusion-influenced geminate recombination [17]. Super photoacids capable of ESPT in a wide array of nonaqueous solvents expanded the range of tunable kinetic regimes. Agmon and coworkers predicted theoretically [46] and one demonstrated experimentally [14] that, depending on the ratio of the proton dissociation rate constant and fluorescence lifetimes of R*OH and R*O⁻, several asymptotic regimes are possible for photoacids. For instance, not only the most common power-law $t^{-3/2}$ decay, but also $1/t$ and even exponential growth are seen for R*OH asymptotic behavior! Therefore, excited-state proton transfer serves as unique tool for verifying interesting aspects of modern theoretical chemical kinetics.

13.3.4
ESPT in the Gas Phase

Gas-phase spectroscopic investigations of photoacids and their clusters with various solvent molecules are extremely important because they offer several opportunities unavailable in the liquid phase [16, 19c, 47] Our studies of gas phase spectroscopy and ESPT kinetics of super photoacids expanded the range of molecular systems systematically investigated using commercially available 1- and 2-naphthols. In the gas phase excitation and emission spectra of isolated molecules are so structured that the fine features of the rotation isomers can be clearly identified. We have found extensive spectroscopic indications of vibronic coupling between L_a and L_b states that, by analogy with 1-naphthol, is probably one of the main factors responsible for the ESPT triggering. Such coupling adds more polar L_a character to the initially excited low polar L_b state [16a].

Another important advantage of the molecular beams is the possibility of selective generation and investigation of the photoacid–solvent clusters of the controlled molecular composition. Clusters of 5CN2 with various numbers of water, ammonia, methanol and DMSO molecules, generated in a molecular beam, have been investigated by resonant two-photon ionization and fluorescence spectroscopies (Fig. 13.12) [16a]. We have observed an interesting correlation between the strength of the photoacid and the ESPT size threshold. In 5CN2–ammonia clusters the threshold for ESPT is either 3 or 4 ammonia molecules, while for

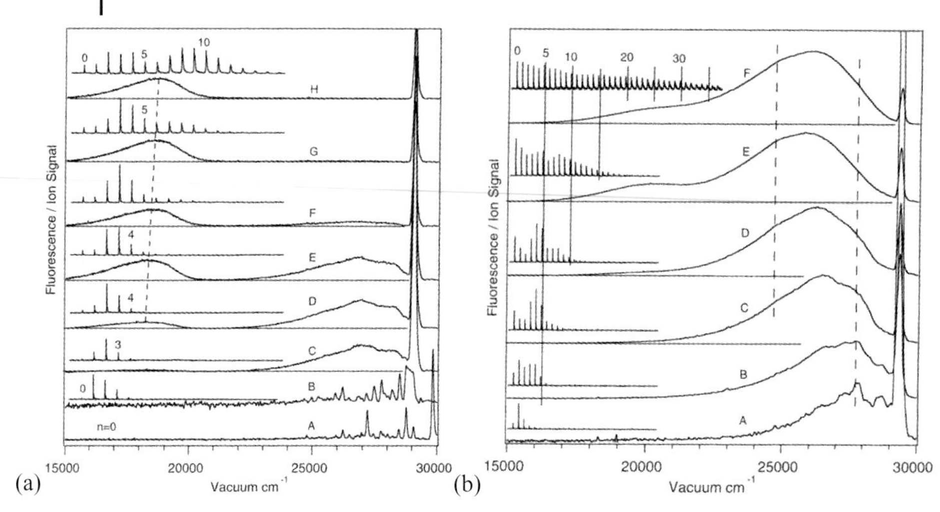

Figure 13.12 Fluorescence spectra of some (a) 5CN2–$(NH_3)_n$ and (b) 5CN2–$(H_2O)_n$ cluster distributions. The mass spectra corresponding to the fluorescence spectra are inset on the left. Red-shifted ESPT emission appears for n (a) > 3 or 4, and (b) > 8–10. Excitation was at 29100 cm^{-1} for spectra C–H, and ionization was at 36350 cm^{-1}. The largest, rightmost peak in each spectrum is predominantly scattered laser excitation light. The free molecule was excited at the *trans*-rotamer origin, as was the $n = 1$ cluster. The dashed line(s) in (a) indicates the blue shift of the ESPT emission with increasing cluster size, and in (b) helps to compare the position and shape of the molecular emission. Reproduced from Ref . [16a] by permission of the American Chemical Society.

1-naphthol-ammonia complexes it is 4. A more pronounced effect of increased photoacidity is observed in 5CN2–$(H_2O)_n$ clusters. For this strong photoacid ESPT appears at $n \approx 10$, as compared to 25–30 in 1-naphthol. The relatively slow fluorescence rise time of the excited 5CN2 anion in both cases and several other factors may indicate the nonadiabatic model of the ESPT in which dynamic solvent relaxation after photoexcitation is a key aspect of the EPST process [16c].

In contrast to room-temperature solutions and the above two cases no ESPT is observed in 5CN2 clusters with methanol, methanol–water and DMSO. At the same time the steady-state emission of R*OH demonstrated significant bathochromic shift with increasing cluster size. We explain the absence of ESPT by a simple thermodynamic factor. In bulk methanol and DMSO the protolytic photodissociation is endergonic. So, the difference in ESPT reactivity between bulk and clusters could be attributed to significant cooling of the latter, leading to enthalpic effects being dominant. We expect that the strongest photoacid DCN2 will extend the range of gas-phase clusters where *intermolecular* ESPT is observed.

13.3.5
Stereochemistry

One of our long-range goals is to provide experiments that allow us to construct models for proton transfer in biological media. Certainly one of the most dramatic aspects of biological systems involving proton transfer is their high stereoselectivity. Thus we wondered if we could find stereospecificity in proton transfer in chiral solvents. The nub of the question is represented in Fig. 13.13, in which we see that, if proton transfer involves interaction between solvent molecules, a transition state involving homochiral solvent molecules will have a different geometry than one involving heterochiral molecules. Thus the proton transfer rates should be different in racemic and nonracemic solvents. This is exactly what we found, with a preference for the homochiral transition state of almost 3:1 [48]. We can see that an enzyme, with its highly structured transition state, should lead to high stereoselectivity for proton transfer. Currently we are investigating chiral photoacids to see if there is a rate dependence of the proton transfer on the chirality of the acceptor. Altogether, we mention that the chirality preservation in the transition state is a prerequisite for stereoselectivity in the photoinduced proton transfer.

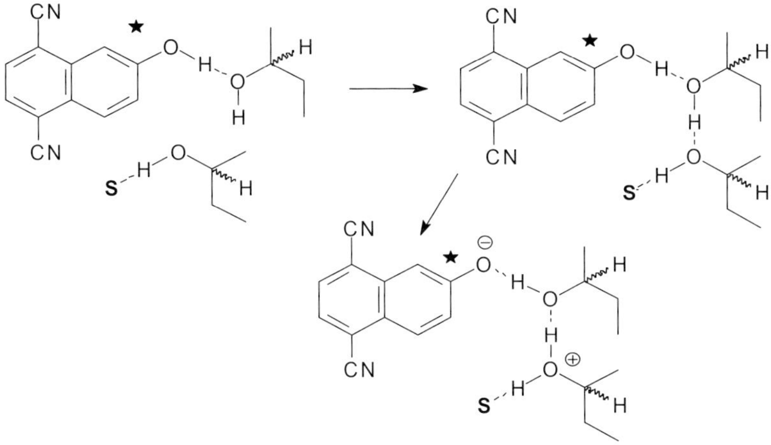

Figure 13.13 Proton transfer in a chiral solvent.

13.4
ESPT in Biological Systems

Despite the paucity of examples of excited-state proton transfer in biology, we believe that ESPT has great potential for the development of tools to examine *in vitro* and *in vivo* proton transfer. Almost by definition, biological media represent nonhomogeneous reaction media which serve to test theories of proton transfer and present opportunities to develop methods for examining the dynamics of proteins. Several examples of biological photoacids have begun to emerge:

Our attention was initially drawn to a medically significant group of hydroxyquinolines, the camptothecin family of topoisomerase I inhibitors [49]. 10-Hydroxycamptothecin (10HCT) and its Mannich derivative topotecan exhibit fluorescence behavior that is, on the basis of our previous studies, entirely rationalizable, if previously unrecognized, as involving prototropic behavior [50]. Whether such behavior also mirrors differences in bioavailability is still a matter of conjecture. The search for new pharmaceuticals and biological mechanisms is beyond the scope of this chapter. Nevertheless, we were intrigued by the prospect that 10-hydroxycamptochecin and topotecan could exhibit excited-state proton transfer in biological media and thus provide insight into the action of proton transfer probes in biology. Indeed, the emission of 10HCT is different from that of the parent 6HQ, and very similar to 6HQNO showing ESPT in pure methanol, isoemissive fluorescence spectra in methanol–water mixtures, and exhibiting no excited-state protonation at nitrogen (Fig. 13.14). Clearly, the additional electron-withdrawing pyridone group acts to diminish the excited-state basicity of the quinoline nitrogen [51], and acid–base photochemistry of 10HCT resembles that of 6HQNO (Fig. 13.5). In water-rich solutions the protolytic photodissociation rate was more than 85 1 ns^{-1}, clearly placing 10HCT among other super photoacids. Despite the presence of several acid–basic groups, the excited-state proton transfer behavior is "normal", exhibiting the typical nanosecond power-law asymptotic dependence common for such systems. In the shorter time domain, an intermediate with the lifetime of 3 ps was detected in a time-resolved pump–probe signal of R*OH form. We may associate it with the recently suggested "loose" hydrogen-bonded complex, therefore making ESPTa three-step process [52].

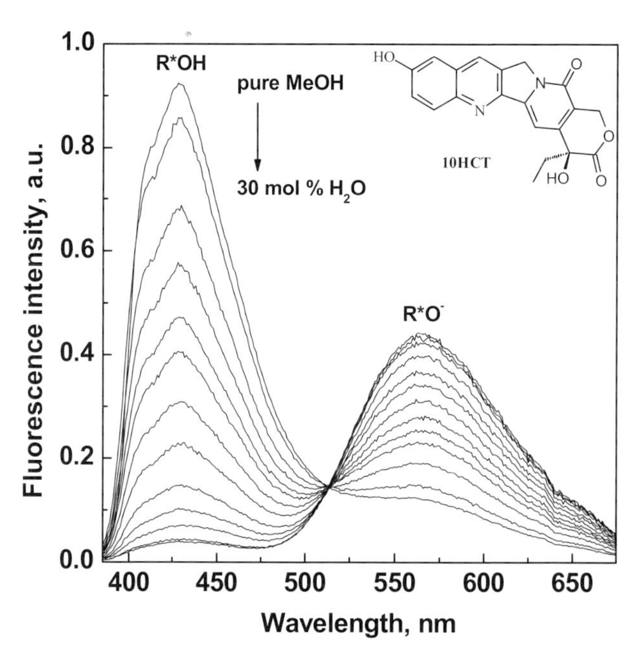

Figure 13.14 Emission from 10HCT.

13.4.1
The Green Fluorescent Protein (GFP) or "ESPT in a Box"

The green fluorescent protein (GFP) of the jellyfish *Aequorea Victoria* has attracted great interest as a biological fluorescence marker and as one of the few demonstrated examples of excited-state proton transfer in nature [53]. The wild-type chromophore *p*-hydroxybenzylidenediazolone is formed without co-factors via post-translational cyclization followed by an autoxidation of a tripeptide unit of the 238 amino acid polypeptide sequence (Fig. 13.15) [54]. Single-crystal X-ray diffraction reveals that the chromophore is threaded through the center of an 11-stranded β-barrel forming a coaxial helix which stabilizes the chromophore from the surrounding medium, including water [55, 56]. Additional noncovalent coupling of the chromophore to the protein backbone is facilitated via an extended hydrogen-bonded network [57].

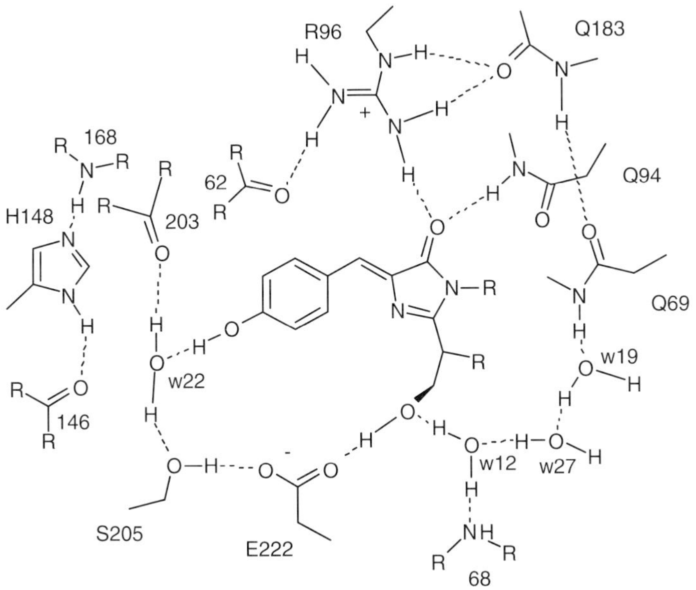

Figure 13.15 GFP active-site structure. Adapted from Ref. [57]. Hydrogens added for emphasis.

At room temperature, wtGFP (wild-type GFP) exhibits two main absorption peaks with maxima at 398 nm (band A) and 478 nm (band B). Several groups have studied the excited state dynamics of wtGFP [58–63]. These led to the conclusion that the two visible absorption bands of the wtGFP correspond to protonated and deprotonated ground-state conformations, while upon photoexcitation the acid form A rapidly deprotonates to form the B form. In addition, ultrafast kinetics suggests the presence of a third intermediate I which involves a proton transfer without conformational relaxation to the B form [64].

Although the vast majority of users of GFP and its derivatives can use these probes without resort to the mechanistic details, the unique barrel structure of the chromophore "box" makes this an ideal system to study proton transfer in a geometrically rigorously well-defined environment. While previous time-resolved studies of GFP have focused on emission decay at short times up to 150 ps [58–63] Huppert et al have focused on fluorescence at longer times (up to 10 ns). Using the TCSPC technique with a dynamic range of about 4 decades and extending the monitoring range of the emission to much longer times, they find that the fluorescence decay of R*OH is nonexponential up to 10 ns. Using a reversible geminate recombination model to fit the fluorescence data of wtGFP samples, they have concluded that the proton is transferred to the proton accepting moieties (hydroxy groups and waters) that are close to the chromophore. As the proton hops from one proton acceptor site to another, its motion can be approximated by a random walk in three dimensions, and hence a diffusion constant can be assigned to such motion. The proton can also recombine to its original site, the hydroxy group of the wtGFP chromophore, and thus repopulate the protonated form, ROH^*, giving rise to the asymptotic power law [65].

Since the excited-state pK_a^* of the GFP chromophore is less than zero already in the protein, then this chromophore, by analogy with the hydroxycamptothecin outlined above, also falls into the category of biological "super" photoacids. Isolated synthetic GFP and its derivatives do not fluoresce outside the β-barrel at room temperature due to very effective internal conversion related to isomerization along the double bond [59–63]. Deep understanding of the photophysics of these dyes as well as synthesis of their fluorescence derivatives capable of ESPT are our immediate goals. We anticipate further studies in biological proton transfer will be facilitated by these powerful new photophysical tools.

13.5 Conclusions

The studies performed to date suggest that "super" photoacids have an important role to play in unraveling the surprisingly complex dynamics of proton transfer, particularly in organized media for which diffusion is less straightforward. Such studies would be facilitated by photoacids with even stronger acidities but without side reactions. Particularly tantalizing is the possibility of time-resolved proton transfer studies of relevance to proton-catalyzed biological reactions, which demands coupling of high-photoacidity with well-characterized protein environments. Such studies are necessarily still in their infancy.

Acknowledgements

Support of this research by the U.S. National Science Foundation, under Grant CHE-0456892, is gratefully acknowledged.

References

1 Rehm, D.; Weller, A. *Isr. J. Chem.* **1970**, *8*, 259–271.

2 (a) Förster, T. *Z. Elektrochem.* **1950**, *54*, 42; (b) Weller, A. *Z. Elektrochem.* **1952**, *56*, 662–668.

3 Agmon, N. *Chem. Phys. Lett.* **1995**, *244*, 456–462.

4 Tolbert, L. M., Solntsev, K. M. *Acc. Chem. Res.* **2002**, *35*, 19–27.

5 For reviews, see (a) Ireland, J. F.; Wyatt, P. A. H. *Adv. Phys. Org. Chem.* **1976**, *12*, 131–221; (b) Schulman, S. G. *Fluorescence and Phosphorescence Spectroscopy*, Pergamon Press, Elmsford, NY, 1977; (c) Huppert, D.; Gutman, M.; Kaufman, K. J. in *Laser Studies of Proton Transfer, Advances in Chemical Physics*, Jortner, J.; Levine, R.D., Rice, S.A., Eds.; Wiley, New York, 1981, Vol. 47; (d) Shizuka, H. *Acc. Chem. Res.* **1985**, *18*, 141–147; (e) Arnaut, L. G.; Formosinho, S. J. *J. Photochem. Photobiol. A* **1993**, *75*, 1–20; (f) Agmon, N. *J. Phys. Chem. A* **2005**, *109*, 13–35.

6 Tolbert, L. M. *Acc. Chem. Res.* **1986**, *19*, 268–273.

7 (a) Flegel, M.; Lukeman, M.; Huck, L.; Wan, P. *J. Am. Chem. Soc.* **2004**, *126*, 7890–7897; (b) Lukeman, M.; Veale, D.; Wan, P.; Munasinghe, V. R. N.; Corrie, J. E. T. *Can. J. Chem.* **2004**, *82*, 240–253; (c) Shukla, D.; Lukeman, M.; Shi, Y.; Wan, P. *J. Photochem Photobiol. A* **2002**, *154*, 93–105; (d) Budac, D.; Wan, P. *Can. J. Chem.* **1996**, *74*, 1447–1464; (e) Wan, P.; Shukla, D. *Chem. Rev.* **1993**, *93*, 571–584; (f) Budac, D.; Wan, P. *J. Org. Chem.* **1992**, *57*, 887–894.

8 Solntsev, K. M.; Huppert, D.; Tolbert, L. M.; Agmon, N. *J. Am. Chem. Soc.* **1998**, *120*, 7981–7982.

9 Weller, A. *Prog. React. Kinet.* **1961**, *1*, 189–214.

10 Lewis, F. D.; Crompton, E. M. *J. Am. Chem. Soc.* **2003**, *125*, 4044–4045.

11 (a) Pines, E.; Tepper, D.; Magnes, B.-Z.; Pines, D.; Barak, T. *Ber. Bunsen-Ges. Phys. Chem.* **1998**, *102*, 504–510; (b) Pines, E.; Fleming, G. R. *Chem. Phys.* **1994**, *183*, 393–402.

12 Webb, S. P.; Philips, L. A.; Yeh, S. W.; Tolbert, L. M.; Clark, J. H. *J. Phys. Chem.* **1986**, *90*, 5154–5164.

13 Agmon, N.; Rettig, W.; Groth, C. *J. Am. Chem. Soc.* **2002**, *124*, 1089–1096.

14 Solntsev, K. M.; Huppert, D., Agmon, N. *Phys. Rev. Lett.* **2001**, *86*, 3427–3430.

15 (a) Tolbert, L. M.; Haubrich, J. E. *J. Am. Chem. Soc.* **1994**, *116*, 10593–10600; (b) Tolbert, L. M.; Haubrich, J. E. *J. Am. Chem. Soc.* **1990**, *112*, 8163–8165.

16 (a) Knochenmuss, R.; Solntsev, K. M.; Tolbert, L. M. *J. Phys. Chem. A* **2001**, *105*, 6393–6401; (b) Knochenmuss, R.; Fischer, I.; Luhrs, D.; Lin, Q. *Isr. J. Chem.* **1999**, *39*, 221–230.

17 (a) Pines, E.; Huppert, D.; Agmon, N. *J. Chem. Phys.* **1988**, *88*, 5620–5630; (b) Agmon, N.; Pines, E.; Huppert, D. *J. Chem. Phys.* **1988**, *88*, 5631–5638.

18 Krissinel, E. B.; Agmon, N. *J. Comput. Sci.* **1996**, *17*, 1085–1098.

19 (a) Bardez, E. *Isr. J. Chem.* **1999**, *39*, 319–332; (b) Bardez, E.; Fedorov, A.; Berberan-Santos, M. N.; Martinho, J. M. G. *J. Phys. Chem. A* **1999**, *103*, 4131–4136; (c) Tanner, C.; Manca, C.; Leutwyler, S. *Science* **2003**, *302*, 1736–1739.

20 Poizat, O.; Bardez, E.; Buntinx, G.; Alain, V. *J. Phys. Chem. A* **2004**, *108*, 1873–1880.

21 Kim, T. G.; Topp, M. R. *J .Phys. Chem. A* **2004**, *108*, 10060–10065.

22 Pérez Lustres, J. L. personal communication.

23 Moreira, P. F., Jr.; Giestas, L.; Yihwa, C.; Vautier-Giongo, C.; Quina, F. H.; Maçanita, A. L.; Lima, J. C. *J. Phys. Chem. A* **2003**, *107*, 4203–4210.

24 Solntsev, K. M.; Clower, C. E.; Tolbert, L. M.; Huppert, D. *J. Am. Chem. Soc.* **2005**, *127*, 8534–8544.

25 (a) Maklashina, E.; Cecchini, G. Arch. *Biochem. Biophys.* **1999**, *369*, 223–232; (b) Rothery, R. A.; Blasco, F.; Weiner, J. H. *Biochemistry*, **2001**, *40*, 5260–5268; (c) McMillian, M. K.; Li, L.; Parker, J. B.; Patel, L.; Zhong, Z.; Gunnett, J. W.; Powers, W. J.; Johnson, M. D. *Cell Biol. Toxicol.* **2002**, *18*, 157–173; (d) Sim-

kovic, M.; Frerman, F. E. *Biochem. J.* **2004**, *378*, 633–640; (e) Reyes, R.; Martinez, J. C.; Delgado, N. M.; Merchant-Larios, H. *Arch. Androl.* **2002**, *48*, 209–219; (f) *Handbook of Fluorescent Probes and Research Products*, 9th Edn., Haughland, R. P.; Ed., Molecular Probes, Eugene, OR, 2003, Ch. 15.2.

26 (a) Kowalik, J.; VanDerveer, D.; Clower, C.; Tolbert, L. M. *Chem. Commun.* **1999**, 2007–2008; (b) Clower, C.; Solntsev, K. M.; Kowalik, J.; Tolbert, L. M.; Huppert, D. *J. Phys. Chem. A* **2002**, *106*, 3114–3122.

27 Seiler, P.; Wirz, J. *Tetrahedron Lett.* **1971**, 1683–1686.

28 Boule, P.; Richard, C.; David-Oudjehani, K.; Grabner, G. *Proc. Indian Acad. Sci., Chem. Sci.* **1997**, *109*, 509–519.

29 Hino, T.; Endo, T. *Macromolecules* **2004**, *37*, 1671–1673.

30 Mansueto, E. S.; Wight, C. A. *J. Am. Chem. Soc.* **1989**, *111*, 1900–1901.

31 Kamlet, M. J.; Abboud, J.-L. M.; Abraham, M. H.; Taft, R. W. *J. Org. Chem.* **1983**, *48*, 2877–2887.

32 Solntsev, K. M.; Huppert, D.; Agmon, N. *J. Phys. Chem. A* **1999**, *103*, 6984–6997.

33 Pines, E.; Pines, D.; Ma, I.-Z.; Fleming, G. R. *ChemPhysChem* **2004**, *5*, 1315–1327.

34 Pines, E.; Pines, D.; Barak, T.; Magnes, B.-Z.; Tolbert, L. M.; Haubrich, J. *Ber. Bunsen.-Ges. Phys. Chem.* **1998**, *102*, 511–517.

35 Lewis, F. D.; Sinks, L. E.; Weigel, W.; Sajimon, M. C.; Crompton, E. M. *J. Phys. Chem. A* **2005**, *109*, 2443–2451.

36 Meot-Ner, M. *J. Am. Chem. Soc.* **1992**, *114*, 3312–3322.

37 (a) Lee, J.; Robinson, G. W.; Webb, S. P.; Philips, L. A.; Clark, J. H. *J. Am. Chem. Soc.* **1986**, *108*, 6538–6542; (b) Yao, S. H.; Lee, J.; Robinson, G. W. *J. Am. Chem. Soc.* **1990**, *112*, 5698–5700; (c) Krishnan, R.; Lee, J.; Robinson, G. W. *J. Phys. Chem.* **1990**, *94*, 6365–6367.

38 Agmon, N.; Huppert, D.; Masad, A.; Pines, E. *J. Phys. Chem.* **1991**, *95*, 10407–10413.

39 Tolbert, L. M.; Harvey, L.C.; Lum, R. C. *J. Phys. Chem.* **1993**, *97*, 13335–13340.

40 (a) Gertner, B. J.; Peslherbe, G. H.; Hynes, J. T. *Isr. J. Chem.* **1999**, *39*, 273–281; (b) Ando, K.; Hynes, J. T. *J. Mol. Liq.* **1995**, *64*, 25–37; (c) Ando, K.; Hynes, J. T.; in *Structure, Energetics and Reactivity on Aqueous Solutions*, Cramer, C. J.; Truhlar, D. G., Eds.; ACS Books, Washington DC, 1994.

41 Solntsev, K. M.; Huppert, D.; Agmon, N.; Tolbert, L. M. *J. Phys. Chem. A* **2000**, *104*, 4658–4669

42 (a) Carmeli, I.; Huppert, D.; Tolbert, L. M.; Haubrich, J. E. *Chem. Phys. Lett.* **1996**, *260*, 109–114; (b) Cohen, B.; Huppert, D. *J. Phys. Chem. A* **2000**, *104*, 2663–2667.

43 Peon, J.; Polshakov, D.; Kohler, B. *J. Am. Chem. Soc.* **2002**, *124*, 6428–6438.

44 (a) Koifman, N.; Cohen, B.; Huppert, D. *J. Phys. Chem. A* **2002**, *106*, 4336–4344; (b) Genosar, L.; Leiderman, P.; Koifman, N.; Huppert, D. *J. Phys. Chem. A.* **2004**, *108*, 309–319.

45 Nunes, R.; Arnaut, L. G.; Solntsev, K. M.; Tolbert, L. M.; Formosinho, S. *J. Am. Chem. Soc.* **2005**, *127*, 11890–11891.

46 Gopich, I. V.; Solntsev, K. M.; Agmon, N. *J. Chem. Phys.* **1999**, *110*, 2164–2174.

47 Knochenmuss, R.; Fisher, I. *Int. J. Mass. Spectrom.* **2002**, *220*, 343–257.

48 Solntsev, K. M.; Tolbert, L. M.; Cohen, B.; Huppert, D.; Hayashi, Y.; Feldman, Y. *J. Am. Chem. Soc.* **2002**, *124*, 9046–9047.

49 (a) Mi, Z.; Burke, T. G. *Biochemistry* **1994**, *33*, 12540–12545; (b) Mi, Z.; Burke, T. G. *Biochemistry* **1994**, *33*, 10325–10336; (c) Burke, T. G.; Malak, H.; Gryczynski, I.; Mi, Z.; Lakowicz, J. R. *Anal. Biochem.* **1996**, *242*, 266–270.

50 Tolbert, L. M.; Nesselroth, S. M. *J. Phys. Chem.* **1991**, *95*, 10331–10336.

51 (a) Ashkenazi, S.; Leiderman, P.; Huppert, D.; Solntsev, K. M.; Tolbert, L. M. in *Femtochemistry and Photobiology*, Martin and Hynes, J. T., Eds., Elsevier, Amsterdam 2004, pp. 201–206; (b) Solntsev, K. M.; Sullivan, E. N.; Tolbert, L. M.; Ashkenazi, S.; Leiderman, P.; Huppert, D. *J. Am. Chem. Soc.* **2004**, *126*, 12701–12708.

52 Rini, M.; Pines, D.; Magnes, B.-Z.; Pines, E.; Nibbering, E. T. J. *J. Chem. Phys.* **2004**, *121*, 9593–9610.

53 Zimmer, M. *Chem. Rev.* **2002**, *102*, 759–781.

54 (a) Tsien, R. Y. *Annu. Rev. Biochem.* **1998**, *67* 509–544, (b) Cubitt, A. B.; Heim, R.; Adams, S. R.; Boyd, A. E.; Gross, L. A.; Tsien, R. Y. *Trends Biochem. Sci.* **1995**, *20*, 448–455.

55 Ormö, M.; Cubitt, A. B.; Kallio, K.; Gross, L. A.; Tsien, R. Y.; Remington, S. J. *Science* **1996**, *273*, 1392–1395.

56 (a) Yang, F.; Moss, L. G.; Phillips, G. N. J. *Nature Biotech.* **1996**, *14*, 1246; (b) Suhling, K.; Siegel, J.; Phillips, D.; French, P. M. W.; Leveque-Fort, S.; Webb, S. E. D.; Davis, D. M. *Biophys. J.* **2002**, *83*, 3589–3595.

57 Brejc, K.; Sixma, T. K.; Kitts, P. A.; Kain, S. R.; Tsien, R. Y.; Ormö, M.; Remington, S. J. *Proc. Natl. Acad. Sci. USA* **1997**, *94*, 2306–2311.

58 Chattoraj, M.; King, B. A.; Bublitz, G. U.; Boxer, S. G. *Proc. Natl. Acad. Sci. USA* **1996**, *93*, 8362–8367.

59 (a) Voityuk, A. A.; Kummer, A. D.; Michel-Beyerle, M.-E.; Rosch, N. *Chem. Phys.* **2001**, *269*, 83–91; (b) Kummer, A. D.; Wiehler, J.; Rehaber, H.; Kompa, C.; Steipe, B.; Michel-Beyerle, M. E. *J. Phys. Chem. B* **2000**, *104*, 4791–4798; (c) Voityuk, A. A.; Michel-Beyerle, M. E.; Rosch, N. *Chem. Phys. Lett.* **1998**, *296*, 269–276; (d) Youvan, D. C.; Michel-Beyerle, M. E. *Nature Biotech.* **1996**, *14*, 1219–1220; (e) Kummer, A. D.; Kompa, C.; Lossau, H.; Pollinger-Dammer, F.; Michel-Beyerle, M. E.; Silva, C. M.; Bylina, E. J.; Coleman, W. J.; Yang, M. M.; Youvan, D. C. *Chem. Phys.* **1998**, *237*, 183–193.

60 Striker, G.; Subramanian, V.; Seidel, C. A. M.; Volkmer, A. *J. Phys. Chem. B* **1999**, *103*, 8612–8617.

61 Winkler, K.; Lindner, J. R.; Subramanian, V.; Jovin, T. M.; Vohringer, P. *Phys. Chem. Chem. Phys.* **2002**, *4*, 1072–1081.

62 Litvinenko, K. L.; Webber, N. M.; Meech, S. R. *Bull. Chem. Soc. Jpn.* **2002**, *75*, 1065–1070.

63 Cotlet, M.; Hofkens, J.; Maus, M.; Gensch, T.; Van der Auweraer, M.; Michiels, J.; Dirix, G. Van Guyse, M.; Vanderleyden, J.; Visser, A. J. W. G.; De Schryver, F. C. *J. Phys. Chem. B* **2001**, *105*, 4999–5006.

64 Wiehler, J.; Jung, G.; Seebacher, C.; Zumbusch, A.; Steipe, B. *ChemBioChem* **2003**, *4*, 1164–1171.

65 Leiderman, P.; Ben-Ziv, M.; Genosar, L.; Huppert, D.; Solntsev, K. M.; Tolbert, L. M. *J. Phys. Chem. B* **2004**, *108*, 8043–8053.

Further Titles of Interest

S.M. Roberts

Catalysts for Fine Chemical Synthesis V 5 – Regio and Stereo-controlled Oxidations and Reductions

2007
ISBN 0-470-09022-7

M. Beller, C. Bolm (Eds)

Transition Metals for Organic Synthesis. Building Blocks and Fine Chemicals

Building Blocks and Fine Chemicals

2004
ISBN 3-527-306137

G. Dyker (Ed.)

Handbook of C-H Transformations

Applications in Organic Synthesis

2005
ISBN 3-527-310746

H. Yamamoto, K. Oshima (Eds.)

Main Group Metals in Organic Synthesis

2004
ISBN 3-527-305084

G. A. Olah, Á. Molnár

Hydrocarbon Chemistry

2003
ISBN 0-471-417823